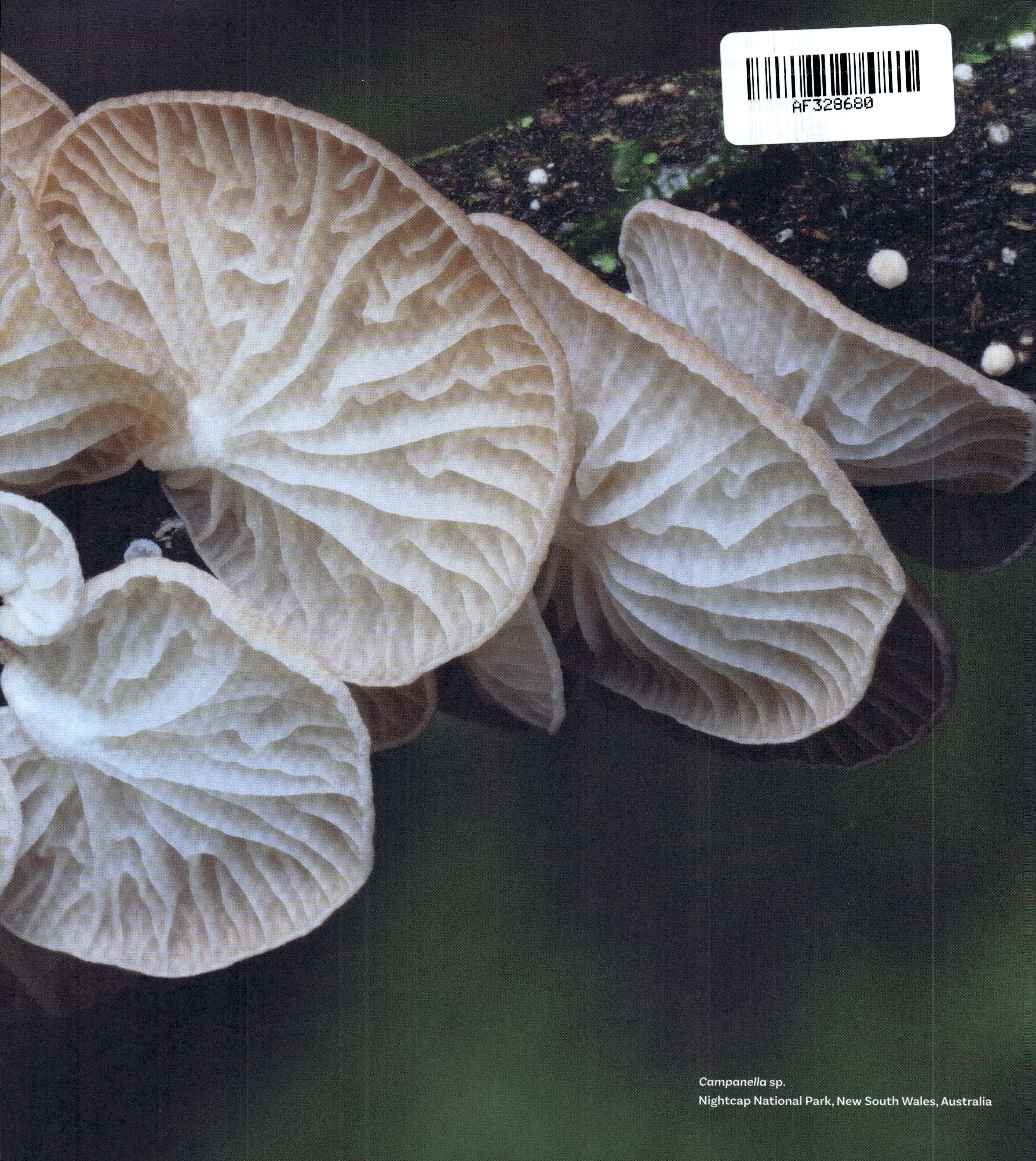

Campanella sp.
Nightcap National Park, New South Wales, Australia

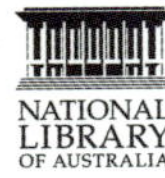 A catalogue record for this book is available from the National Library of Australia

ISBN: 9781486318186 (hbk)
ISBN: 9781486318193 (epdf)
ISBN: 9781486318209 (epub)

Published in print in Australia, New Zealand and the Americas, and in all other formats throughout the world, by CSIRO Publishing.

CSIRO Publishing
36 Gardiner Road, Clayton VIC 3168
Private Bag 10, Clayton South VIC 3169
Australia

Telephone: +61 3 9545 8400
Email: publishing.sales@csiro.au
Website: www.publish.csiro.au
Sign up to our email alerts: publish.csiro.au/earlyalert

Published in print only, in the United Kingdom, Europe, Middle East, Africa and Asia, by CABI.
ISBN 9781836991335

CABI
Nosworthy Way
Wallingford
Oxfordshire OX10 8DE
UK

CABI
200 Portland Street
Boston
MA 02111
USA

Tel: +44 (0)1491 832111
E-mail: info@cabi.org
Website: www.cabi.org

Tel: +1 (617)682-9015
E-mail: cabi-nao@cabi.org

CABI is a trading name of CAB International

Front cover: *Coprinopsis pulchricaerulea* (photo by Stephen Axford)
Back cover: (top to bottom) *Heimiomyces* sp.; *Cruentomycena viscidocruenta*; *Gliophorus graminicolor*
(photos by Stephen Axford)

Edited by Janet Walker
Cover design and layout by Cath Pirret Design
Index by Indexicana
Printed in Malaysia by Papercraft

CSIRO Publishing publishes and distributes scientific, technical and health science books, magazines and journals from Australia to a worldwide audience and conducts these activities autonomously from the research activities of the Commonwealth Scientific and Industrial Research Organisation (CSIRO). The views expressed in this publication are those of the author(s) and do not necessarily represent those of, and should not be attributed to, the publisher or CSIRO. The copyright owner shall not be liable for technical or other errors or omissions contained herein. The reader/user accepts all risks and responsibility for losses, damages, costs and other consequences resulting directly or indirectly from using this information.

CSIRO acknowledges the Traditional Owners of the lands that we live and work on and pays its respect to Elders past and present. CSIRO recognises that Aboriginal and Torres Strait Islander peoples in Australia and other Indigenous peoples around the world have made and will continue to make extraordinary contributions to all aspects of life including culture, economy and science. CSIRO is committed to reconciliation and demonstrating respect for Indigenous knowledge and science. The use of Western science in this publication should not be interpreted as diminishing the knowledge of plants, animals and environment from Indigenous ecological knowledge systems.

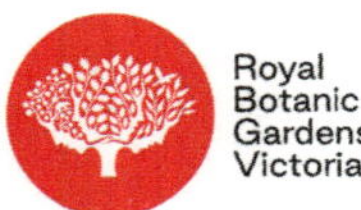 With thanks to Royal Botanic Gardens Victoria for their support of this publication.

The paper this book is printed on is in accordance with the standards of the Forest Stewardship Council® and other controlled material. The FSC® promotes environmentally responsible, socially beneficial and economically viable management of the world's forests.

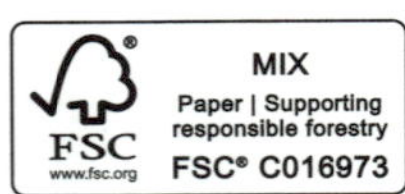

Mar25_01

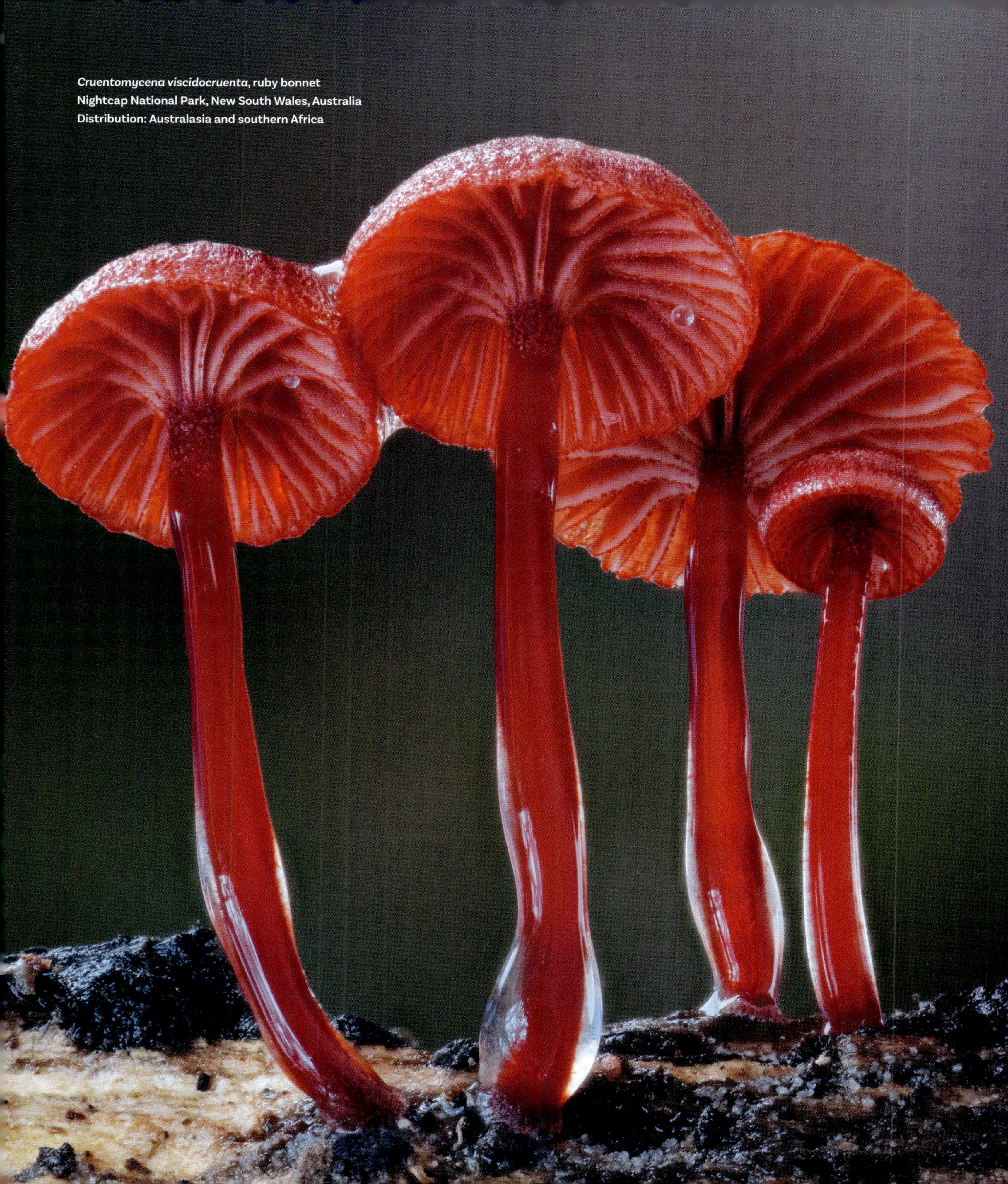

Cruentomycena viscidocruenta, ruby bonnet
Nightcap National Park, New South Wales, Australia
Distribution: Australasia and southern Africa

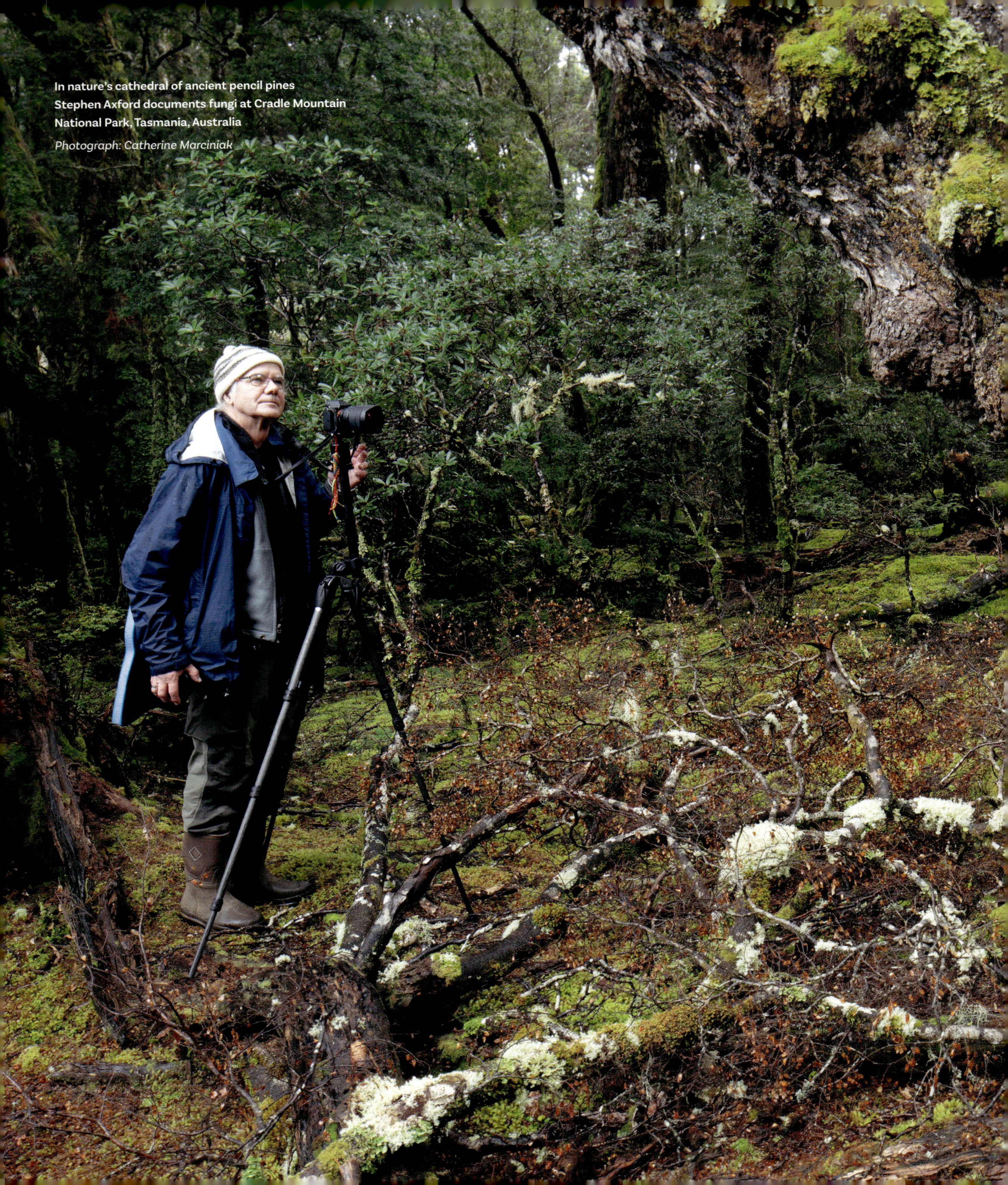

In nature's cathedral of ancient pencil pines
Stephen Axford documents fungi at Cradle Mountain
National Park, Tasmania, Australia
Photograph: Catherine Marciniak

Planet Fungi

A PHOTOGRAPHER'S FORAY

Catherine Marciniak,

Stephen Axford &

Tom May

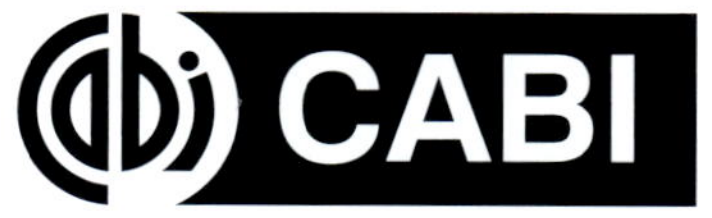

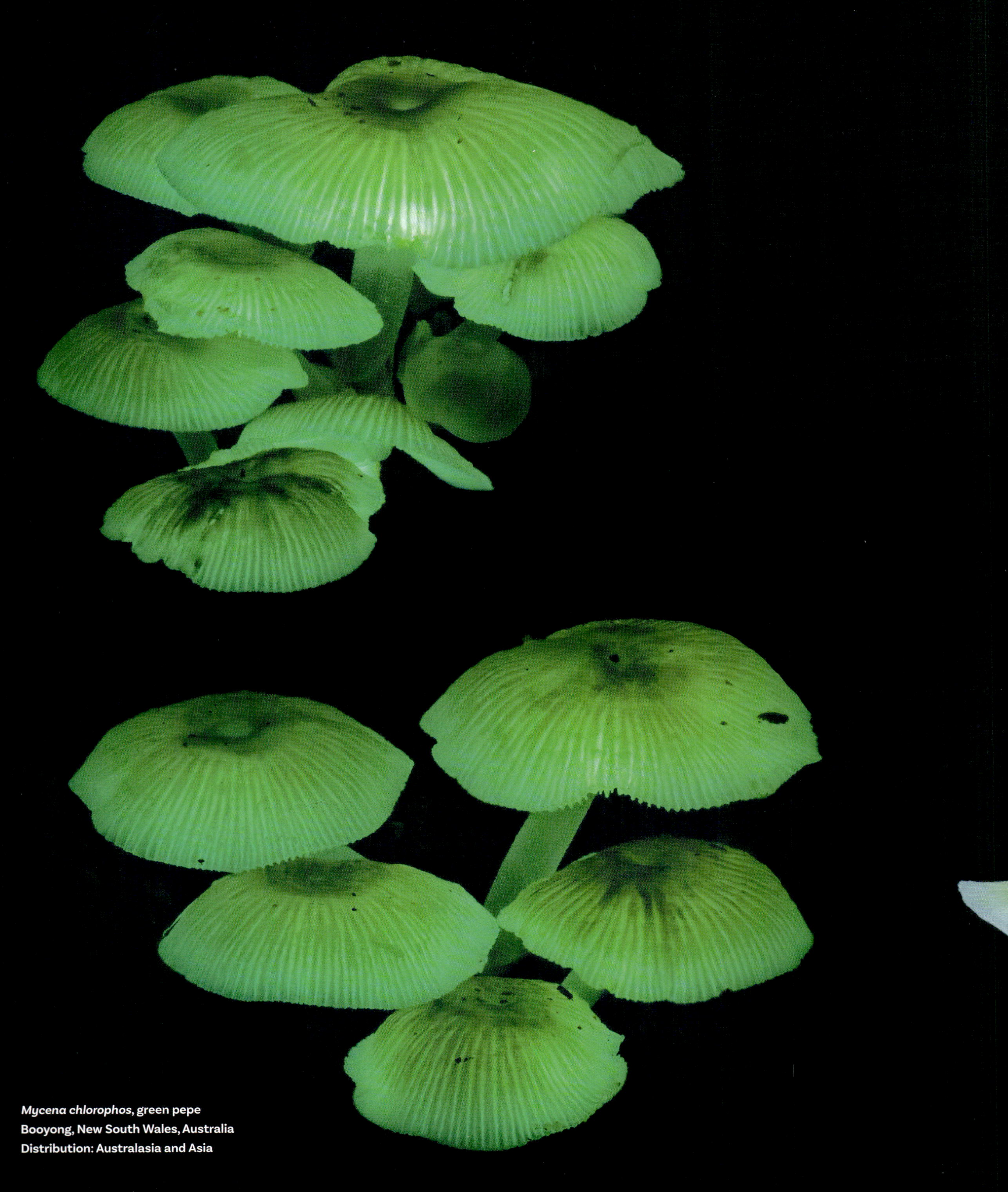

Mycena chlorophos, green pepe
Booyong, New South Wales, Australia
Distribution: Australasia and Asia

Contents

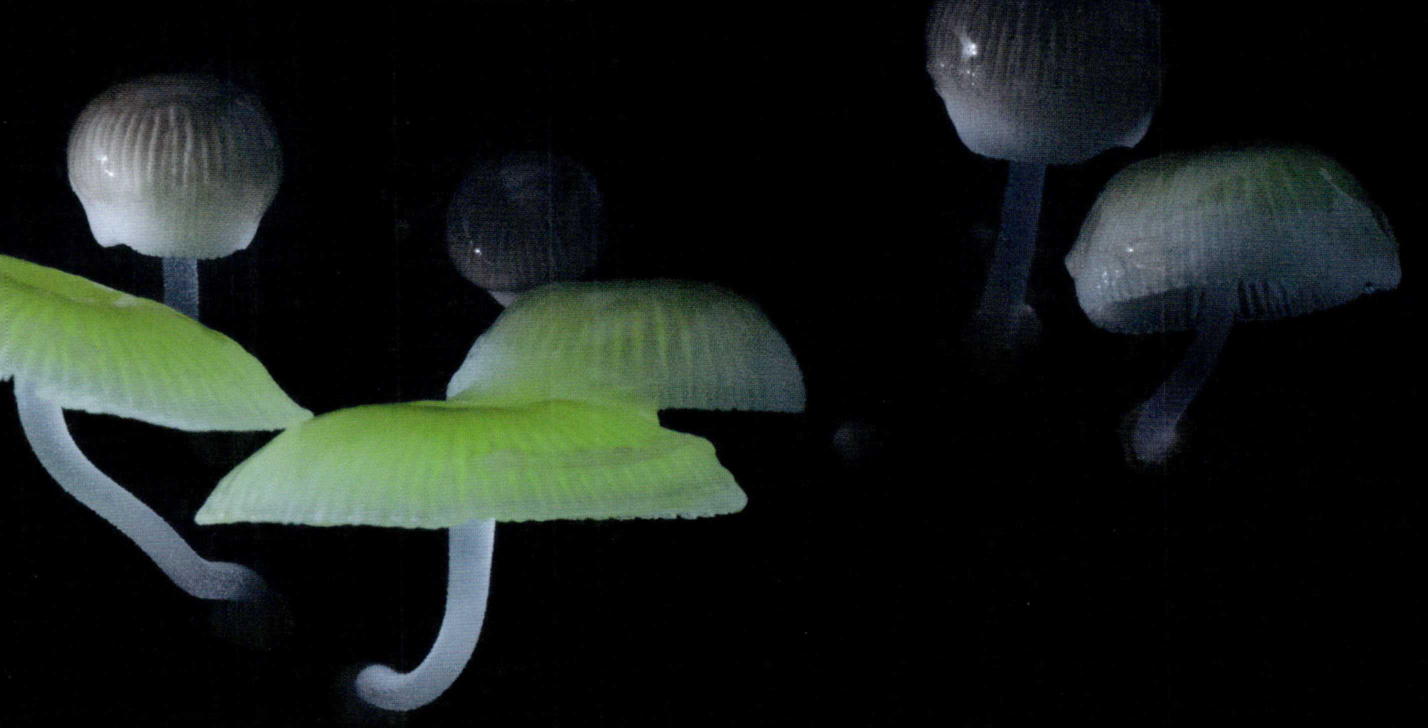

Authors' notes

All photographs are the work of Stephen Axford unless otherwise credited.

Only around 155,000 of the 2-5 million species of fungi scientists think exist have been described and named to the taxonomic level of species. Therefore, only some of the fungi in the images in this book have species names, some are compared to known species ('cf.'), others are identified only to genus or are unknown species.

The identifications are based on the author's and field mycologists' experience and morphological characteristics observed during the field trips and/or in the images. The exceptions are species identified as new discoveries, for which DNA sequences have been obtained from specimens of the illustrated fungi.

Disclaimer

Foraging for mushrooms should be treated with caution. The information in this book should not be relied upon as recommending or promoting the edibility of any fungus or any specific dietary practice. It is essential to be absolutely sure of the identity of any species of fungi before eating it. If there is any doubt at all about its edibility, do not eat the fungus. There is a reason why some mushrooms have the nickname death cap.

Roridomyces austrororidus, **austral dripping bonnet**
Waratah, Tasmania, Australia
Distribution: Australasia and South America

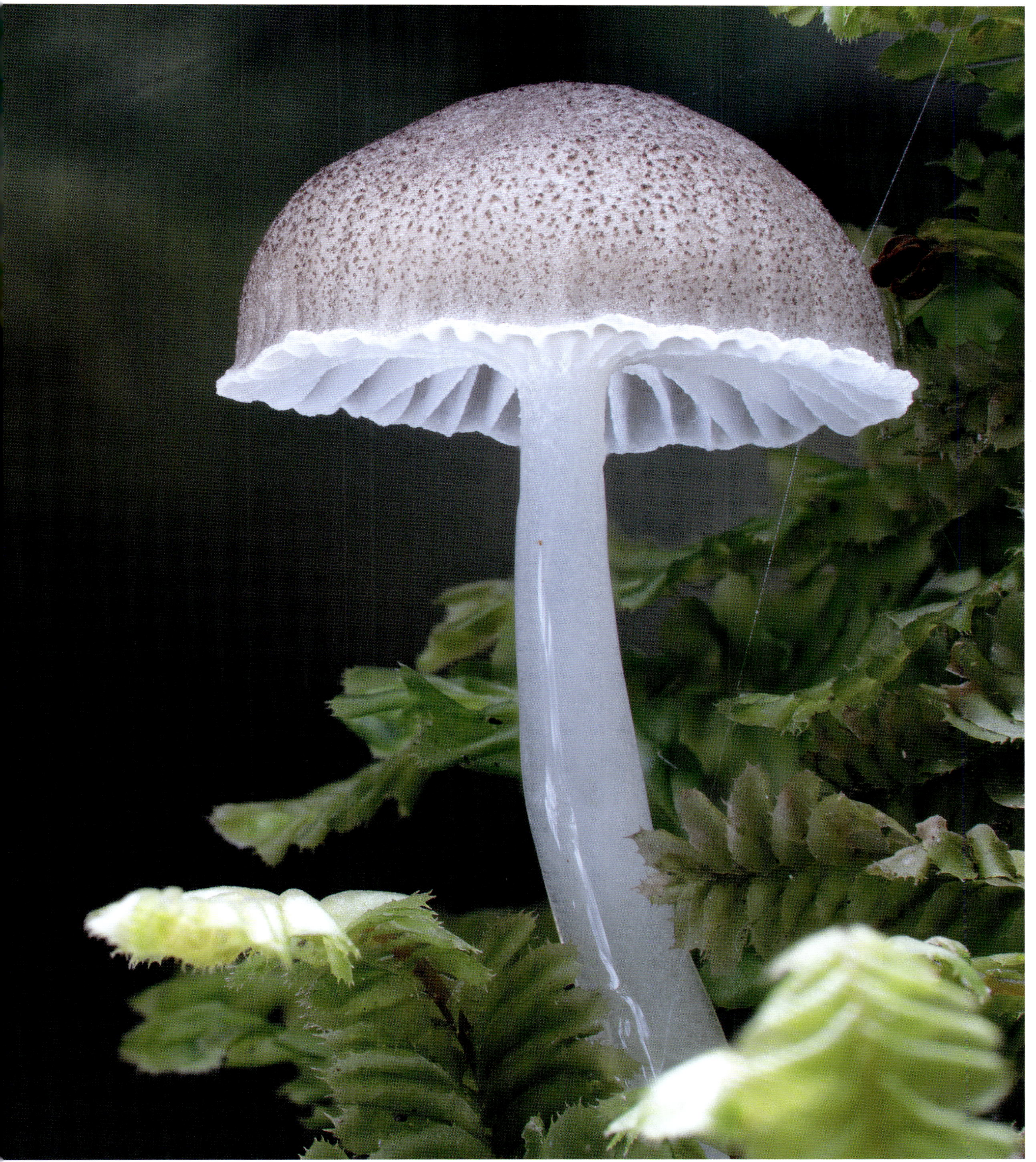

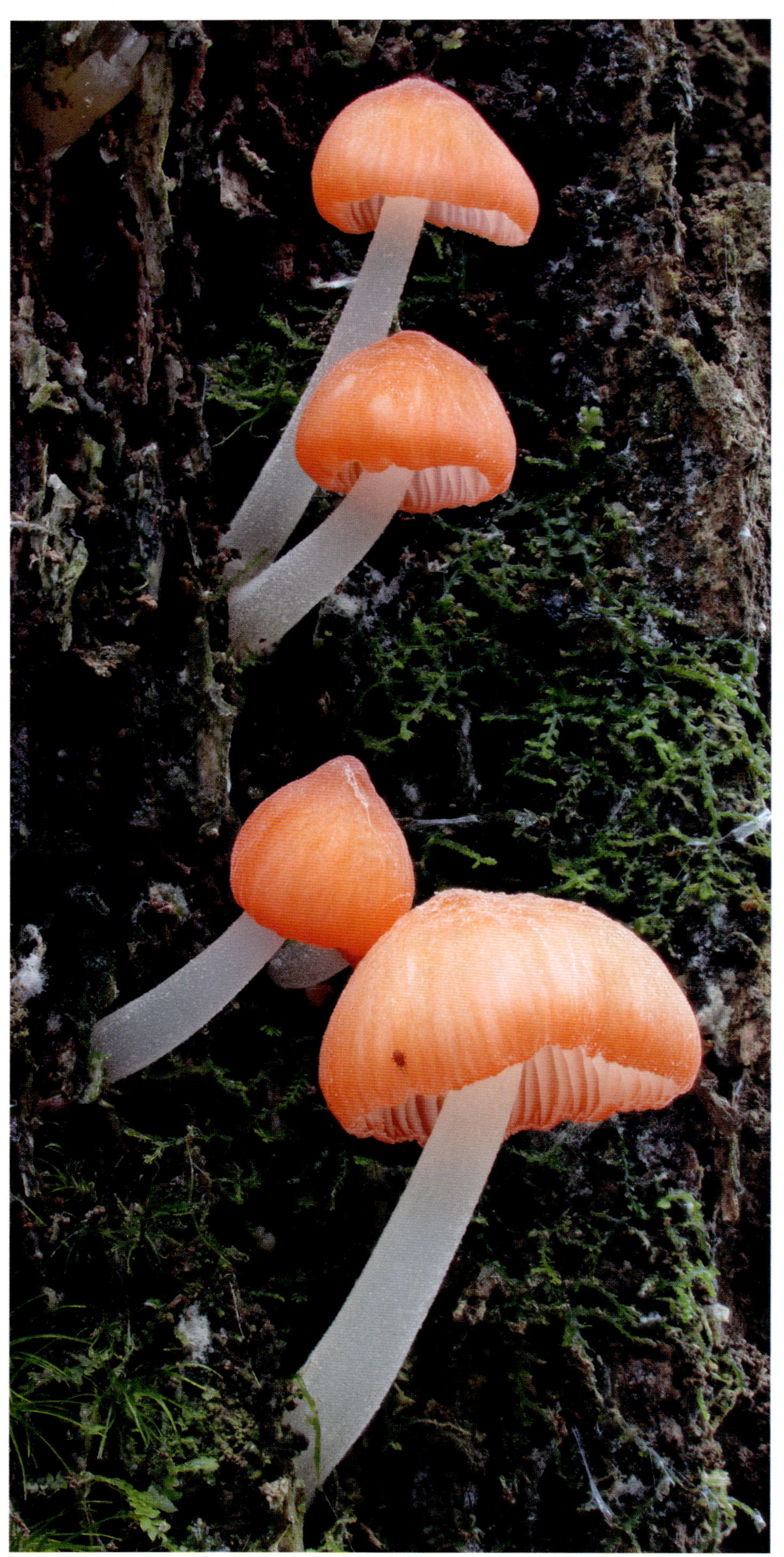

Atheniella adonis, scarlet bonnet
Tinjure-Milke-Jaljale Rhododendron
Conservation Area, Chauki, Nepal
Distribution: worldwide, mostly
in the Northern Hemisphere

Acknowledgements

We want to express our immense gratitude to all the wonderful contributors and supporters who have helped bring to life our vision of creating this book.

A special thanks to the experts who diligently reviewed the scientific content and shared their remarkable stories: Giuliana Furci, Matthew Smith, Donald Pfister (Patagonia), Peter Mortimer, Samantha Chandranath Karunarathna, Gautam Baruah, Arun Kumar Dutta (China, India and Myanmar), Roger Shivas, Donovan Teal (entomopathogenic fungi), Cecile Gueidan, Gintaras Kantvilas (lichens) and Teresa Lebel (*Coprinopsis pulchricaerulea* investigation).

We would also like to thank our dear friend Annie Benzie for her unwavering support and valuable feedback on story ideas and style.

Furthermore, we would like to acknowledge the exceptional work of Fungimap in Australia and the Fungi Foundation internationally. They have been instrumental in shedding light on the incredible importance of the Kingdom of Fungi to life as we know it.

We are grateful to Giuliana Furci, whose influence has inspired and motivated us to follow in her footsteps.

In addition, we would like to extend our gratitude to Sony Australia and New Zealand for their support with camera equipment courtesy of their Digital Imaging Advocate Program.

We acknowledge the Traditional Owners and custodians throughout the world. We recognise the deep cultural and scientific knowledge of fungi held by Indigenous peoples for millennia.

Lastly, we would like to express our heartfelt appreciation to all our scientific collaborators and fungi friends for their invaluable mentorship and guidance in our fungi advocacy, film and photographic work. Your support has meant the world to us.

About the authors

Catherine Marciniak, our lead writer, was raised in the Blue Mountains in Australia. The forest has always been her sanctuary, nurturing an enduring love of nature. She is an award-winning documentary filmmaker, photographer and journalist. In a career spanning four decades she has produced over 80 hours of television, multi-screen exhibitions and hundreds of digital articles covering a wide range of topics. She co-wrote the iMax documentary *FUNGI – Web of Life* and directed, wrote and filmed *Follow the Rain*, a magical journey into the weird and wonderful world of fungi. In her role as a Features Reporter with the Australian Broadcasting Corporation, Catherine was a two-time finalist in the Walkley Awards for Excellence in Journalism.

Stephen Axford has worn many hats in his life, from the hardhat of a miner to the corporate hat of a computer systems designer, and now the pioneering hat of the fungi hunter. He is renowned worldwide for his expertise in macro fungi field photography, with his exquisite images featured in hundreds of publications, books and exhibitions globally. His time-lapses of fungi growing are showcased in award-winning science documentaries including *Planet Earth 2*, *The Kingdom: How Fungi Made the World*, *Fantastic Fungi*, *Hostile Planet*, *Follow the Rain* and the IMAX documentary *FUNGI – Web of Life*. He is the presenter and co-writer of Planet Fungi's video productions. Stephen captures the intricate beauty of fungi in remote forests worldwide, showcasing nature's marvels through his lens.

Dr Tom May, scientific advisor to the book, has worked as a mycologist at Royal Botanic Gardens Victoria, in Melbourne, Australia, for almost three decades. Tom has combined a distinguished research career with a passion for communicating and educating about fungi. He has held many roles in professional and citizen science organisations, including the inaugural President of Fungimap. Tom has published widely in scientific and popular journals, co-authoring FunKey, an interactive key to mushroom genera, and the book *Wild Mushrooming in Australia*. His contributions to science and citizen science have been recognised with the Nancy T. Burbidge Medal, honouring those who have made a long-standing and significant contribution to Australasian systematic botany, and the Australian Natural History Medallion.

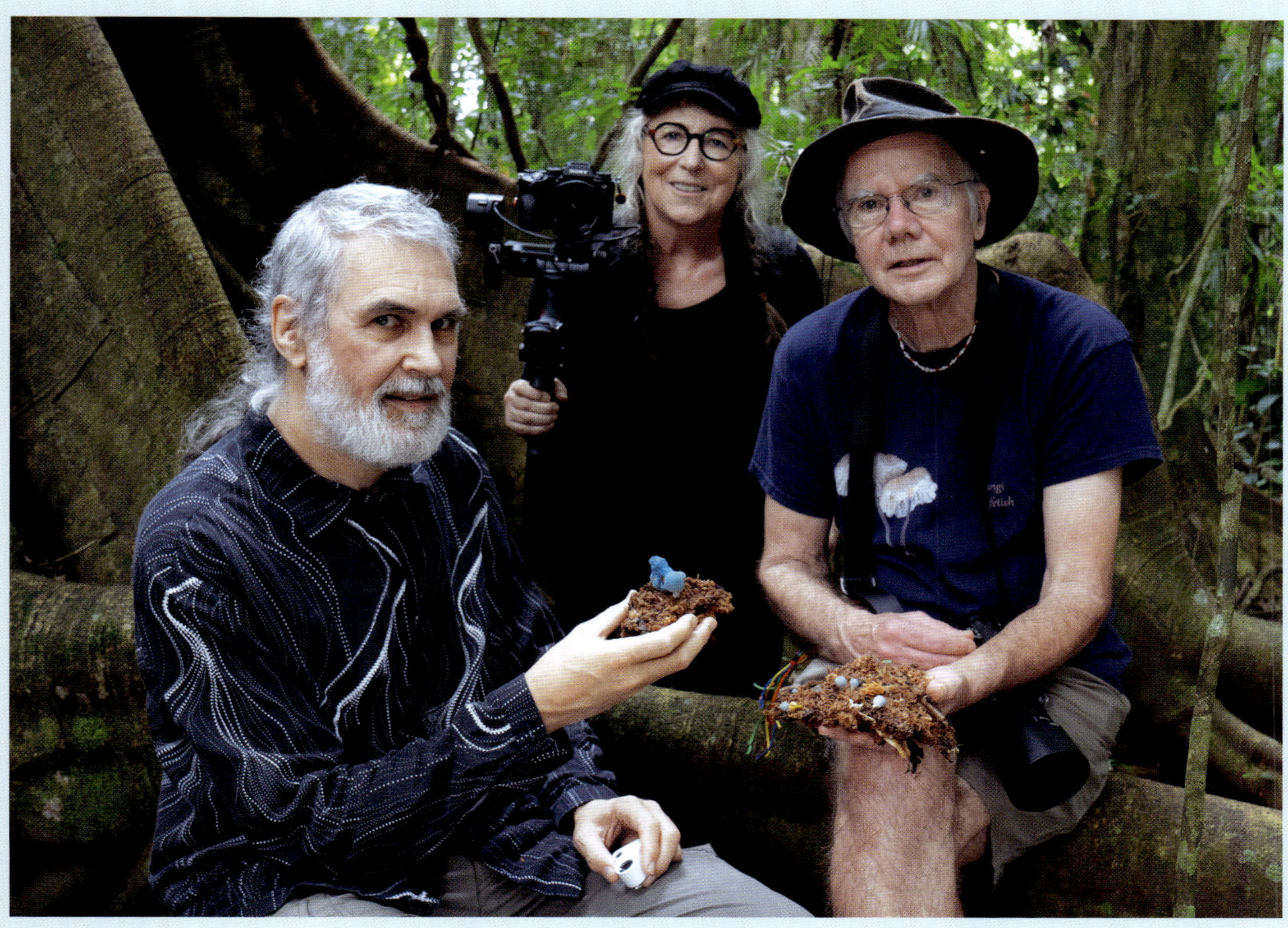

From left to right: : Dr Tom May, Catherine Marciniak and Stephen Axford

Photograph: Catherine Marciniak

How fungi changed my view of the world

Confronting mortality was the catalyst for Stephen Axford's passion for fungi.

In 1998, Stephen's wife of 13 years, Pat Flannagan, was diagnosed with terminal breast cancer. It was a five-year journey of hospitals, a deepening love and saying goodbyes as Stephen cared for her until she died. Only months later, Stephen discovered he also had a life-threatening illness.

'Against all odds, I recovered, but in caring for Pat and facing death twice, I discovered an inner strength, a courage I didn't know I had.

'After being a troubled and rebellious youth, I realised I liked the man I had become.

'This helped me to rethink my life. I was ready to take some risks, to reinvent myself.'

The wild places of Australia's east coast and their ancient landscapes became Stephen's sanctuary. He began walking long distances ... a solo adventurer.

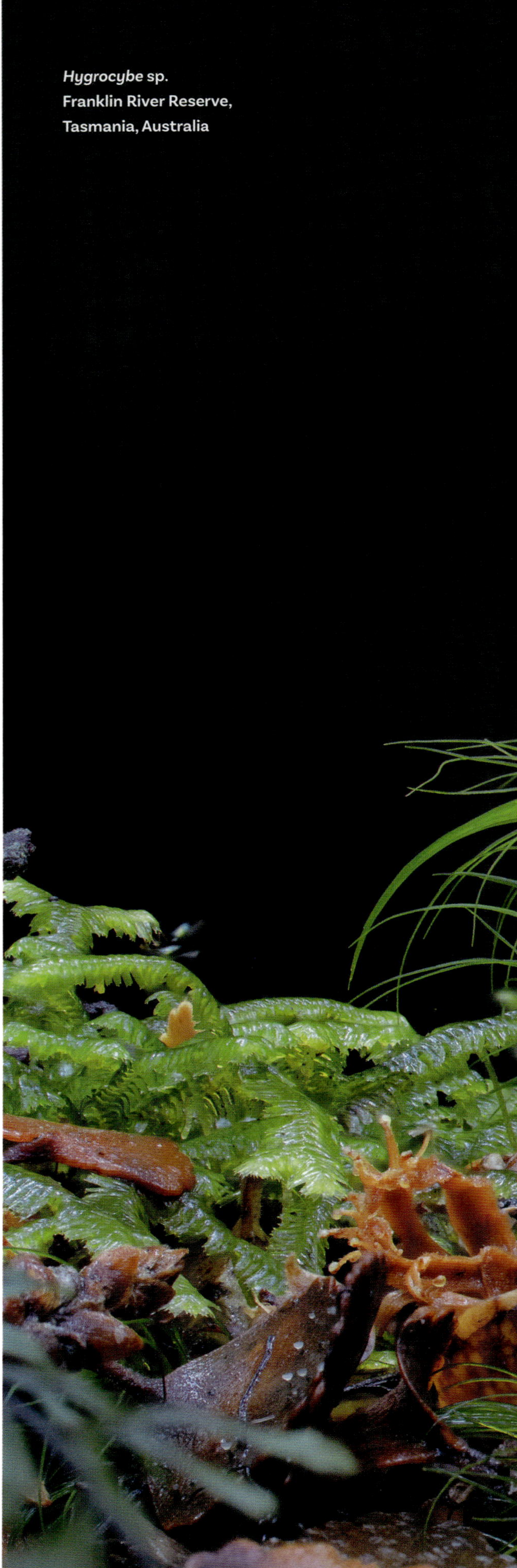

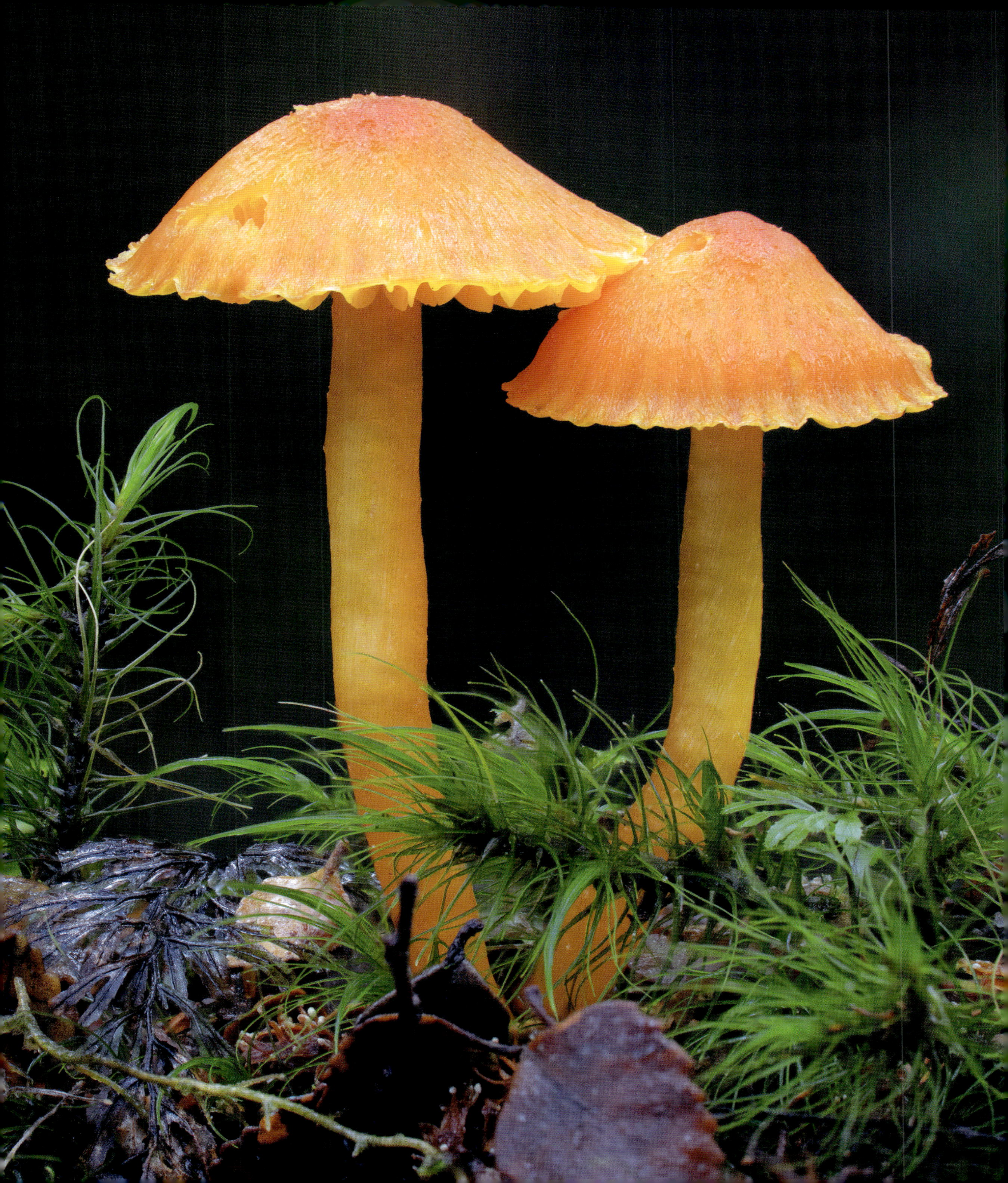

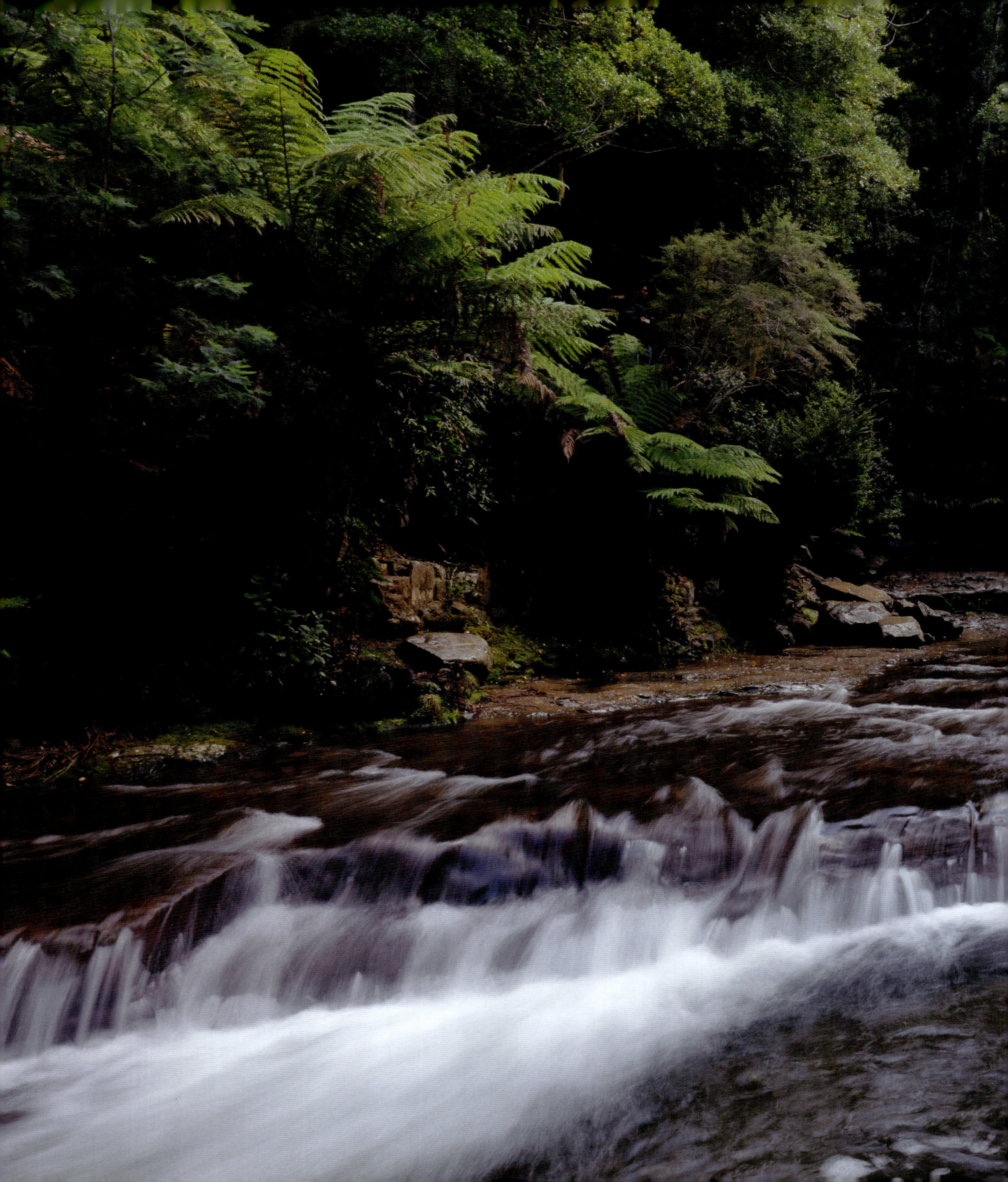

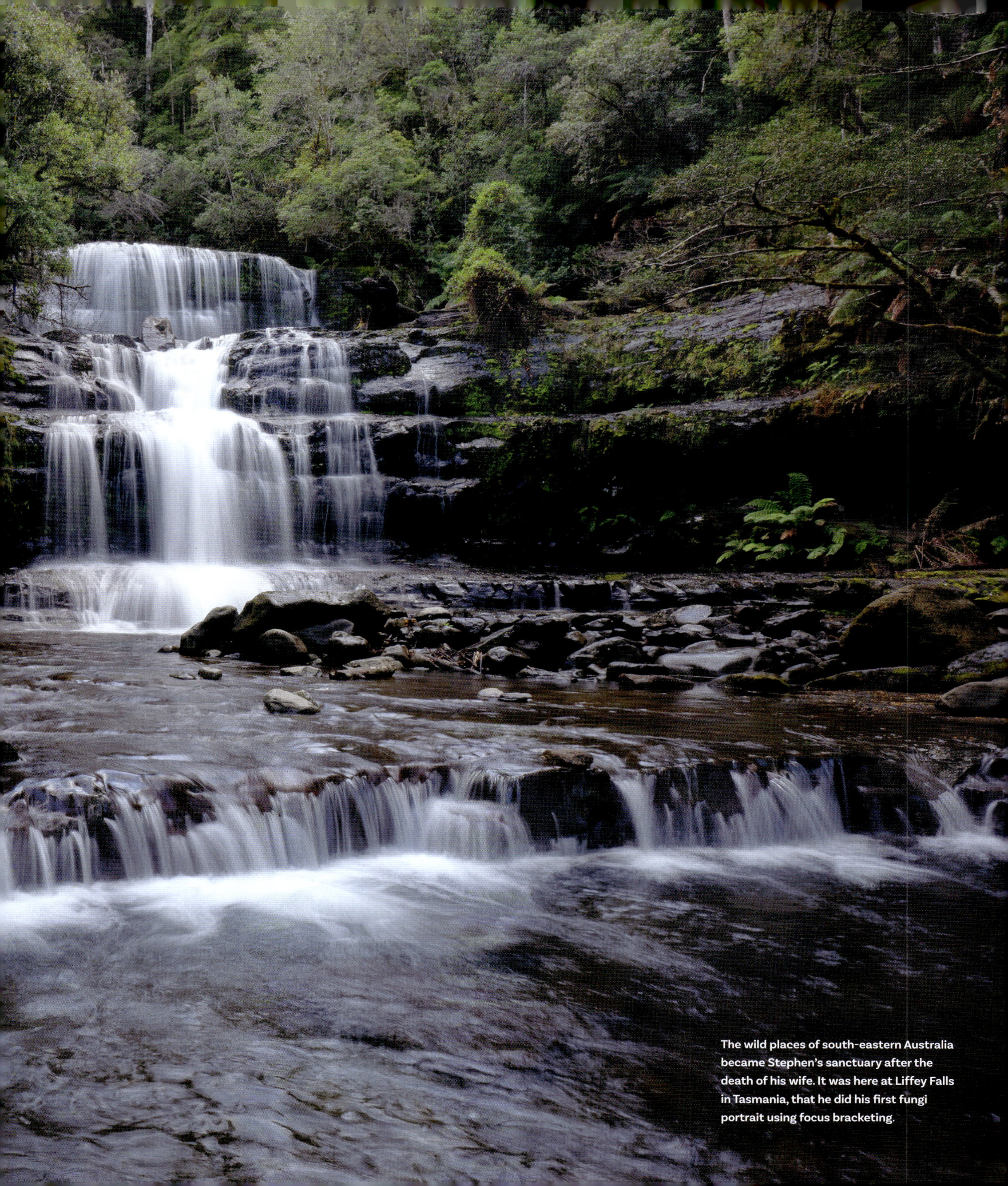

The wild places of south-eastern Australia became Stephen's sanctuary after the death of his wife. It was here at Liffey Falls in Tasmania, that he did his first fungi portrait using focus bracketing.

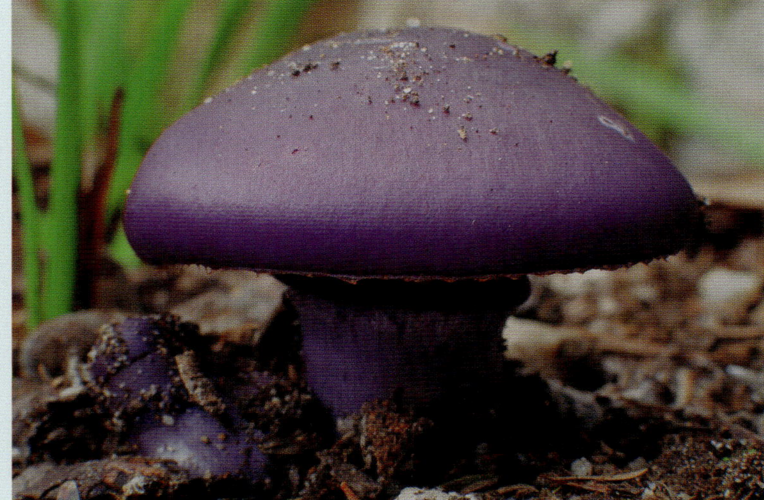

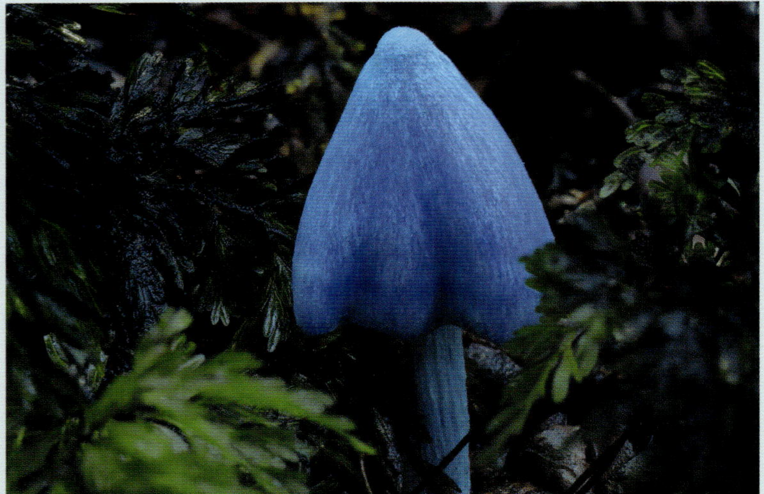

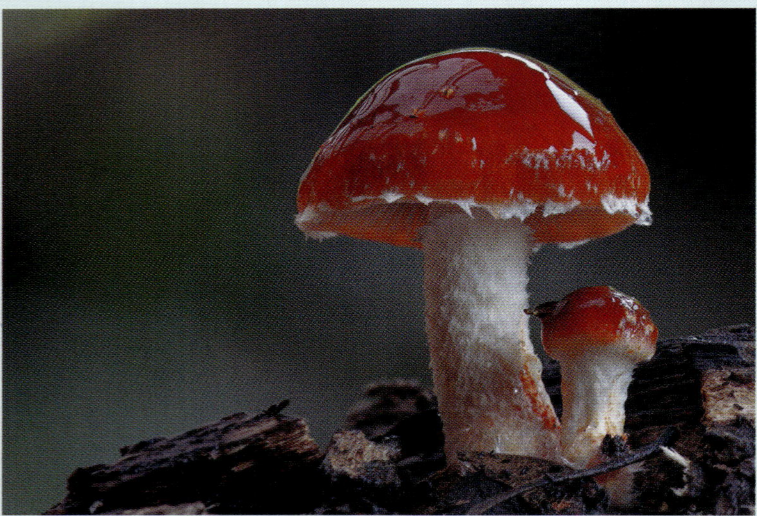

'It was the endless discovery that was so alluring, from observing a small fern frond unfolding to a fairy wren busy in the industry of constructing a home to waves crashing against rocks on a wild coastline. The miniature and the majestic transported me into a place of calm.'

Walking along a coastal heathland track on Wilsons Promontory on Australia's southern coastline, Stephen had a chance encounter with a brilliant purple mushroom.

'I didn't even know there were purple mushrooms, and if you asked me what a mushroom was, I could probably tell you it wasn't a plant or an animal, but that's about all I knew.'

TOP
***Cortinarius archeri*, emperor cortinar**
Wilsons Promontory, Victoria, Australia
Distribution: Australia

This image of an immature *C. archeri* is the mushroom photograph that started it all in May 2002.

MIDDLE
***Entoloma hochstetteri*, blue pinkgill**
Arnold River Dam Walk, Dobson, New Zealand
Distribution: New Zealand

BOTTOM
***Leratiomyces ceres*, chip cherries**
Booyong, New South Wales, Australia
Distribution: worldwide

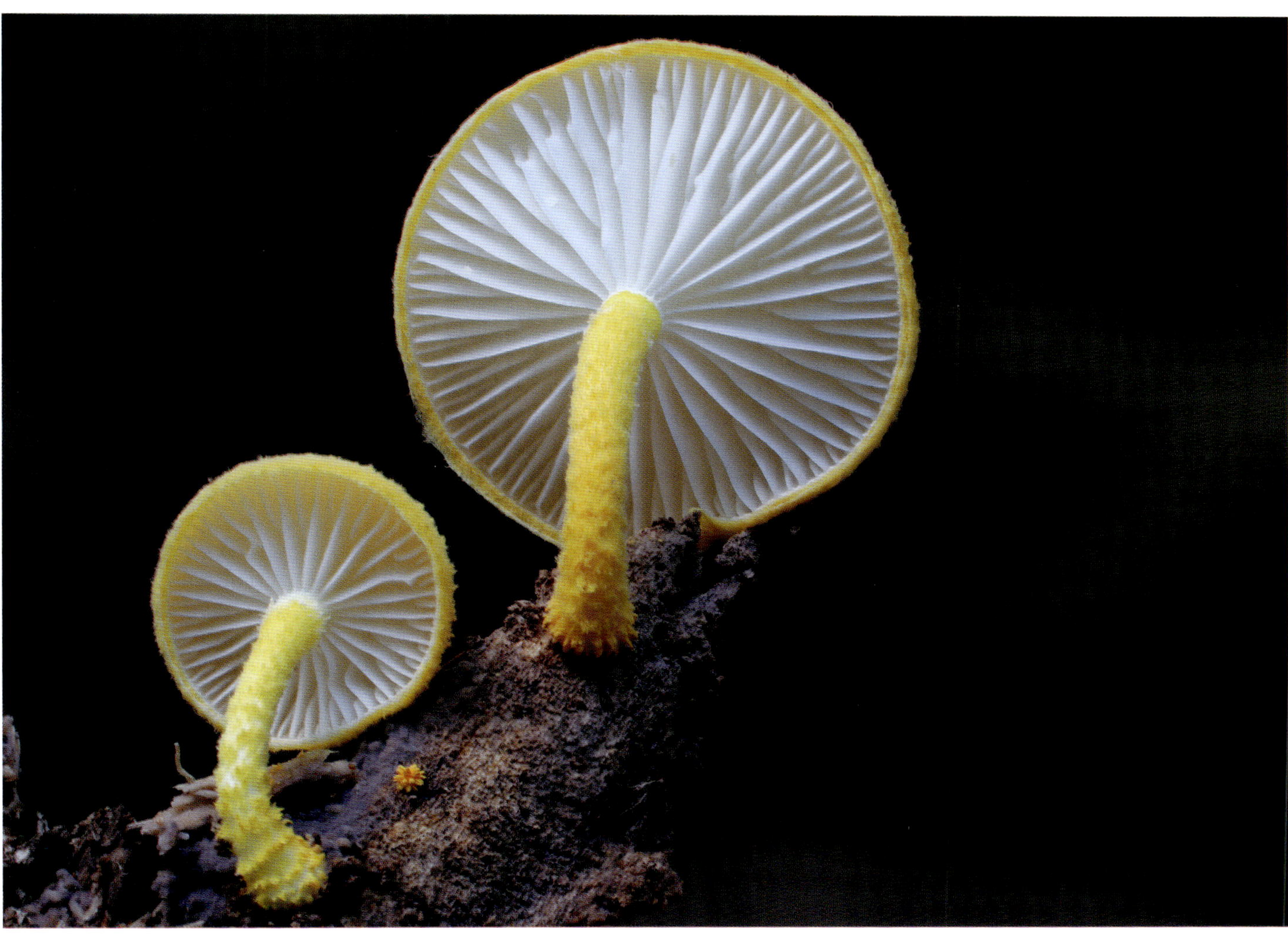

Stephen has never studied botany or ecology. He was a computer software engineer, a self-confessed geek with a growing interest in photography.

Armed with the latest Canon 4-megapixel point-and-shoot camera, he documented the purple mushroom, and in that moment, his view of the forest changed forever.

'Instead of looking up, I looked down, scouring the ground for hidden treasures.'

Cyptotrama asprata, golden-scruffy collybia
Booyong, New South Wales, Australia
Distribution: worldwide in tropical regions

It was the familiar fungal friends Stephen saw first. Those resembling the button mushrooms on our dinner plate or the fairy ring champignons erupting in the lawn. They have a cap (pileus), a stem (stipe) and gills (lamellae).

'But then I began to see fungal organisms that didn't look anything like a classic mushroom.

'Mushrooms that look like coral in the sea or shrivelled fingers on a dead man's hand.

'Mushrooms with eyelashes and others that mimic 1960s designer chairs.'

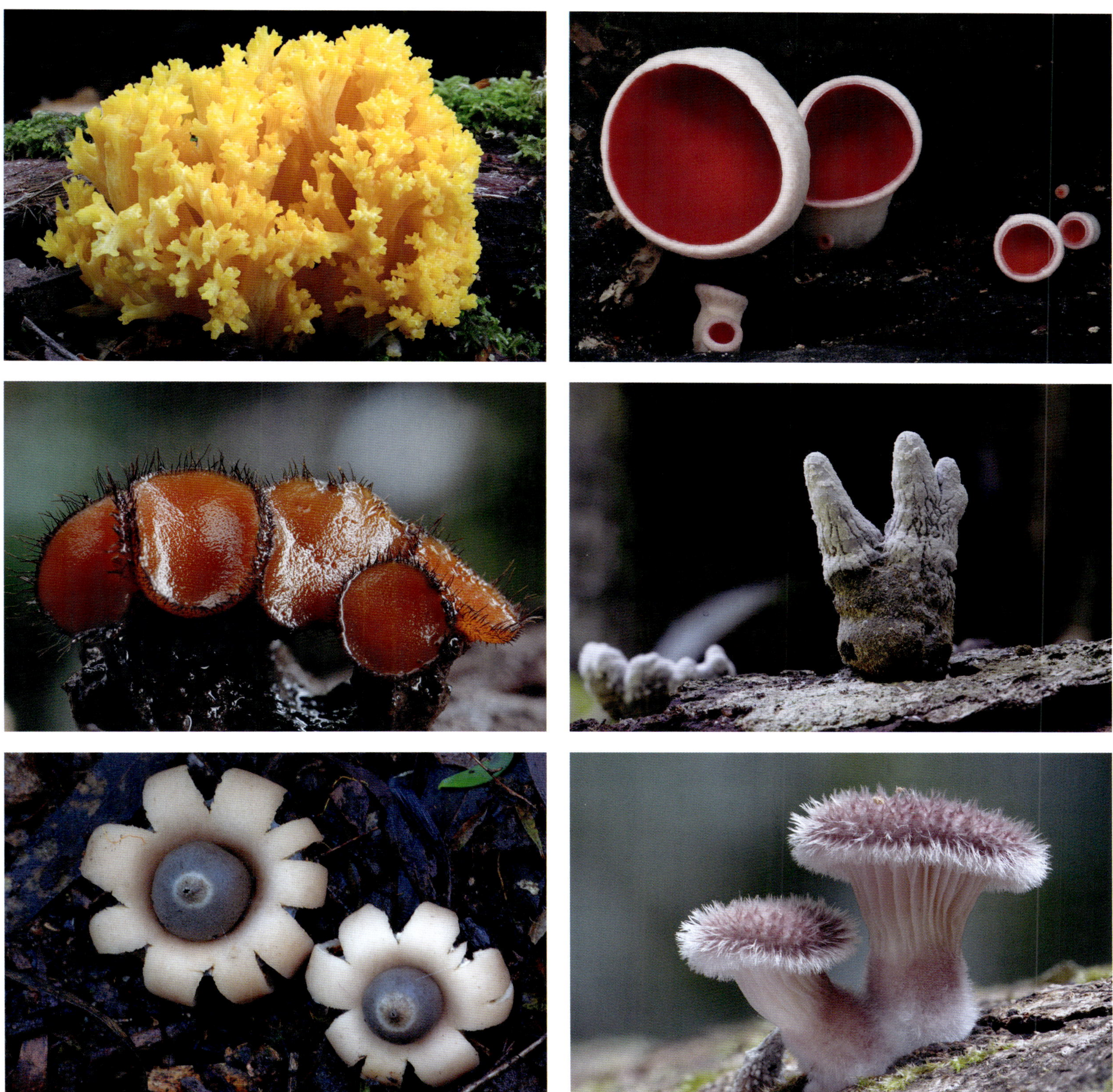

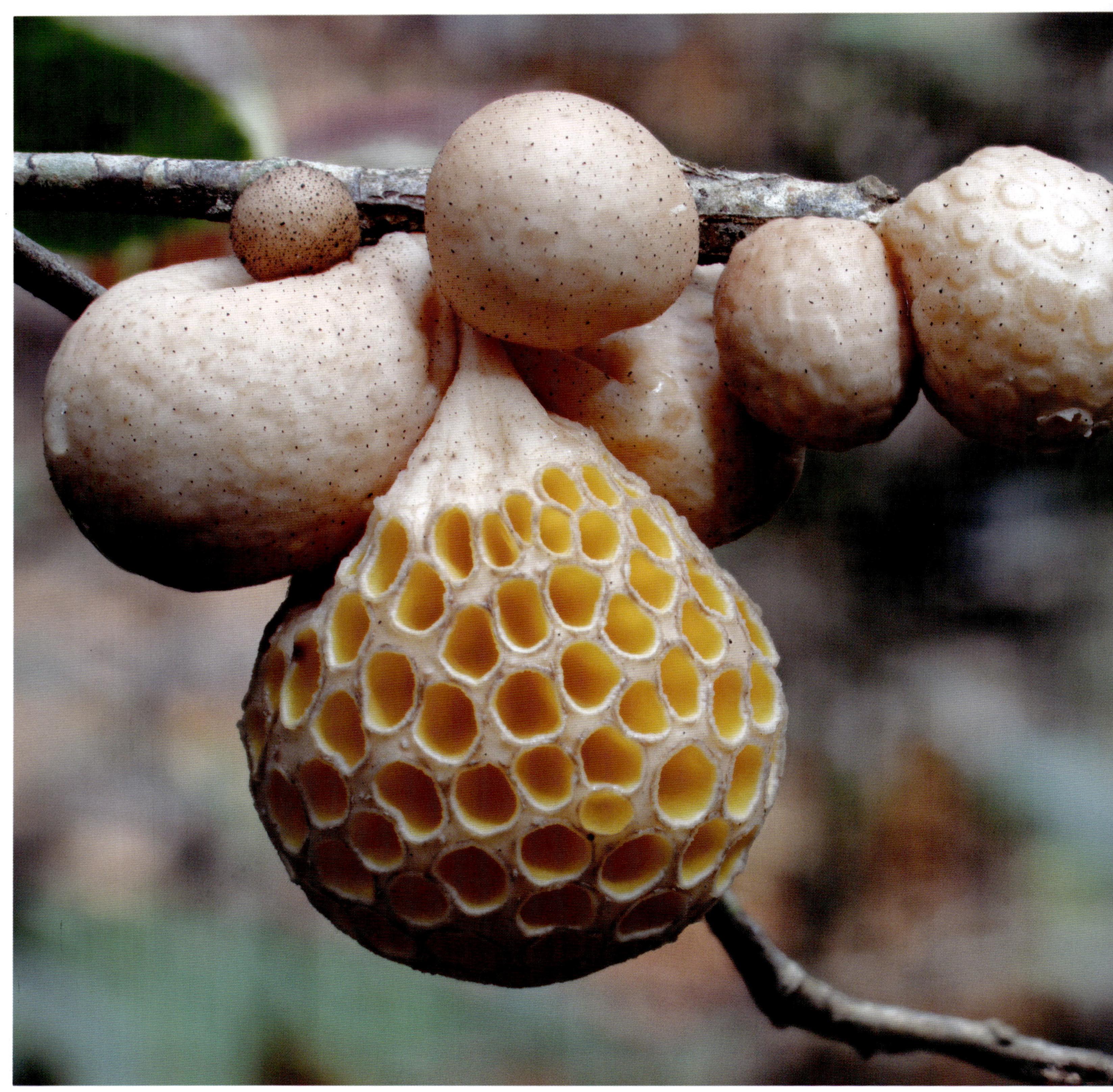

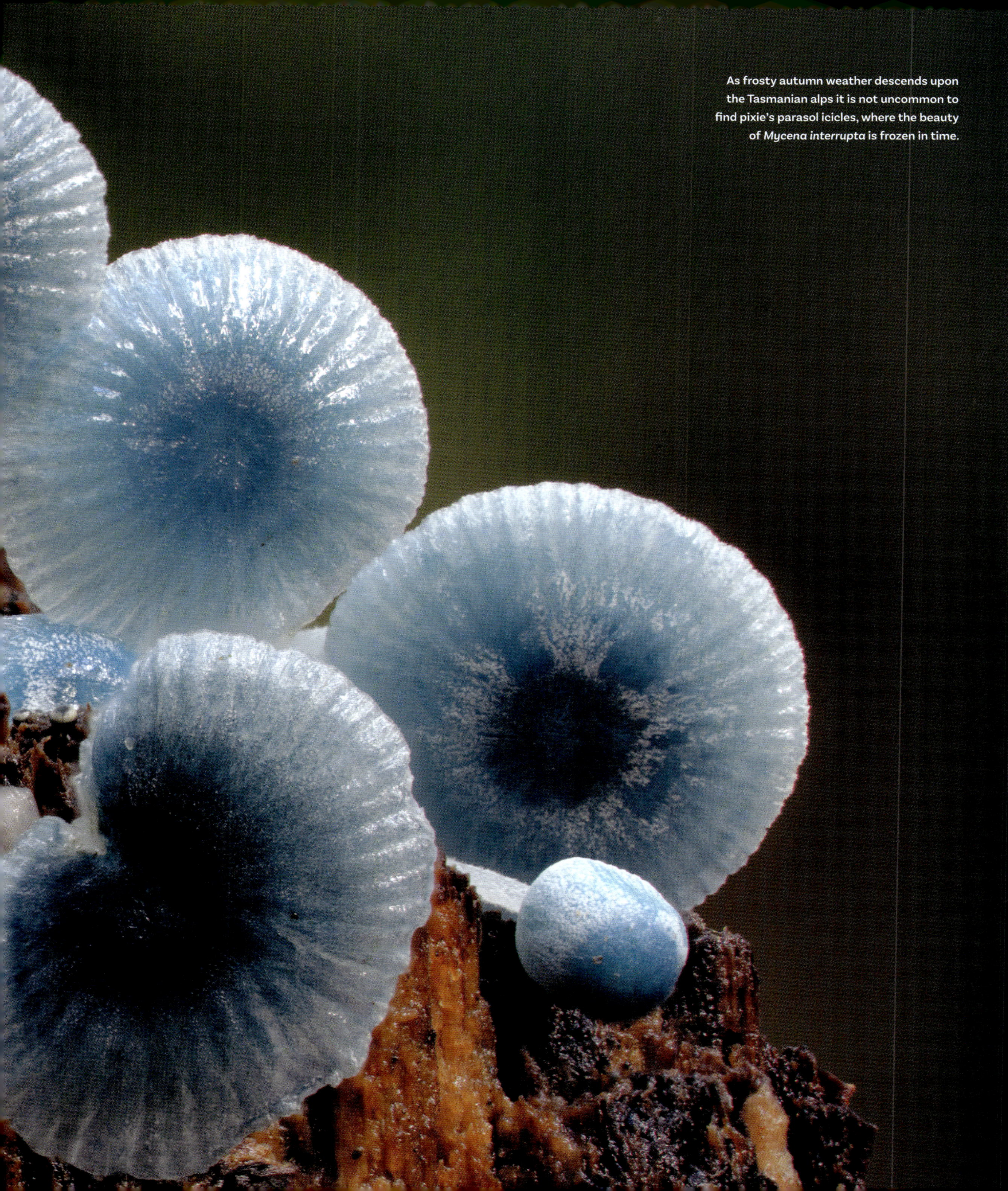

As frosty autumn weather descends upon the Tasmanian alps it is not uncommon to find pixie's parasol icicles, where the beauty of *Mycena interrupta* is frozen in time.

'A seemingly endless diversity of shapes and colours made me realise I had a very narrow idea of what a mushroom is.

'I caught the fungi hunter's obsession, the thrill of discovery and the compulsion to find more.'

Spied growing amid roots of trees, in piles of dead branches and even sprouting from decaying leaves, these early finds helped Stephen hone his skills as a fungi photographer.

His creative mission was to capture the beauty of these mushrooms in all their intricate detail.

His intellectual mission was to try find out how these charismatic organisms fit into the world of living things. He did not realise how challenging that would prove to be.

This was long before the mass myco-awakening of today.

For millennia, humans had been researching and naming the life that surrounds us, but fungi are amazingly understudied compared to animals and plants. They were until recently a neglected area of scientific study.

In part this is because historically fungal organisms were thought to be plants.

Cyttaria septentrionalis
Gloucester Tops National Park, New South Wales, Australia
Distribution: Australia, growing on Antarctic Beech
on the tops of a few mountain areas in southern
Queensland and northern New South Wales.

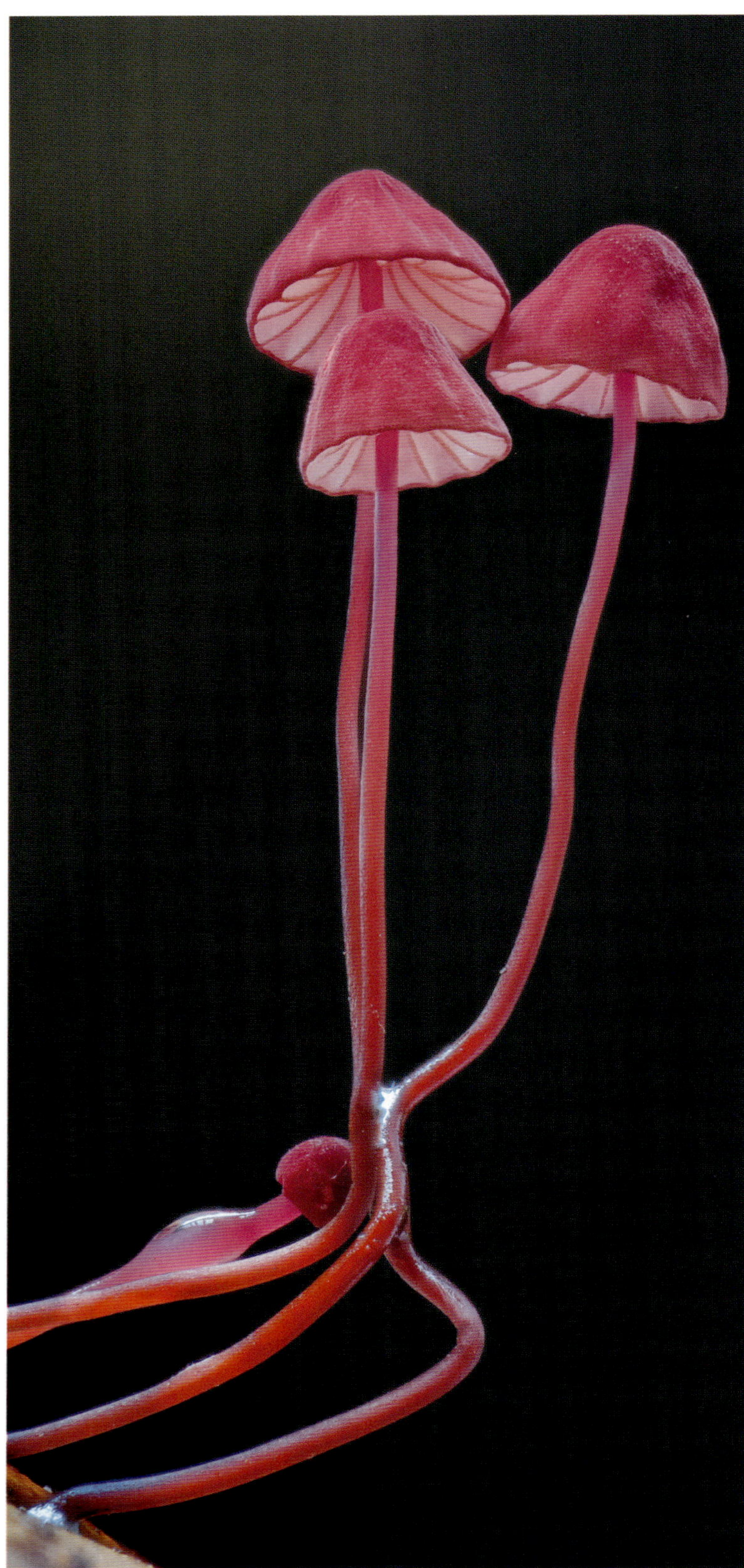

LEFT
Marasmius cf. *haematocephalus*
Booyong, New South Wales, Australia

OPPOSITE
Xylaria sp.
Booyong, New South Wales, Australia

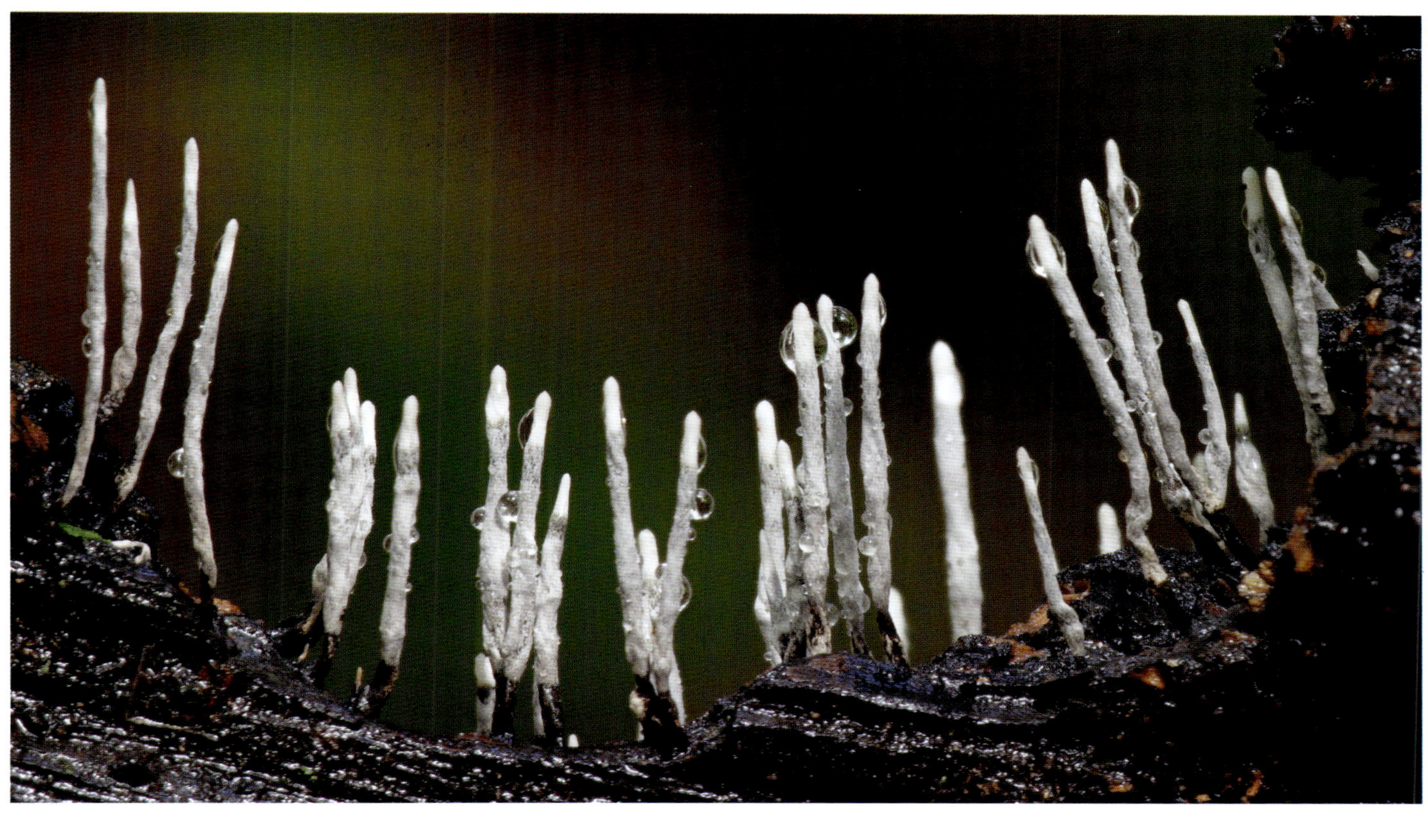

Rewind a couple of centuries, before the developments of molecular biology and DNA sequencing. The science of taxonomy, the classification of organisms in the natural world, was in its infancy

In the 18th and 19th centuries there was a surge in scientific exploration and discovery.

European explorers, enabled by safer navigation and shipbuilding, were on a quest to find new lands, new trade routes and markets.

It was also the Age of Enlightenment, an intellectual and cultural movement in Europe focused on reason rather than religion and the sciences as a source of new knowledge.

When the explorers set sail on their pioneering adventures, scientists went with them, resulting in a boom time for cataloguing plants and animals.

A system of classification was designed by Carl Linnaeus, a Swedish botanist, physician and zoologist known as the 'father of modern taxonomy'.

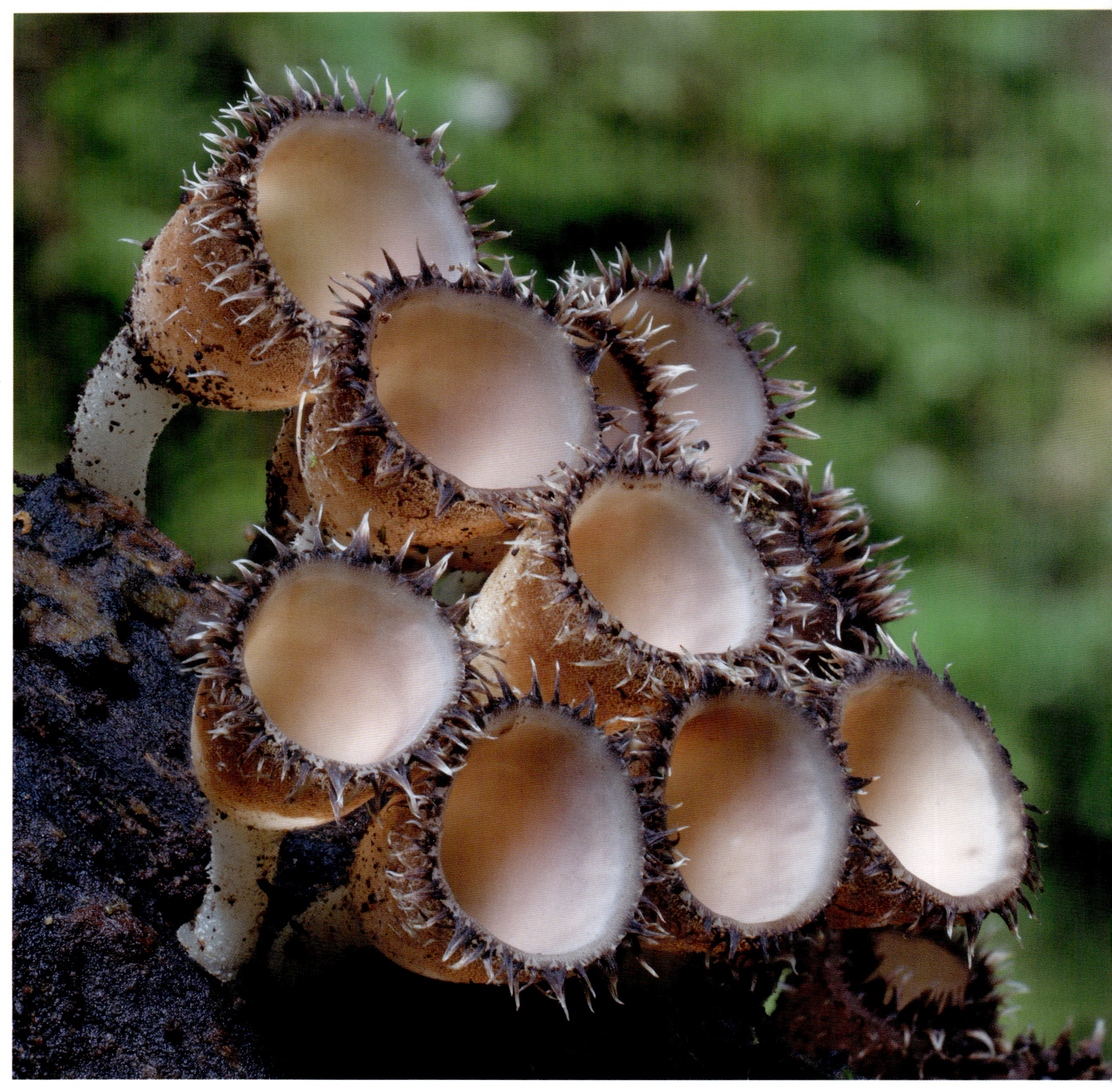

Virologist Charles H. Calisher observed that Linnaeus 'liked things "neat" and he was not satisfied with the unwieldy names used at that time for biological entities ("Physalis annua ramosissima, ramis angulosis glabris, foliis dentato-serratis" do not exactly roll off the tongue).'

'He brilliantly and consistently applied to all sorts of living things what we call "binomial nomenclature".'[1]

In his 1753 publication *Species Plantarum,* Linnaeus detailed a system in which each organism has a two-part name comprising the name of a genus followed by a species epithet.[2]

It mirrors the way many societies refer to people by a first name and a surname.

The living world was divided into two groups based on the perceptions of scientists and naturalists dating back to the time of Aristotle, the kingdom of plants were the organisms that are stationary, and the kingdom of animals those that are mobile.

The window into the fungal world at that time was primarily mushrooms, which are stationary, so fungi were lumped into the kingdom of plants.

***Cookeina insititia*, fringed goblet**
Nightcap National Park, New South Wales, Australia
Distribution: Asia and Australasia

It was not until 1969, the same year that Neil Armstrong made a small step onto the surface of the Moon, that mycology made a long overdue giant leap into its rightful place in the sciences. In a classification reshuffle, a unique Fungi kingdom was recognised, firmly establishing that fungi were neither plants nor animals.

Three decades later, in 2002, when Stephen had his first fungal 'a ha! moment', interest in mycology was starting to gain momentum.

In Australia, there were some illustrated field guides for fungi. The citizen science group Fungimap, the brainchild of mycologist Dr Tom May, was targeting the mapping of specific species and harnessing the power of citizen scientists to increase knowledge of Australian fungi.[3]

It was the perfect time for someone like Stephen to enter the realm of the fungal kingdom.

Armed with a camera instead of a sketch pad, he became part of a global modern-day naturalist phenomenon: people with time, a love of nature and the tools to document it.

It is Stephen's specialisation in fungus macro photography and his skill in transcending the limitations of a single focal plane that sets him apart.

Early on, Stephen blended his inner scientist and inner artist, as he mastered the art of focus stacking. This is a technique that was only made possible with the development of digital cameras.

With the camera on a tripod, several frames are taken of the same shot. This is like a CT scan where each image has a different slice of the mushroom in focus. These multiple images are then seamlessly merged into a single frame.

Today, in-camera and post-production software can merge hundreds of images into one frame in minutes but in the early 2000s it had to be done manually.

Each image took Stephen hours to post-produce.

Caloboletus sp.
Diqing Tibetan Autonomous Prefecture, Yunnan, China

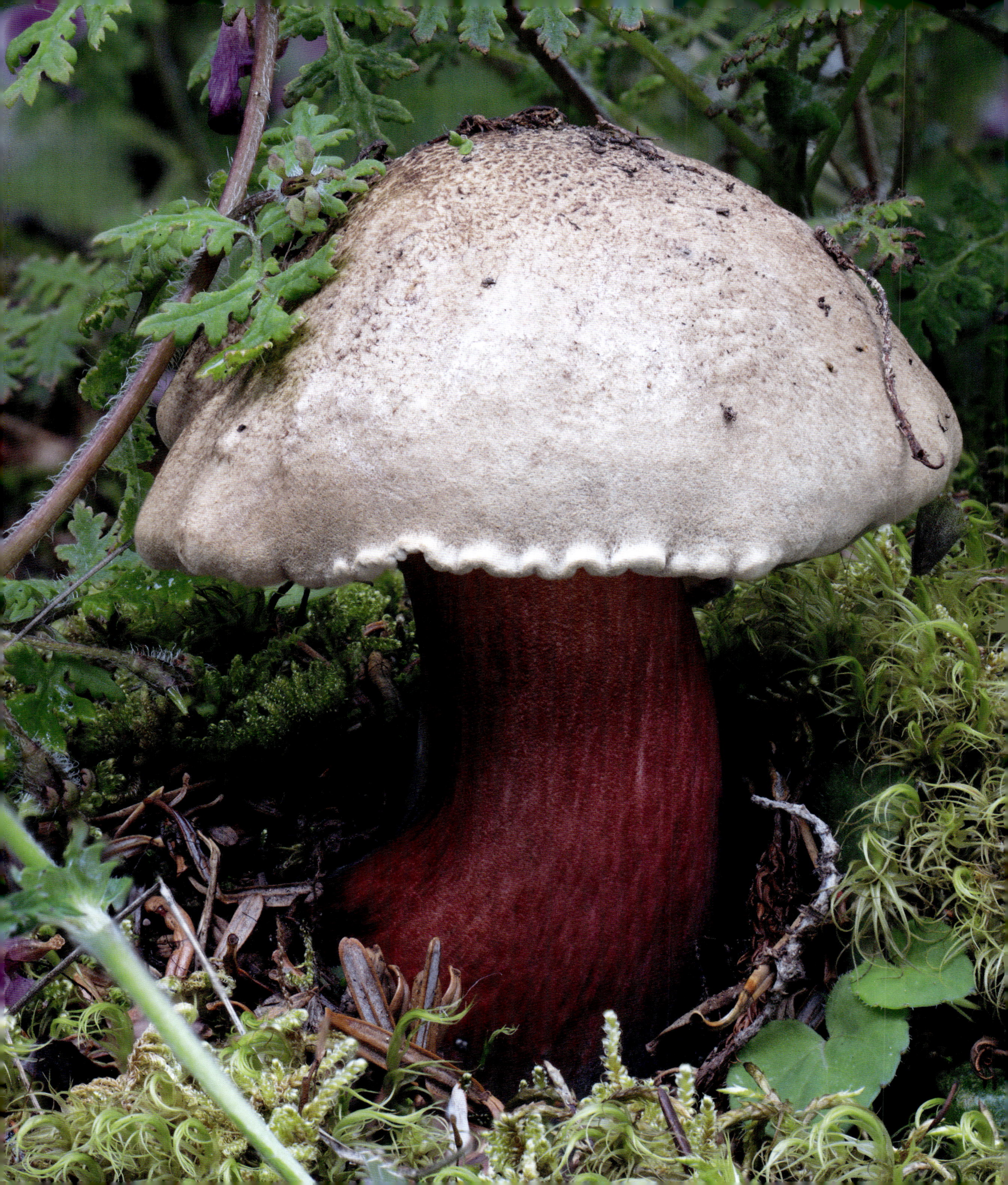

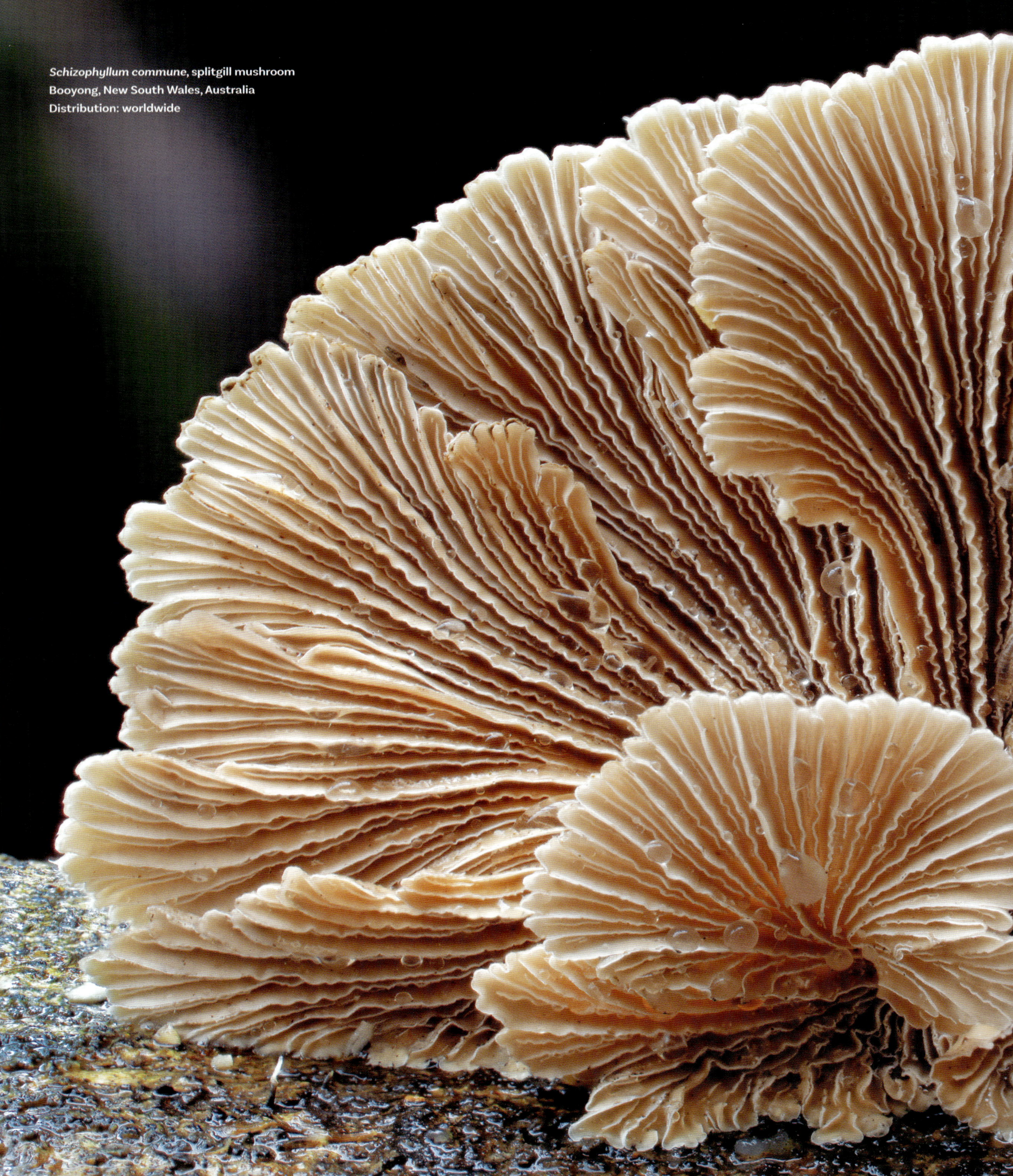

Schizophyllum commune, splitgill mushroom
Booyong, New South Wales, Australia
Distribution: worldwide

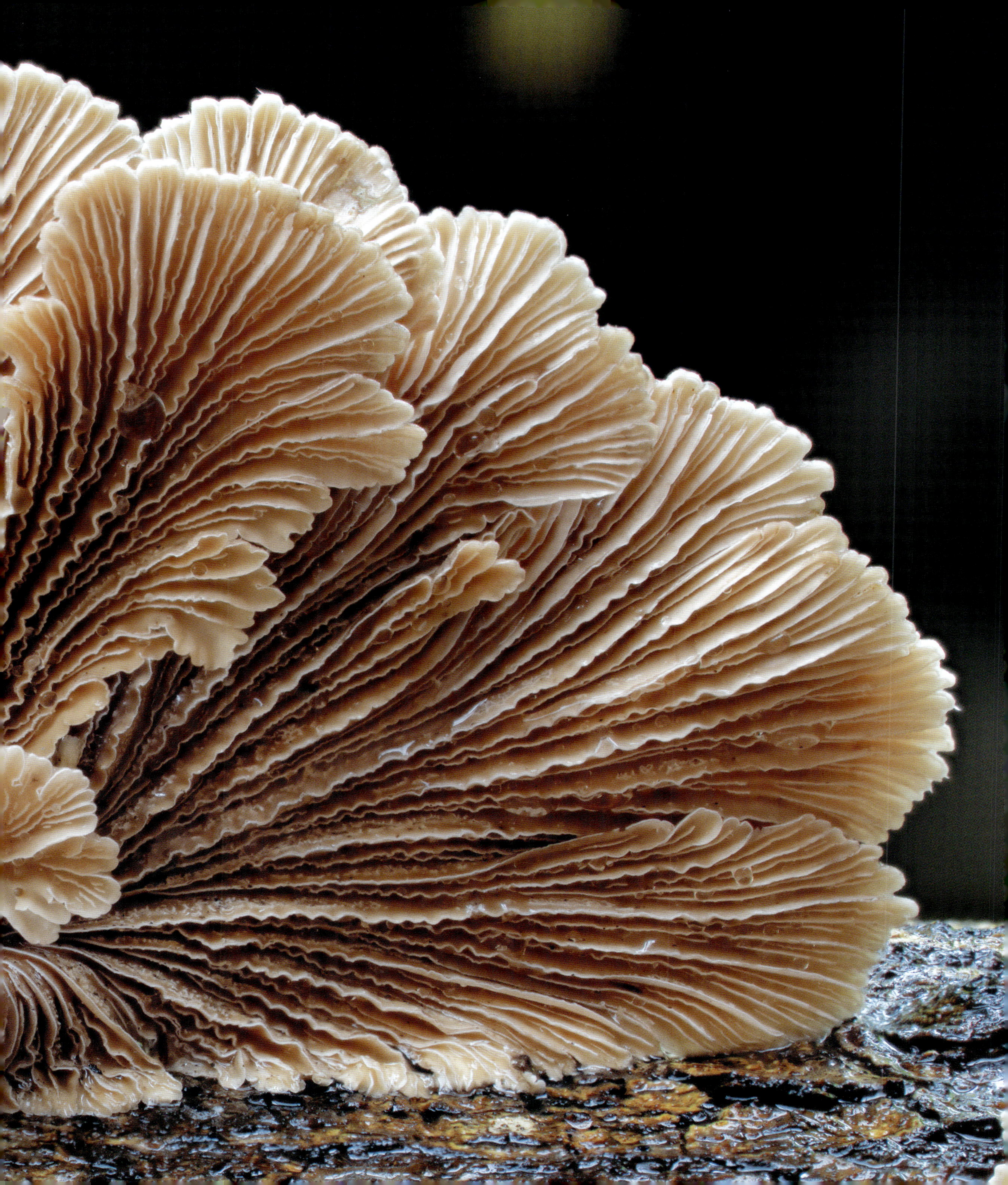

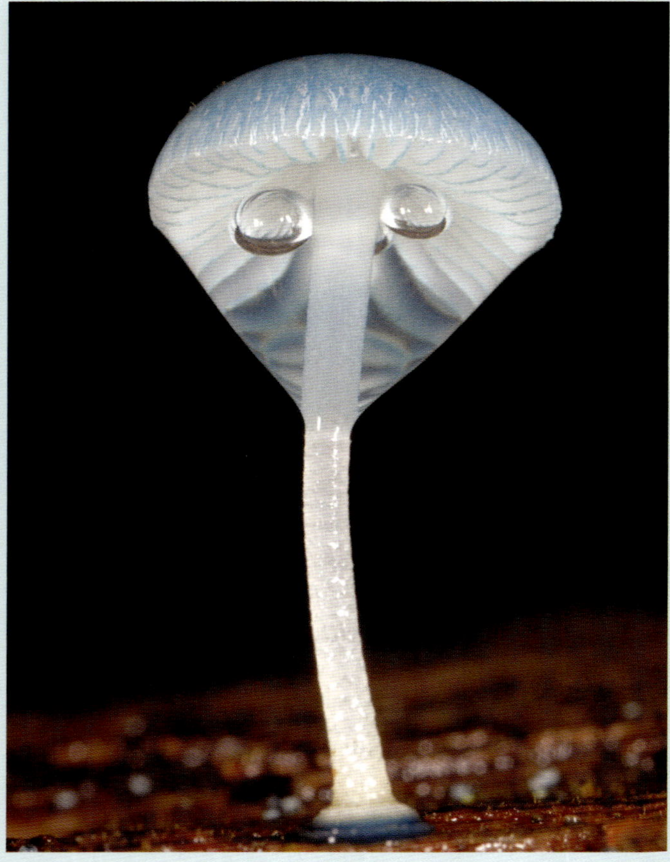

The first image Stephen focus stacked was of a tiny mushroom with two air bubbles in a drop of water. It was just 8 millimetres (mm) high and 4 mm across the cap.

The common name for this striking blue mushroom *Mycena interrupta* is pixie's parasol.

It is one of a group of very similar species found in the Gondwanaland rainforests of South America, New Zealand and Australia.[4]

This one was on a fallen log on the banks of the Liffey Falls walking trail in Tasmania, Australia.

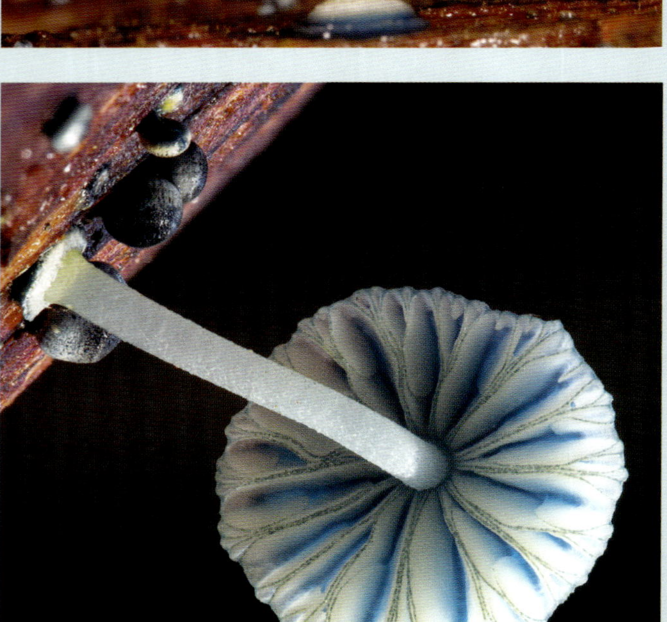

TOP
The first photograph that Stephen manually focus-bracketed was a tiny *Mycena interrupta* at Liffey Falls in Tasmania.

BOTTOM
Mycena interrupta, pixie's parasol
Trowutta Arch, Tasmania, Australia
Distribution: Australasia

OPPOSITE
Mycena interrupta grows on wood and is attached to the substrate by a flat white disk rimmed with dark blue, with caps ranging from as small as 0.4 cm (centimetres) to as large as 2 cm. It is often found in photogenic groups.

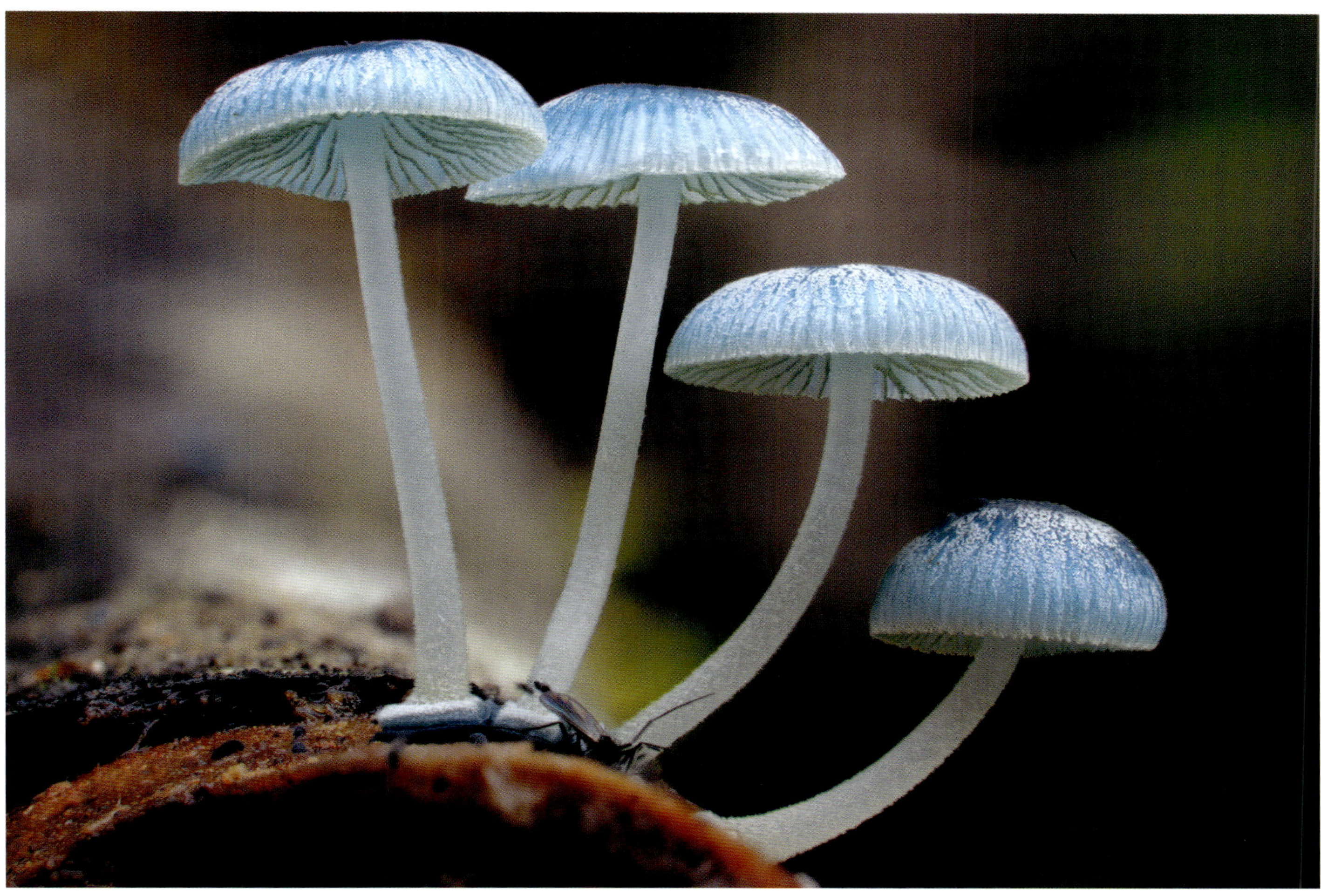

'I had carried an unusual lens in with me, the Canon MP-E 65 mm. It's a specialist macro lens and won't focus on anything that is more than 31 cm away.'

'It doesn't have a focus ring so the only way to get the image sharp is to change the distance between the camera and the subject. In the field I use it handheld but that has its challenges for focus stacking, as every frame is slightly different.'

Using an early version of Adobe Photoshop, Stephen layered three photographs on top of each other, resizing and skewing each frame to correct perspective shifts. Then with the mask tool, the most focused area of each frame is painted onto the final image.

The result is an ethereal portrait, capturing the interplay of soft light and razor-sharp details. It also reveals a visual narrative, a moment of interaction between fungus and the gentle raindrops of a winter shower.

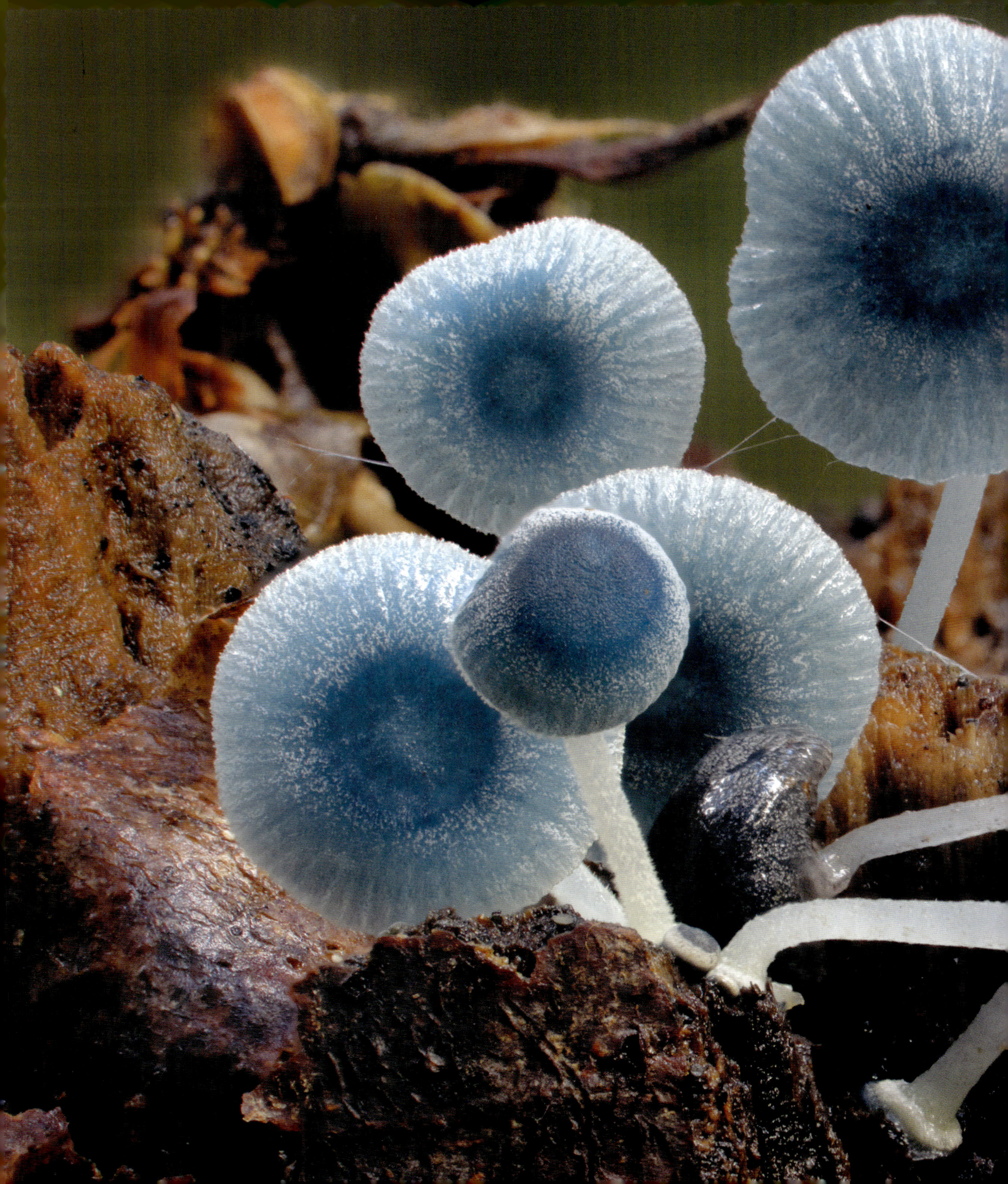

'I spent two hours studying and photographing this one mushroom and the ecosystem that surrounded it. People call it forest bathing these days. Losing oneself in the minutiae of life, it's good for the soul.'

Stephen became captivated by pixie's parasol's vibrant hues, ranging from brilliant turquoise to deep sapphire.

This seduction by colour dominates his early fungi photography, and as the technology of focus stacking developed, his fungal portraits became living sculptures, teeming with textures, patterns and intricate details impossible to see with the naked eye.

They are images that evoke a sense of wonder, transcending mere documentation.

Podoserpula pusio, pagoda fungus
Queenstown, Tasmania, Australia
Distribution: Australasia and Madagascar

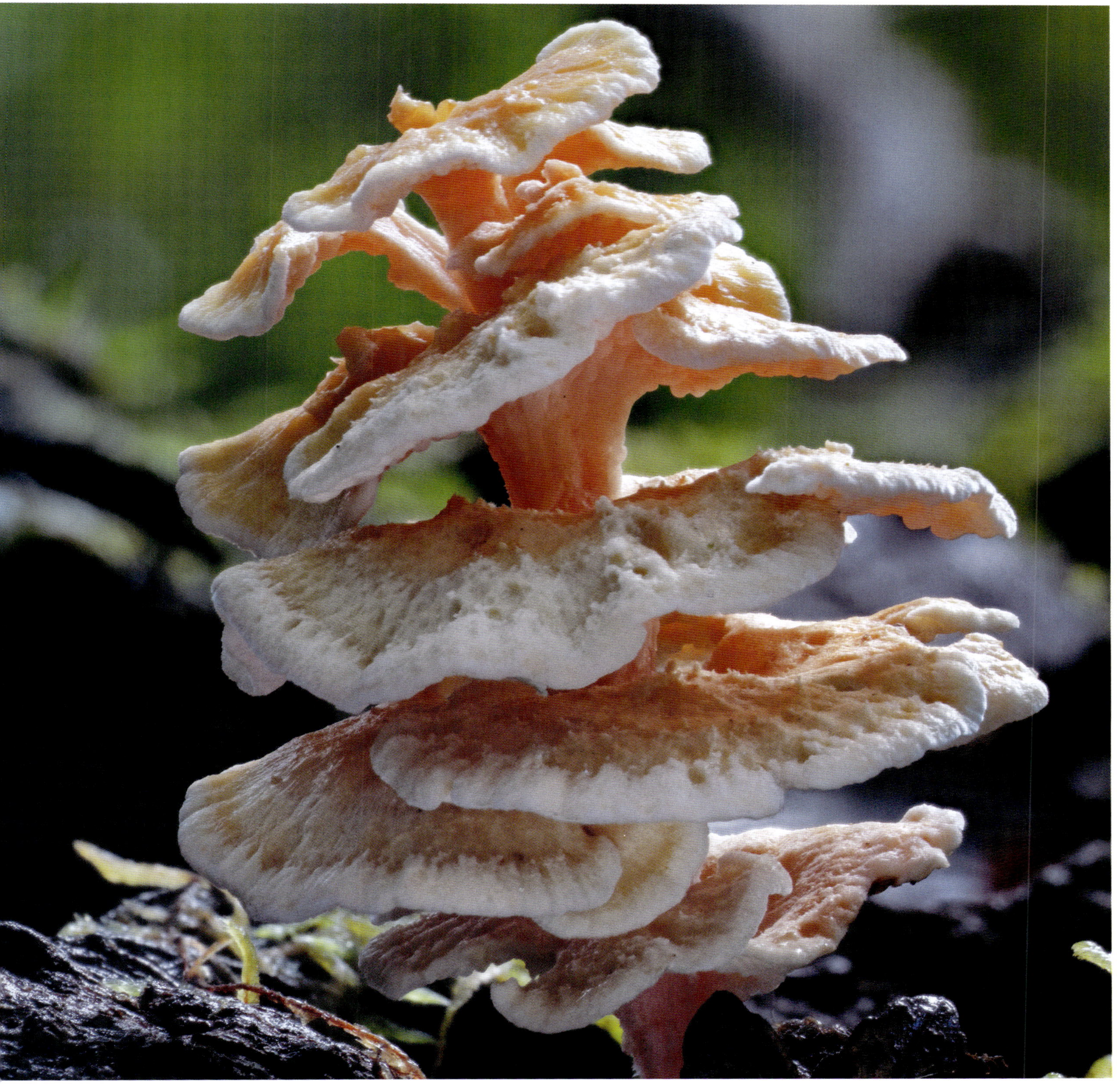

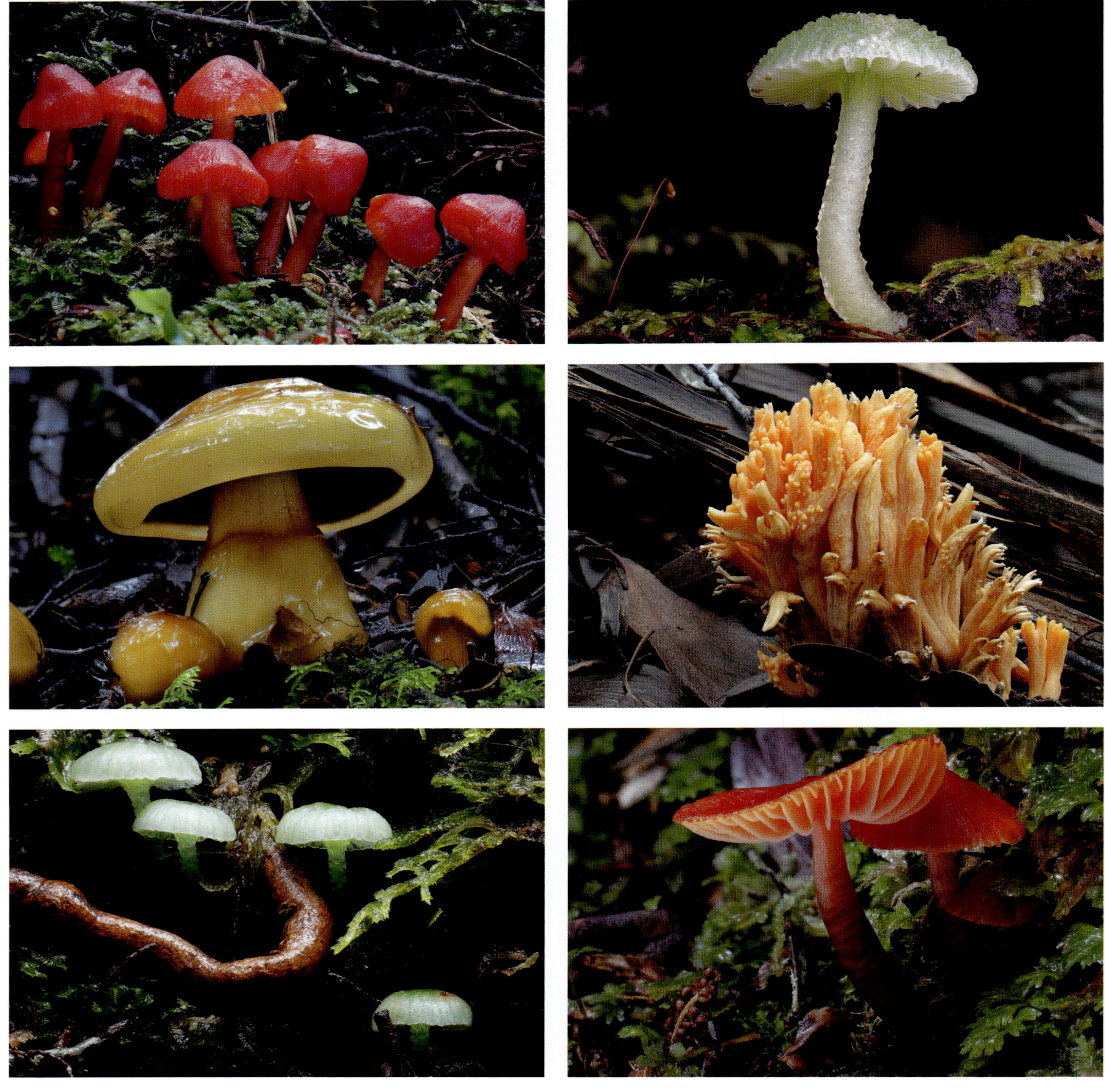

'I started to set myself fungal challenges, like photographing the unusually coloured mushrooms of the forest.

'I was very excited when I found my first Gliophorus graminicolor, a stunning green mushroom that grows in the ancient rainforests of Tasmania in Australia and New Zealand. It's quite small, only 10 to 20 mm across the cap, and they are very hard to see, moss-green mushrooms growing in moss-green moss.

'I knelt and took the photograph, which I thought was a very special single specimen. Then I stood up and realised I'd been kneeling on a whole patch of green mushrooms hiding in plain view.

'It was like someone opened my eyes.'

OPPOSITE TOP LEFT
Hygrocybe sp.
Julius River, Tasmania, Australia

OPPOSITE MIDDLE LEFT
Cortinarius sinapicolor
Frenchmans Cap Track, Tasmania, Australia
Distribution: Australia

OPPOSITE BOTTOM LEFT
Gliophorus graminicolor
Franklin River Nature Trail, Tasmania, Australia
Distribution: Australasia

OPPOSITE TOP RIGHT
Gliophorus graminicolor
Whakapapanui Walk, Tongariro National Park, New Zealand
Distribution: Australasia

OPPOSITE MIDDLE RIGHT
Ramaria sp.
Cradle Mountain, Tasmania, Australia

OPPOSITE BOTTOM RIGHT
Hygrocybe sp.
Julius River, Tasmania, Australia

Humans for millennia have observed that seasonal rains summon countless fungal characters to the forest stage.

So Stephen followed the rain to document the fungi that emerged. As a result, his catalogue of fungi images from uncharted wild places expanded to thousands in just a few years.

The more his lens unveiled, the more questions took root.

Why are some mushrooms so transient, here one day and gone the next, while others are found under the same ancient tree, year after year?

Why are there are so many different colours, shapes and sizes?

As he began to research, the first revelation emerged.

These mushrooms that had so thoroughly charmed him were but the fleeting reproductive bodies of an intricate and hidden living web – a mycelium.

This is the part of a fungus that grows and feeds. It is composed of a network of hyphae – thread-like structures, interweaving in soil and within organic matter.

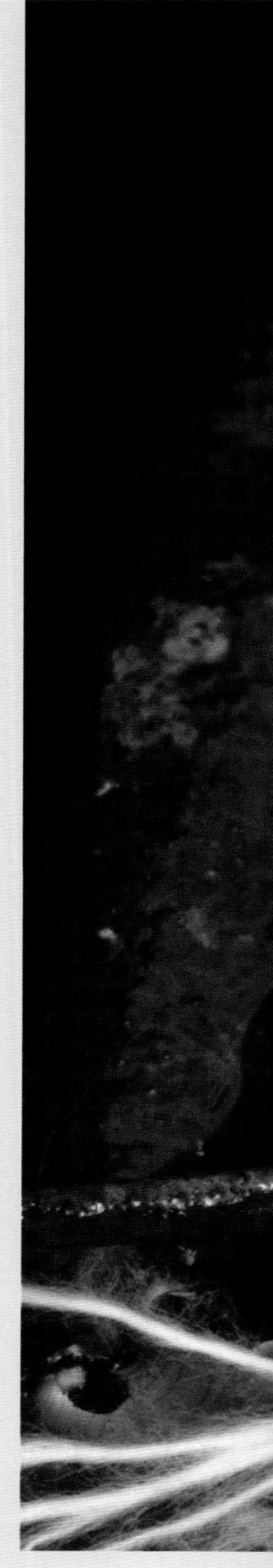

Hyphal threads branching and bundling together to form a mycelium.

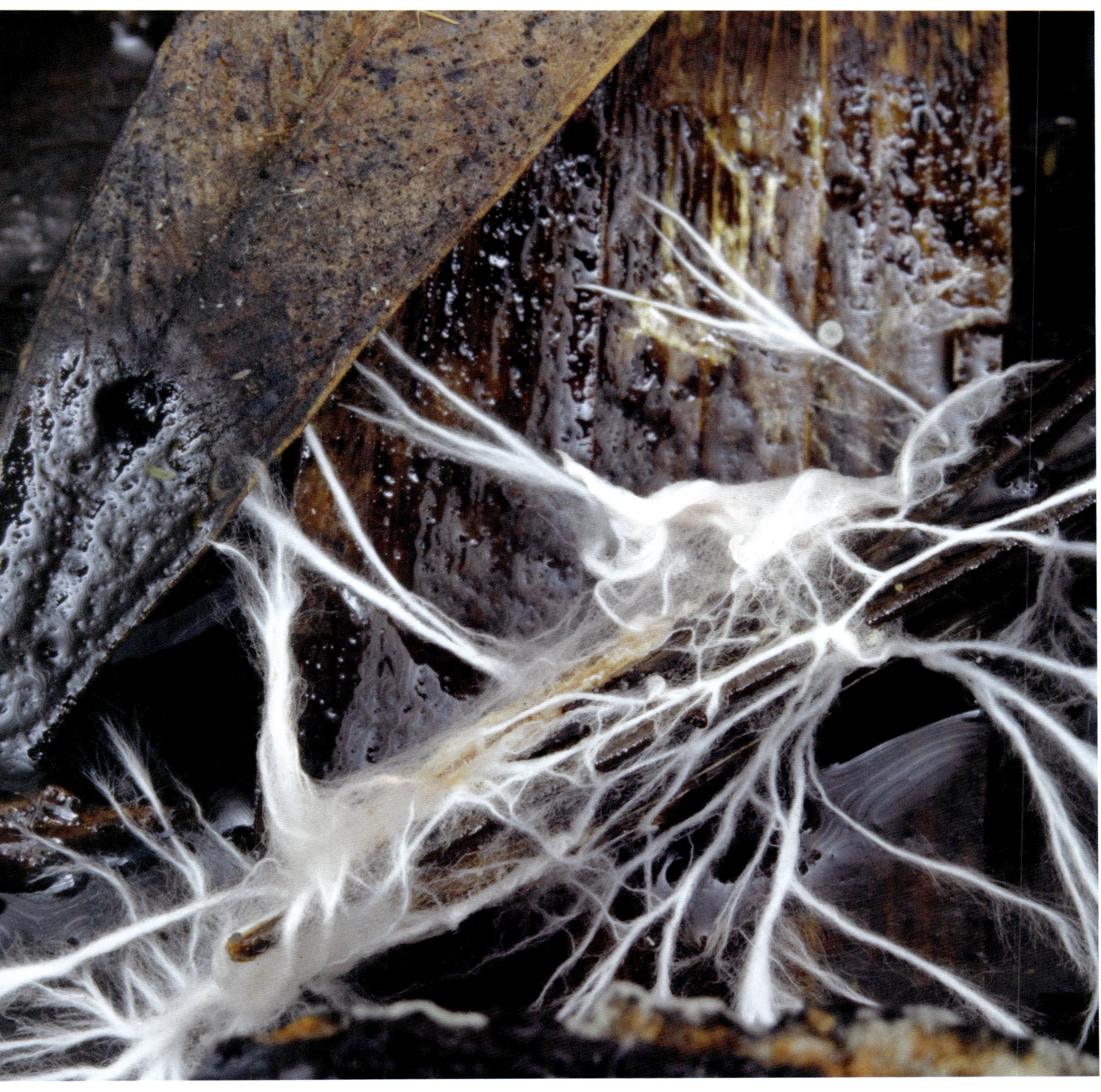

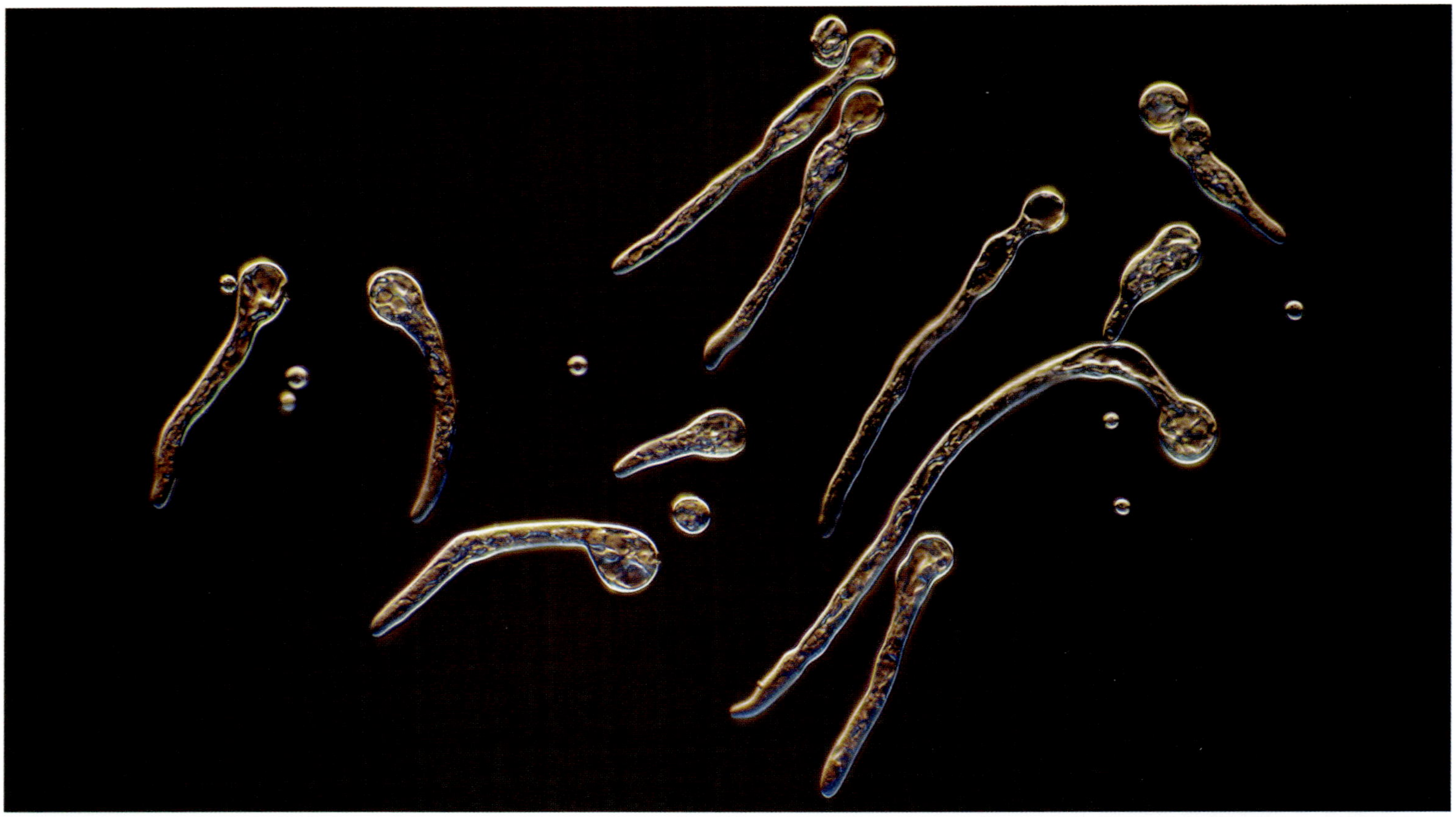

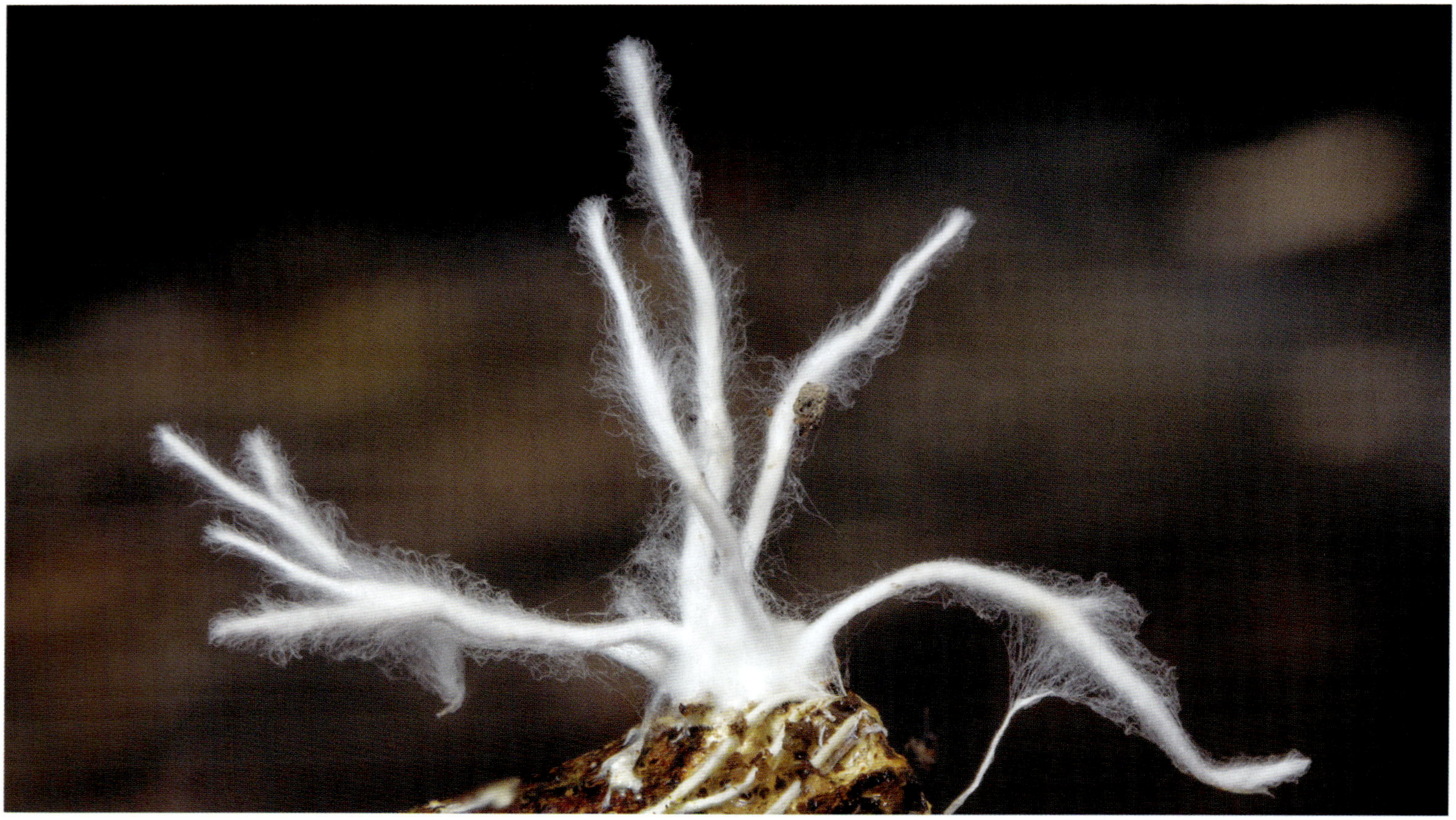

Picture life as a spore, drifting from a mushroom's embrace. It settles on the ideal canvas for germination and sends forth a short hypha known as a germ tube.

Astonishingly fine, individual hyphae measure no more than a hundredth of a millimetre in diameter.

From this humble beginning, millions of branches radiate, branch and reconnect in a complex three-dimensional network.

At the tip of each hypha lies a world of potential, the power of unlimited growth, which allows it to explore its substrate for nourishment and the opportunity to reproduce.

In the fungal kingdom, sex radically differs from that of plants and animals.

Instead of a joining of males with females, compatibility is determined by mating types, a genetic code guiding the potential of union.

To mate, the hyphal tips of two mycelia with different but compatible mating types merge.

Some fungi have two mating types, while others boast thousands.

OPPOSITE TOP
The moment of germination when hyphal threads emerge from spores.
Photograph: Wim van Egmond

OPPOSITE BOTTOM
Mycelium inside a fallen branch becomes visible at it grows towards its next meal.

'I can still recall the first time I encountered the fungus Schizophyllum commune. *At first glance, it looked like a scruffy specimen with a drab furry surface, but as soon as I turned it over and saw the sculptural lacework of its splitting gills, I was astonished by its beauty.*

'I did some further research on this species and realised that when immersing myself in the fungal world, I was going to have to rethink what I knew about gender and sex completely.

*'*Schizophyllum commune *has an incredible 23,000 different mating types.*[5]

'It is fascinating to think that in this species, any individual can partner with 99 per cent of its own kind, including some of its siblings, as long as they are a different mating type.'[6]

***Schizophyllum commune*, splitgill mushroom**
Booyong, New South Wales, Australia
Distribution: worldwide

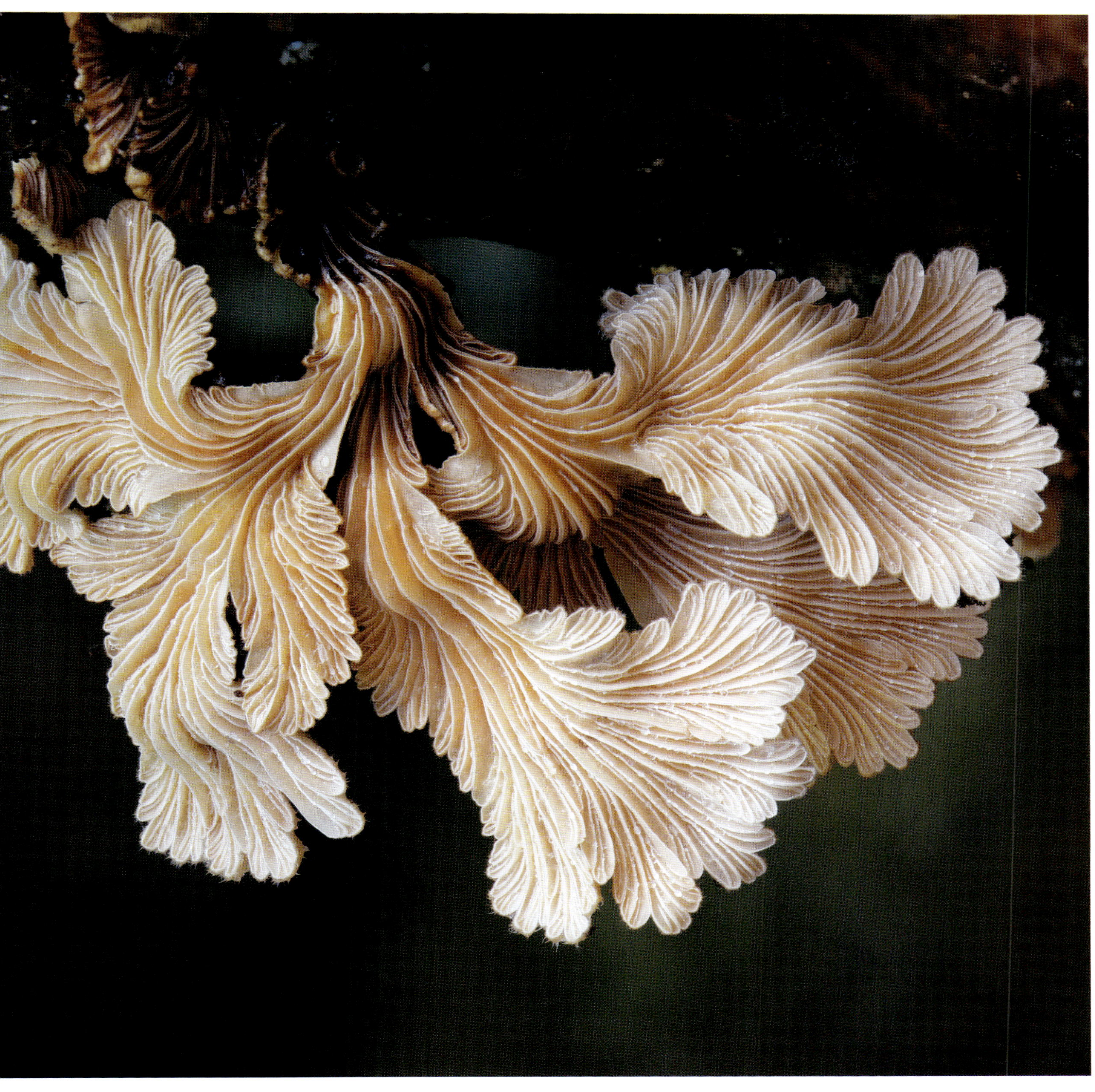

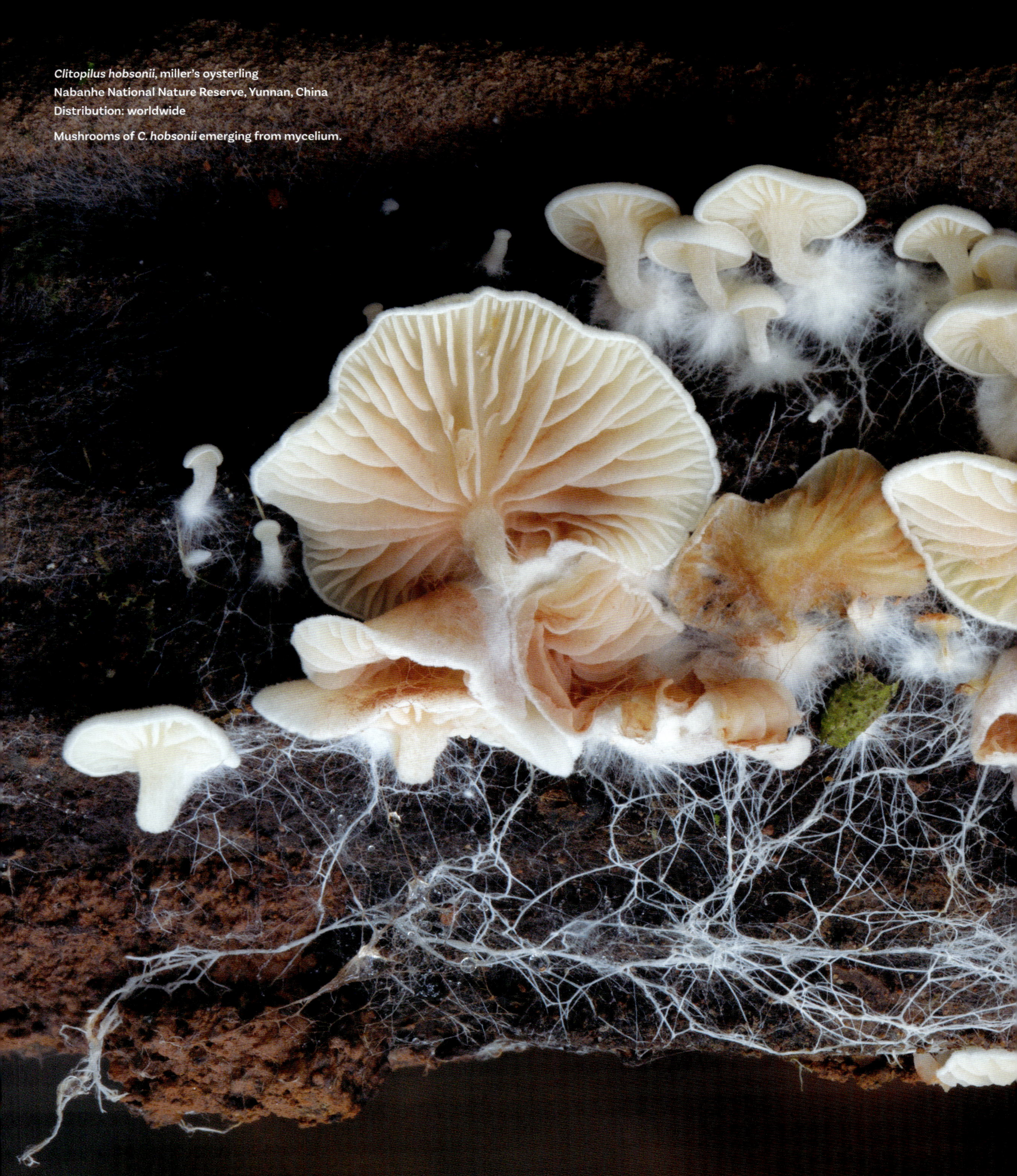

Clitopilus hobsonii, miller's oysterling
Nabanhe National Nature Reserve, Yunnan, China
Distribution: worldwide

Mushrooms of *C. hobsonii* emerging from mycelium.

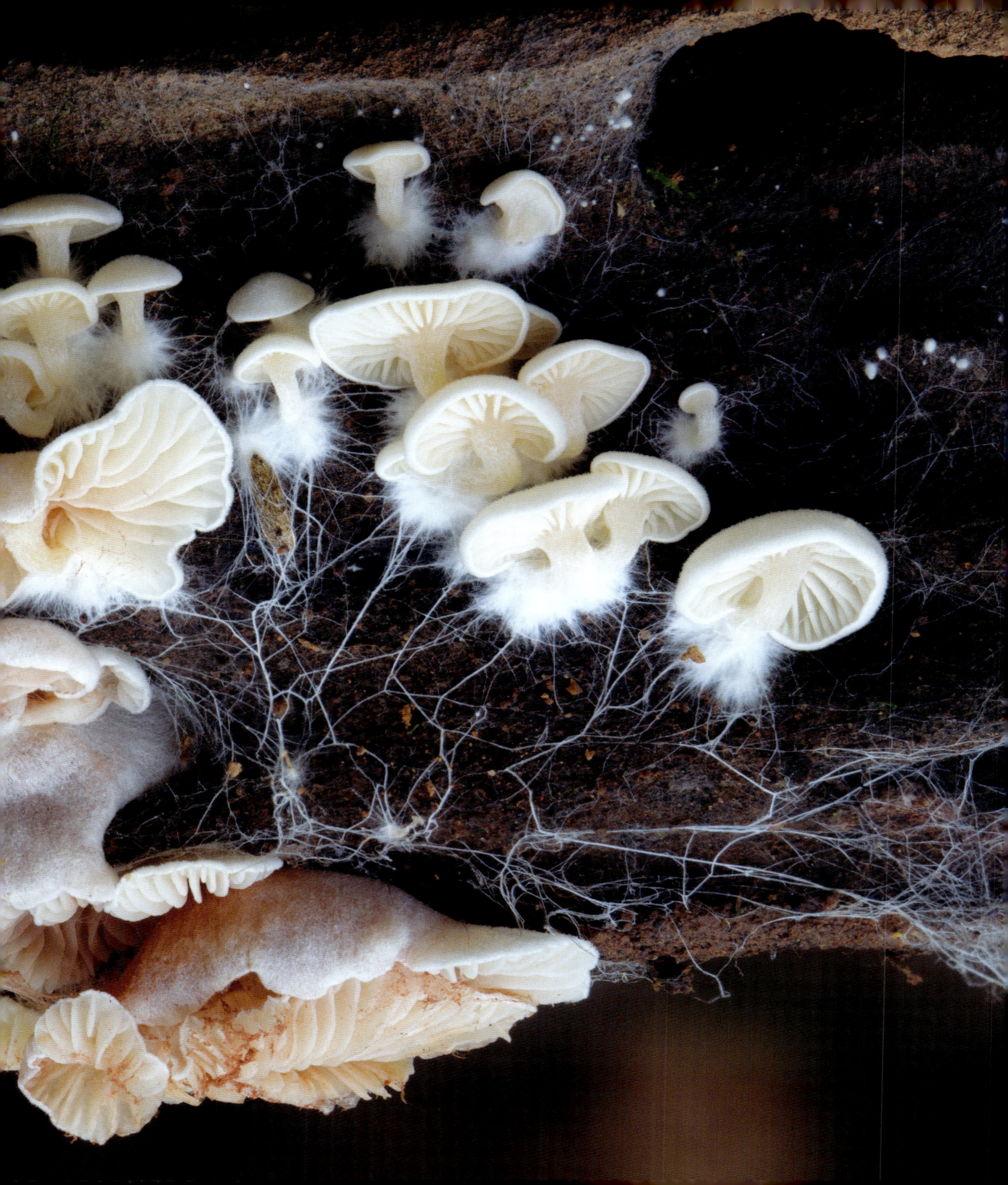

But mycelia merging is just the start of fungal sex.

The individual nuclei from both spores remain intact until the mycelium produces a sporing body, the mushroom we see nudge its way into the world.

It is only in the final moment of reproduction that the nuclei derived from the parent spores unite within the mushroom.

From these fused nuclei, genetic material is shuffled and exchanged and then packaged into new nuclei ready for the manufacture of a new generation of spores.

Millions, sometimes billions of spores, are produced, each with the potential to land in the perfect place to germinate, grow a hypha, expand and start the endless and beautiful cycle over again.

Stephen's second revelation was how few species of fungus have names.

In the 1990s, scientists estimated there were 250,000 species of fungus.[7]

By 2023 that number had grown to anywhere between 2 and 5 million species, with some scientists speculating up to 11 million species.[8,9,10]

Yet only around 155,000 have names.[7]

The estimate of undiscovered diversity is staggering and imprecise because much is still unknown about the fungal kingdom, particularly for the world of microfungi.

Microfungi are tiny fungi that are not easily visible compared to the mushrooms, puffballs and bracket fungi that we notice when walking through the forest.

OPPOSITE
Geastrum triplex, collared earthstar
Dorrigo, New South Wales, Australia
Distribution: worldwide

Raindrops are the trigger for jets of spores
to be released from the spherical inner sac
of *G. triplex*.

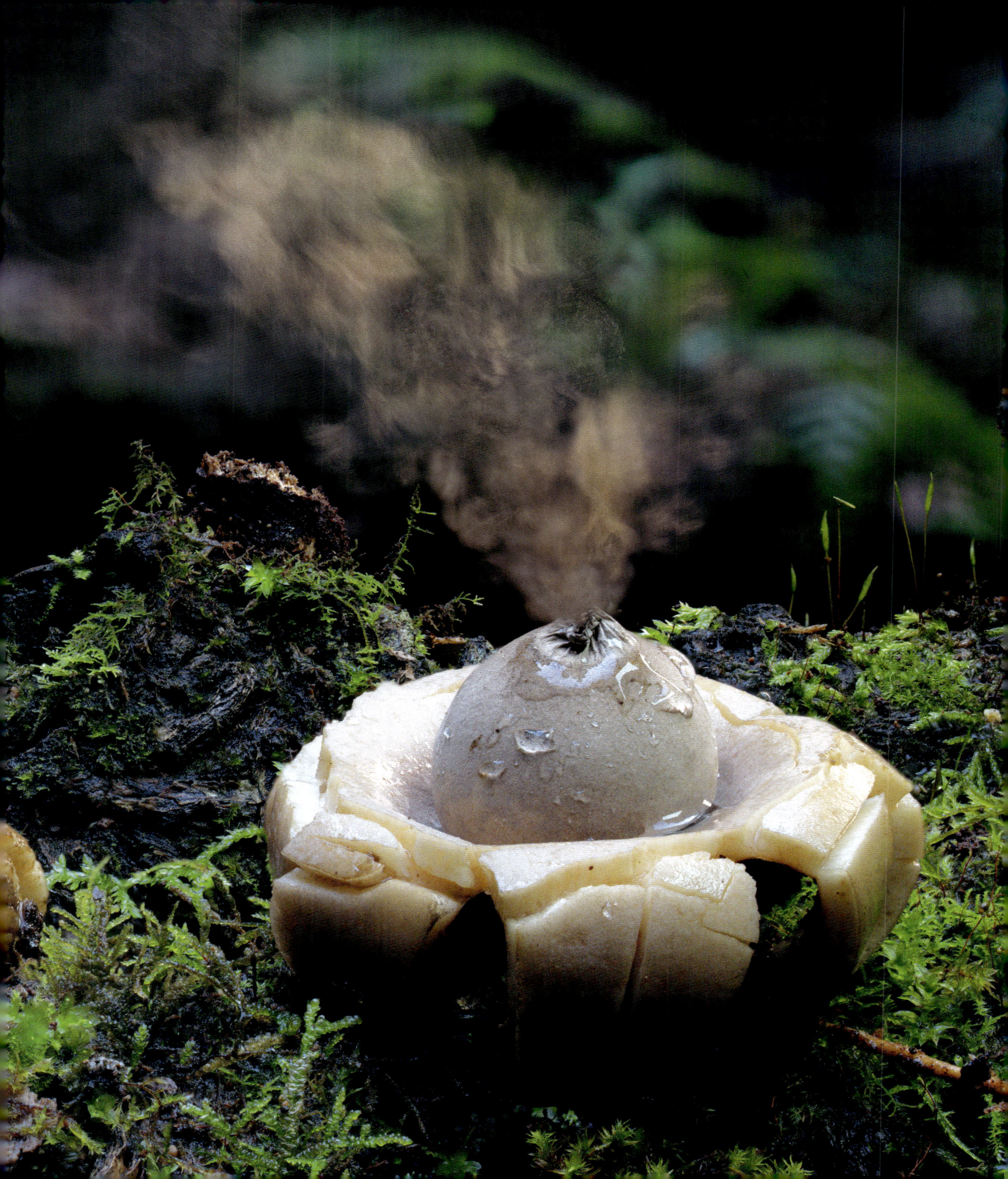

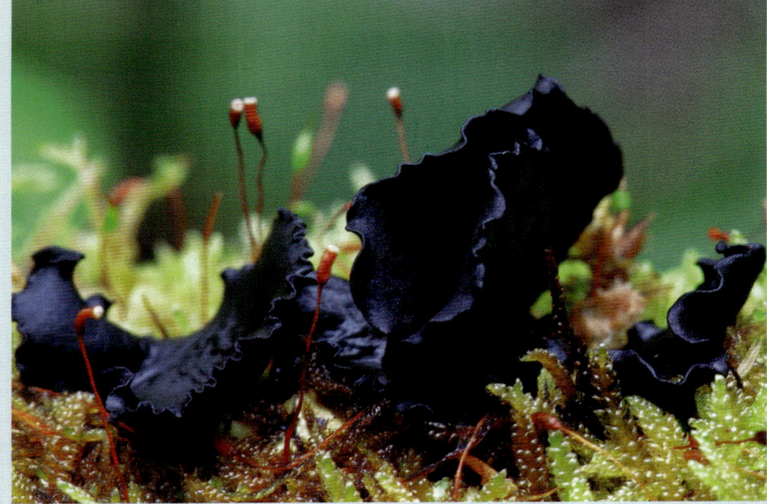

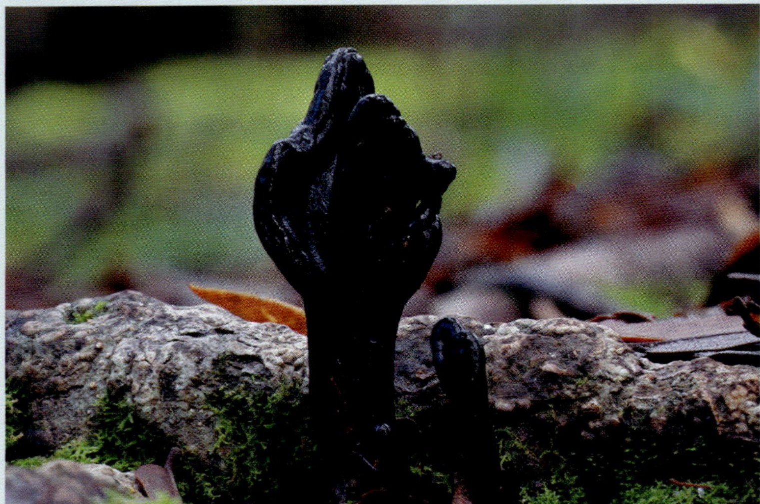

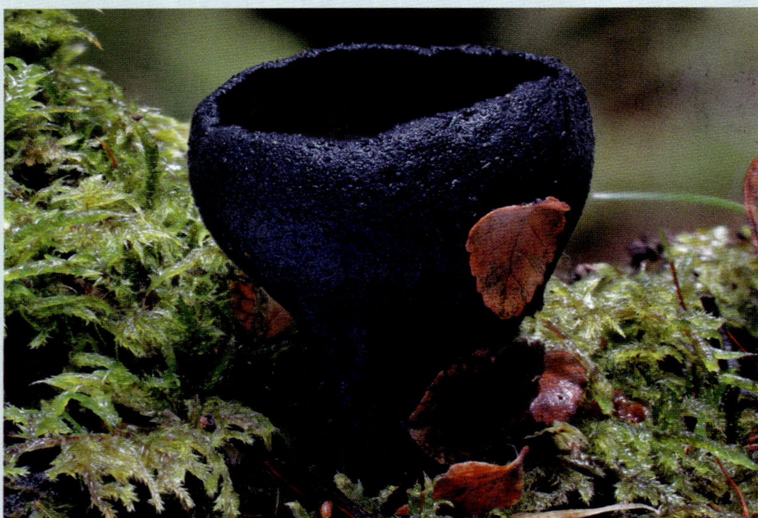

A 2006 study searched the lush rainforests of Australia for fungi in the decaying leaf litter. Inside just six leaves, hundreds of species of microfungi were found, a hidden mass of biodiversity, many new to science.[11]

A few years later, scientists embarked on a global project to research fungi in soil. In a mere handful, thousands of unique species were documented, revealing that the earth beneath our feet harbours a bustling ecosystem of fungi.[12]

TOP
Cordierites frondosa
Mt Macedon, Victoria, Australia
Distribution: Asia and Australasia

MIDDLE
Geoglossum **sp.**
Strahan, Tasmania, Australia

BOTTOM
Pseudoplectania tasmanica
Philosopher Falls, Tasmania, Australia
Distribution: Australia

OPPOSITE
Tiny mushrooms from an unidentified
species of fungus emerge from the
mycelium inside a leaf.
Lake Marian Track, Fiordland National Park, New Zealand

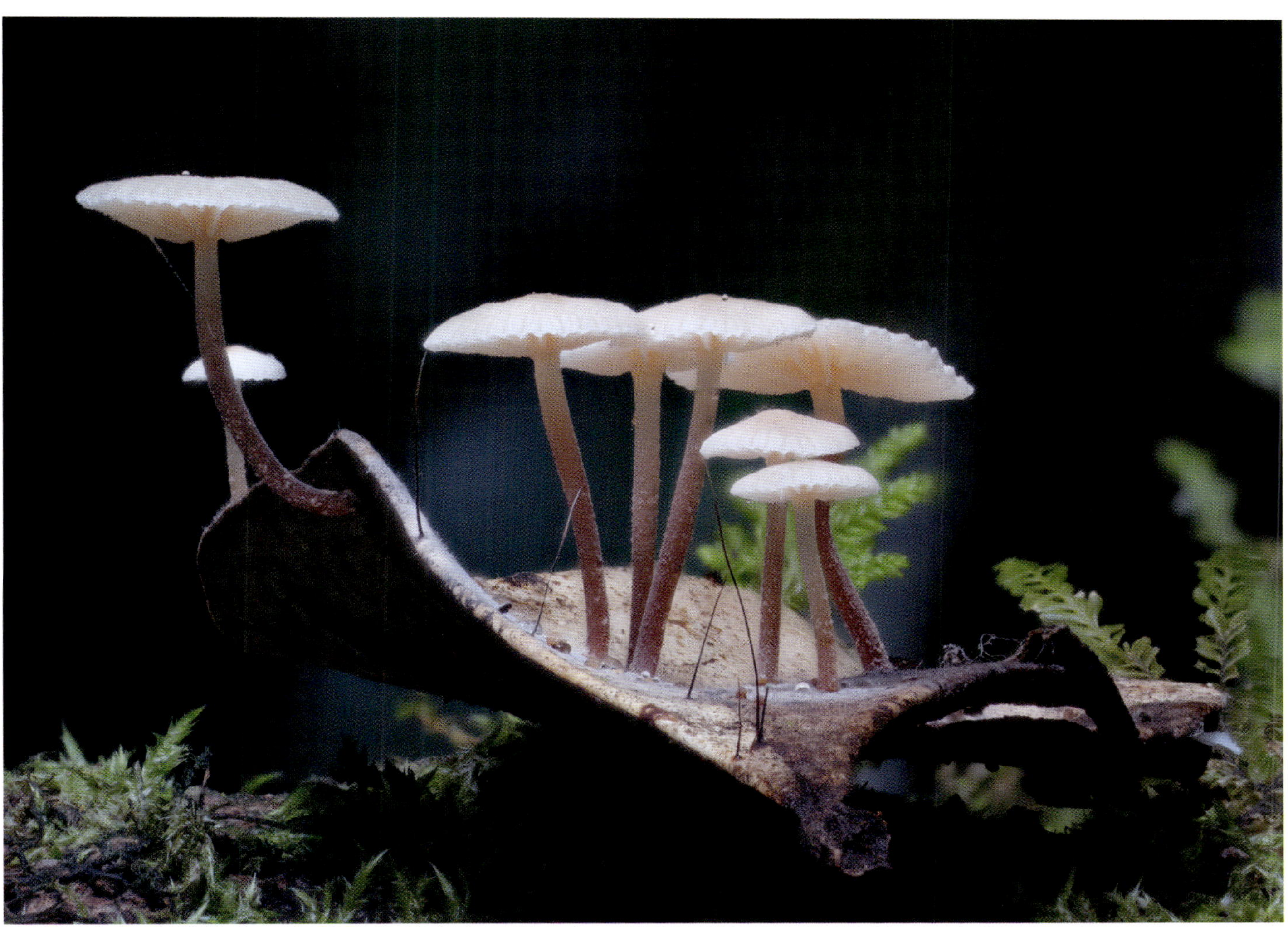

The collective realisation dawned: fungus, once underestimated and neglected, was emerging as profoundly instrumental in shaping life on our planet.

Stephen found himself contributing to a new frontier of scientific discovery.

Within its mycelial tapestry, he had discovered not just a passion but a purpose: to assist in expanding our knowledge of these wonderful organisms.

'A simple gift – a little purple mushroom – became the key that unlocked for me the doors to an entire planet.'

Fast forward to 2011.

Fungi identification is easier. Many field guides and dedicated fungi Facebook and Instagram pages are mushrooming across the internet.

A digital revolution is in full swing. DIY websites and social media are now user-friendly.

Professional and amateur photographers are uploading their images to a global audience.

The fungi galleries on Stephen's website receive millions of views every year.

He is developing relationships with mycologists who are intrigued to see the specimens he is unearthing in forests where fungi surveys have never been done.

This is also the year that I met Stephen. A shared love blossomed, and we became partners in life and playmates in the fungal kingdom.

Creatively, it was a marriage of visual storytelling skills, Stephen's photography and my documentary filmmaking.

It was on a sultry spring afternoon a year into our relationship that we embarked on a life-altering cinematic journey. The setting was magical; a tropical thunderstorm gathered and intensified over a mountain peak in the subtropics of New South Wales, Australia.

With only our cameras and fingers to depress the shutter, we counted off seconds, clicking through hundreds of images over four hours, in our very first attempt at time-lapse videography.

OPPOSITE
Auriscalpium vulgare, earpick fungus
Honghe Amoshan Nature Reserve, Yunnan, China
Distribution: Northern Hemisphere

The favourite substrate of *A. vulgare* is fallen conifer cones.

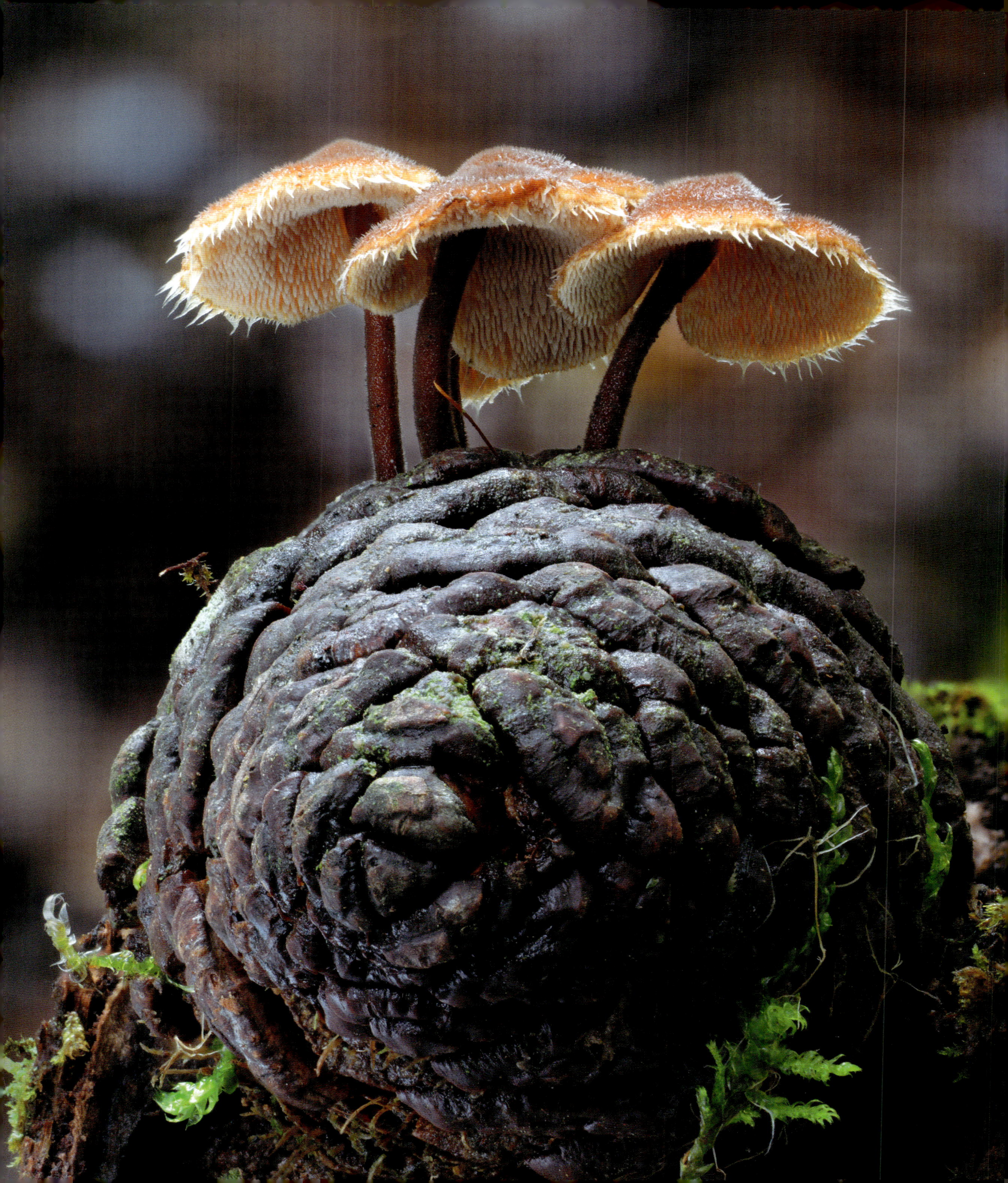

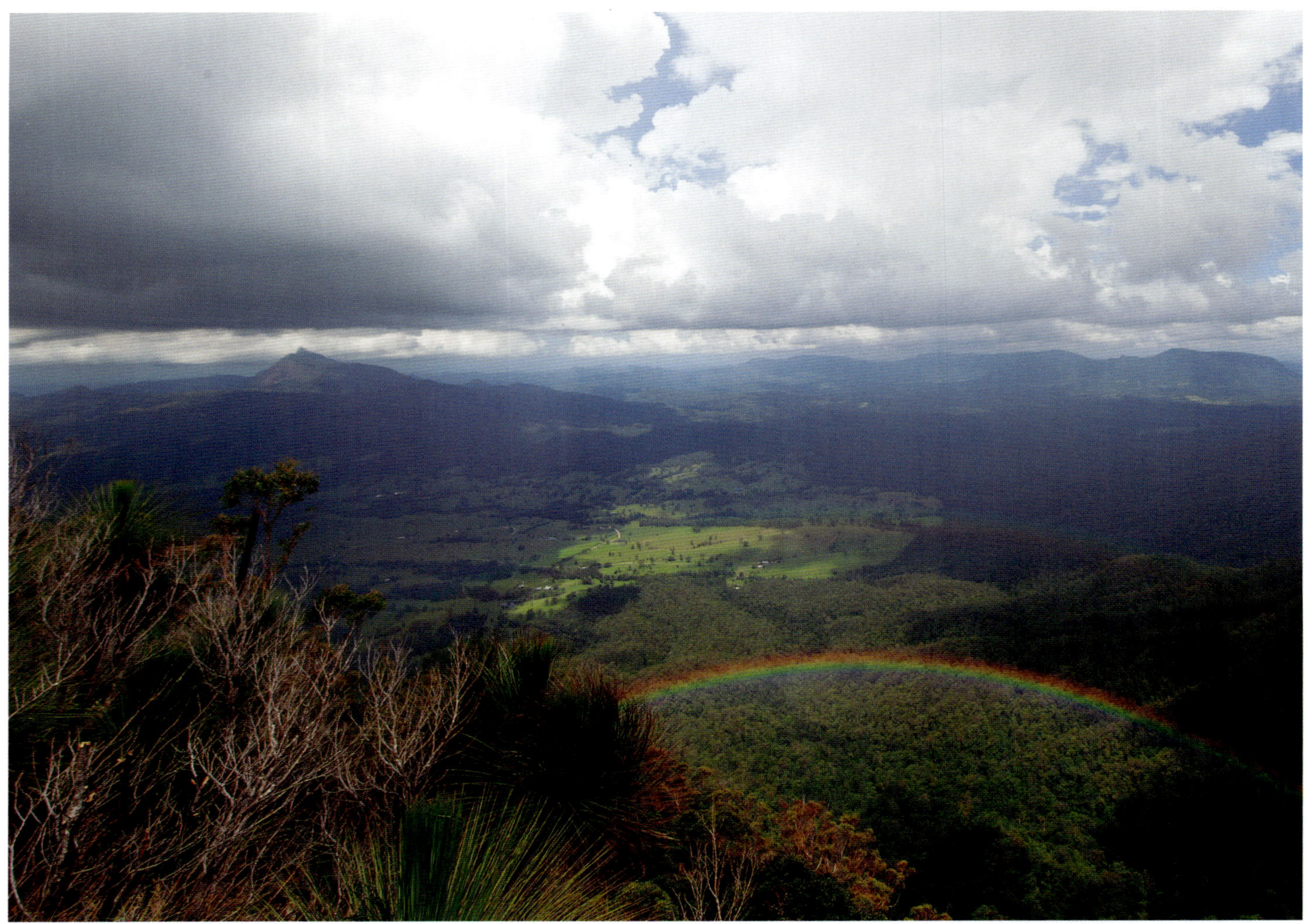

Creating a time-lapse involves capturing a series of still images at set intervals and then playing them back in sequence at a faster speed to create the illusion of time passing quickly.

That day the storm washed over us, the sun emerged and a rainbow graced the bottom of our frame.

It is still one of the most stunning landscape time-lapses we have created.

On the journey home, a creative vision was hatched. A shower cubicle in our laundry was surrendered to the cause. It was transformed into a cave with rotting logs, blackout curtains and our art.

Our quest: to capture the elusive dance of time within the reproductive life of forest fungi.

The inaugural time-lapse featured the enchanting *Mycena chlorophos*, a tropical bioluminescent fungus that glows in the dark.

Conditions were ideal – it was very wet, very warm and very humid.

Over 662 frames spanning three days, the fungus blossomed from bud to glorious neon-green mushrooms, illuminating the constructed darkness of their makeshift studio.

The shower recess has evolved into our very own fungarium, a shipping container with four time-lapse studios.

RIGHT

Mycena chlorophos was the perfect actor for Stephen's first fungi time-lapse. These are frames 100, 239 and 662, of a time-lapse that runs for 26 seconds when played at 25 frames per second.

OPPOSITE

'Wollumbin' means 'cloud catcher' in the Bundjalung language of the Indigenous peoples of the New South Wales north coast. It is their name for the majestic remnants of the volcanic plug at the centre of an ancient caldera in the Tweed Valley.

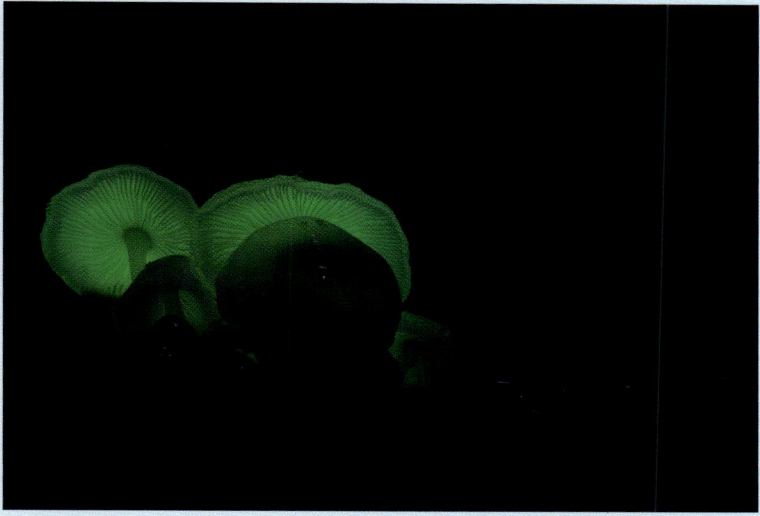

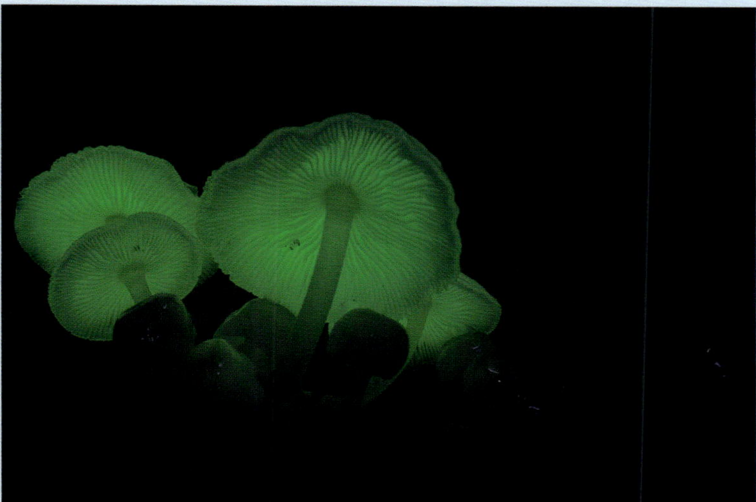

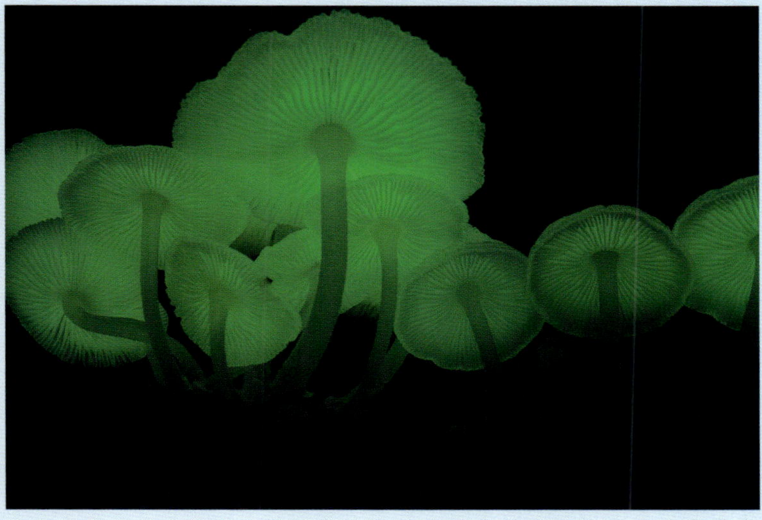

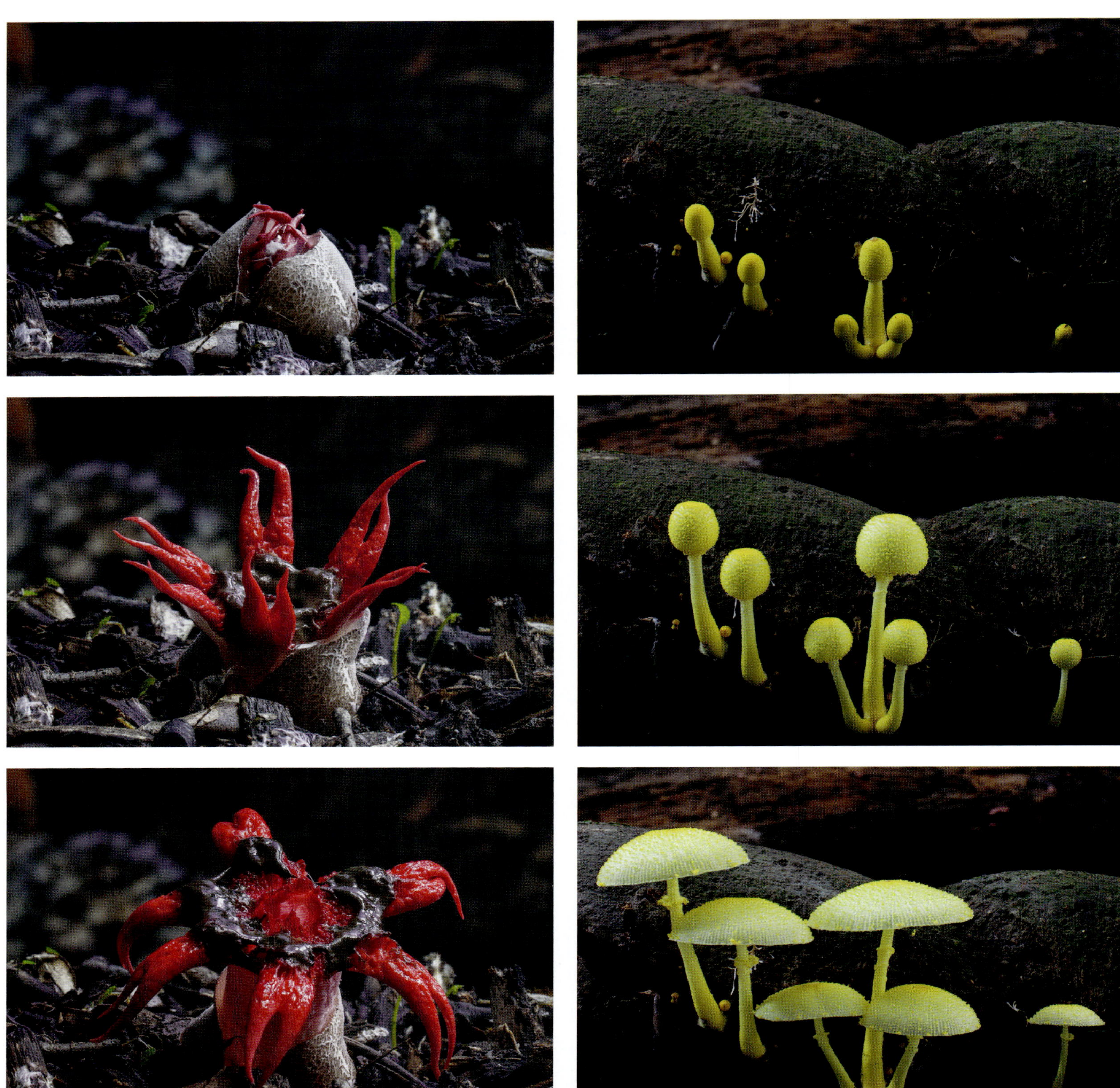

It sounds glamorous, but in truth, the fungarium is a damp place with decaying wood and shocking conditions for camera equipment.

Nevertheless, Stephen spends hours there unveiling the secret lives and diversity of fungal species.

Combining these time-lapses with documentary filmmaking, we craft narratives of epic fungi adventures, sharing the allure of fungi and raising awareness about the vital ecological roles played by these organisms.

Studying fungi and sharing what we learn has become an intellectual pursuit and lifelong passion.

The creations are our ode to fungi and the natural world.

OPPOSITE LEFT

To time-lapse the stinkhorn fungus *Aseroe rubra* emerging from its egg requires considerable patience. It can take weeks for a crack to appear in the egg, but once the birthing process begins the mushroom can unfurl in just a few hours.

These are frames 100, 132 and 203 of a time-lapse that runs for 16 seconds when played at 25 frames per second.

OPPOSITE RIGHT

If you see a yellow mushroom growing from the soil in your houseplant, it is most likely to be *Leucocoprinus birnbaumii,* known worldwide as the flowerpot fungus.

It is the perfect specimen to observe how mushrooms inflate with water, expanding into the mature structure of cap (pileus) and stem (stipe).

These are frames 174, 297 and 426 of a time-lapse that runs for 23 seconds when played at 25 frames per second.

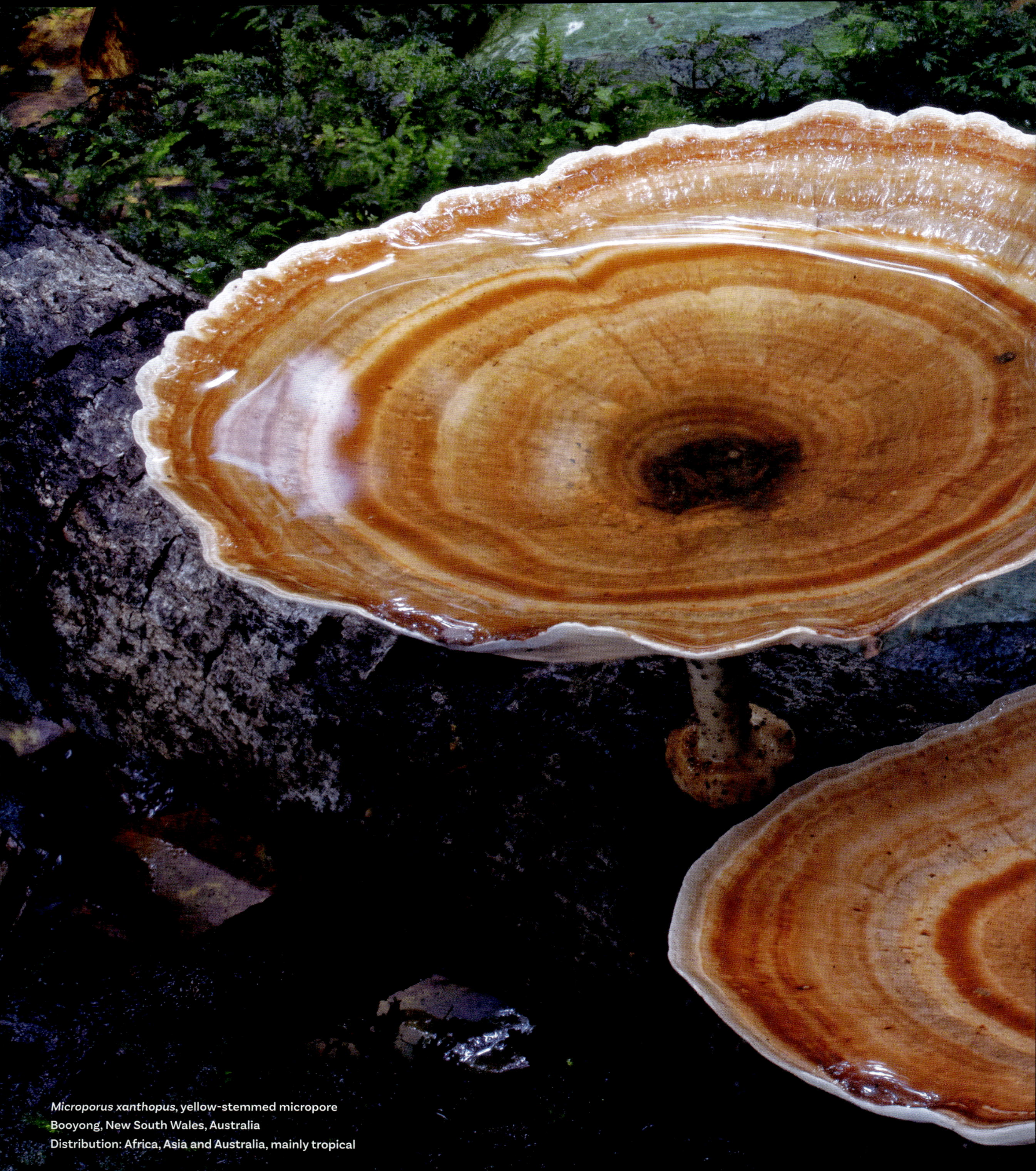

Microporus xanthopus, yellow-stemmed micropore
Booyong, New South Wales, Australia
Distribution: Africa, Asia and Australia, mainly tropical

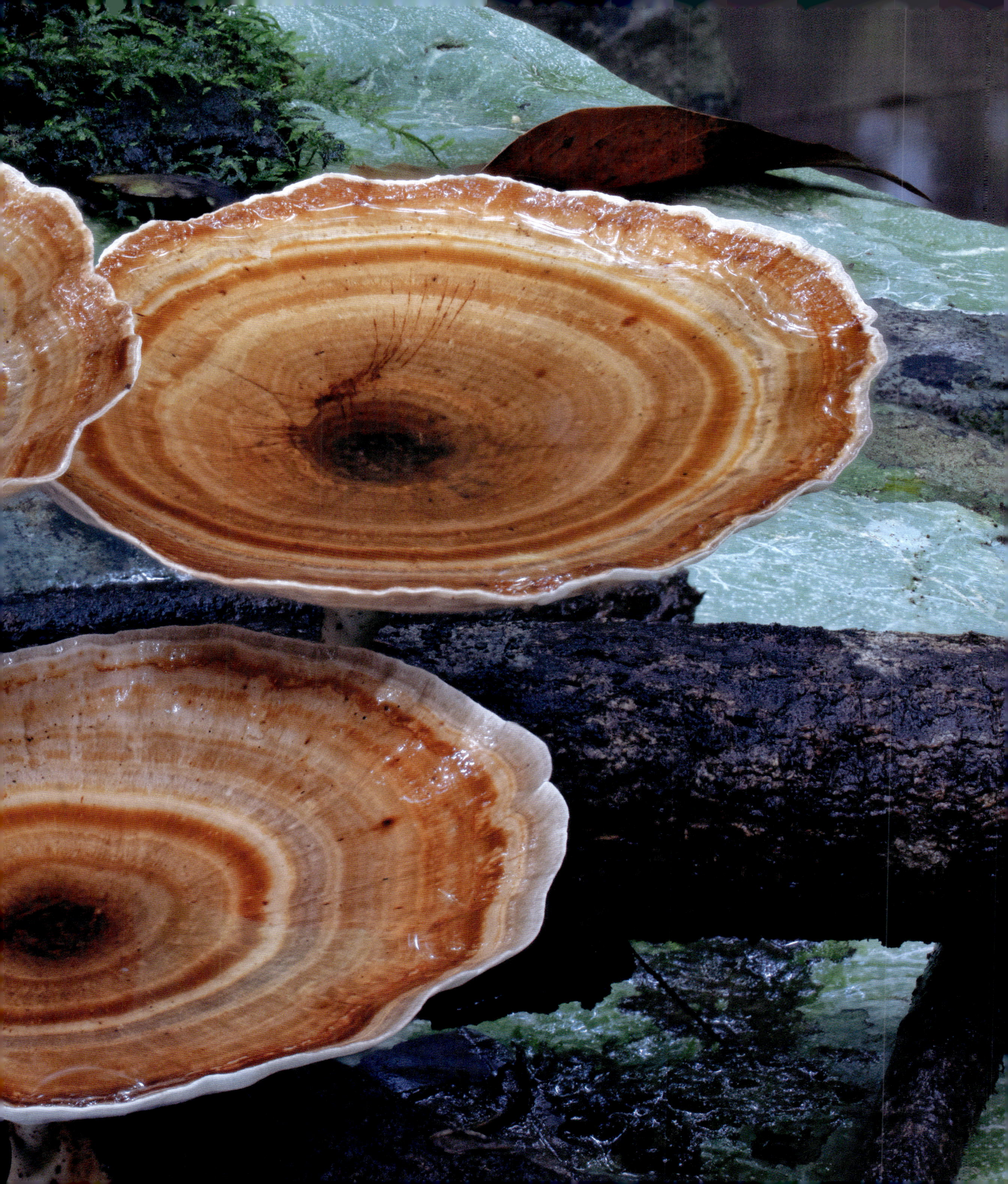

Traders, undertakers and executioners

A silent dance plays out as fungi, in all their different guises, feast on the inanimate and the living in the theatre of nature.

As the rains settle in, misty mountains and swollen rivers herald the start of the fungi season.

It is a time of year fungi hunters eagerly anticipate.

One of the many joys of a passion for fungi is venturing time and time again into the same forests.

Observing the faithful return of the same mushrooms under the same trees, year after year.

Witnessing a fallen log gradually disappear as a succession of fungal characters take their turn to break down its structure.

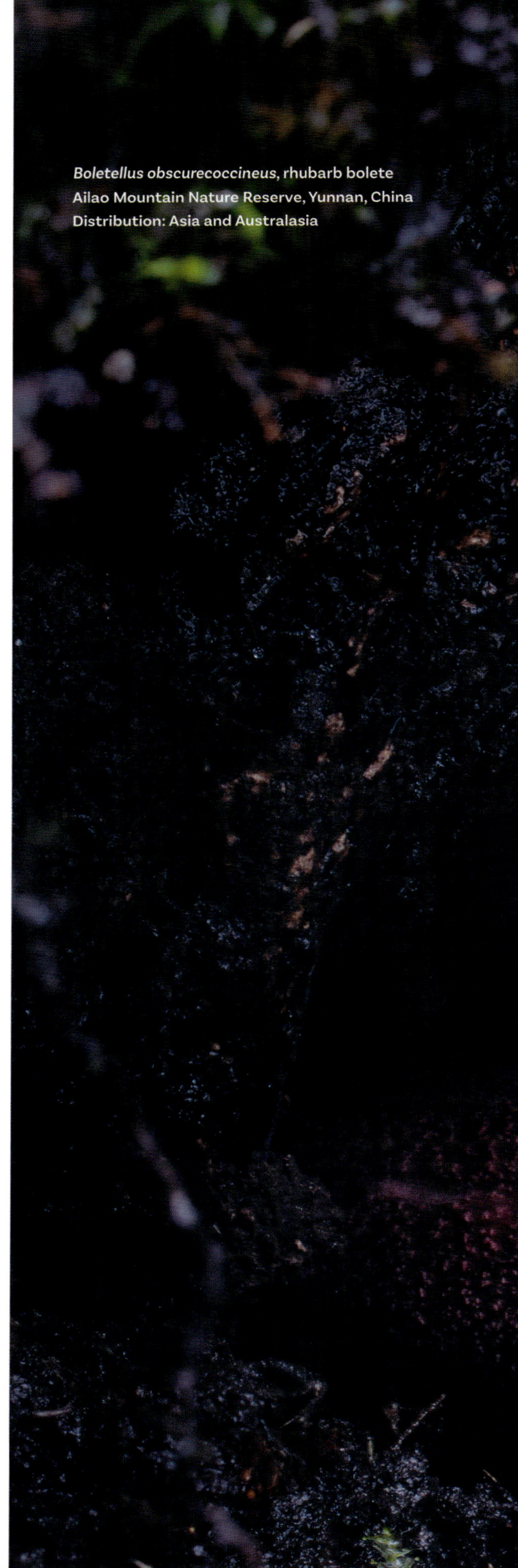

Boletellus obscurecoccineus, rhubarb bolete
Ailao Mountain Nature Reserve, Yunnan, China
Distribution: Asia and Australasia

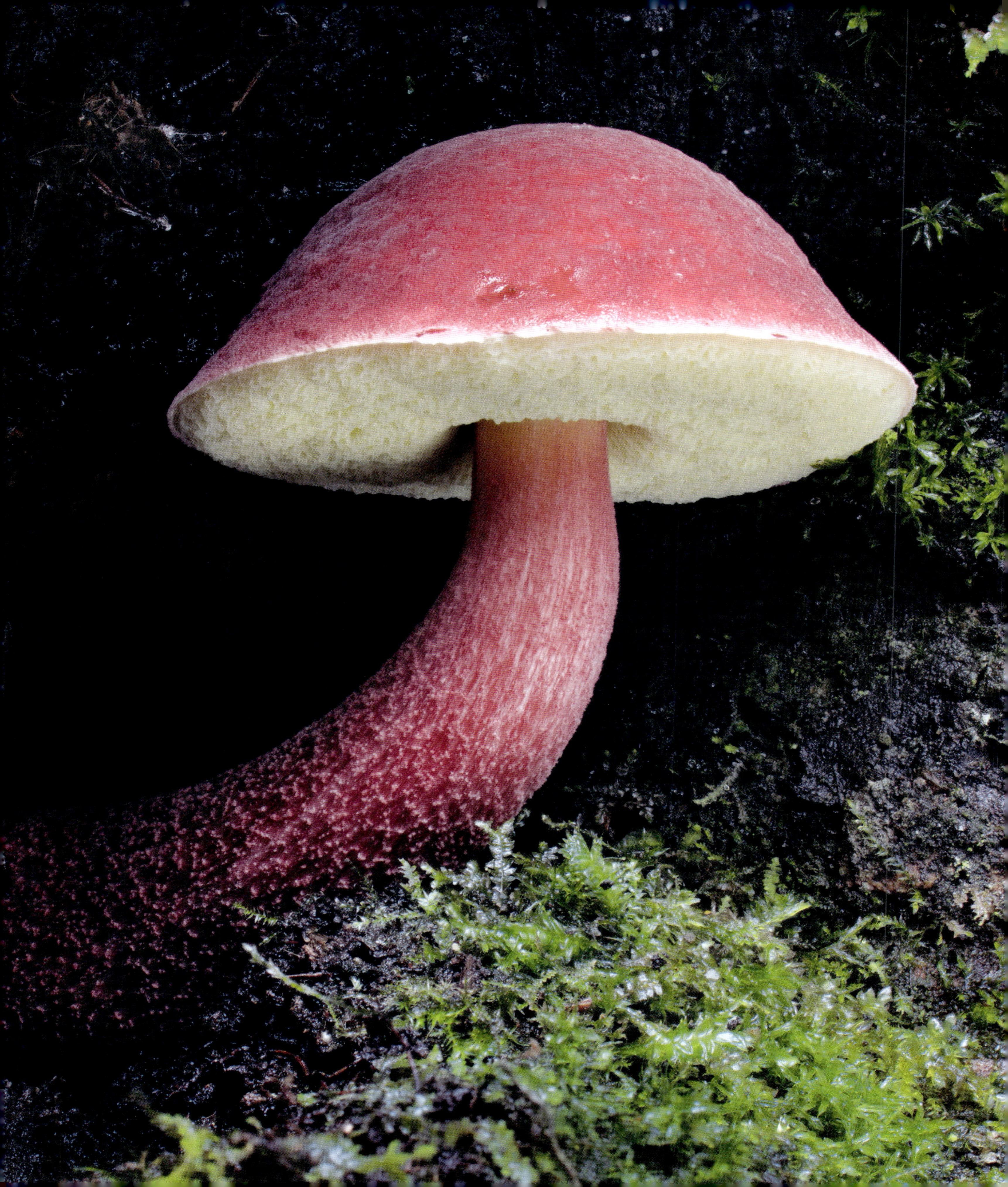

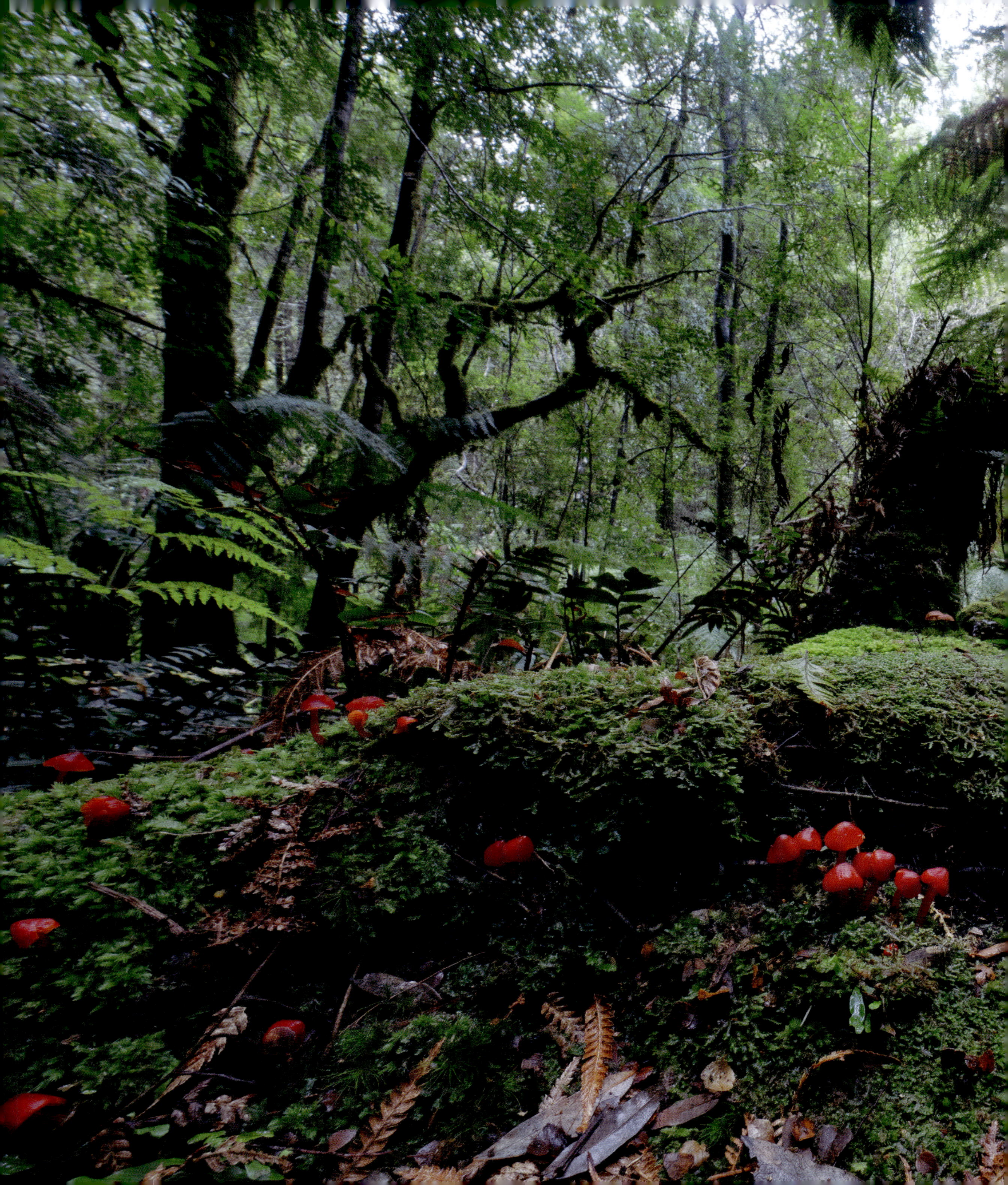

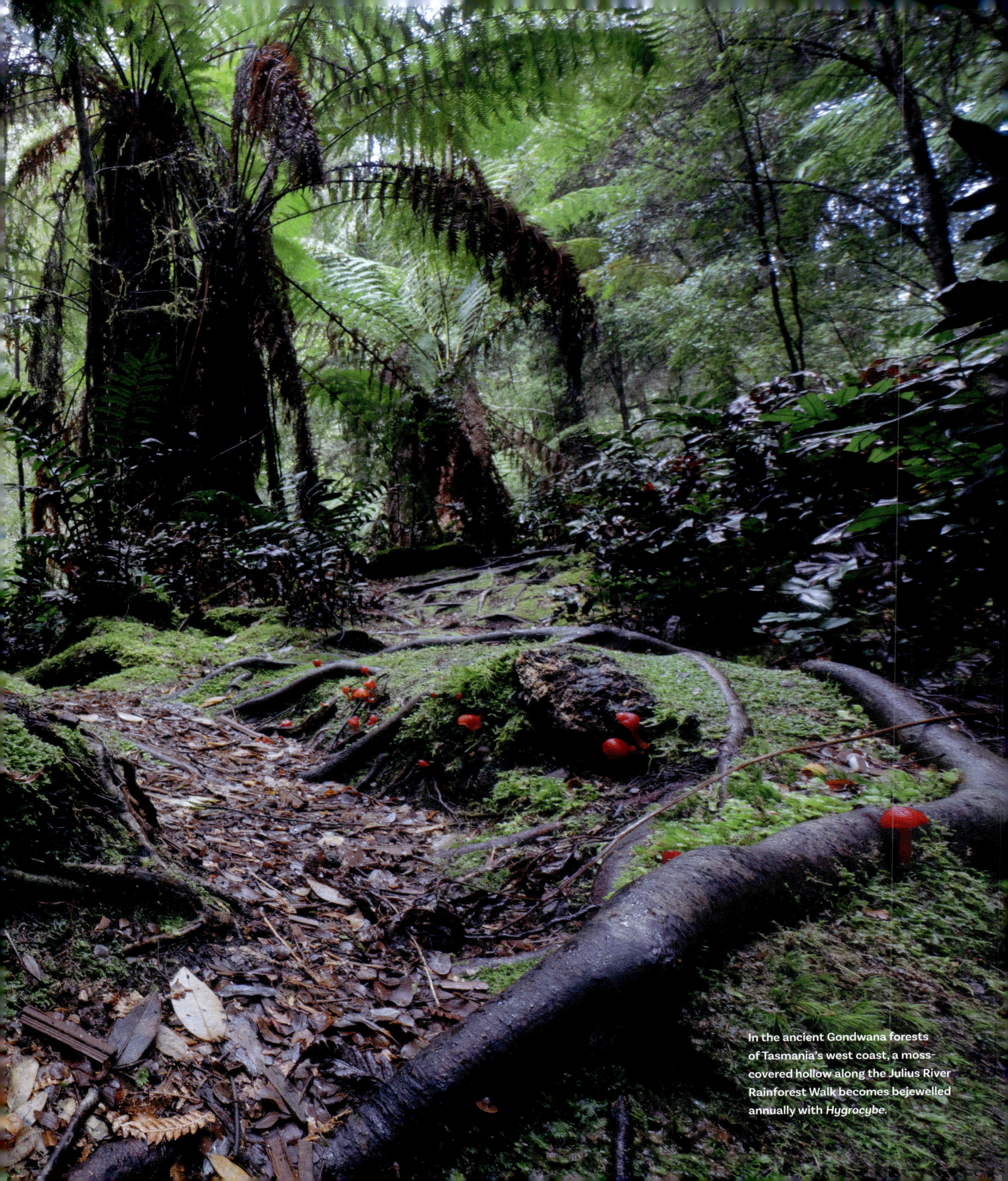

In the ancient Gondwana forests of Tasmania's west coast, a moss-covered hollow along the Julius River Rainforest Walk becomes bejewelled annually with *Hygrocybe*.

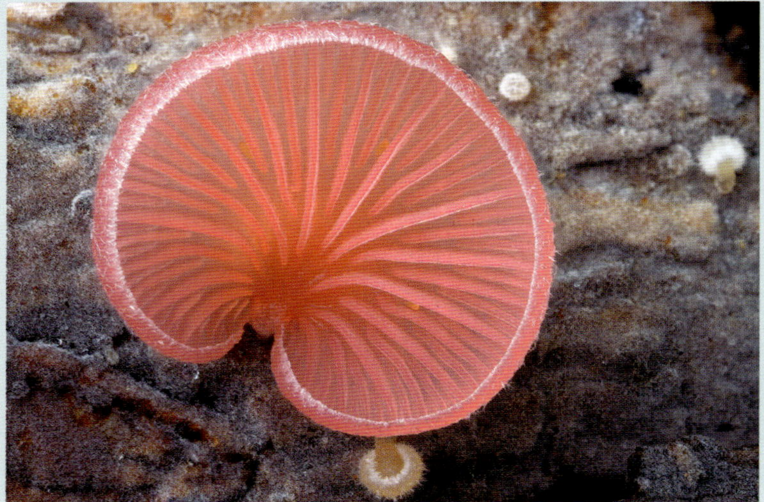

'In Tasmania, in Australia, one of my all-time favourite trails is at Julius River in the ancient cool temperate rainforests of the Tarkine.

'Every year that I make the pilgrimage during fungi season, I know I will find a cluster of red Hygrocybe, emerging in the moss under the beech, leatherwood and sassafras trees.

'Closer to home in the subtropics, fallen trees are recycled back into soil in just a few years.'

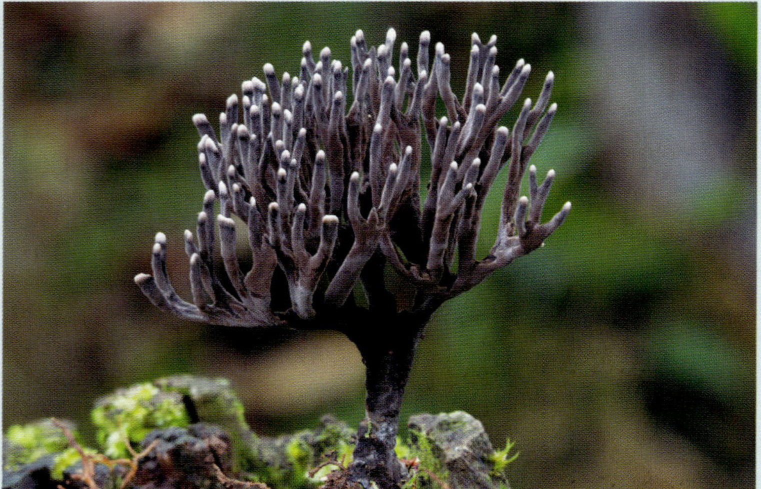

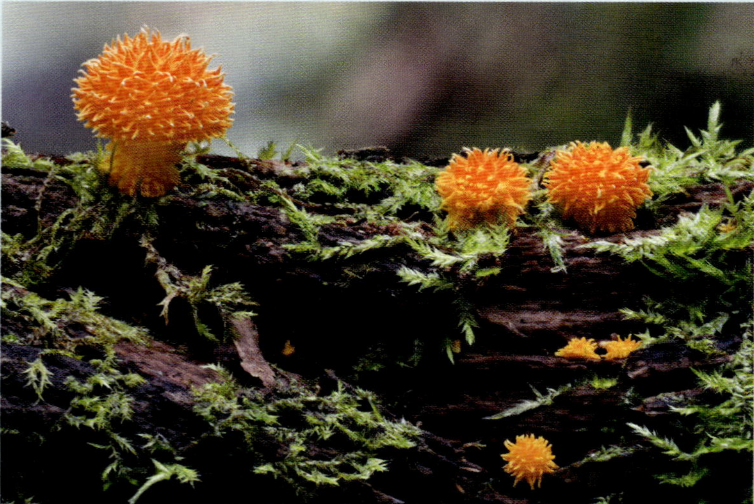

TOP
Crepidotus sp.
Booyong, New South Wales, Australia

MIDDLE
Xylaria sp.
Nabanhe National Nature Reserve, Yunnan, China

BOTTOM
Cyptotrama asprata, golden-scruffy collybia
Nightcap National Park, New South Wales, Australia
Distribution: worldwide, except Europe

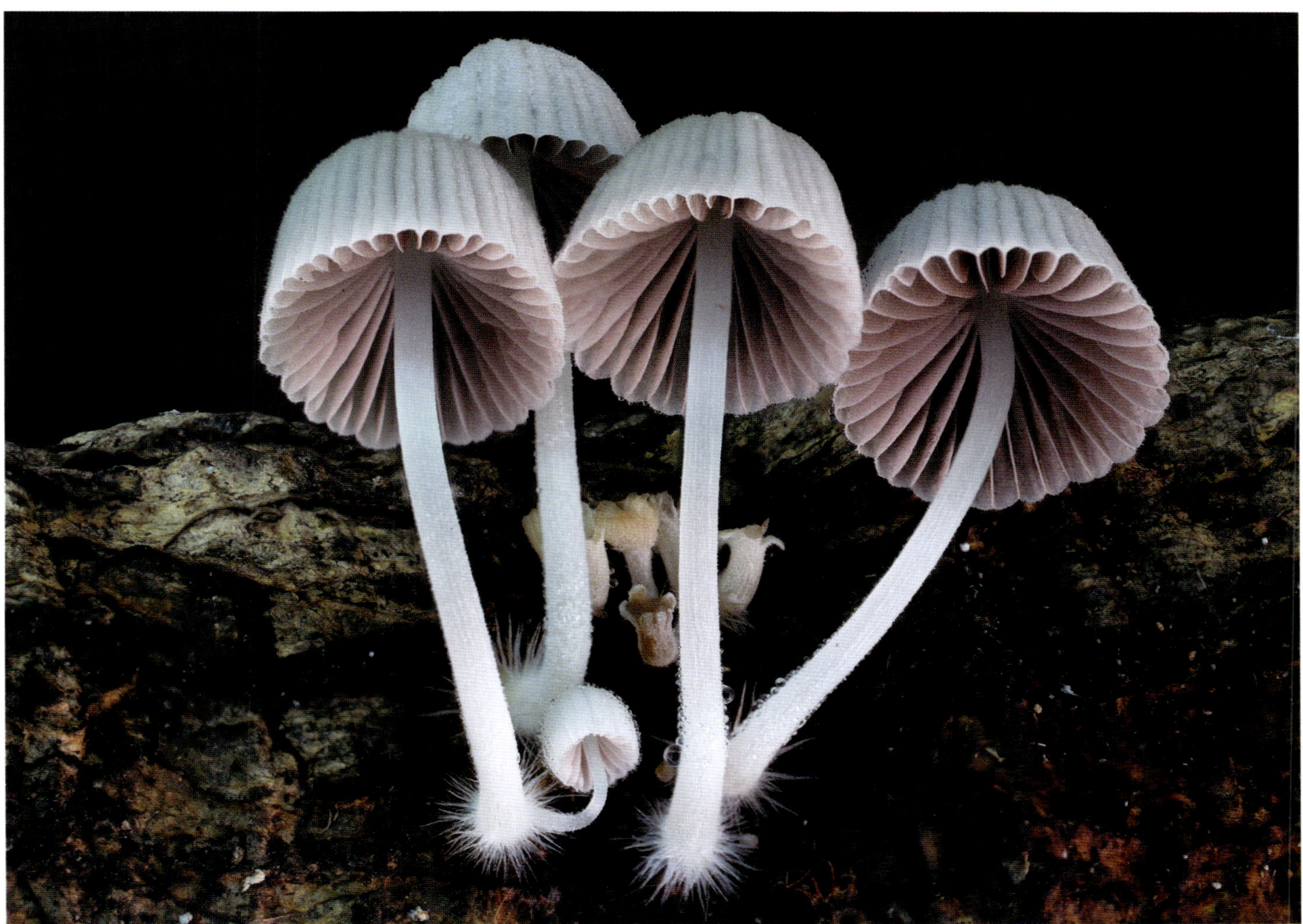

'Over that time, I can see dainty pink bonnets of a Coprinellus, *the gnarled powdery branches of Xylaria, followed by stunning pink fans of a Crepidotus inhabit the rotting log.'*

In these cyclical expeditions, Stephen becomes a storyteller, using the fleeting appearance of mushrooms to observe the unique roles and eating habits of different fungal organisms.

Coprinellus disseminatus, **fairy inkcap**
Booyong, New South Wales, Australia
Distribution: worldwide

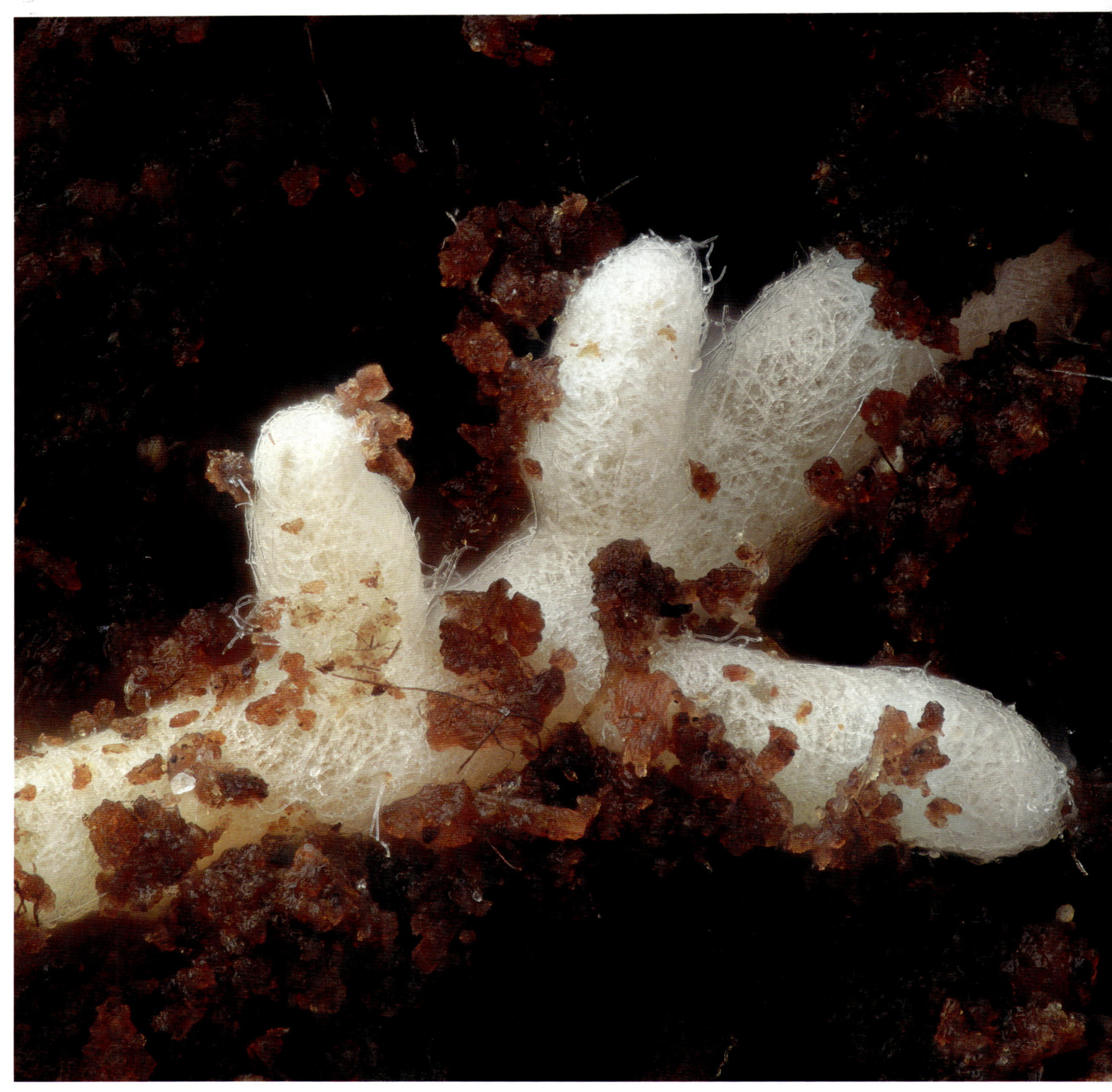

First, there is the extraordinary tale of mycorrhizal fungi, an intimate connection between the underground network of mycelium, comprising its microscopic threads of hyphae intermingling with plant roots.

Picture them as nature's traders, where the web of hyphae creates a subterranean commodity transport system.

The mycelium seeks out, absorbs and moves water and essential minerals such as nitrogen and phosphorous from the soil to nourish the plants.

Yet this is no free ride. The tree must offer up some of the sugars it produces from photosynthesis, sending these back down the hyphae to satiate the fungal appetite.[1,2,3]

In the underground theatre of most landscapes, there are two distinct mycorrhizal fungi. Arbuscular mycorrhizal fungi infiltrate a plant's roots, filling its cells with a dense mycelial mass and thereby extending the roots' reach. These fungi support the growth of 80 per cent of the world's plants.[3,4,5]

Meanwhile, the fine threads of ectomycorrhizal fungi envelop the tip of a plant's root, sheathing it with a mycelial sock. They then extend and weave through the soil, foraging far and wide into tiny cracks and crevices too minuscule for individual plant roots.

These ectomycorrhizal fungi underpin the growth of most of the planet's towering tree species.[1,2,6]

Hyphae of ectomycorrhizal fungi enveloping the roots tips of a plant.

Photograph: Wim van Egmond

But relying on a single ectomycorrhizal fungus would be a risky gamble for any tree. This is an arrangement that is both collaborative and competitive, where fungus and tree are selective, choosing partners based on the offerings each brings to the trade deal at any given moment.[5,6]

Trees can form alliances with dozens of ectomycorrhizal fungi. This diversity, this flexible arrangement, is the secret to the forest's resilience – and so the forest thrives and adapts to change.[6,7]

As we explore the landscapes of our planet, we are constantly amazed by the incredible diversity of ectomycorrhizal fungi that reveal themselves through their stunning and complex sporing bodies.

We can easily spot the typical mushroom-shaped sporing bodies such as a bright yellow *Amanita*, adorned with a frilly petticoat. However, ectomycorrhizal fungi also produce less conventional shapes – vibrant flames and corals, black trumpets, scaly vases, earth balls and many more.

Stephen and I have come to learn that the importance of these fungi goes beyond their beauty. The soil in many of these forests is often poor in nutrients and so they become a hotspot for ectomycorrhizal fungi that are key players in shaping the landscape and promoting biodiversity.

OPPOSITE TOP
Amanita sp.
Yumthang Valley, Sikkim, India

OPPOSITE BOTTOM
Clavulinopsis sp.
Philosopher Falls, Tasmania, Australia

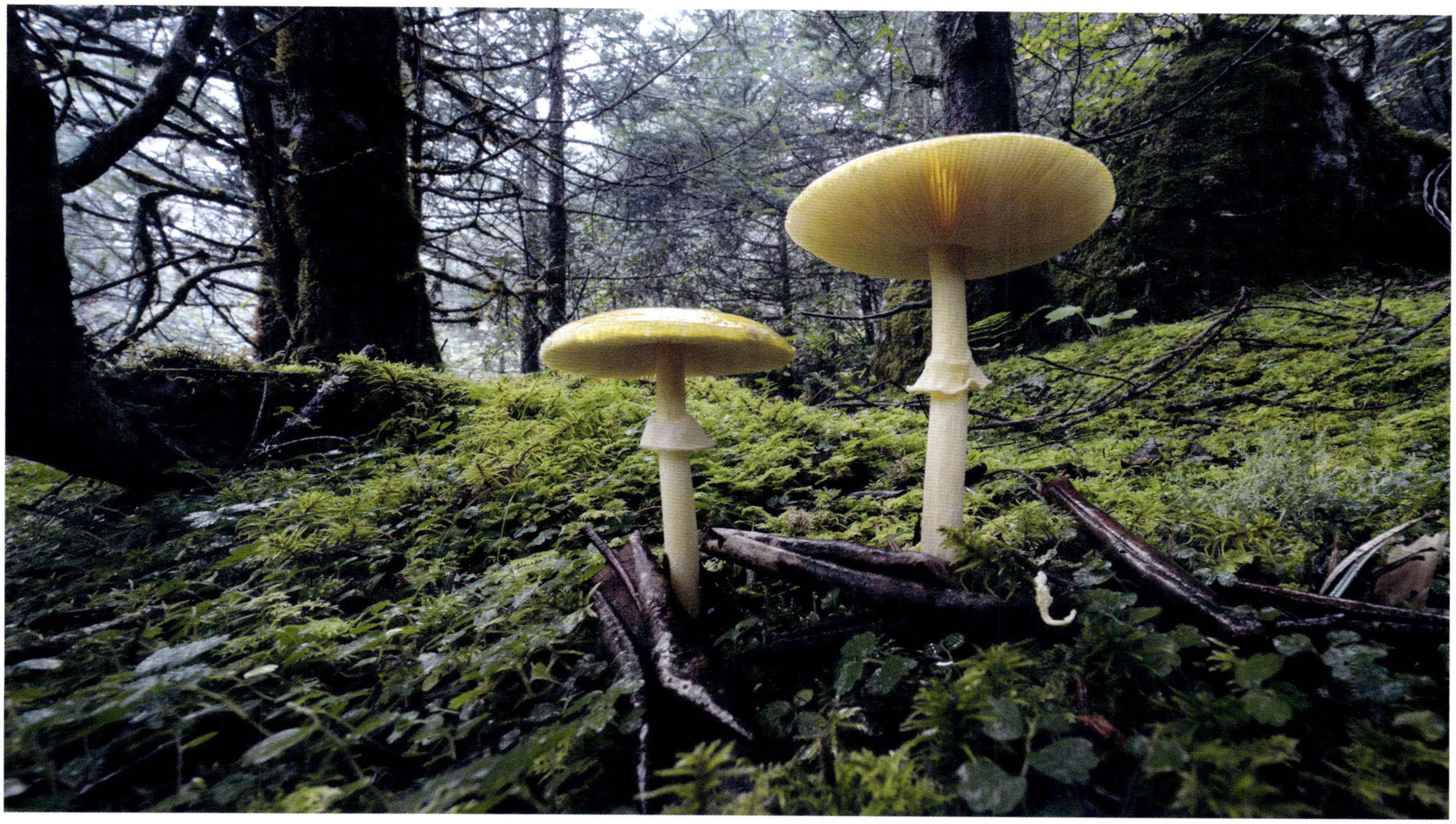

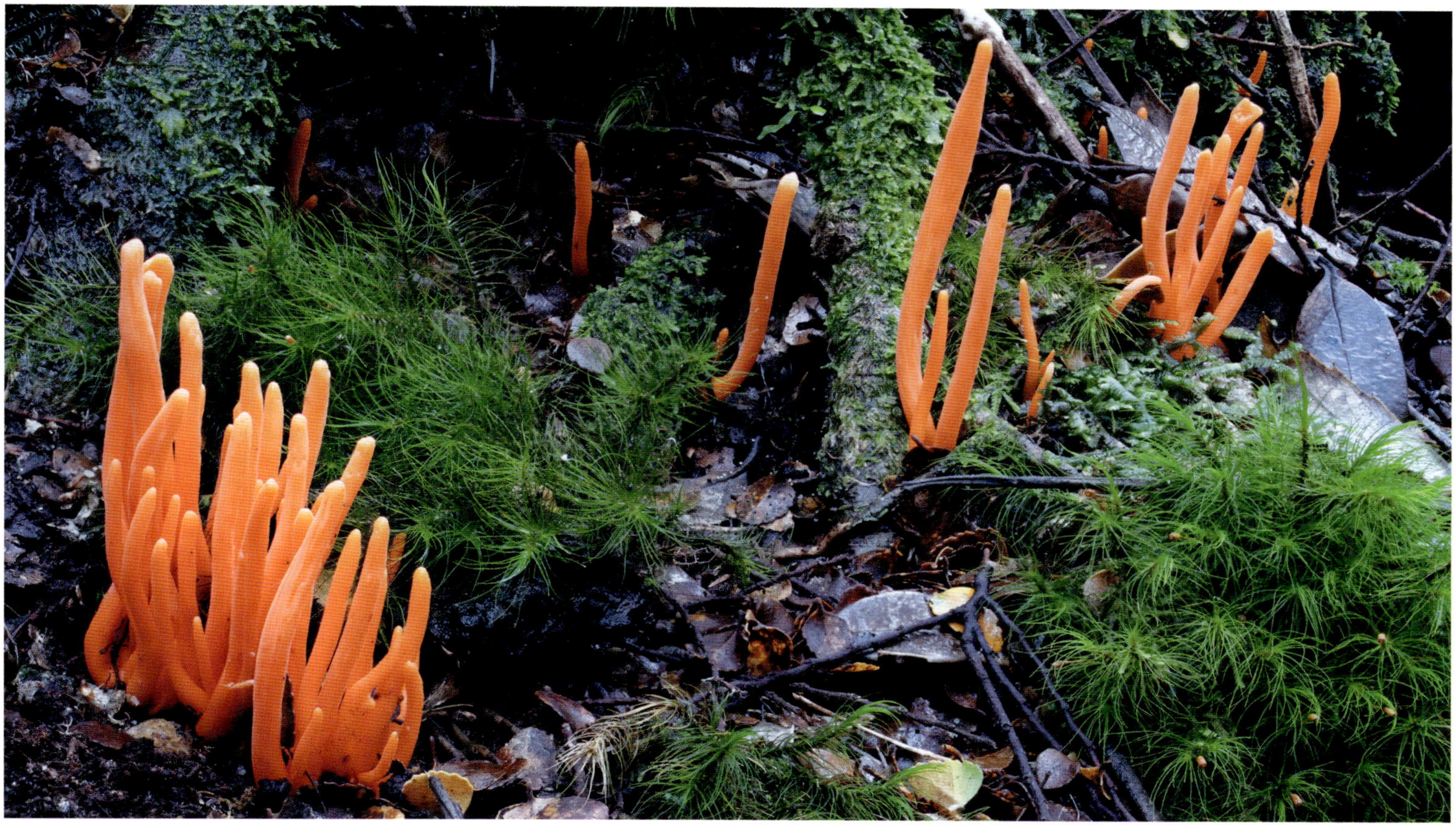

‘We are just starting to realise that these fungi are critical to the balance and health of our planet.

‘Their mycelium is omnipresent, the hidden foundation beneath our feet, connecting different organisms and facilitating a complex exchange of nutrients.

‘Without mycelia, these forests would not exist, life as we know it would not exist, and we humans would not exist.

‘I couldn't believe I didn't know this.’

OPPOSITE TOP LEFT
Laccaria amethystina, amethyst deceiver
Nabanhe National Nature Reserve, Yunnan, China
Distribution: Northern Hemisphere

OPPOSITE MIDDLE LEFT
Craterellus cornucopioides, horn of plenty
Nabanhe National Nature Reserve, Yunnan, China
Distribution: most continents except Africa and
South America

OPPOSITE BOTTOM LEFT
Cantharellus sp.
Ialong, Meghalaya, India

OPPOSITE TOP RIGHT
Scleroderma sp.
Mawlynnong, Meghalaya, India

OPPOSITE MIDDLE RIGHT
Cortinarius sp.
Saint Arnaud Loop Walk, Saint Arnaud, New Zealand

OPPOSITE RIGHT LEFT
Boletaceae sp.
Booyong, New South Wales, Australia

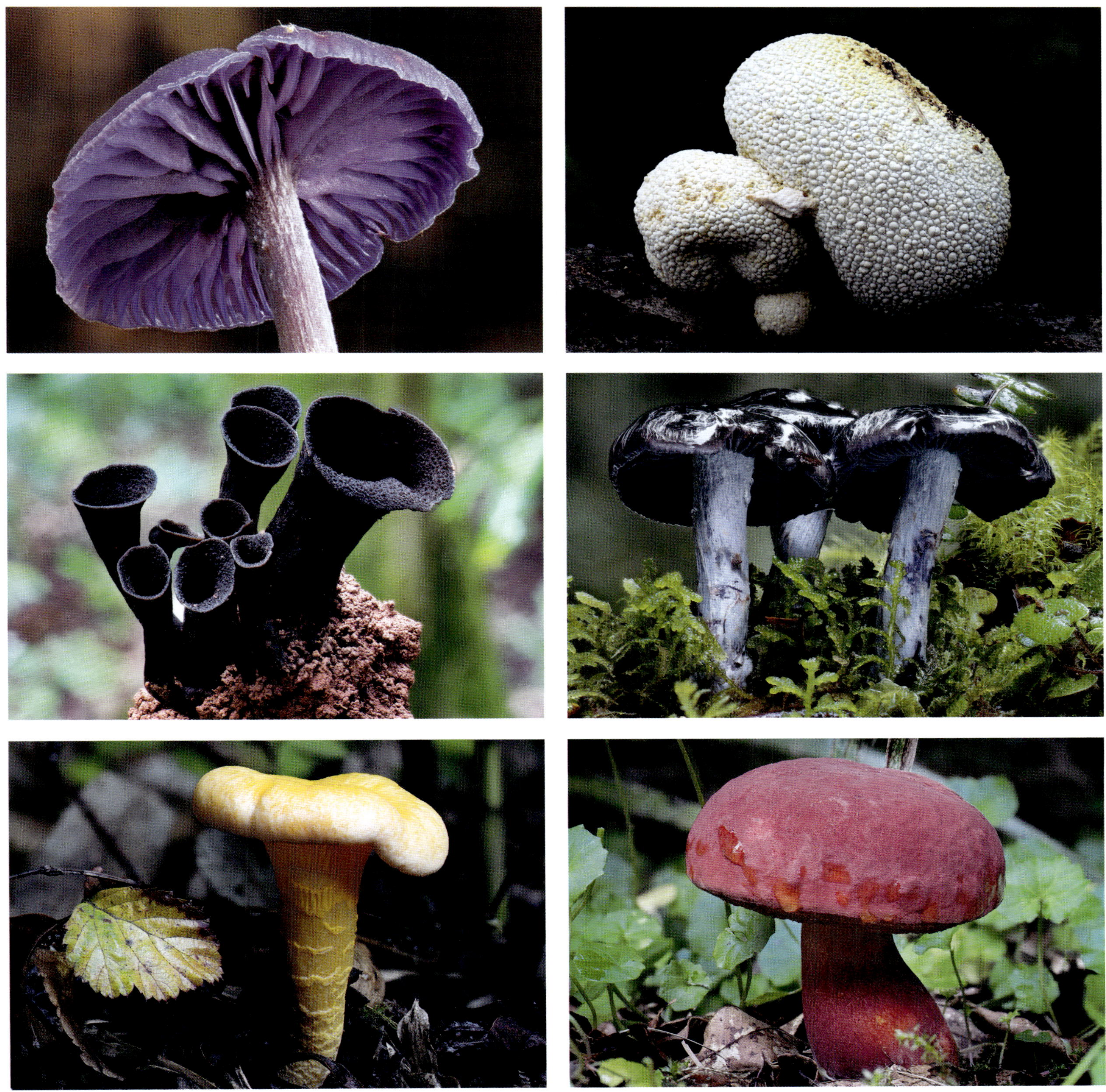

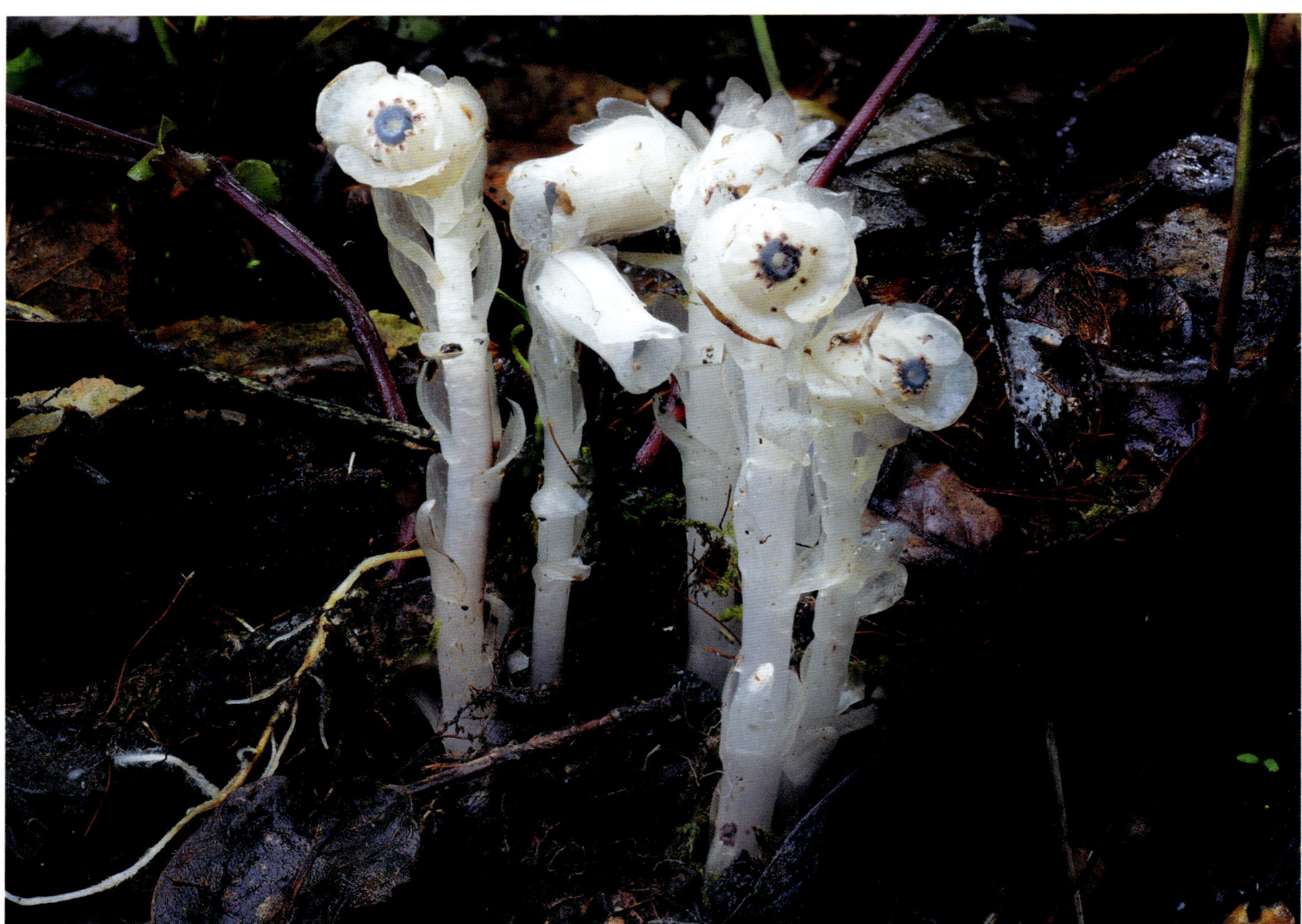

But just like human trade agreements, not everyone plays by the same rules. The kingdoms of plants and fungi, with their millions of different species, are extremely complex, and there are many adaptions of mutualism. Some plant species have even learned how to hijack mycorrhizal trade agreements.

Take, for instance, the plant *Monotropa uniflora*. We were thrilled to stumble upon this botanical opportunist in the high-altitude rhododendron forests of Nepal.

These mycoheterotrophic plants are often found in nutrient-poor soils and in landscapes where sunlight is scarce.

Unlike most plants, mycoheterotrophs do not create their own food via photosynthesis.

Instead, they tap into the mycorrhizal network that connects plants and fungi, siphoning off nutrients and carbon compounds intended for their neighbours, without offering any compensation in return.[8]

'I love that these mycoheterotrophic plants are rebels. They challenge us to expand our thinking, to see that there are no hard fast rules and many adaptions in the complexity of co-evolution.'

Mycorrhizal networks are just one aspect of co-evolution in the forest's fungal theatre.

The second group of fungi we observe in the landscape are the often-overlooked heroes in this drama, those with an insatiable appetite for decay.

TOP
Microporus xanthopus, yellow-stemmed micropore
Booyong, New South Wales, Australia
Distribution: Africa, Asia and Australia, mainly tropical

MIDDLE
Cantharellus sp.
Corinna, Tasmania, Australia

BOTTOM
Lycoperdon perlatum, common puffball
Franklin River Nature Trail, Tasmania, Australia
Distribution: worldwide

OPPOSITE
Monotropa uniflora, ghost plant
Tinjure-Milke-Jaljale Complex Conservation Project, Nepal
Distribution: North America and Asia

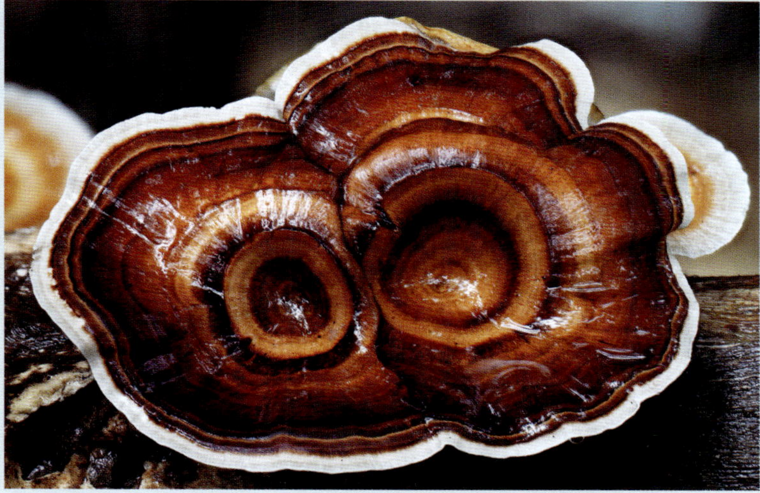

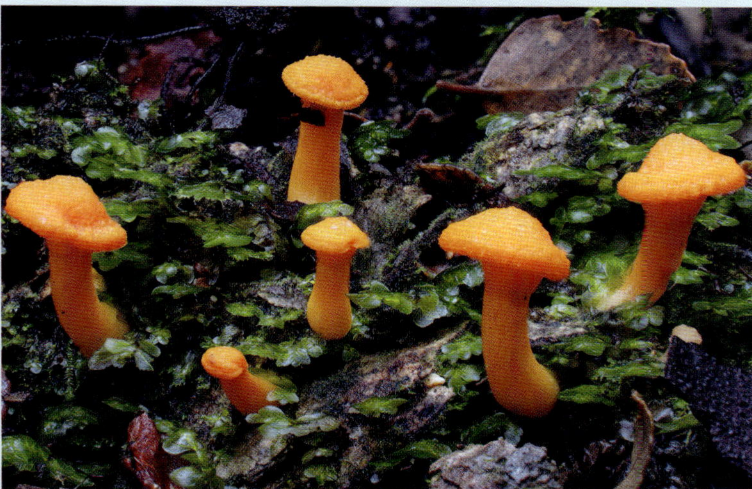

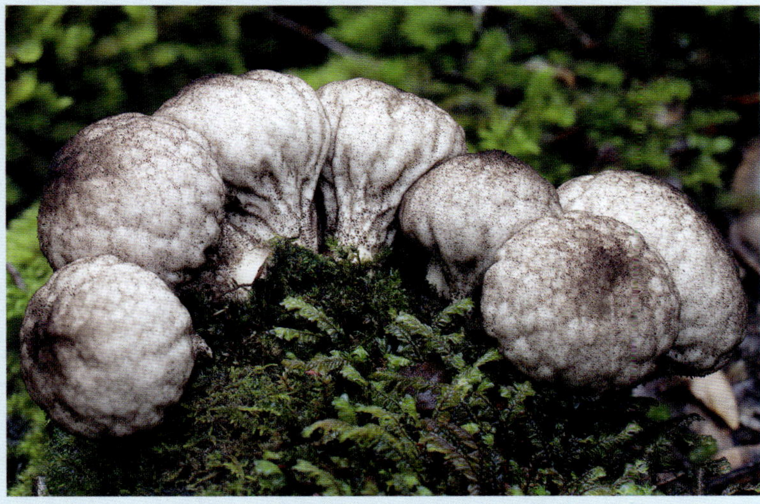

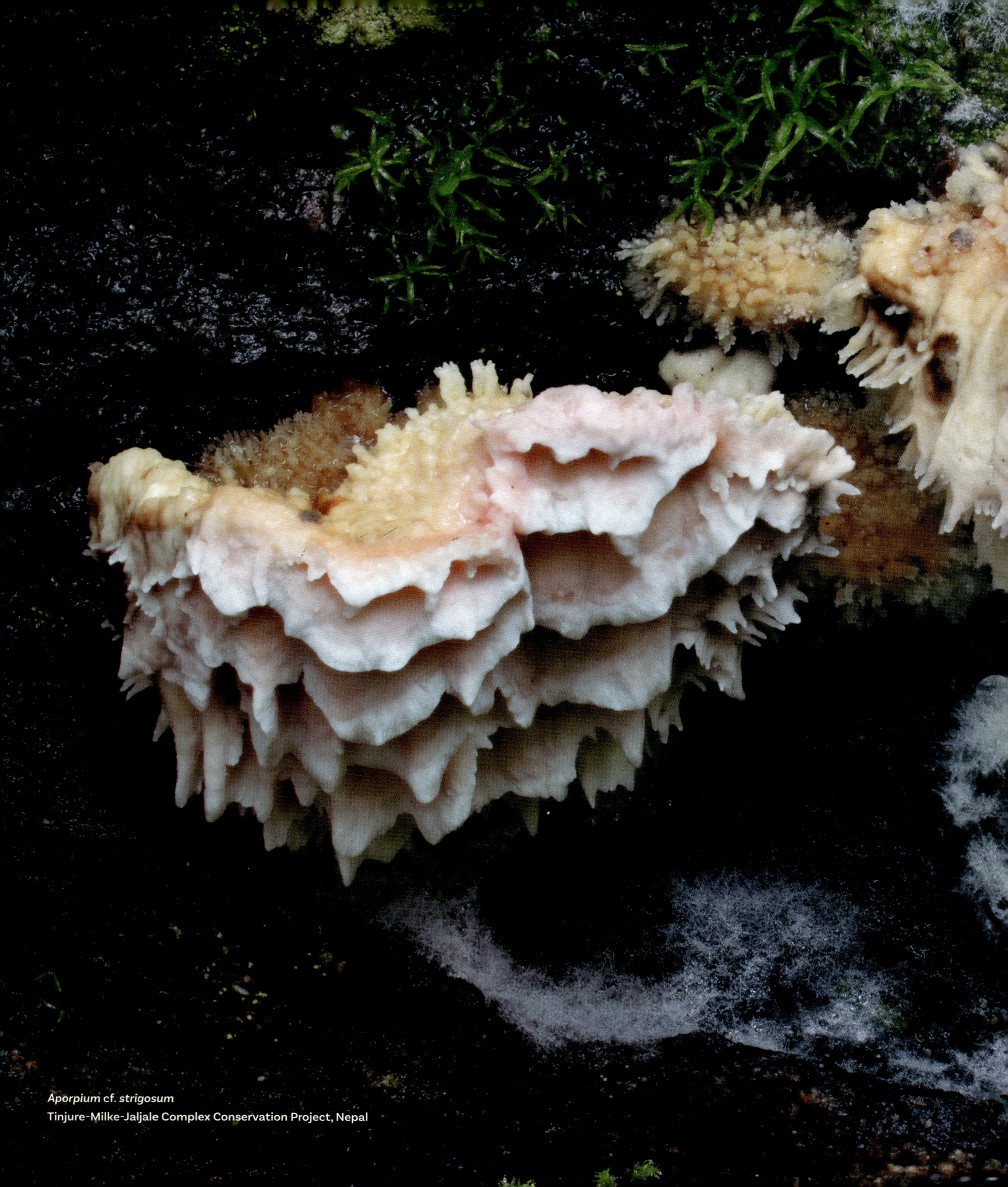

Aporpium cf. strigosum
Tinjure-Milke-Jaljale Complex Conservation Project, Nepal

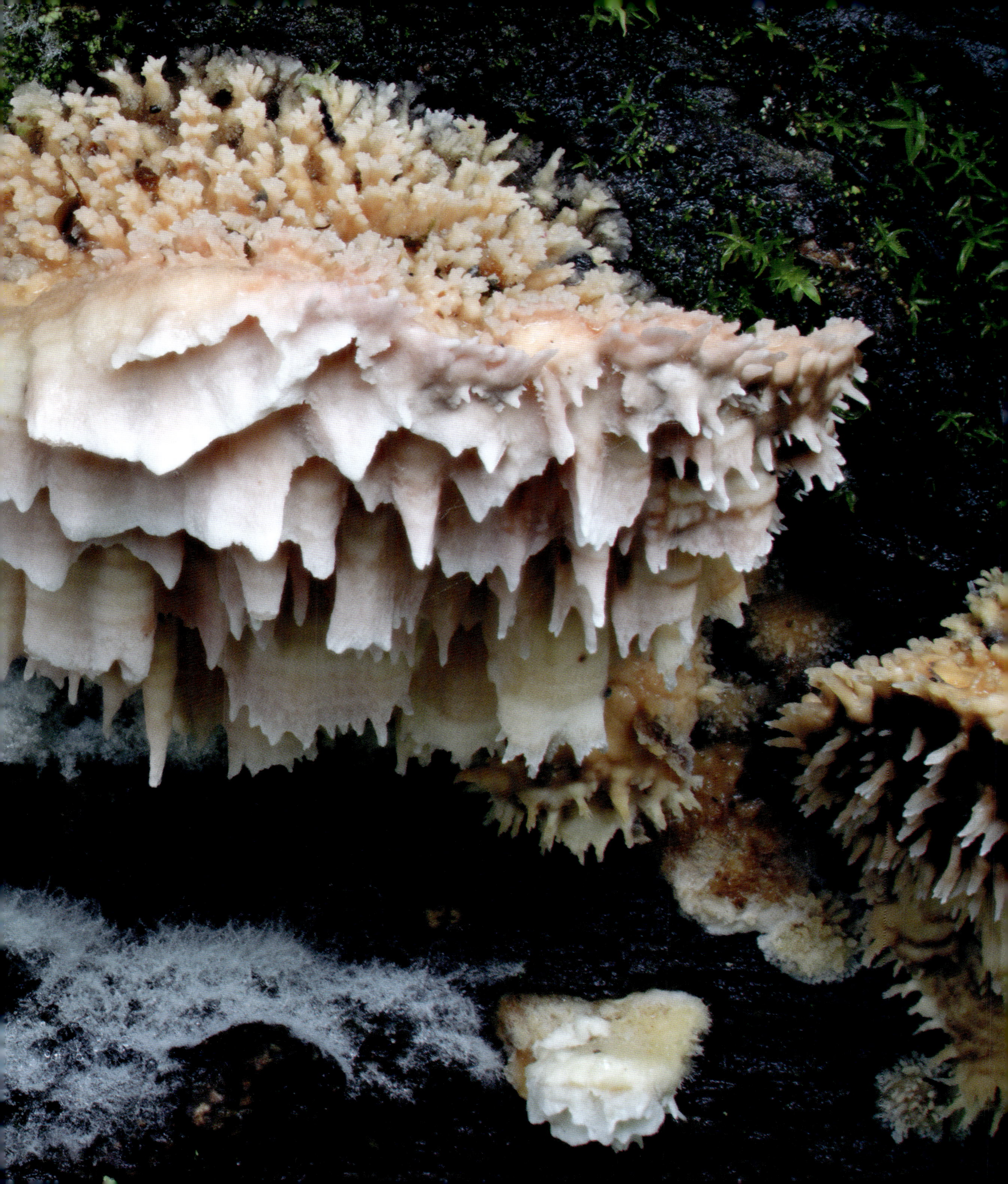

They are the saprotrophic fungi, the undertakers and regenerators of the ecosystem.

Their story dates back many millions of years to the time when plants first ventured from their watery worlds to establish themselves on solid ground.

While primitive plants already used cellulose to form their cell walls, the transition to terrestrial life demanded a sturdier framework. This led to the development of lignin, a tough and almost unbreakable polymer that provides plants with their backbone.[9]

However, lignin posed a massive clean-up challenge when plants die.

Before the saprotrophic fungal, lignin remained impervious to decomposition. Plant matter accumulated as peat, eventually giving rise to the coal seams we unearth today.

Enter the saprotrophic fungal pioneers – 290 million years ago, in the twilight of the Carboniferous period, an ancestral white rot fungus emerged, armed with an enzyme capable of breaking down the seemingly unbreakable.[9]

No longer did coal deposits swell; instead, vast quantities of plant biomass decayed, liberating carbon dioxide into the atmosphere.

This was a turning point in the Earth's carbon cycle, forever altering the course of our planet's ecology.

OPPOSITE
Pseudocolus fusiformis, **stinky squid**
Alstonville, New South Wales, Australia
Distribution: mainly North America, East Asia and Australasia

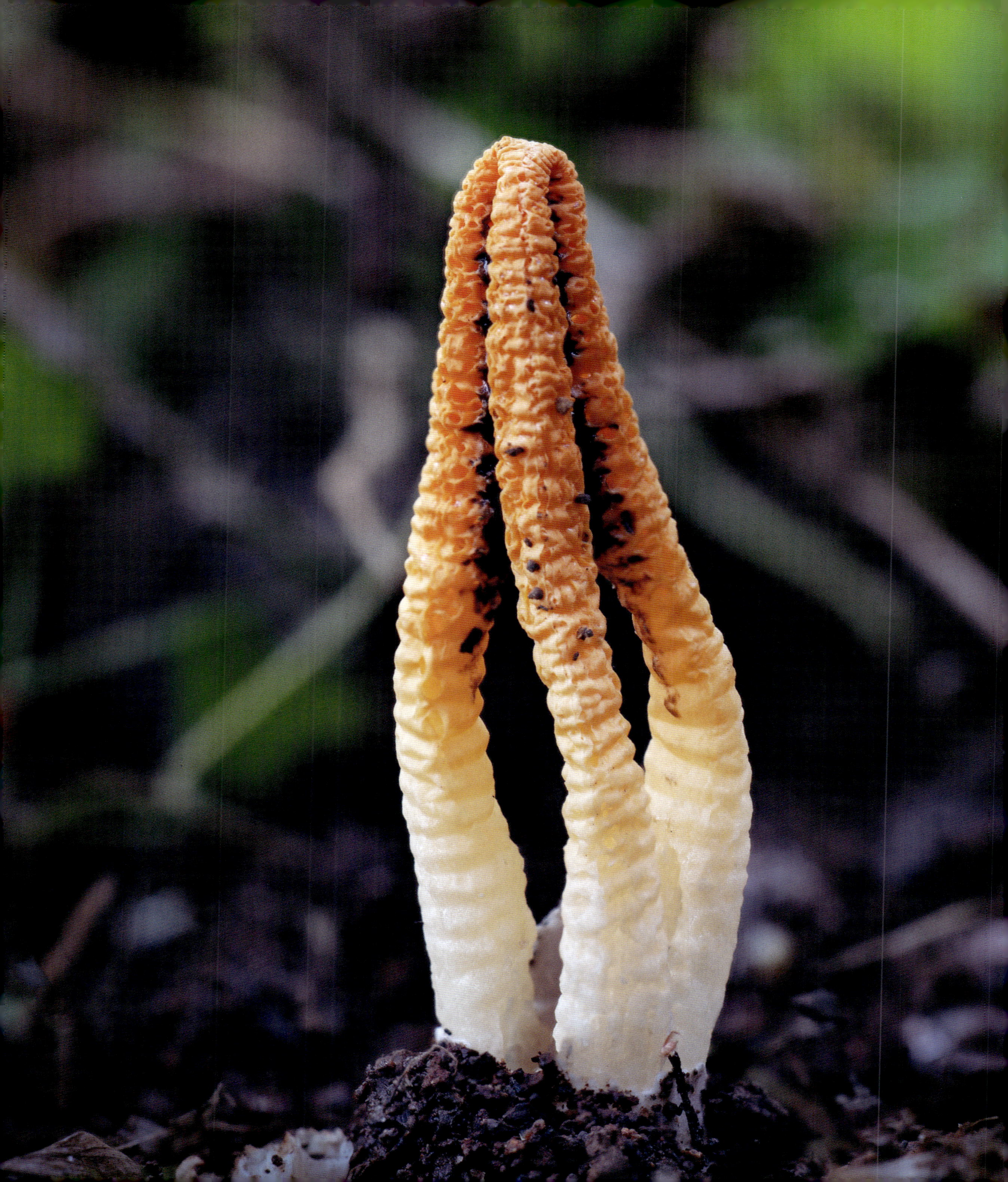

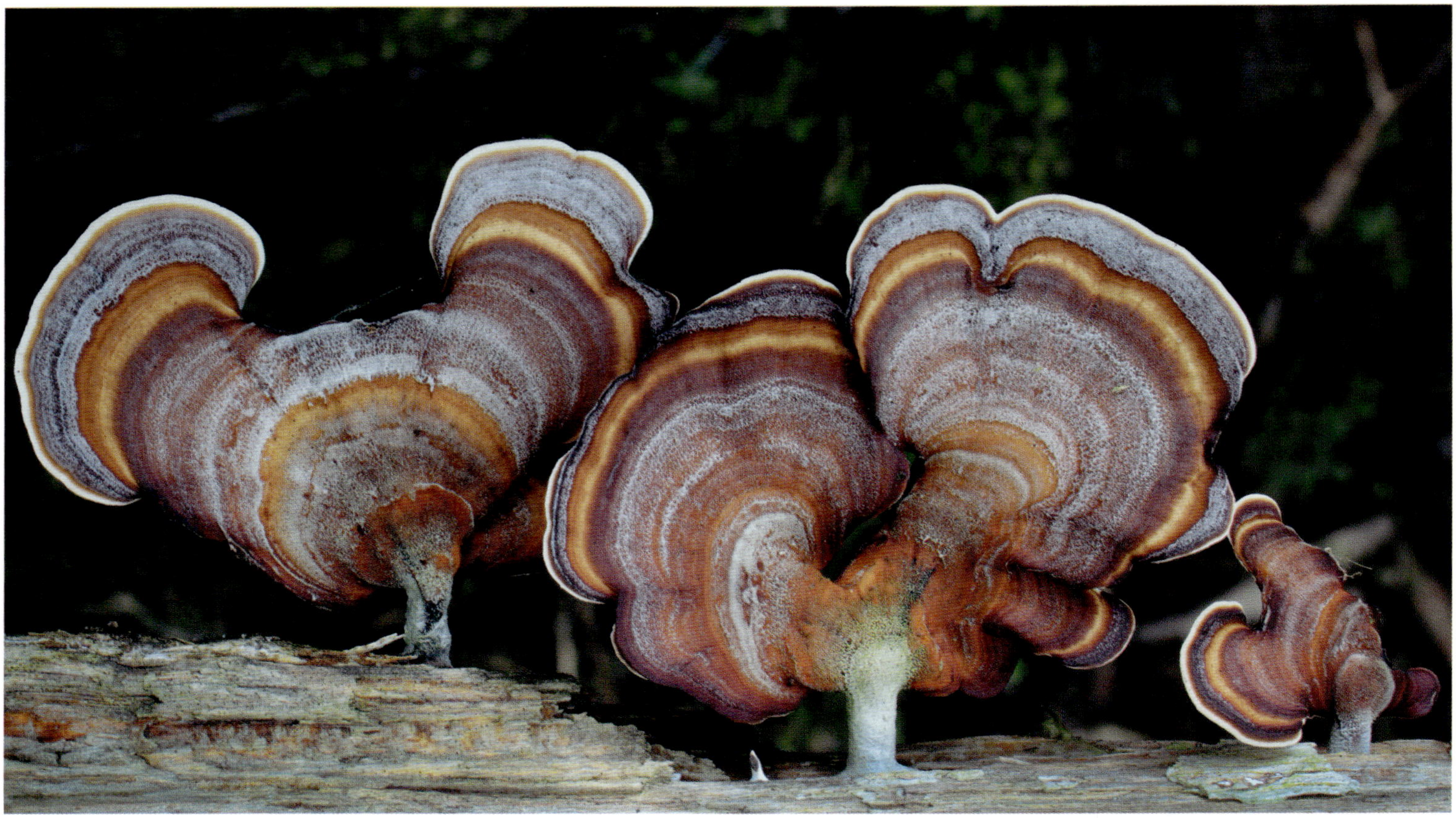

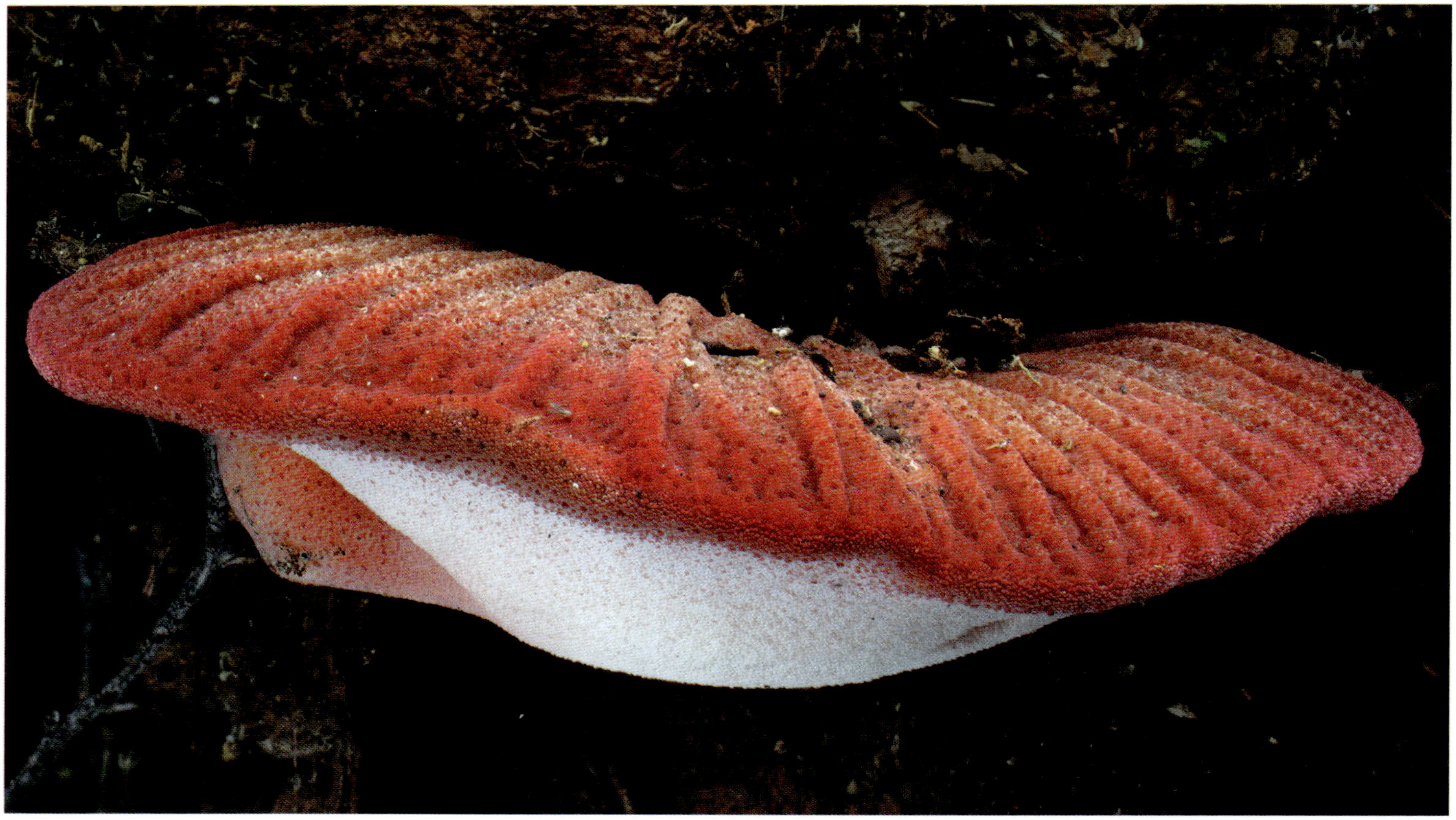

Over time, as fungi evolved and plants diversified, white and then brown rot fungi also multiplied and diversified.

White rot fungi include the magnificent *Microporus affinis* that Stephen photographed in Meghalaya in India. These fungi have unique enzymes that enable them to dismantle the structure of both lignin and cellulose.

In contrast, brown rot fungi such as the highly prized edible mushroom in China, *Fistulina hepatica* or beefsteak polypore, have a different strategy for recycling wood. They break down the cellulose, leaving behind the lignin as a crumbling skeleton.[3,10,11]

These white and brown rot fungi are the characters in many of Stephen's most spectacular photographs.

'Saprotrophic fungi are the unsung champions of decay and the custodians of renewal within our ecosystems.

'If not for these fungi, we would be buried under a build-up of dead wood, leaf matter, and the remains of animals.

'They have taught me to let things rot: to leave a fallen tree or branch on the ground so it can recycle, build soil, feed the critters who inhabit that underworld, nourish the plants.

'We humans have a fear of death. But I like to think that these fungi teach us that death is just part of creating new life.'

OPPOSITE TOP
***Microporus affinis*, dark-footed tinypore**
Mawlynnong, Meghalaya, India
Distribution: mostly Africa, Asia and Australasia

OPPOSITE BOTTOM
Fistulina spiculifera
Philosopher Falls, Tasmania, Australia
Distribution: Australia

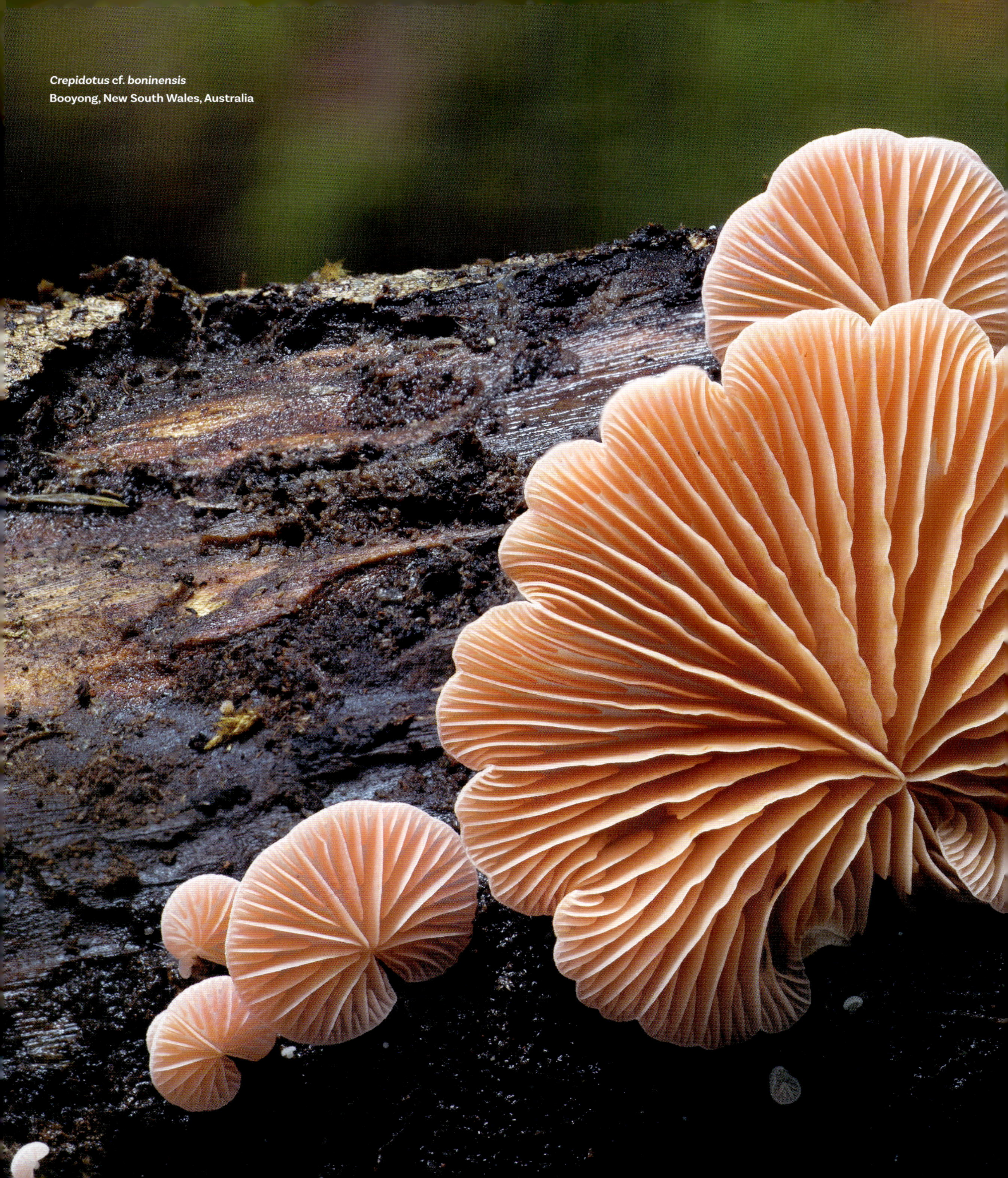

Crepidotus cf. boninensis
Booyong, New South Wales, Australia

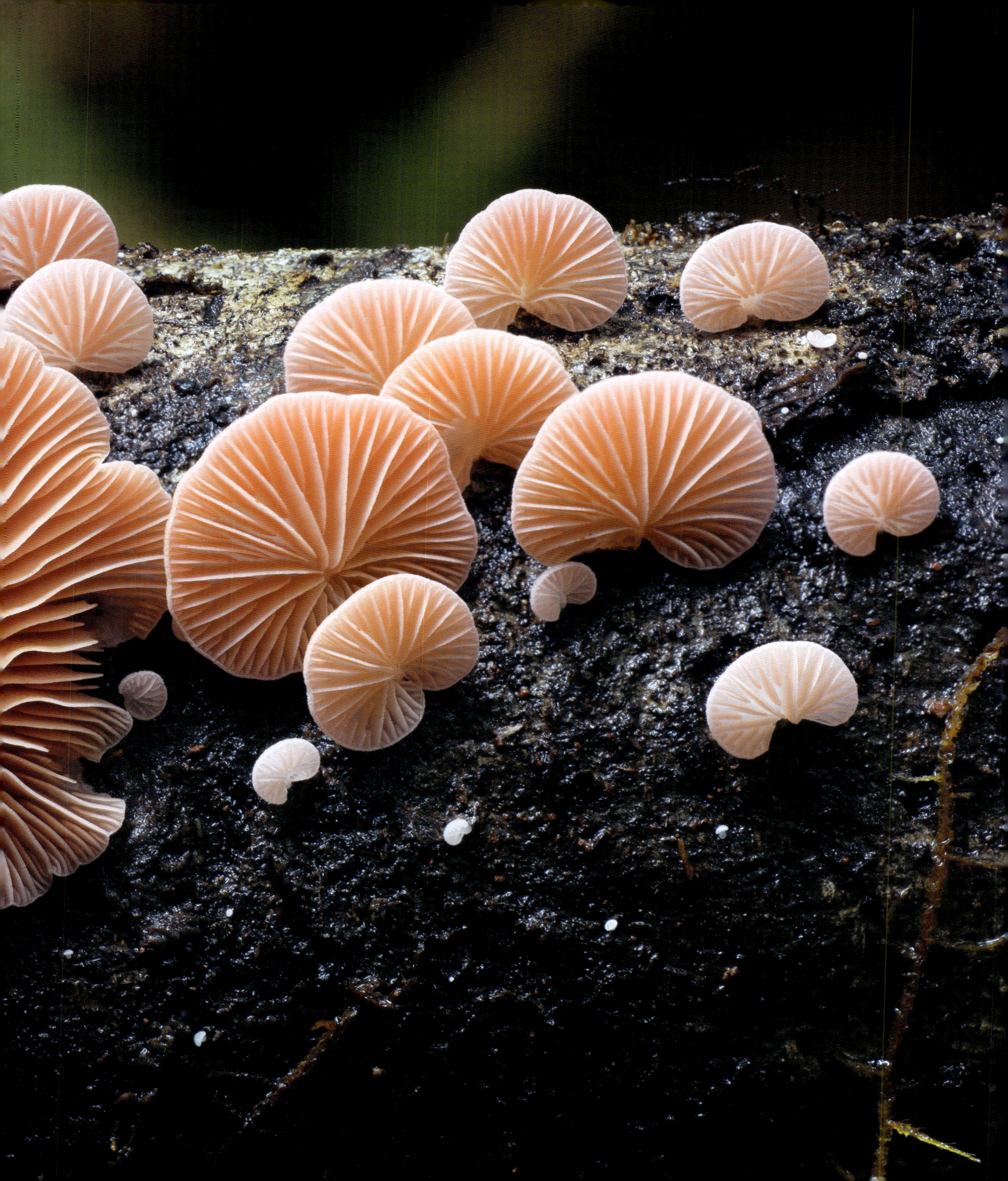

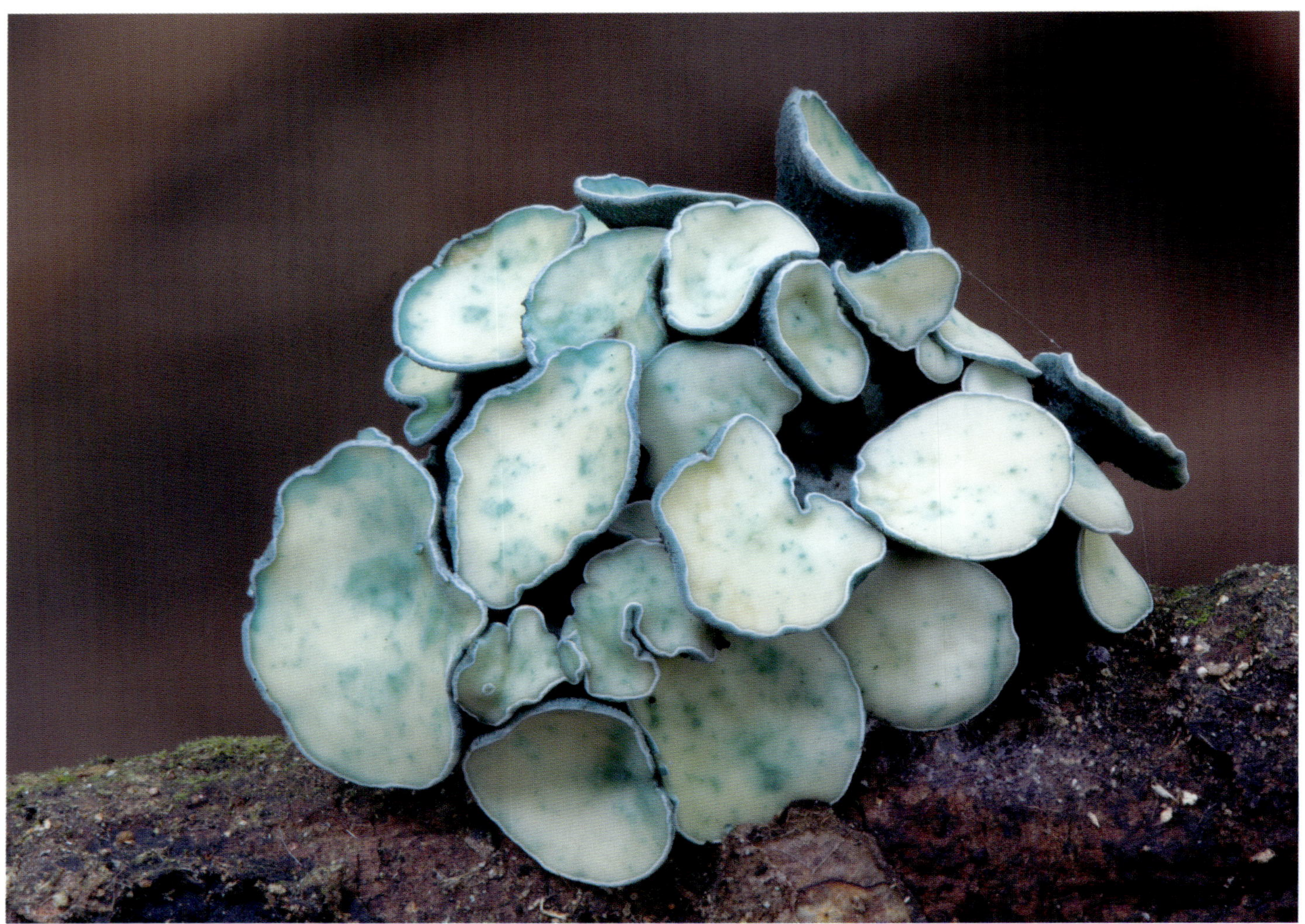

A forest undertaker with a fascinating story is *Chlorociboria aeruginascens*.

This fungal species produces exquisite blue-green cups, and its mycelium creates a pigment called xylindein that dyes the wood in similar shades of blue and green.

In the 14th and 15th centuries, Italian craftsmen skilfully used this tinted wood for ornate inlay designs, and woodworkers still highly value it for its unique beauty.

ABOVE

Chlorociboria cf. *aeruginascens*
Nabanhe National Nature Reserve, Yunnan, China

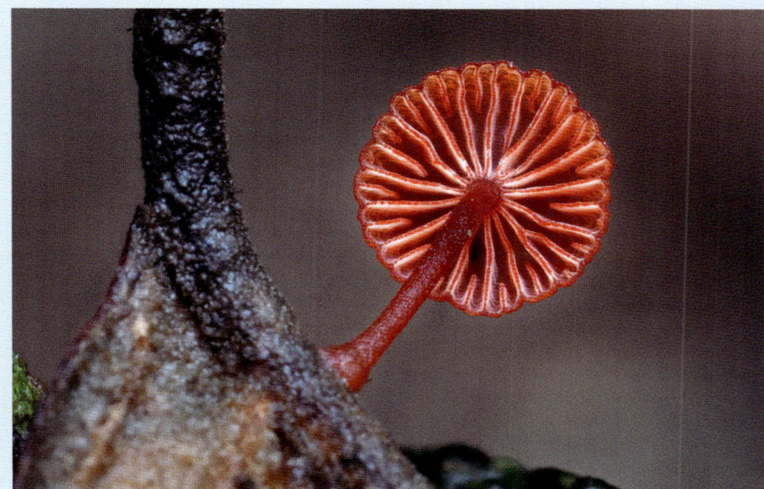

As we explore the wilderness in search of fungi, we come across countless saprotrophic treasures on every dead surface.

Tiny parasols grow out of crumbling leaves.

Capsules open to reveal a nest of eggs, spore packets that splash out in the next shower.

Fallen twigs serve as scaffolding for elaborate fans.

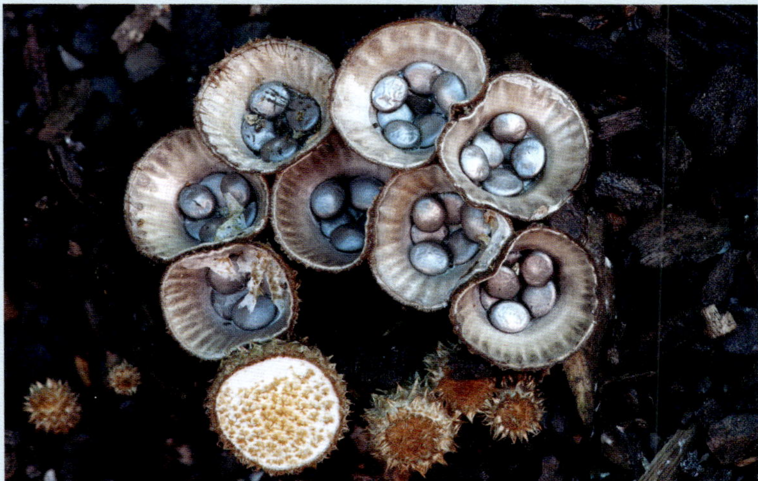

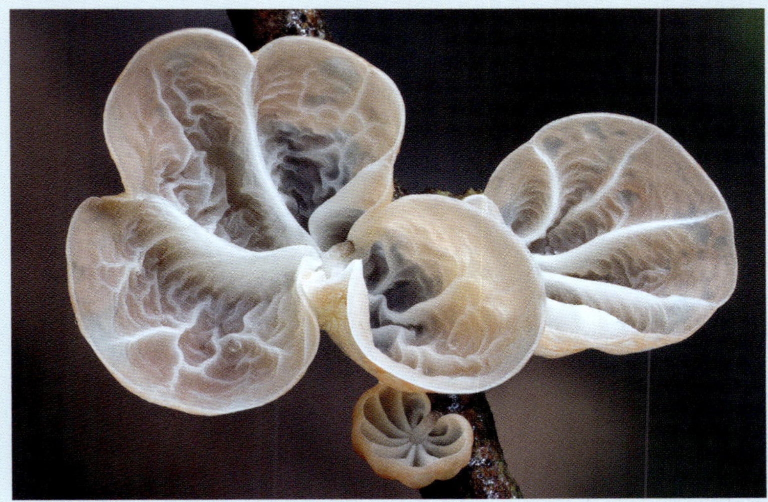

TOP
Cruentomycena sp.
Mawkyrwat, Meghalaya, India

MIDDLE
Cyathus stercoreus, dung-loving bird's nest fungus
Booyong, New South Wales, Australia
Distribution: worldwide

BOTTOM
Campanella sp.
Nightcap National Park, New South Wales, Australia

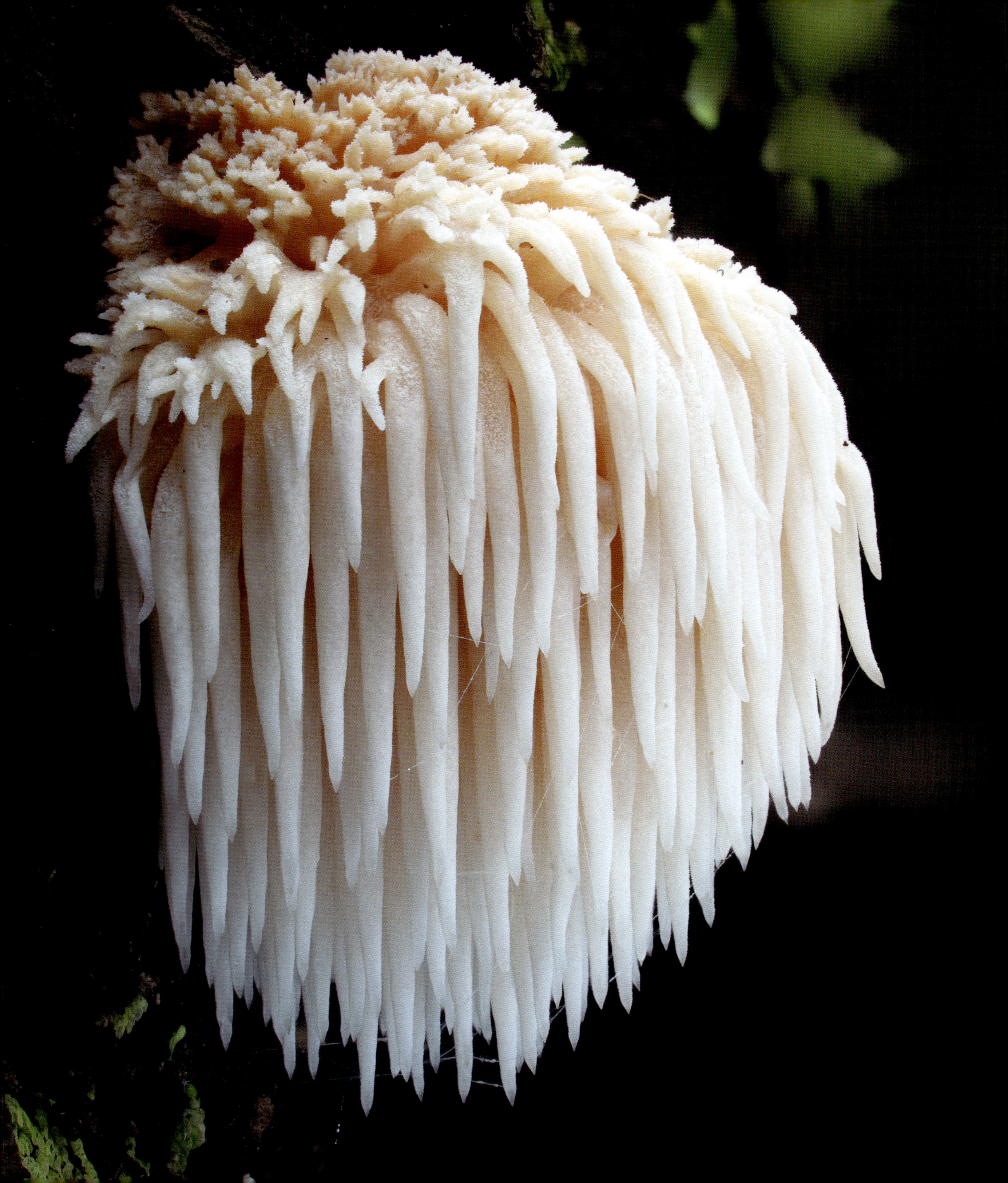

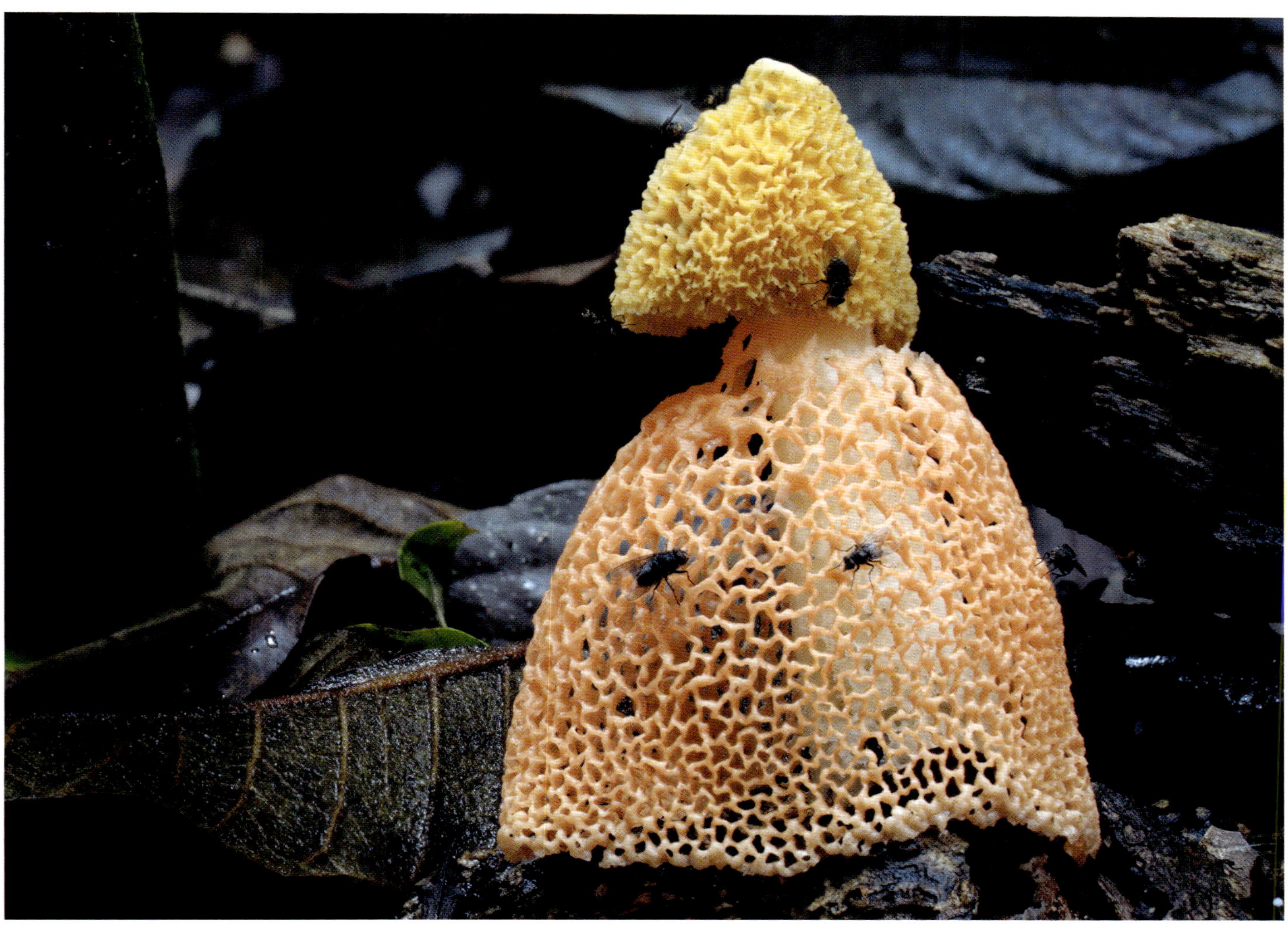

A lion's mane adorns a decaying log.

Bizarre stinkhorns emerge from soft white eggs in the forest debris. We smell them before we
see them – a stench of rotting meat designed to attract flies to spread their spores.

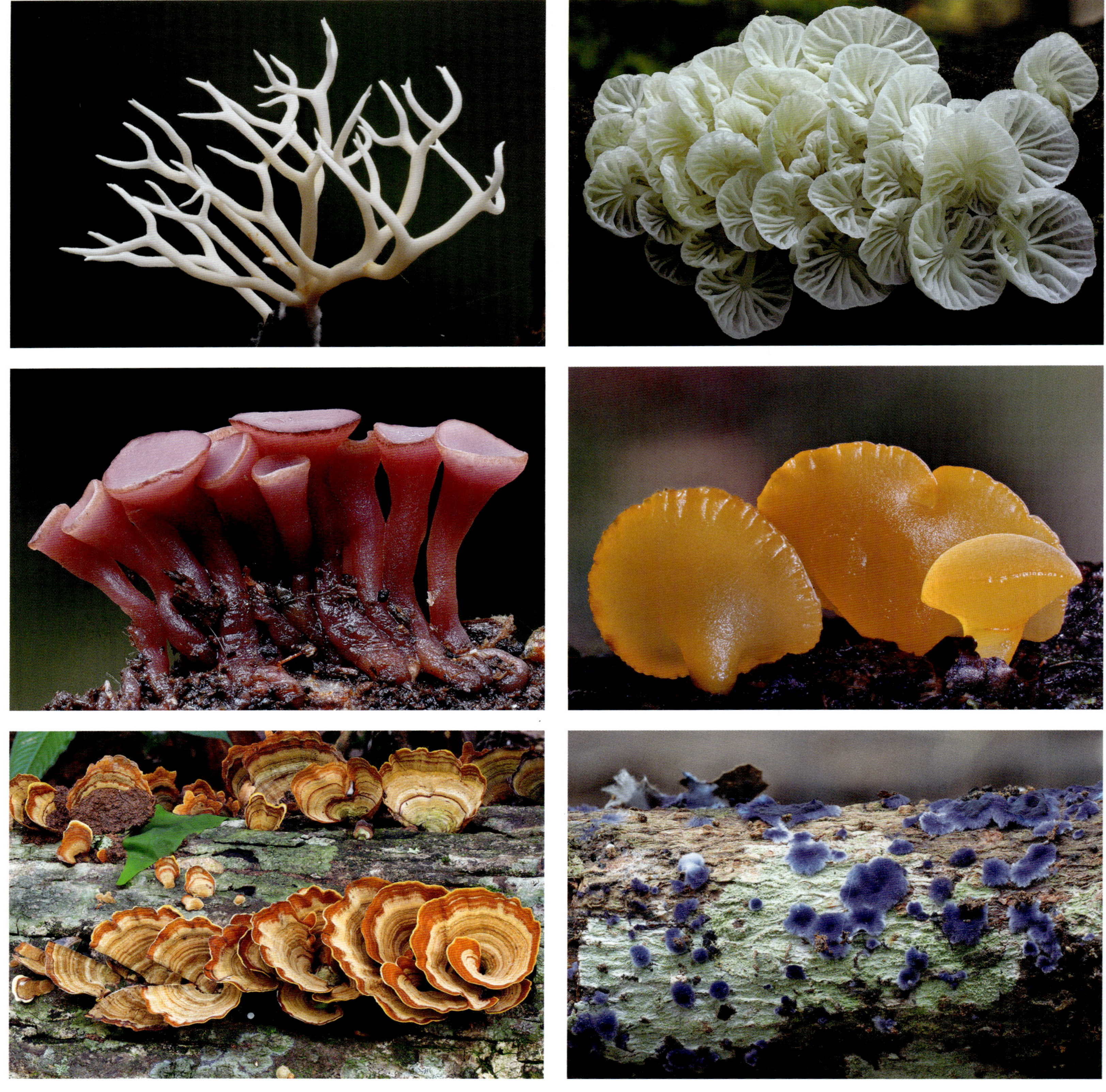

'Mycelium is fascinating, but when you look at it with the naked eye or through a camera lens, it just looks like a bunch of matted threads.

'It is almost impossible to tell what species of fungus it is.

'It is a different story when the sporing bodies emerge. A whole new world opens up. Suddenly, you get to witness the incredible shapes, sizes, colours and forms that make each fungus unique.

'As a fungi photographer, it's a magical moment when that character is revealed.

'It is like seeing a hidden world that's been waiting to be discovered.'

OPPOSITE TOP LEFT
Xylaria sp.
Nightcap National Park, New South Wales, Australia

OPPOSITE TOP RIGHT
Marasmiellus sp.
Nightcap National Park, New South Wales, Australia

OPPOSITE MIDDLE LEFT
Ascocoryne sarcoides, purple jellydisc
Montana Falls, Tasmania, Australia
Distribution: worldwide in temperate regions

OPPOSITE MIDDLE RIGHT
Heterotextus peziziformis, jelly bells
Queenstown, Tasmania, Australia
Distribution: Australasia

OPPOSITE BOTTOM LEFT
Stereum sp.
Nabanhe National Nature Reserve,
Yunnan, China

OPPOSITE BOTTOM RIGHT
Terana caerulea, cobalt crust
Booyong, New South Wales, Australia
Distribution: worldwide

We also encounter mushrooms growing on living trees, often appearing as shelves and brackets.

For a long time, these fungi were believed to be agents of death for old trees but now we know they are actually allies of the forest elders.

As a mature tree undergoes its natural aging process, the hyphae of these fungal agents navigate the inner realms, unleashing enzymes that break down the dead cells of the heartwood. This process unlocks nutrients that were previously trapped within the tree.

These fungi consume the timber used for construction and carpentry, which is why plantation timber industries see them as a threat. The heartwood also serves as a structural support for the tree, therefore hollowing it out can impact its longevity.

These fungal organisms in the forest, however, are increasingly viewed as collaborators in the age-old relationship between living trees and fungi. The hollow cylinders formed in the heartwood can lighten the load on the decaying roots of a veteran tree, helping it to withstand strong winds and potentially prolonging its life.[12]

The hollows created by these fungi also provide a safe haven for creatures like possums, birds and bears.

Meanwhile, the nutrient-rich soil, a product of decay from within, fosters new roots for the tree and acts as a nursery for other plants.

OPPOSITE TOP
Ganoderma **sp.**
Ialong, Meghalaya, India

OPPOSITE BOTTOM
Lenzites betulinus, **gilled polypore**
Mawphlang, Meghalaya, India
Distribution: worldwide

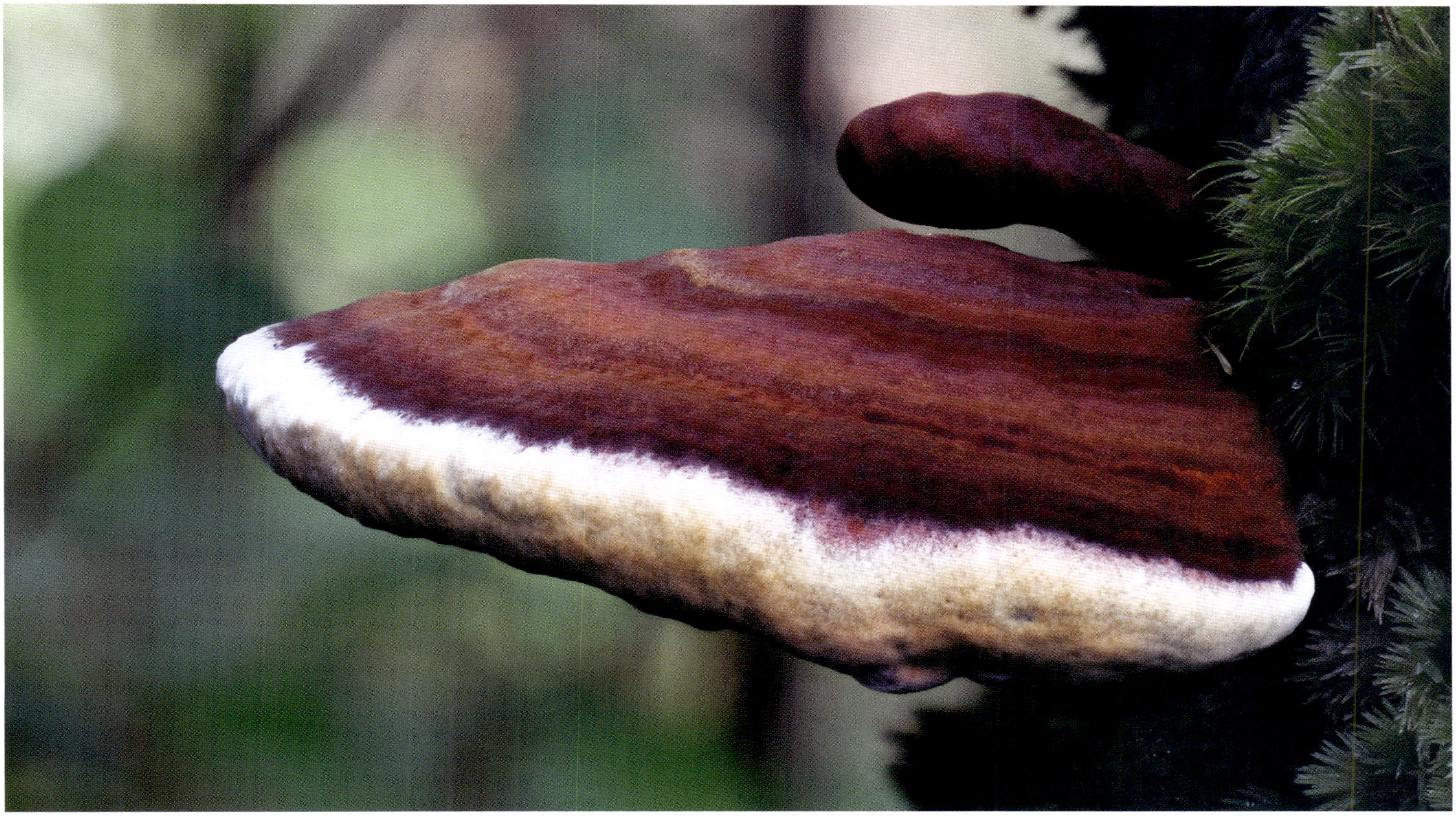

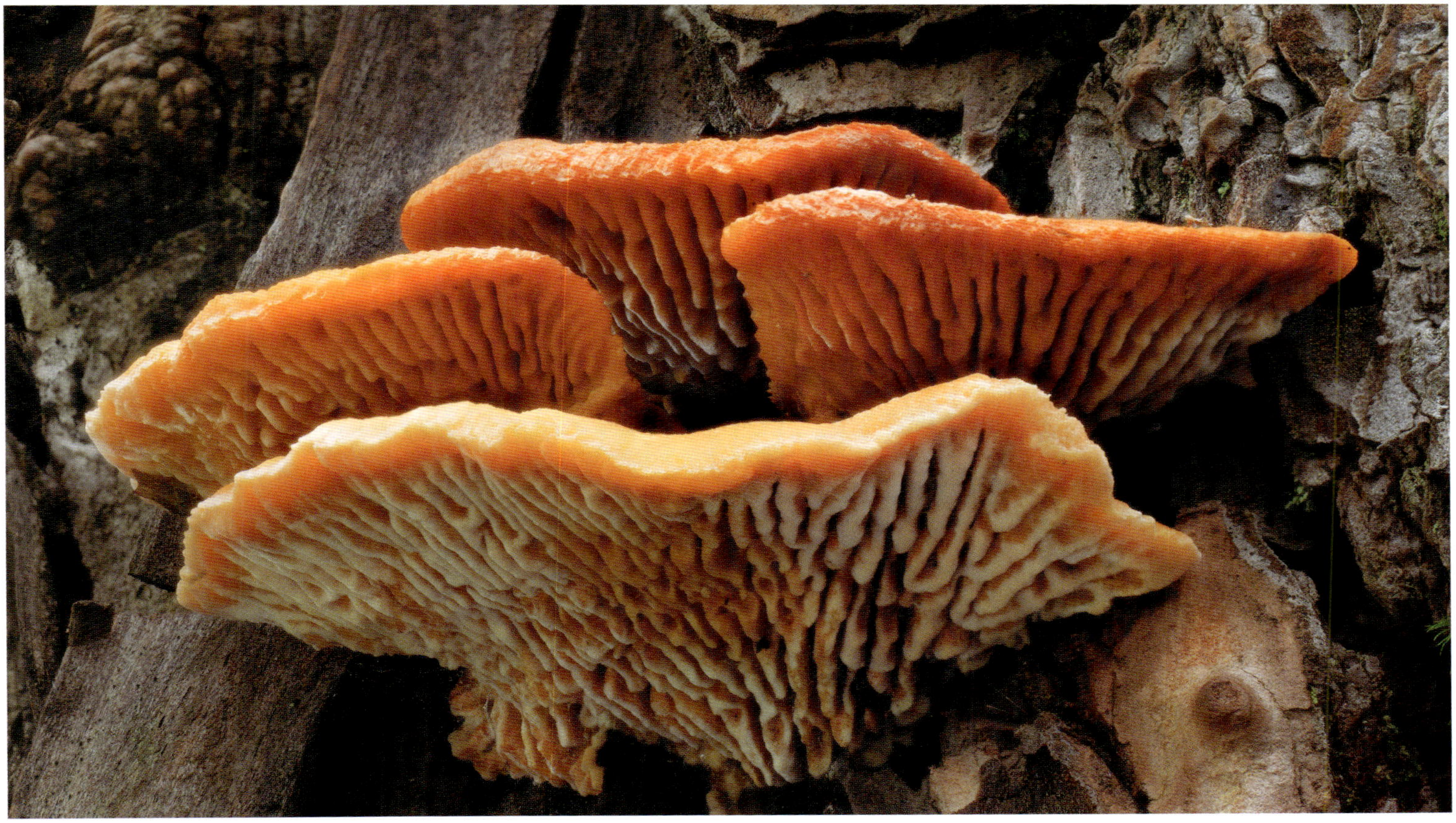

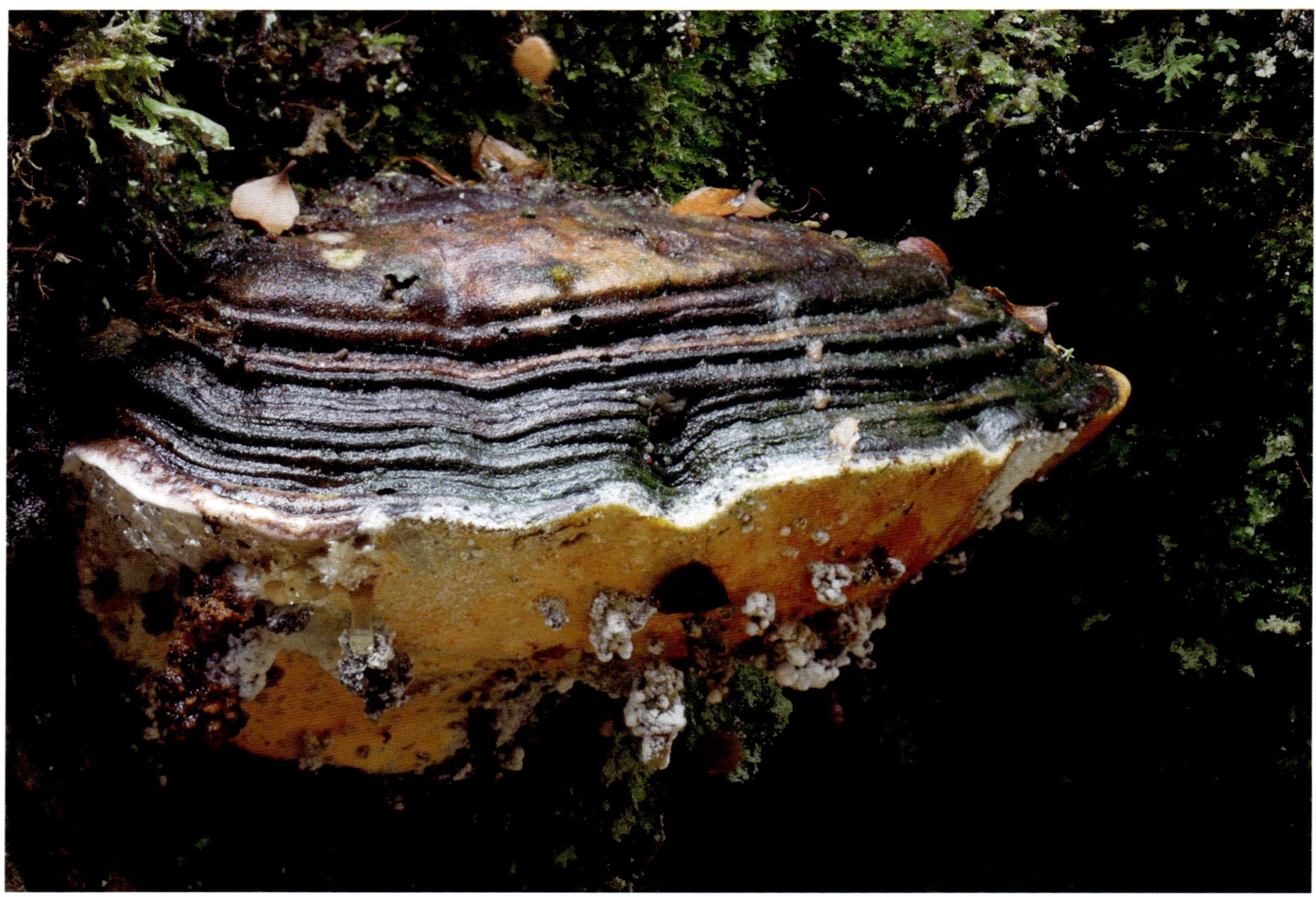

The shelves and brackets are home to a vast array of spiders, mites and insects, creating a whole mini ecosystem hanging from the tree.[13,14]

Together, these fungi, animals and trees are a dynamic, ancient life-support system, reminding us of nature's incredible interconnectedness.

ABOVE
Unknown polypore fungus
Corrina, Tasmania, Australia

OPPOSITE
Rust fungus infecting a plant
Photograph: Alison Pollack

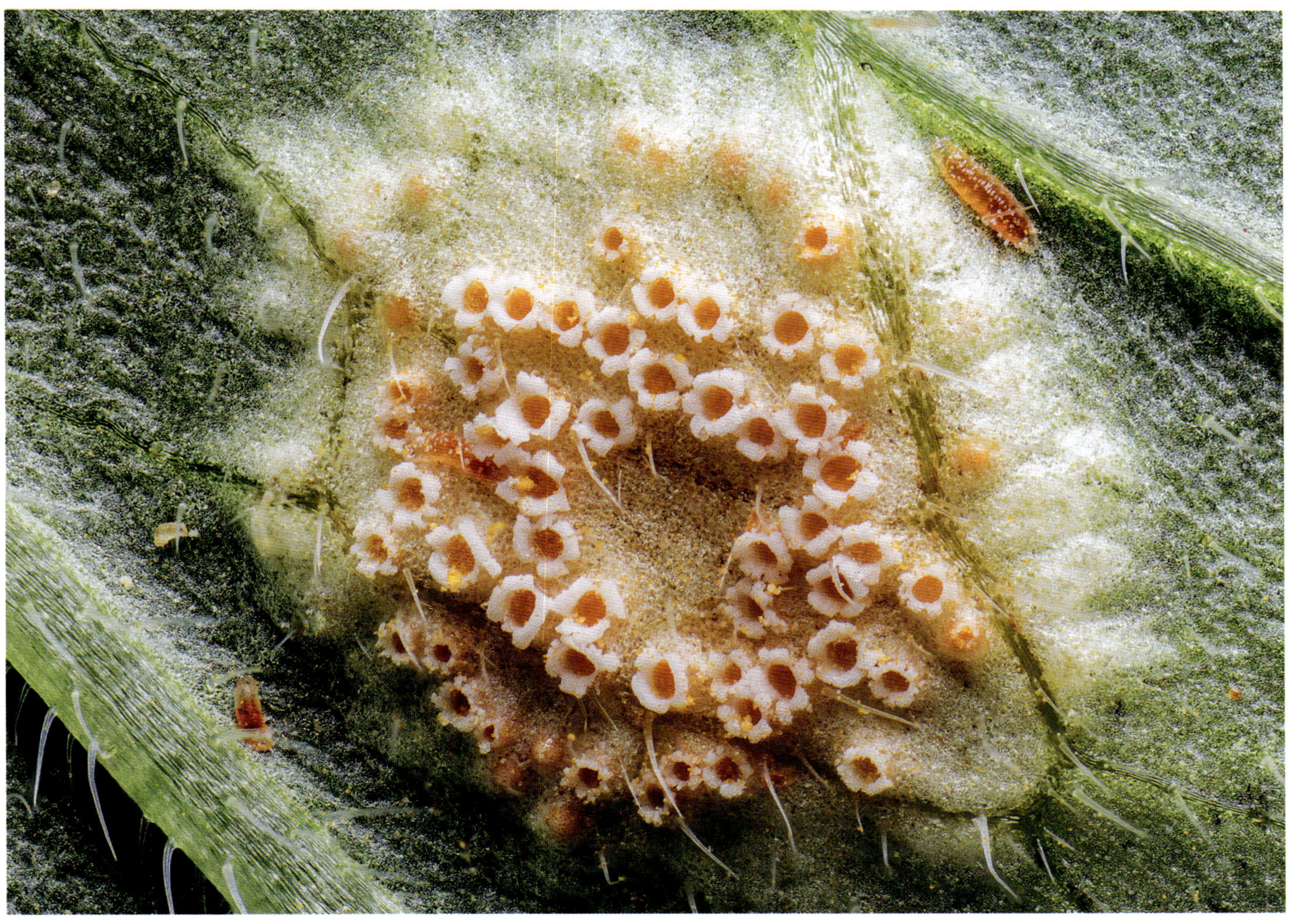

In the theatre of nature, the third ensemble of fungal protagonists we see take the stage are the executioners or parasites.

Some, like rusts and smuts, invade the living cells of plants, extracting nutrients and gradually sapping the life from their hosts. This pathogenic troupe instils fear in humans with the potential devastation they inflict upon crops – fungal diseases that have caused 'starvation, economic collapse, social conflict, warfare, and mass emigration'.[15]

Despite their reputation, it is not in the parasites' interest to eradicate their host.

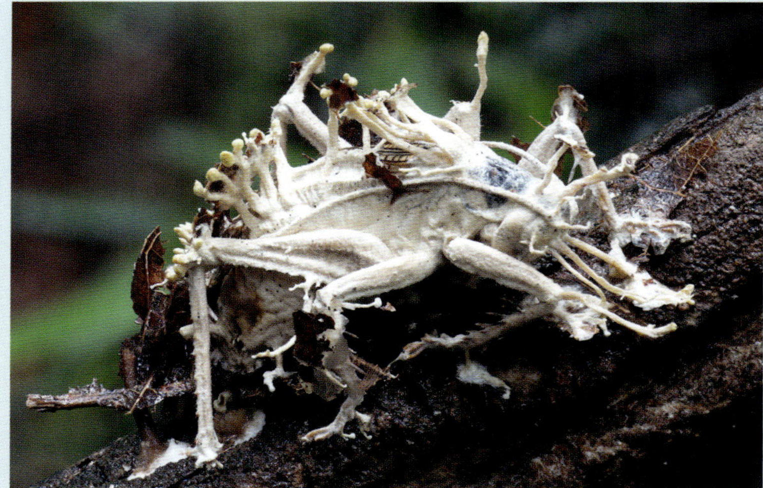

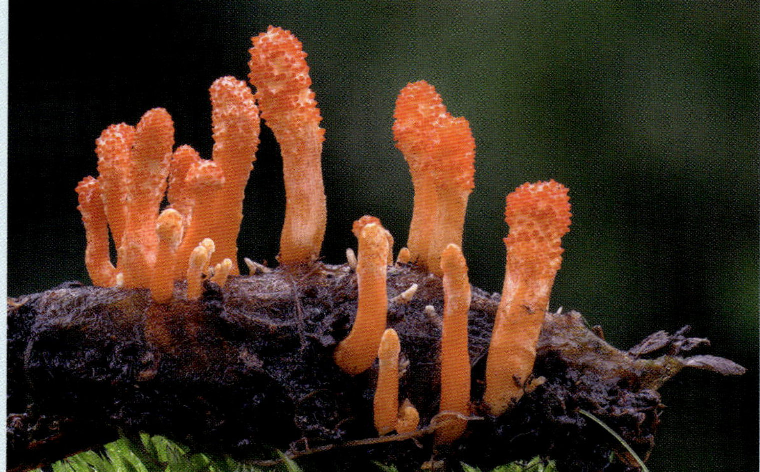

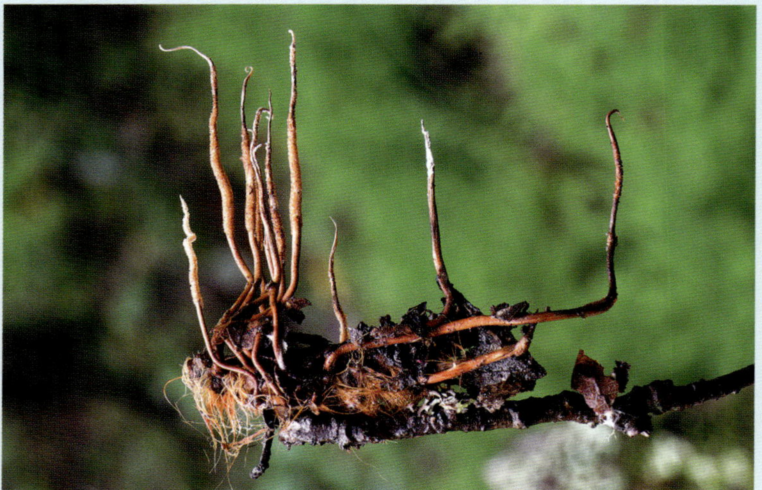

In undisturbed landscapes of forests, heathlands and deserts, healthy plants demonstrate a remarkable ability to repel these pathogenic fungi, so the effects of parasitic fungi in native ecosystems are often contained, confined to mere portions of a plant or those weakened by other forces.

Parasitic fungi that are increasingly taking centre stage in popular culture are the flesh-eating fungi.

These were thrust into the spotlight in the television series and video game 'The Last of Us', in which humans, after contact with a fictional *Cordyceps* fungus, turn into villainous mushroom monsters.

Fortunately, a healthy human is not a suitable host for these fungi. But these transformations do happen in the insect world – all the time.

TOP
Akanthomyces sp.
Honghe Amoshan Nature Reserve, Yunnan, China

Host: a cricket

MIDDLE
Cordyceps militaris, scarlet caterpillar club
Honghe Amoshan Nature Reserve, Yunnan, China
Distribution: worldwide

BOTTOM
Ophiocordyceps crinalis
Honghe Amoshan Nature Reserve, Yunnan, China
Distribution: Asia and North America

Host: lepidopteran larva

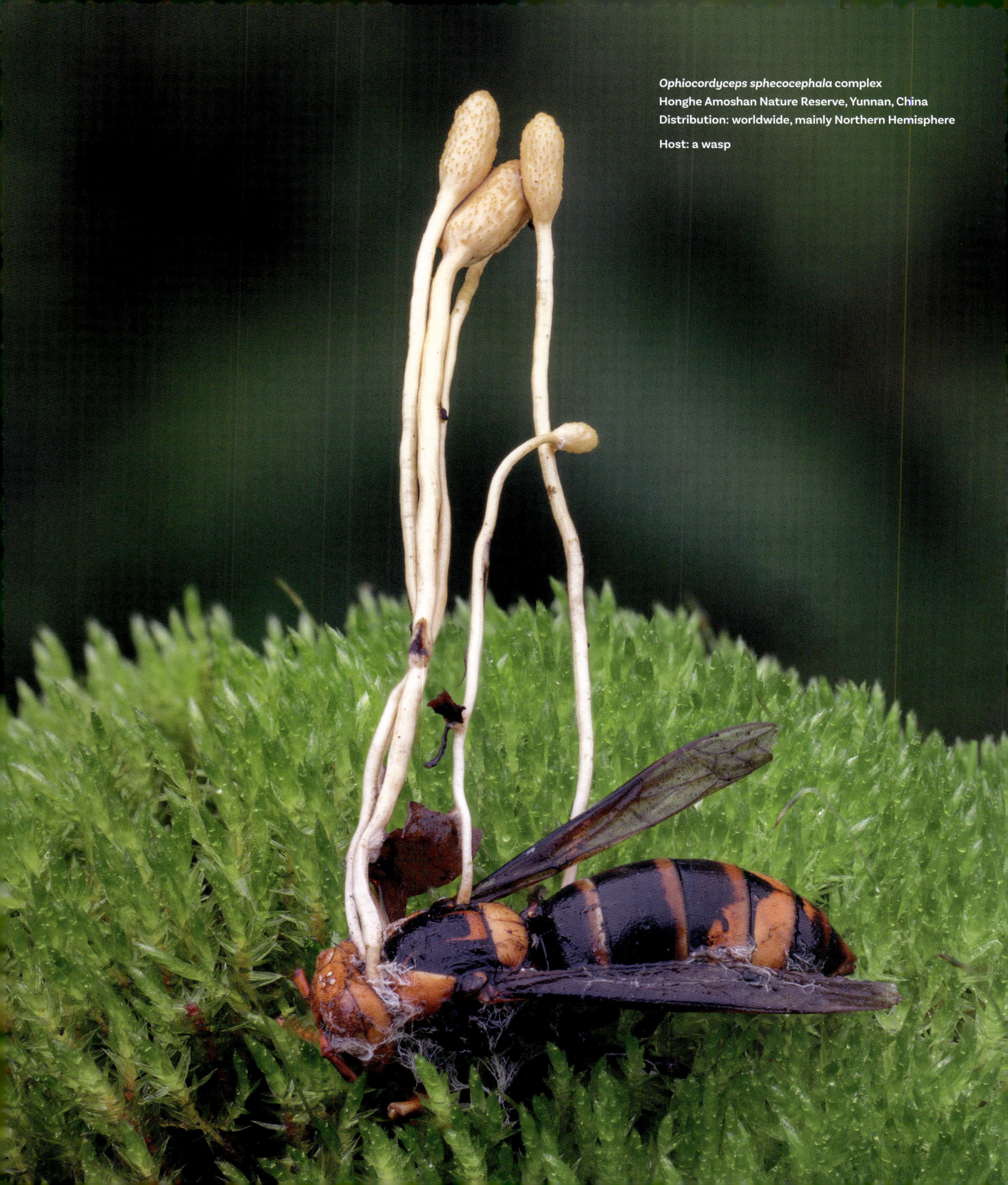

Ophiocordyceps sphecocephala complex
Honghe Amoshan Nature Reserve, Yunnan, China
Distribution: worldwide, mainly Northern Hemisphere

Host: a wasp

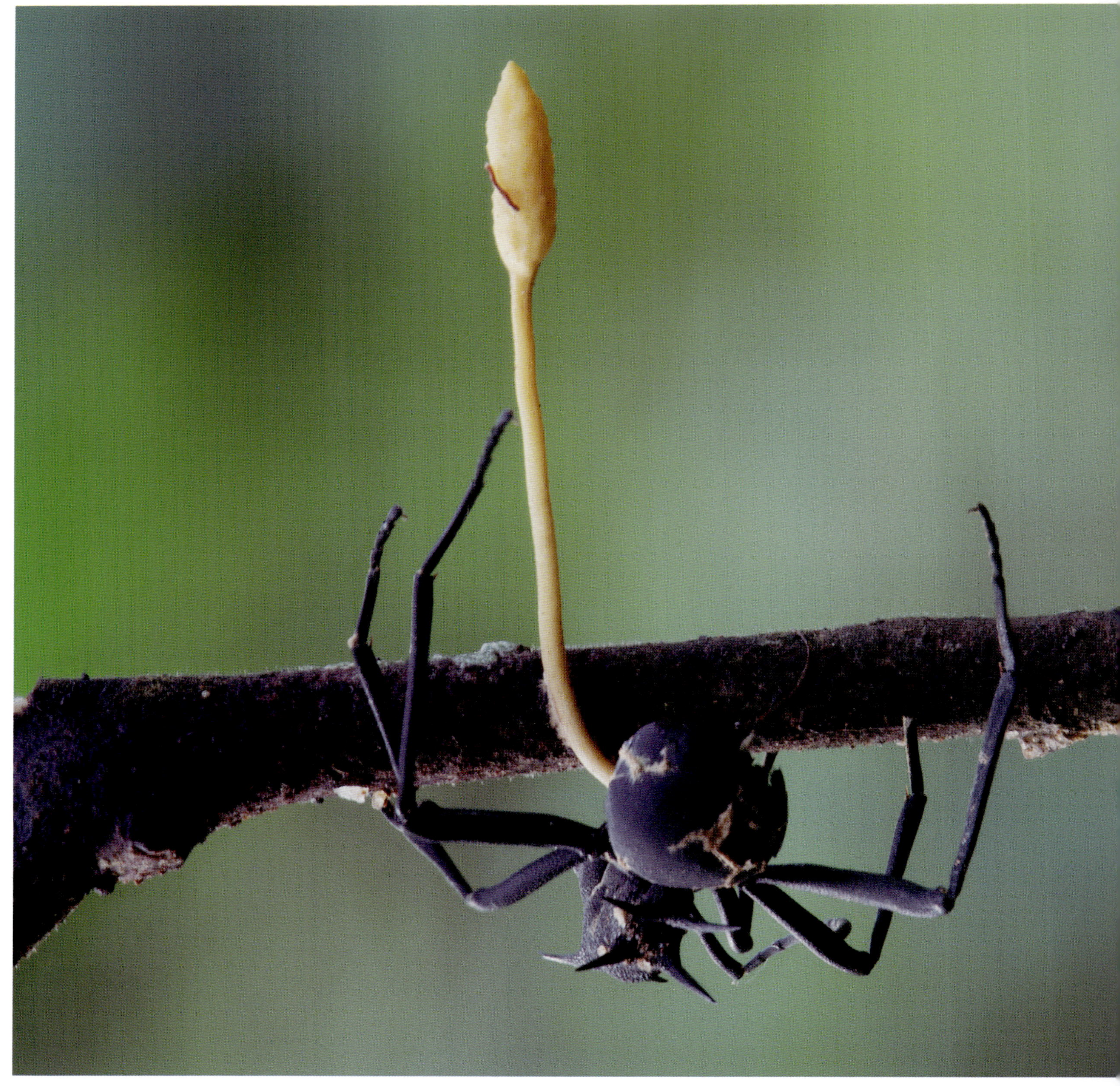

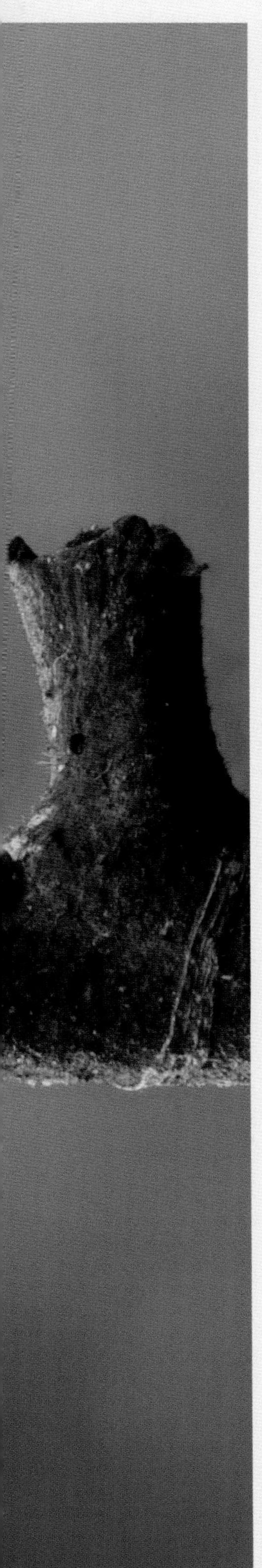

Often highly specialised, many of these fungi have adapted to invading and killing a specific species of insect or spider.

Imagine you are a Hercules ant.

As you go about your industry on the rainforest floor you are unaware that amid decaying foliage and gnarled branches a spore of the zombie-making fungus, *Ophiocordyceps formicarum*, awaits your arrival.[16]

The spore finds a route through your armour, and in a quiet invasion, tendrils of hyphae weave through your anatomy – a microscopic takeover of your body, your muscles and your central nervous system.[16]

Like a puppeteer, the fungus then manipulates your movements, forcing you to leave the nest and climb up nearby foliage. Once you reach a high point, you latch onto a branch or leaf. It is a death grip designed to keep you from falling. You are no longer an ant but a fungus resembling an ant – a vessel with one purpose: reproduction.[16,17]

After your death, a sexual structure covered in spores erupts from your corpse. From this elevated position, spores are spread far and wide.

There are many species of zombie ant fungi on the planet, but they are just one of the fascinating examples of entomopathogenic fungi that feed on the flesh of insects and spiders.

Stephen and I have become obsessed with hunting down these wonderfully creepy fungal organisms, and in doing so, we have discovered a teeming and diverse array of them all around us.

Ophiocordyceps formicarum, zombie ant fungus
The Mushroom Research Centre, Chaing Mai, Thailand
Distribution: Asia

Host: an ant

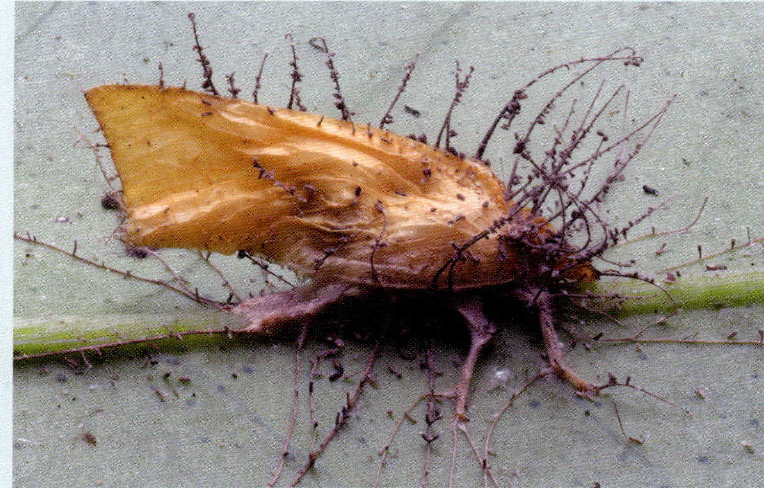

We can find as many on a wild lemon tree in a friend's backyard as we do in the forest, and many are new to science.

These flesh-eating fungi are so small they're hard to see with the naked eye.

However, with the aid of high-resolution photography and intense magnification, Stephen can reveal the gossamer-thin fungal threads holding the prey in place and their surreal reproductive bodies.

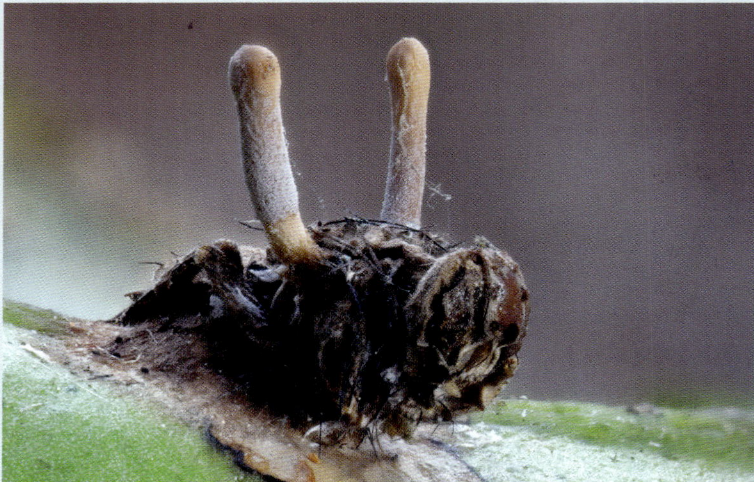

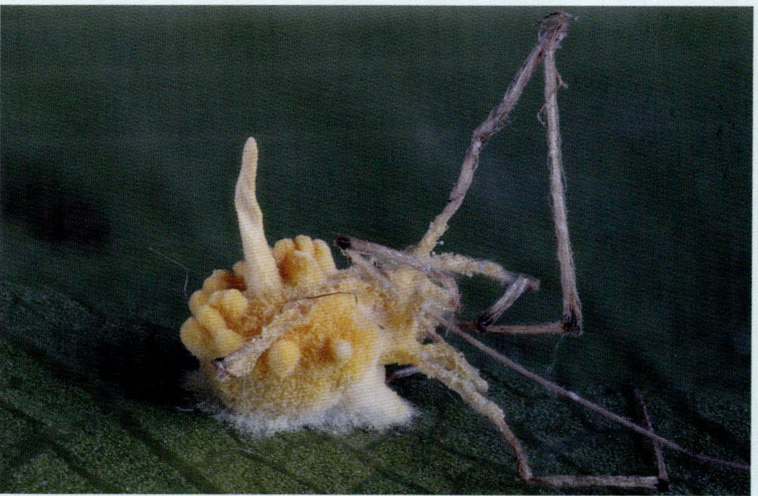

TOP
Hirsutella citriformis
Uki, New South Wales, Australia
Distribution: worldwide, in scattered localities

Host: a planthopper

MIDDLE
Ophiocordyceps dipterigena
Uki, New South Wales, Australia
Distribution: worldwide, mostly in tropical regions

Host: a fly

BOTTOM
Gibellula sp.
Uki, New South Wales, Australia

Host: an unrecognisable invertebrate, possibly a spider

OPPOSITE
Gibellula sp.
Booyong, New South Wales, Australia

Host: a spider

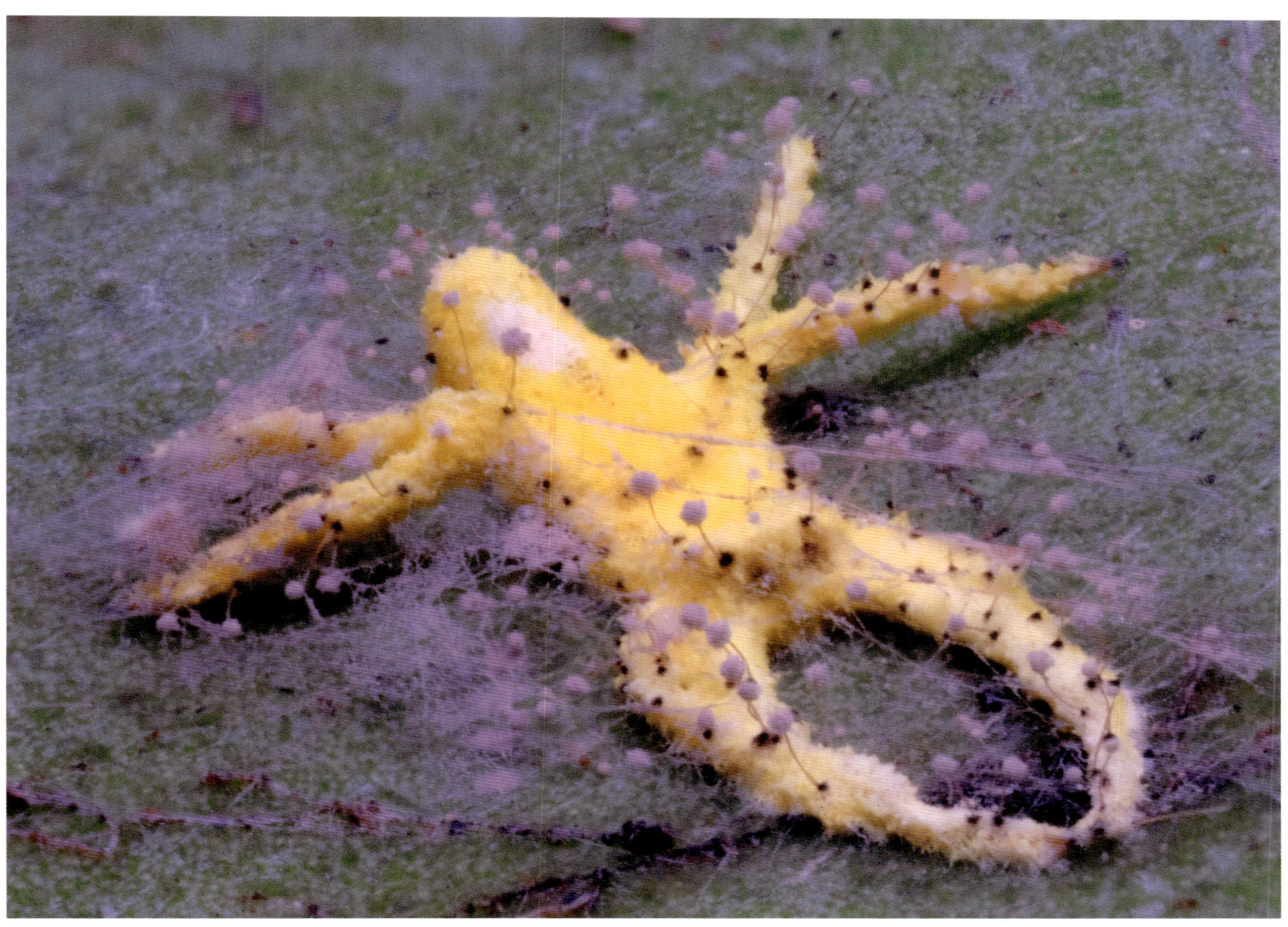

Some flesh-eating fungi look like horns or spiky barbs, while others appear as fluffy fleshy torpedoes erupting into twisting towers.

These otherworldly multifarious designs all have the same goal – to produce and release fungal spores.

'The most common combination we find is a spider carcass covered in yellow fibres with lilac pom poms on stalks.

'It reminds me of the scene in the book Gulliver's Travels, with Gulliver pinned to the ground by the miniature army of Lilliput.'

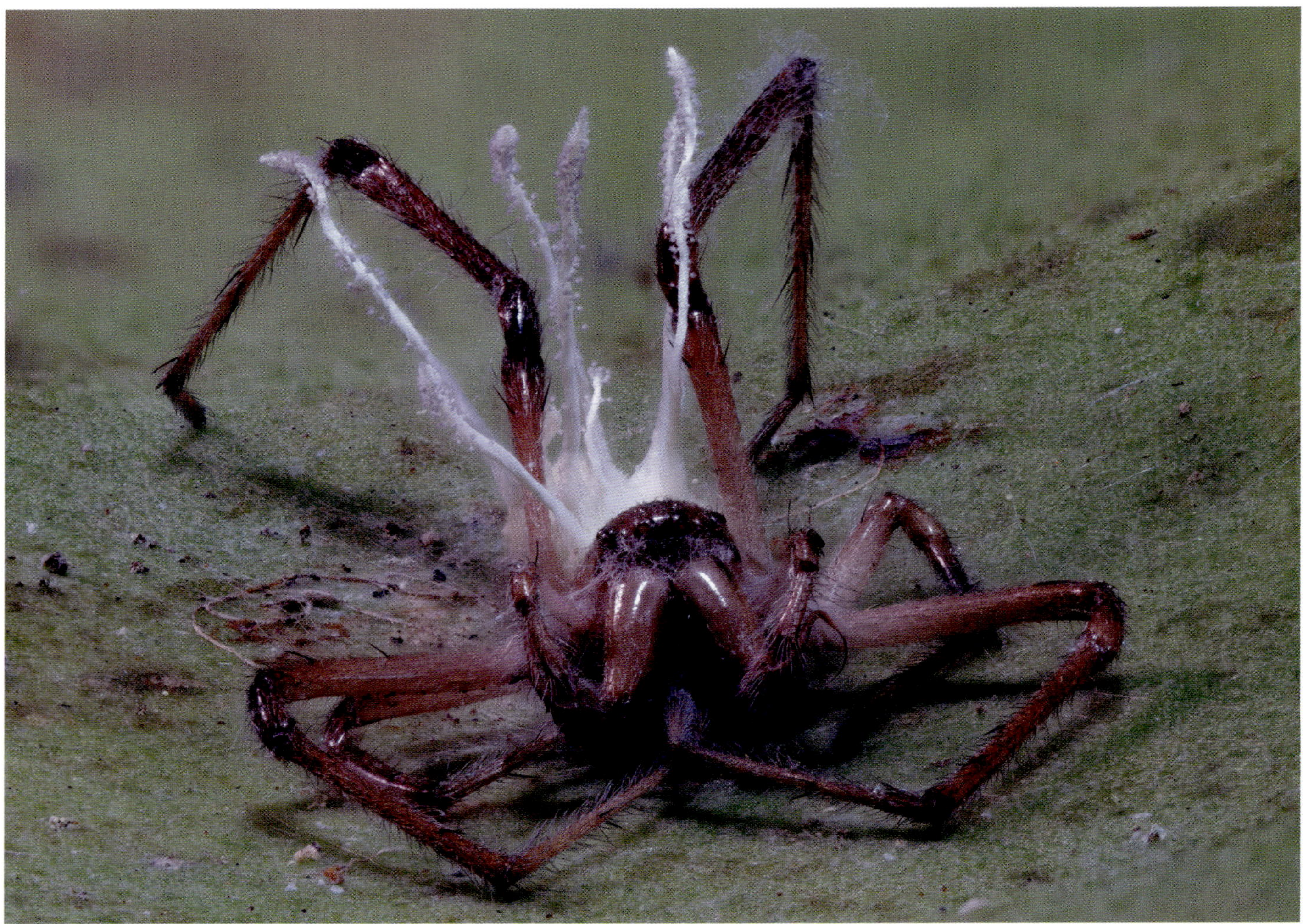

The co-evolution of entomopathogenic fungi and their insect hosts functions as nature's pest control system, maintaining biodiversity and ensuring that no single species dominates to the detriment of others.[17]

They demonstrate that even the smallest players in the natural world can wield significant influence in maintaining the delicate balance of life.

Gibellula sp.
Uki, New South Wales, Australia

Host: a long-legged sac spider

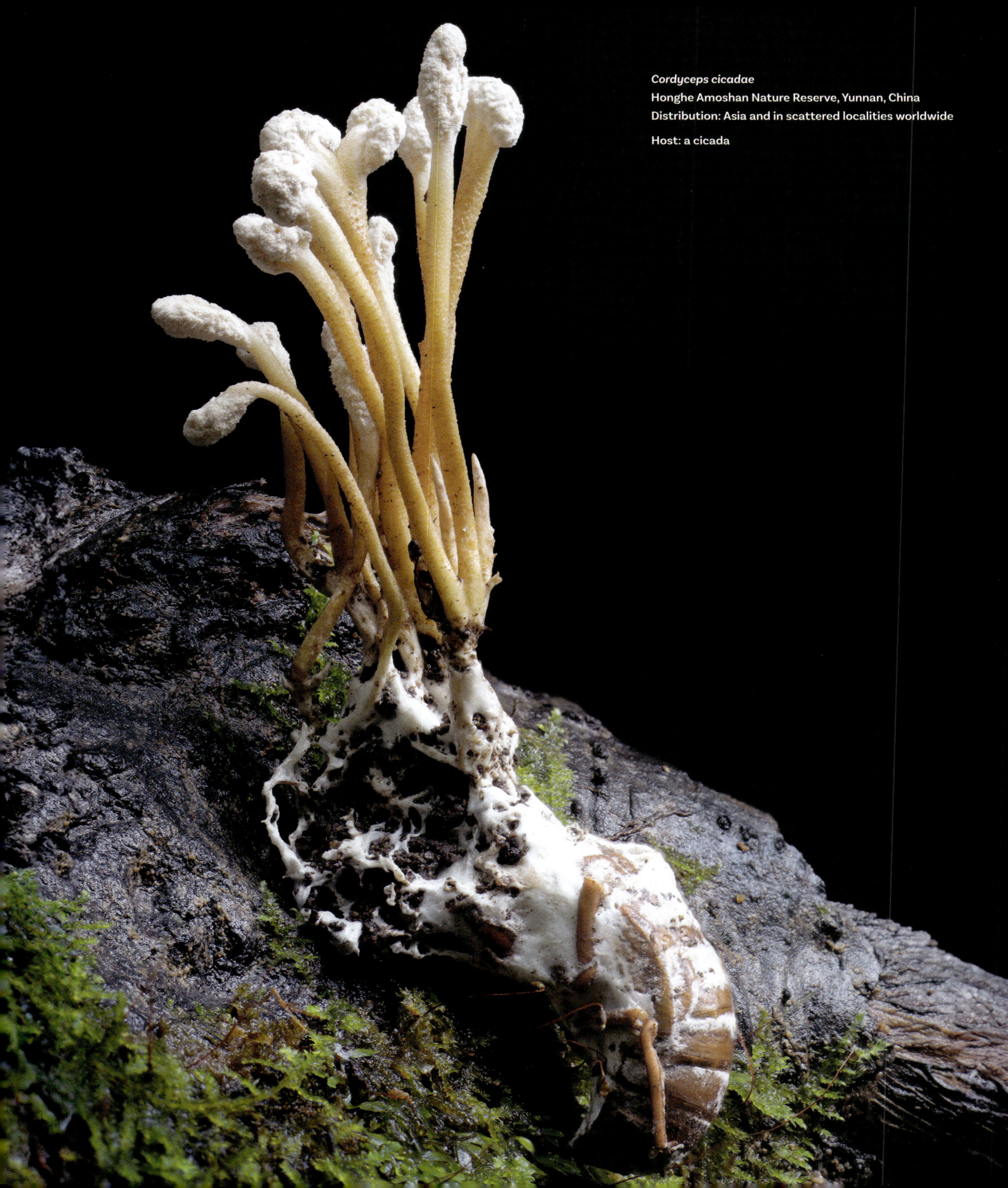

Cordyceps cicadae
Honghe Amoshan Nature Reserve, Yunnan, China
Distribution: Asia and in scattered localities worldwide

Host: a cicada

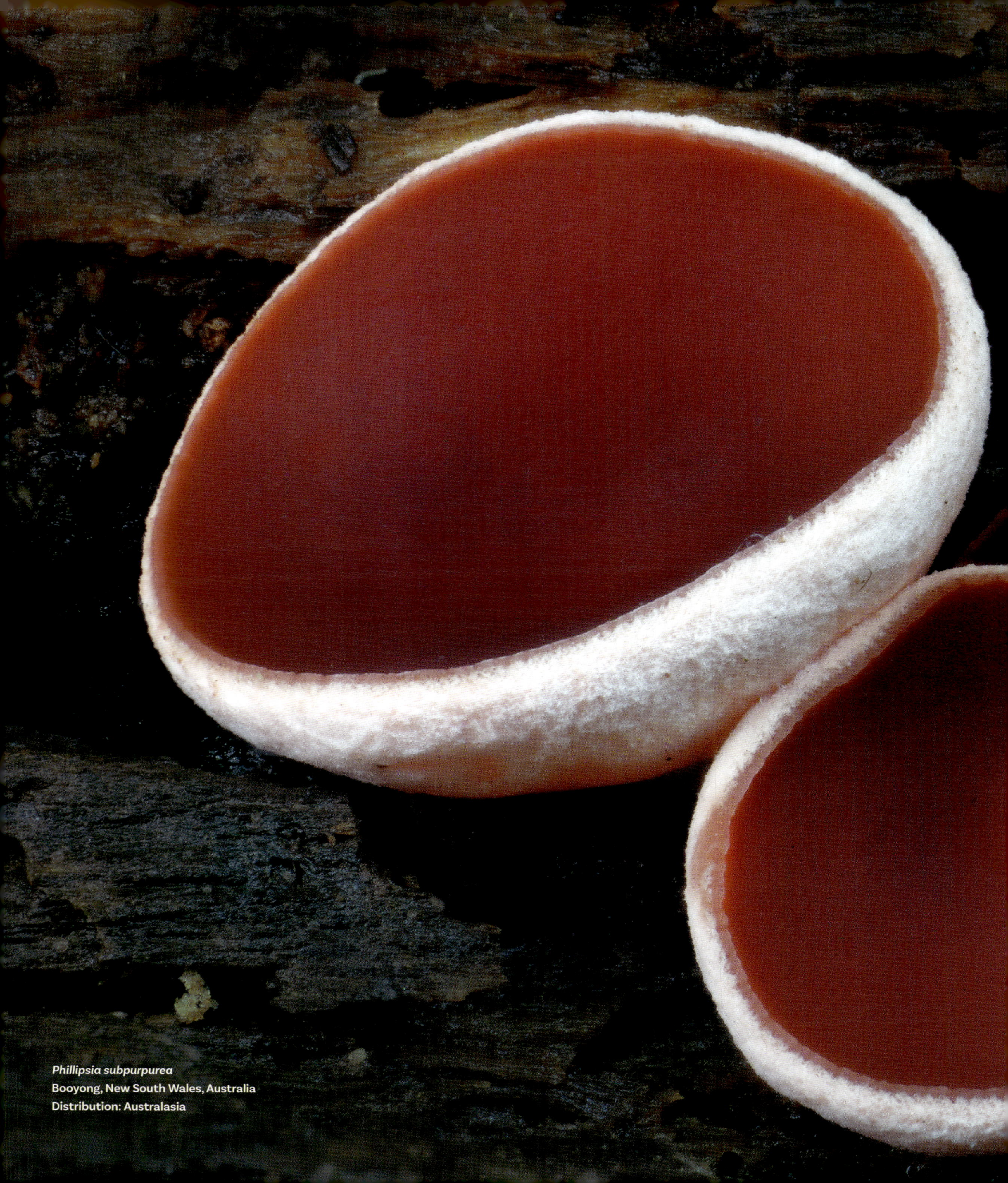

Phillipsia subpurpurea
Booyong, New South Wales, Australia
Distribution: Australasia

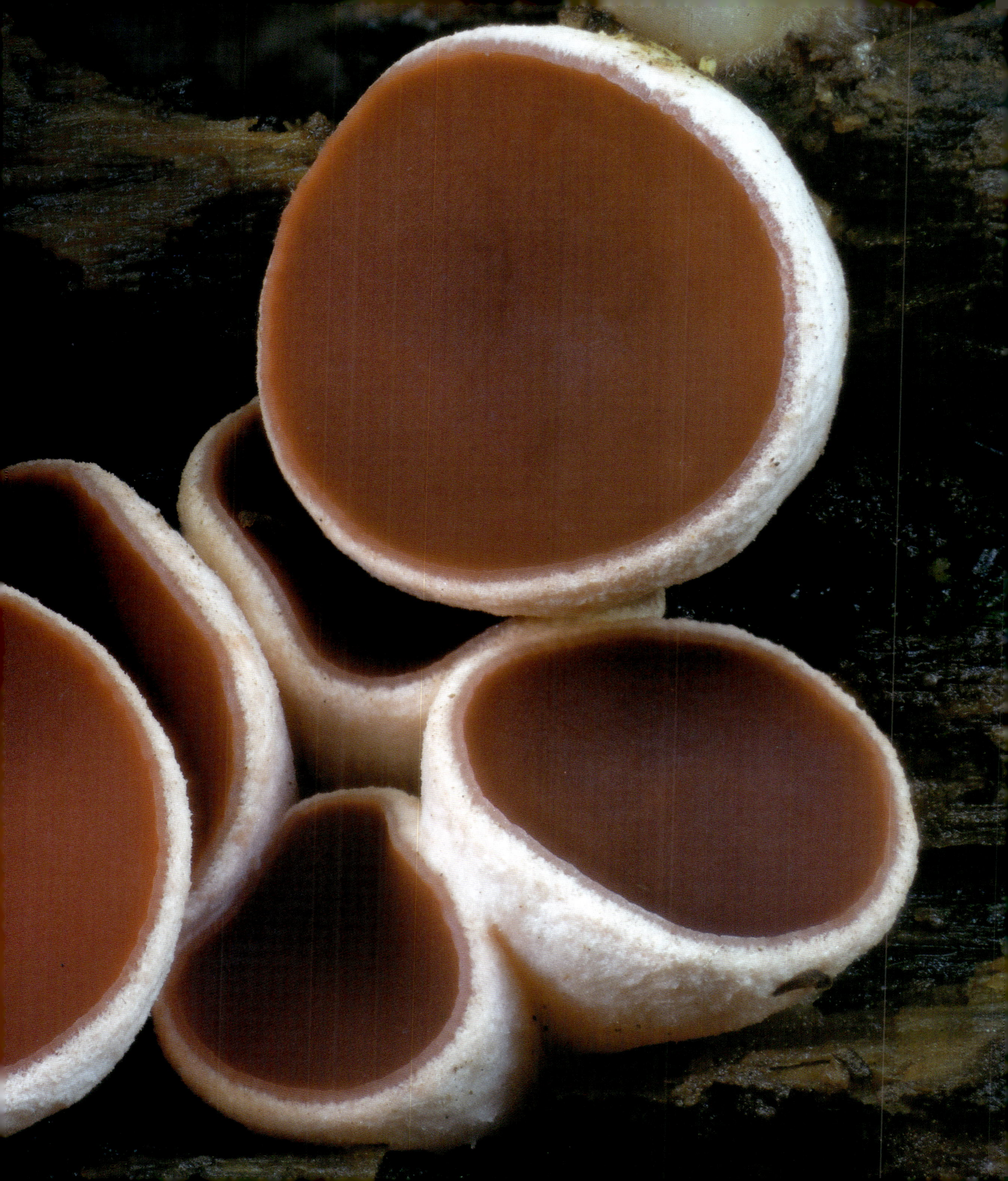

Fungiphilia

In the cloud-shrouded mountains of southern China, a fungal feast awaits.

In late 2014, Stephen received an email from Professor Peter Mortimer, a soil ecologist originally from South Africa and now at Yunnan's Kunming Institute of Botany, part of the Chinese Academy of Sciences.

Professor Mortimer and the head of his department, ethnobiologist Professor Xu Jianchu, lead a team of mycologists involved in pioneering fungal research.

Their activities include annual field trips into the remote forests across southern China, Laos, Myanmar and Thailand.

Stephen's fungi photographs on the internet had caught the eye of Professor Mortimer's father, who shared them with his son.

Impressed and captivated by the beauty captured in the images, he invited us to join the group's next fungi field trip in China - a spontaneous decision, as Professor Mortimer candidly admits in his email, conceived 'without too much forward planning'.

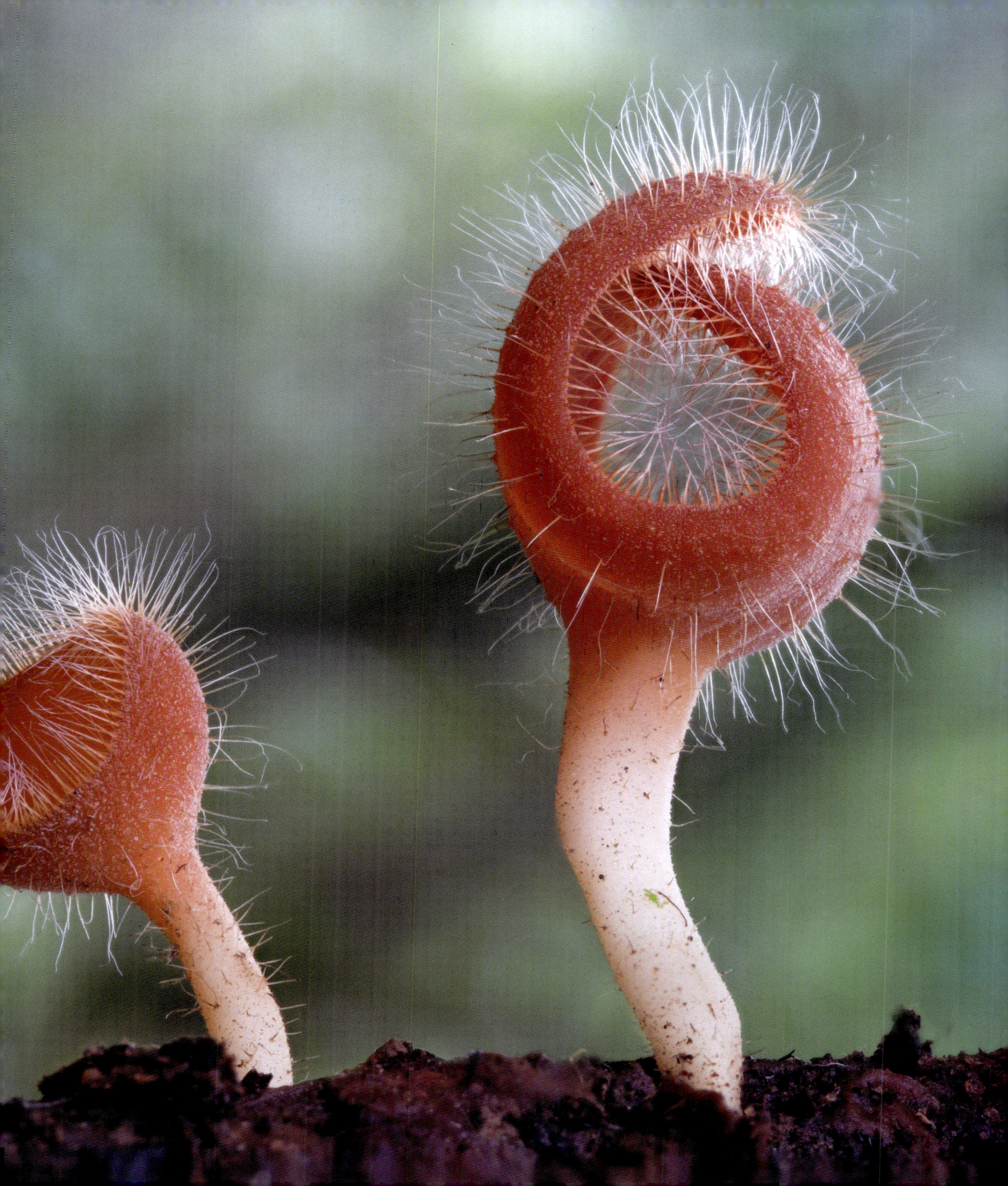

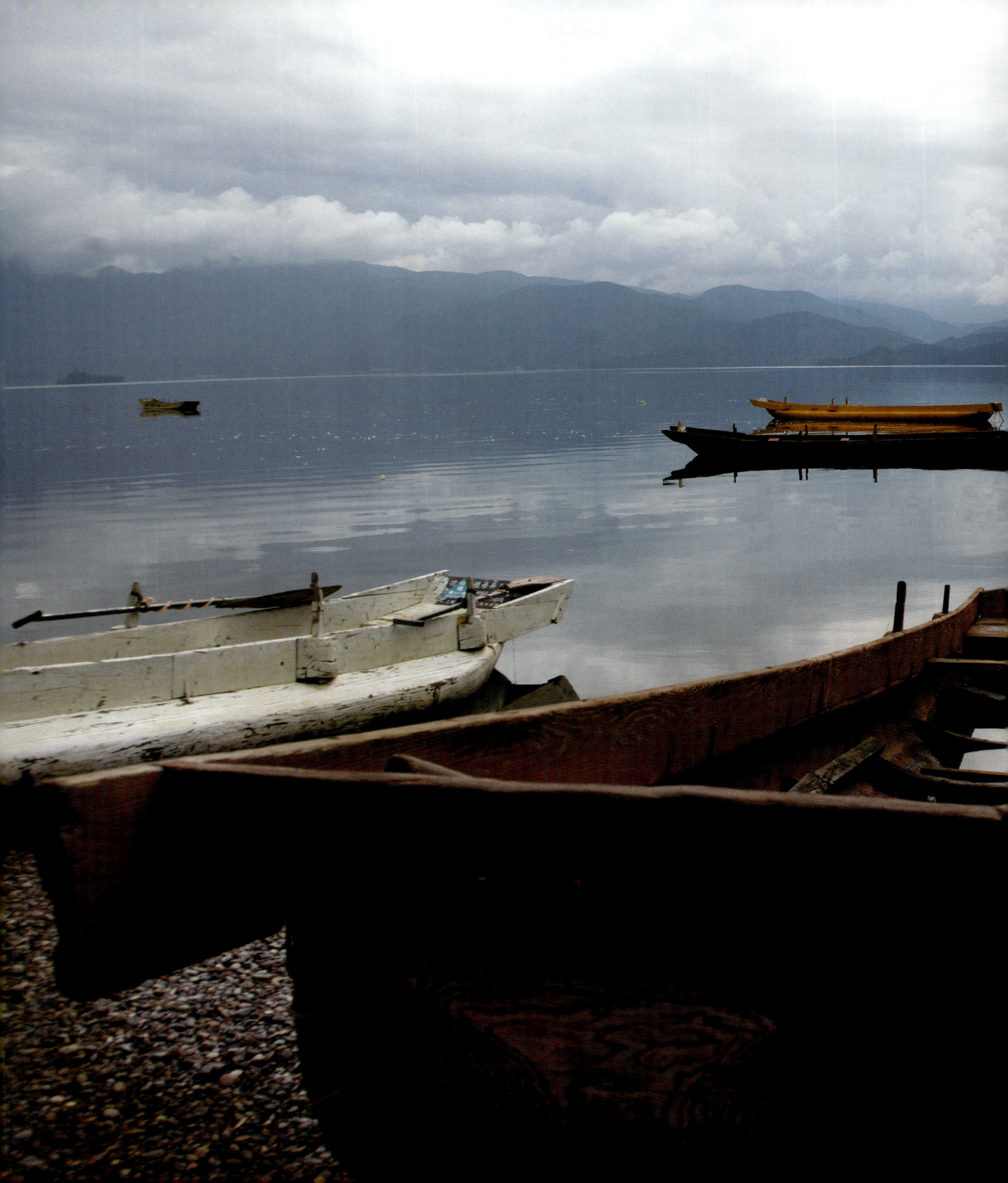

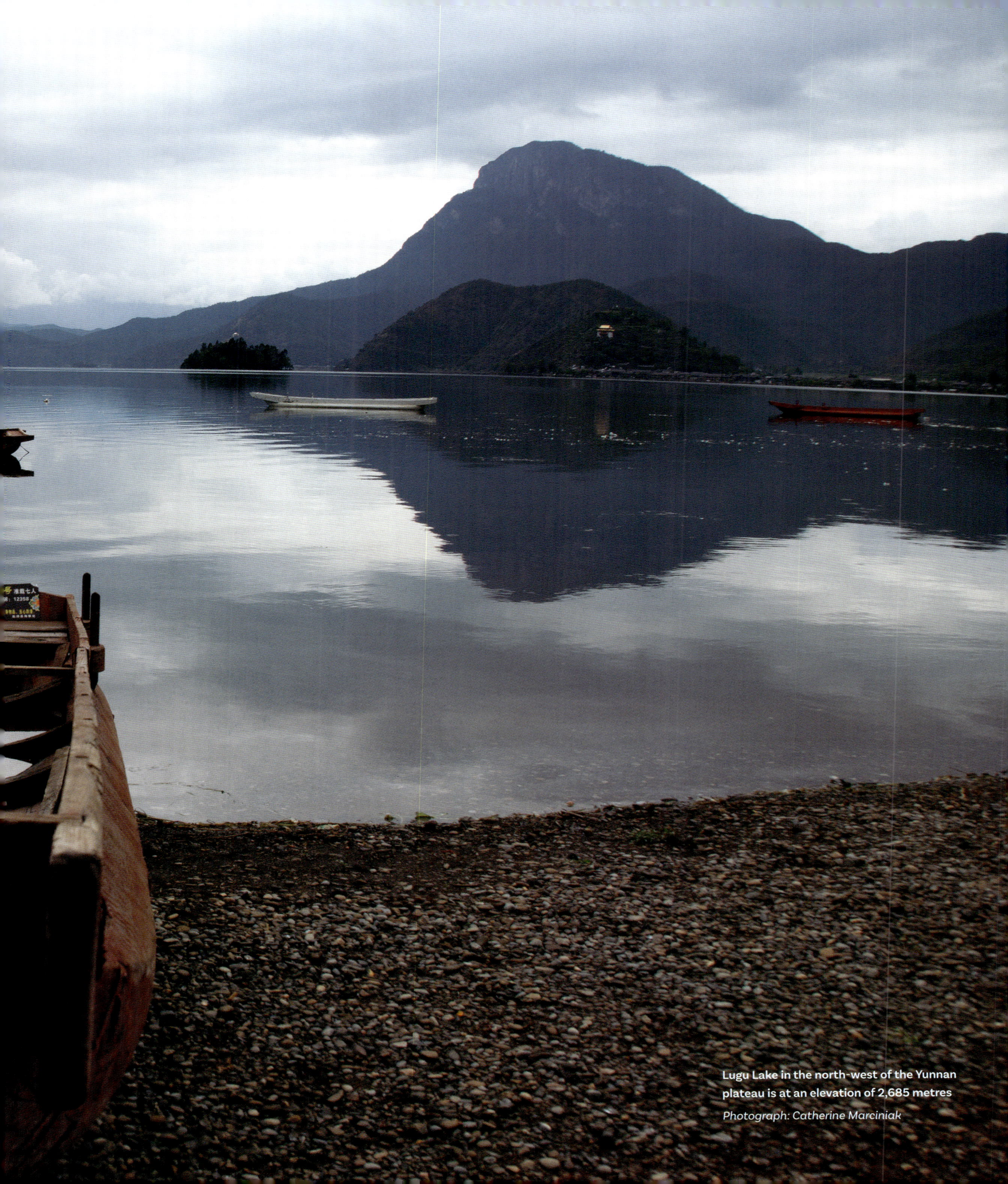

Lugu Lake in the north-west of the Yunnan plateau is at an elevation of 2,685 metres

Photograph: Catherine Marciniak

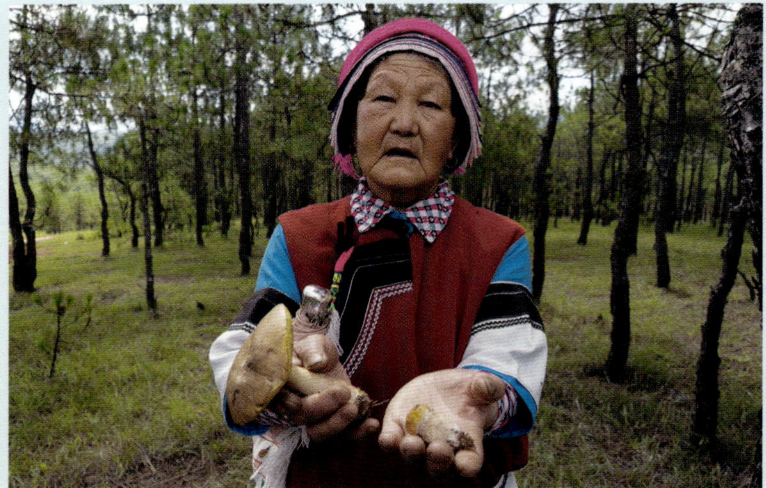

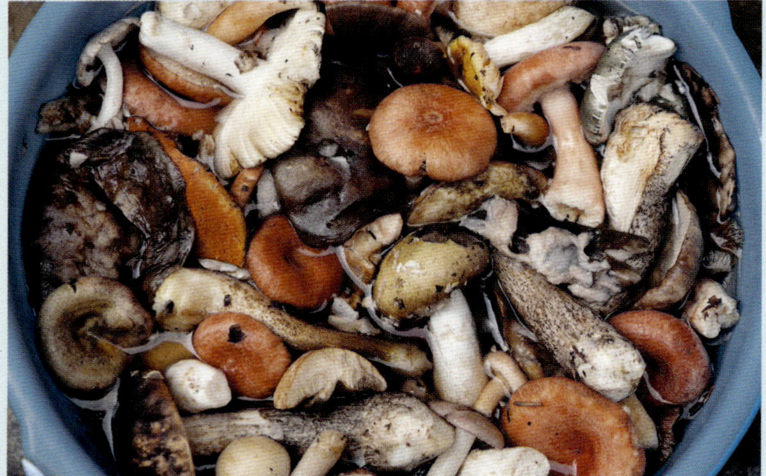

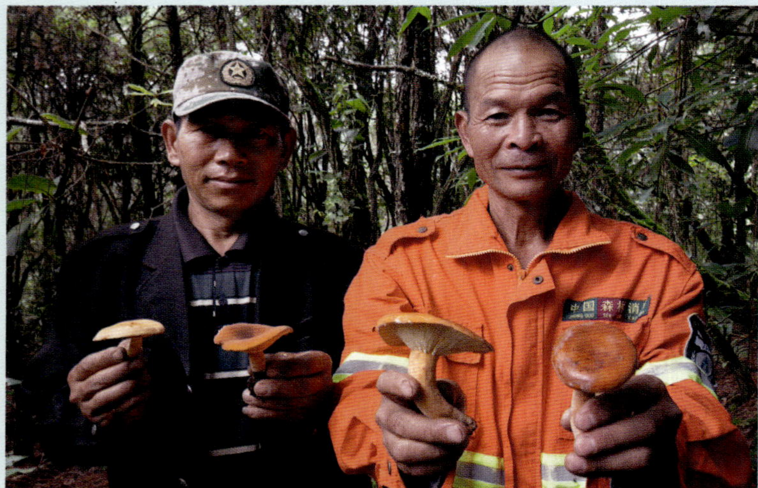

There was no certainty about funding, and the onus of paying for flights to China rested on our shoulders.

Yet the tantalising prospect of gaining access to otherwise restricted forest areas became an irresistible carrot dangling before us.

Admittedly, we are kindred spirits when it comes to risky adventures.

Stephen started climbing volcanoes after his wife Pat passed away, and I had dabbled in rock and mountain climbing in the Himalayas during my youth.

TOP

The village experts on edible fungi are often women. This elder of the Dai tribe near Dali led our fungi hunt in the nearby pine forests. Her tips for preparing boletes varied with each species. Some were eaten raw, some boiled in soup, while others were fried.

Photograph: Catherine Marciniak

MIDDLE

Mushrooms from a day's forage washed and ready for our evening meal. They are mostly russulas and lactarii. Nabanhe National Nature Reserve, Yunnan, China

Photograph: Catherine Marciniak

BOTTOM

The park rangers of Honghe Amoshan Nature Reserve were keen to show us some of the edible varieties in their cloud forest. These are *Lactifluus volemus*. Their common name is weeping milkcap as they ooze white milk when broken.

Photograph: Catherine Marciniak

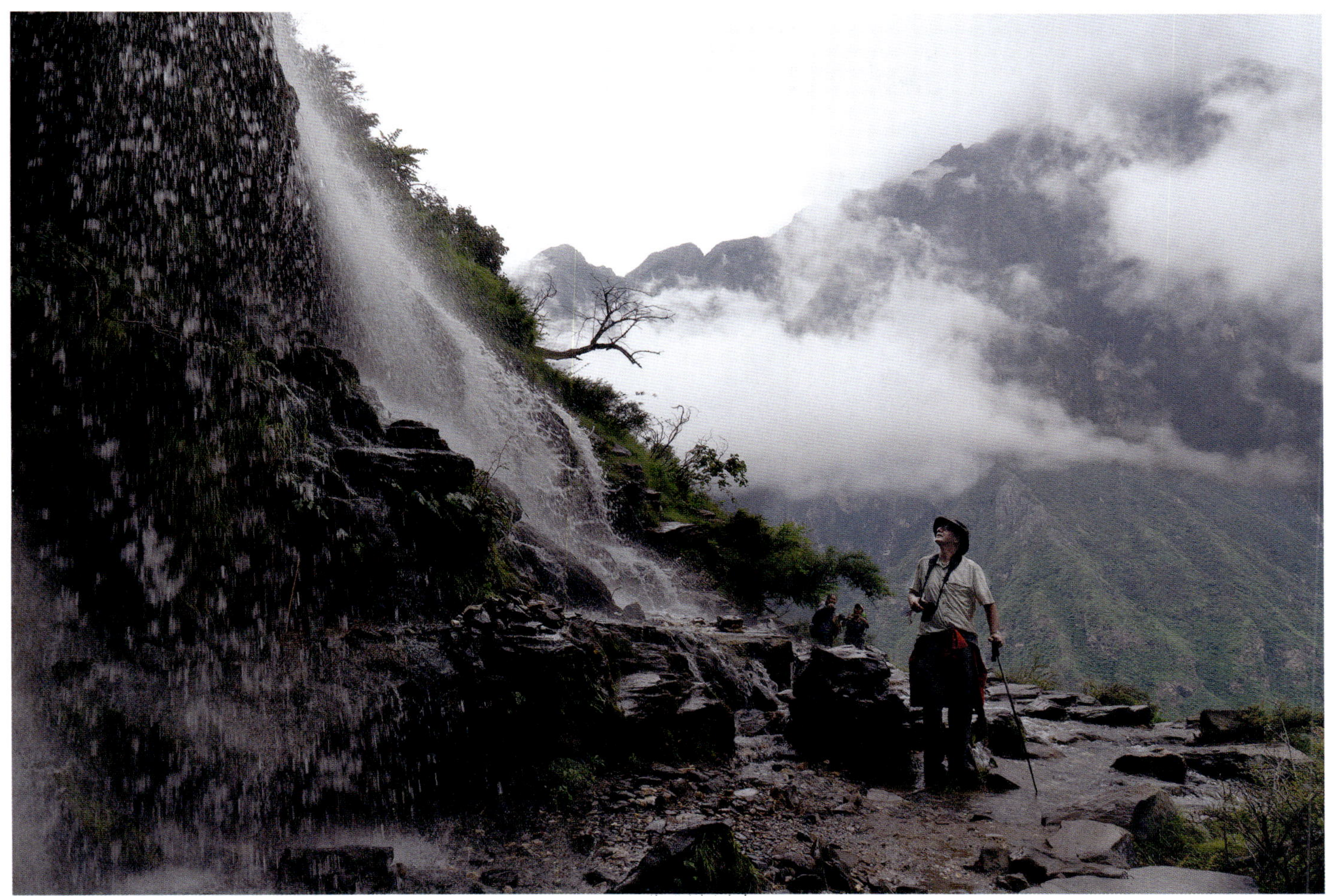

So, in August 2015, as the monsoonal deluge descended on the forests of Yunnan, we arrived in China to join the Kunming Institute of Botany's fungal expedition.

On our first day in Kunming, Stephen experienced an avalanche of fungal revelations.

Yunnan, with its misty Himalayan foothills, crystal clear plateau lakes and lush rainforests is home to 25 of China's ethnic minority groups. Each has unique traditions and a shared fascination with mushrooms.[1]

ABOVE

Fungi hunting amidst the clouds on the Tiger Leaping Gorge hike, Yunnan, China

Photograph: Catherine Marciniak

Professor Mortimer explained that these people embraced over 900 species of fungus in their culinary repertoire.[2,3]

Whole markets and entire cities are dedicated to the celebration of and trade in mushrooms, underscoring the economic juggernaut of a billion-dollar fungal industry, even back in 2015.

'In Australia, we're mainly what is called a fungi-phobic society.

'People are generally scared to eat forest fungi – with good cause. We don't know which are poisonous and which are edible.

'Our supermarkets have a limited selection of edible mushrooms, at best about five species, and these are cultivated not wild-collected.

'Whereas in some towns in Yunnan, streets are lined with mushroom sellers, each offering dozens of varieties.

'We were told that everyone, from an urban businesswoman to a paddy field labourer, possesses an intimate knowledge of mushrooms – a truly fungi-philic community.'

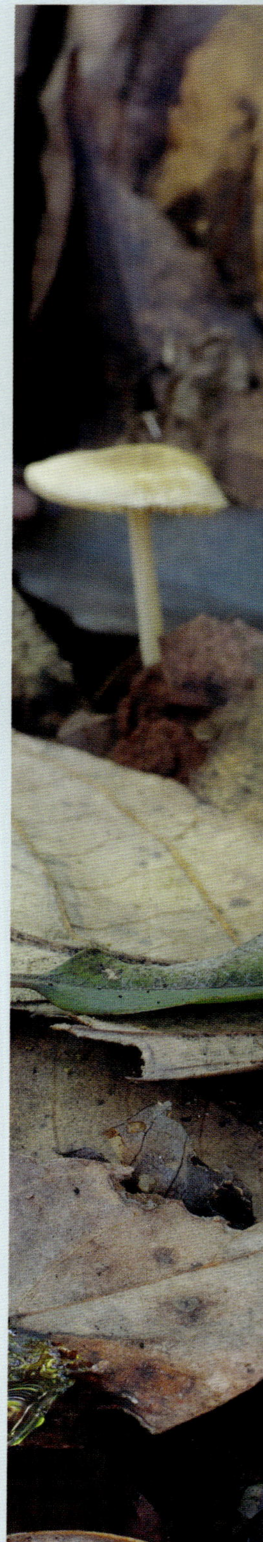

Russula griseocarnosa
Honghe Amoshan Nature Reserve, Yunnan, China
Distribution: Asia

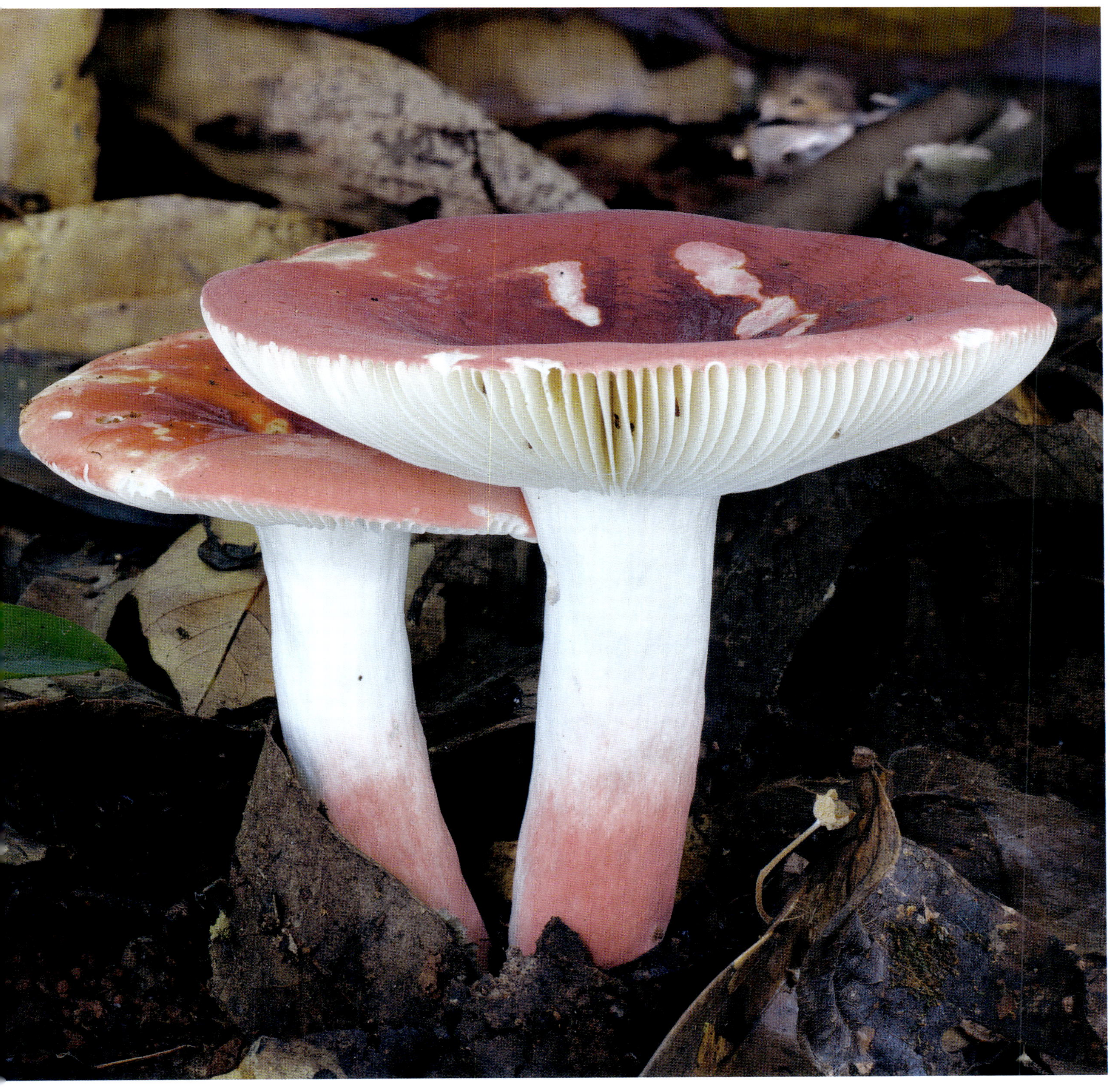

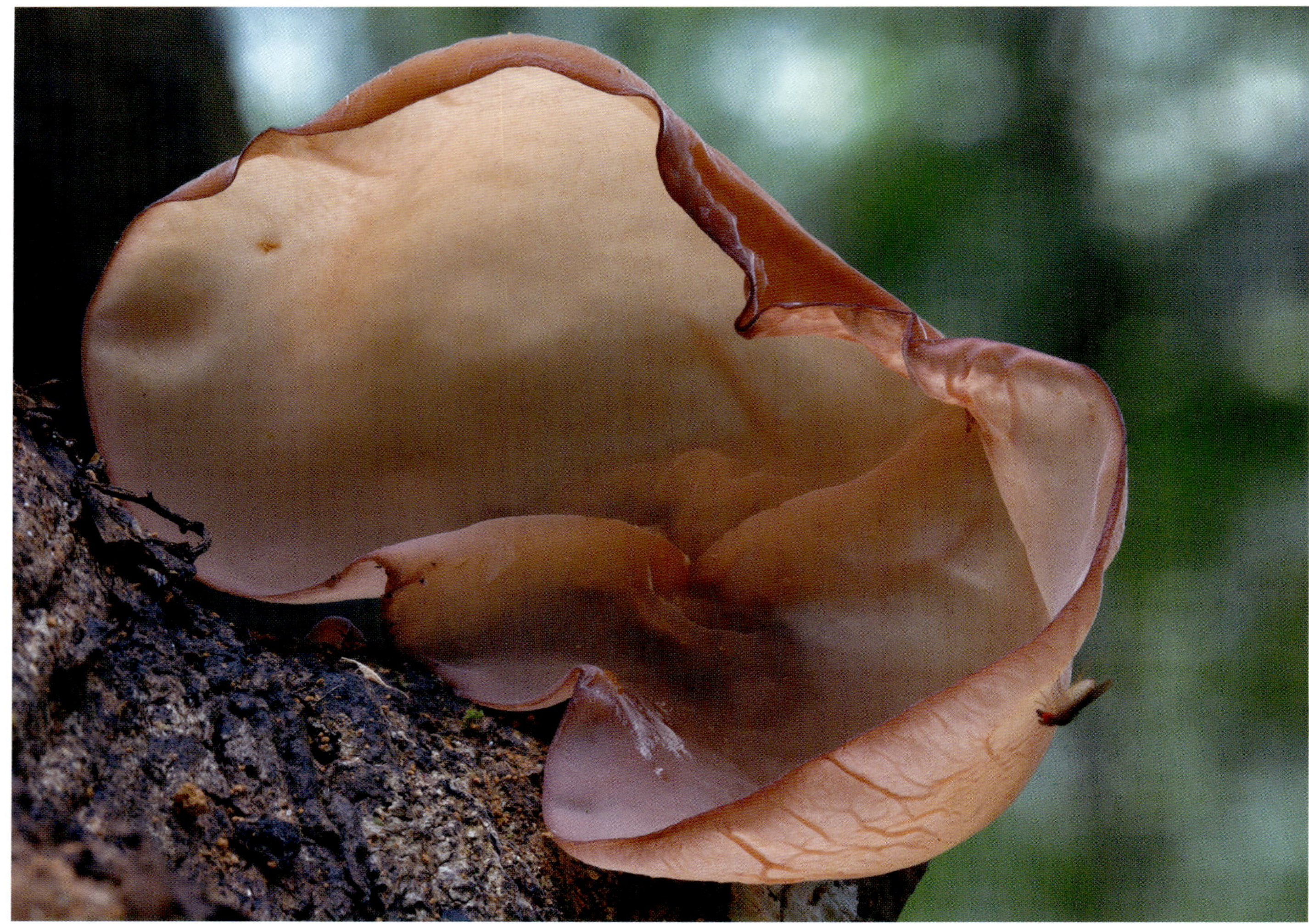

The Kunming Institute of Botany is a research facility that prides itself on attracting brilliant young scientists from Asia, the Middle East and beyond.

During the fungi safari, Professor Mortimer was accompanied by three other researchers. Professor Samantha Chandranath Karunarathna, originally from Sri Lanka, specialises in the taxonomy of wild fungi. Dr Li Huili was researching edible stinkhorns and puffballs. Dr Ye Le is fascinated with entomopathogenic fungi, the organisms that attack and take over the bodies of insects.

ABOVE

Auricularia fibrillifera
Nabanhe National Nature Reserve, Yunnan, China
Distribution: Asia and New Guinea

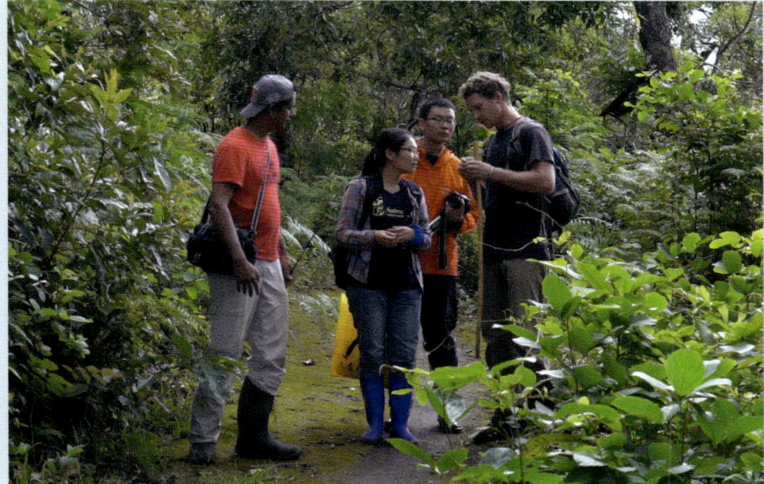

The field trip location was Nabanhe National Nature Reserve of Xishuangbanna Dai Autonomous Prefecture – a biodiversity hotspot.[4]

Although Xishuangbanna covers a mere 0.2 per cent of China, it hosts 16 per cent of its flora, 22 per cent of its mammals and 36 per cent of its birds.[5]

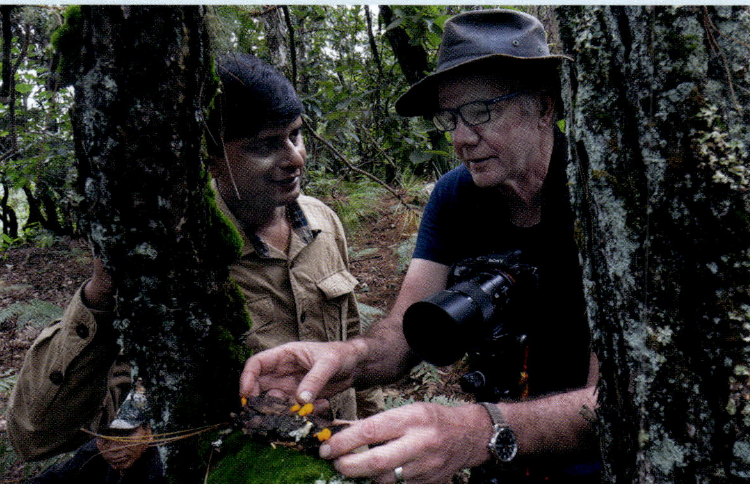

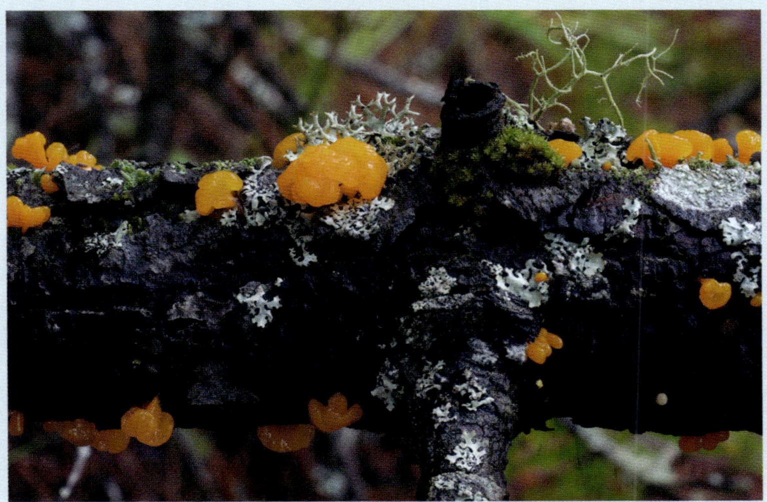

TOP

The fungi field trip in Nabanhe National Nature Reserve of Xishuangbanna, was led by Professor Peter Mortimer. The team members included Samantha Chandranath Karunarathna, Li Huili and Ye Le who in 2015 were PhD candidates.

Photograph: Catherine Marciniak

MIDDLE

Stephen Axford discusses with Samantha Chandranath Karunarathna how he will photograph a specimen of jelly fungus.

Photograph: Catherine Marciniak

BOTTOM

Unidentified jelly fungus
Nabanhe National Nature Reserve, Yunnan, China

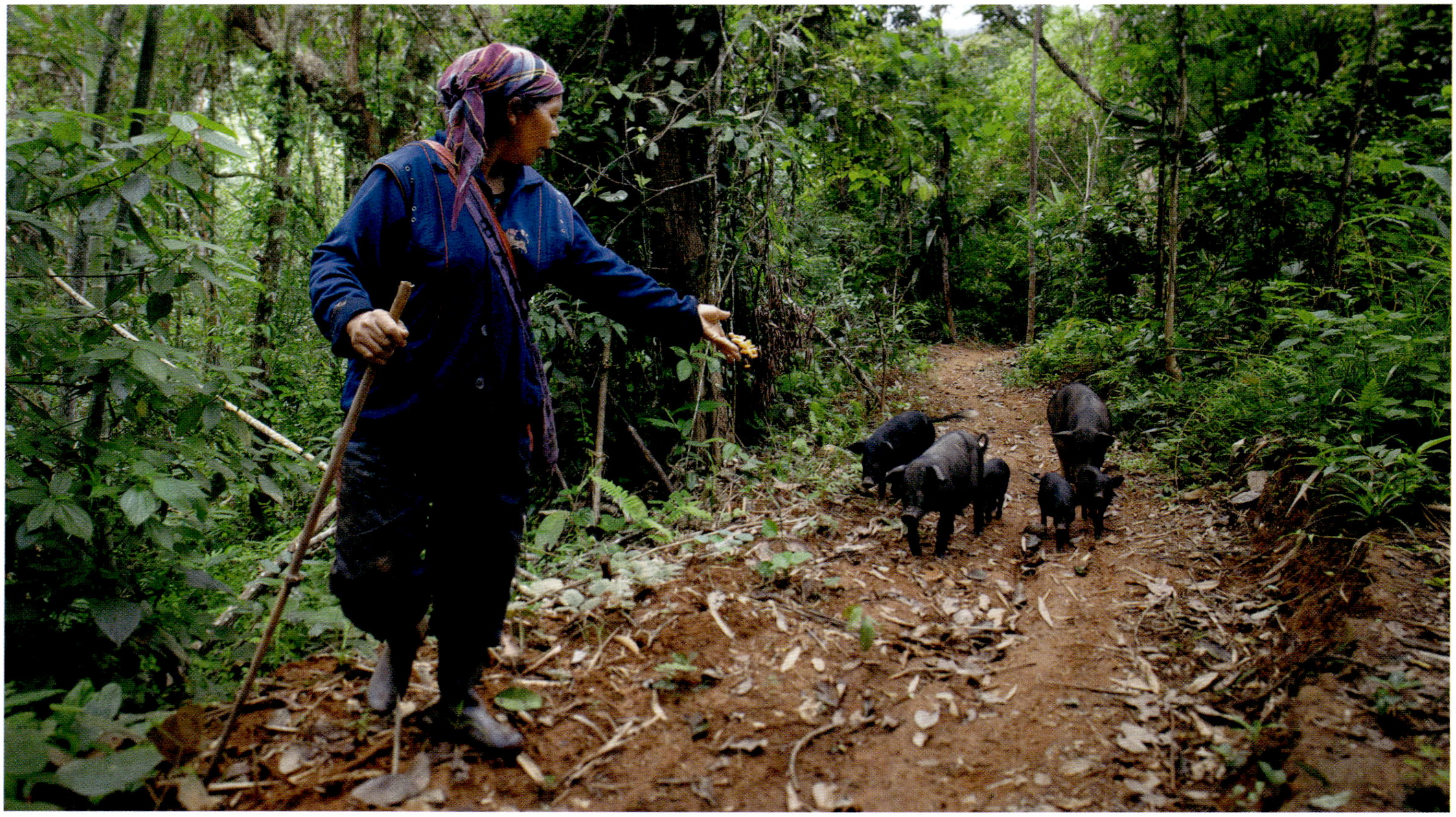

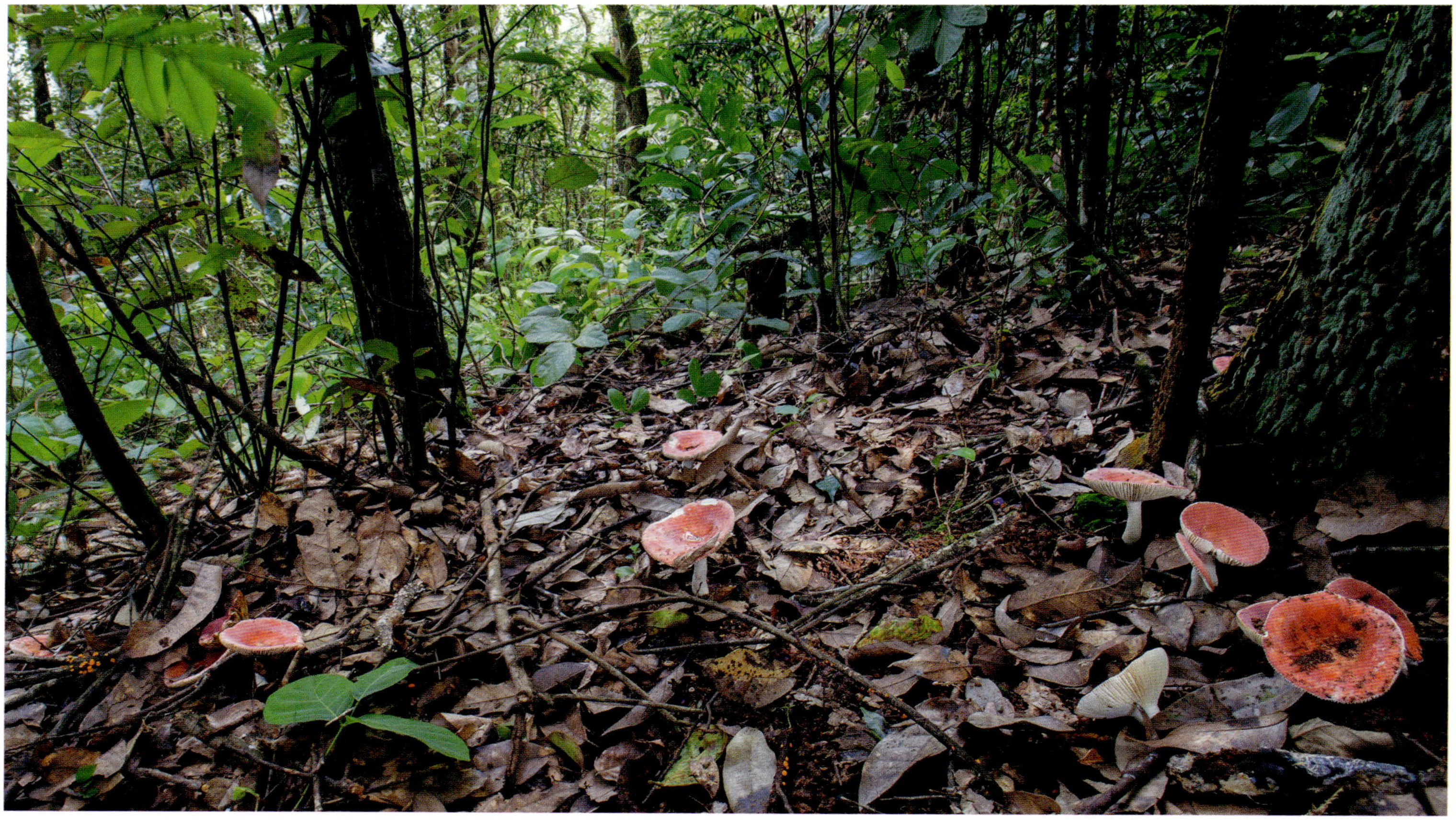

The 26,600 hectares of mountainous subtropical evergreen broad-leaved forest and tropical monsoon forest is a landscape sculpted by deep valleys carved by the upper reaches of the Mekong River.[6]

The indigenous Dai and Bulang people have been stewards of the natural landscape for generations, with the Reserve home to ~4,000 villagers.

Their spiritual philosophy is deeply entrenched with concepts about the interconnectedness of humanity and the environment, where they, humans, are part of the ecosystem.[5,6,7,8]

The forest, revered as the birthplace of rivers and the provider of life-giving water, holds a position of utmost importance.[9]

While they harnessed the lower lands for rice paddies and food gardens, they also traditionally protected the forests on the upper slopes, with some areas revered as 'Holy Hills', sacrosanct and free from all harvesting activity.[9]

This delicate equilibrium between human activity and nature played a significant role in preserving biodiversity in this region.

However, since 1976, the rainforests of Xishuangbanna have been receding at an alarming rate. The primary culprit behind this deforestation is the rapid expansion of rubber plantations.[9,10,11,12]

OPPOSITE TOP

The Dai and Bulang people regard themselves as part of the natural ecosystem of the Nabanhe National Nature Reserve of Xishuangbanna.

Photograph: Catherine Marciniak

OPPOSITE BOTTOM

The Nabanhe National Nature Reserve comprises 26,600 hectares of mountainous subtropical ever-green broad-leaved forest and tropical monsoon forest.

Photograph: Catherine Marciniak

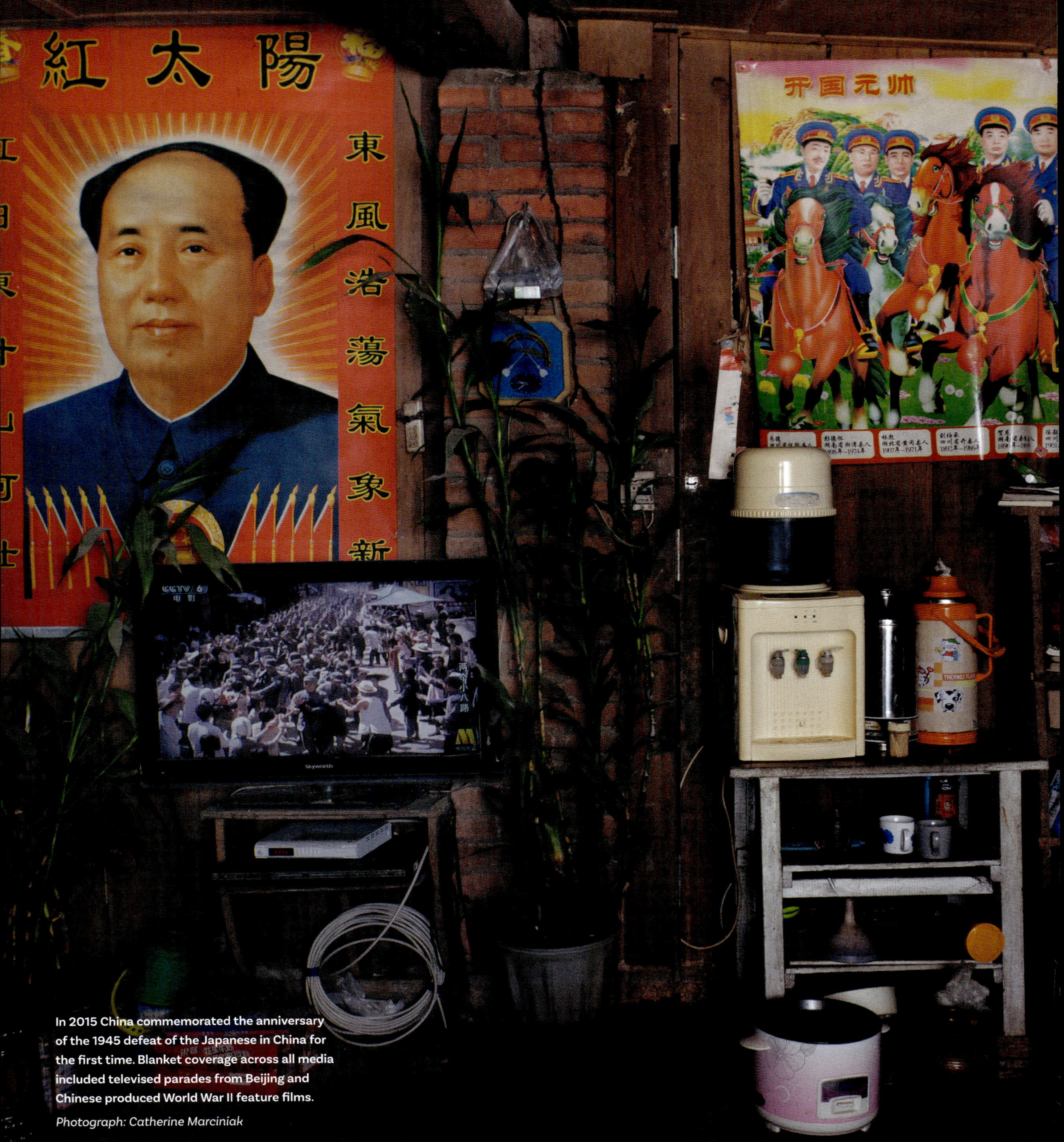

In 2015 China commemorated the anniversary of the 1945 defeat of the Japanese in China for the first time. Blanket coverage across all media included televised parades from Beijing and Chinese produced World War II feature films.

Photograph: Catherine Marciniak

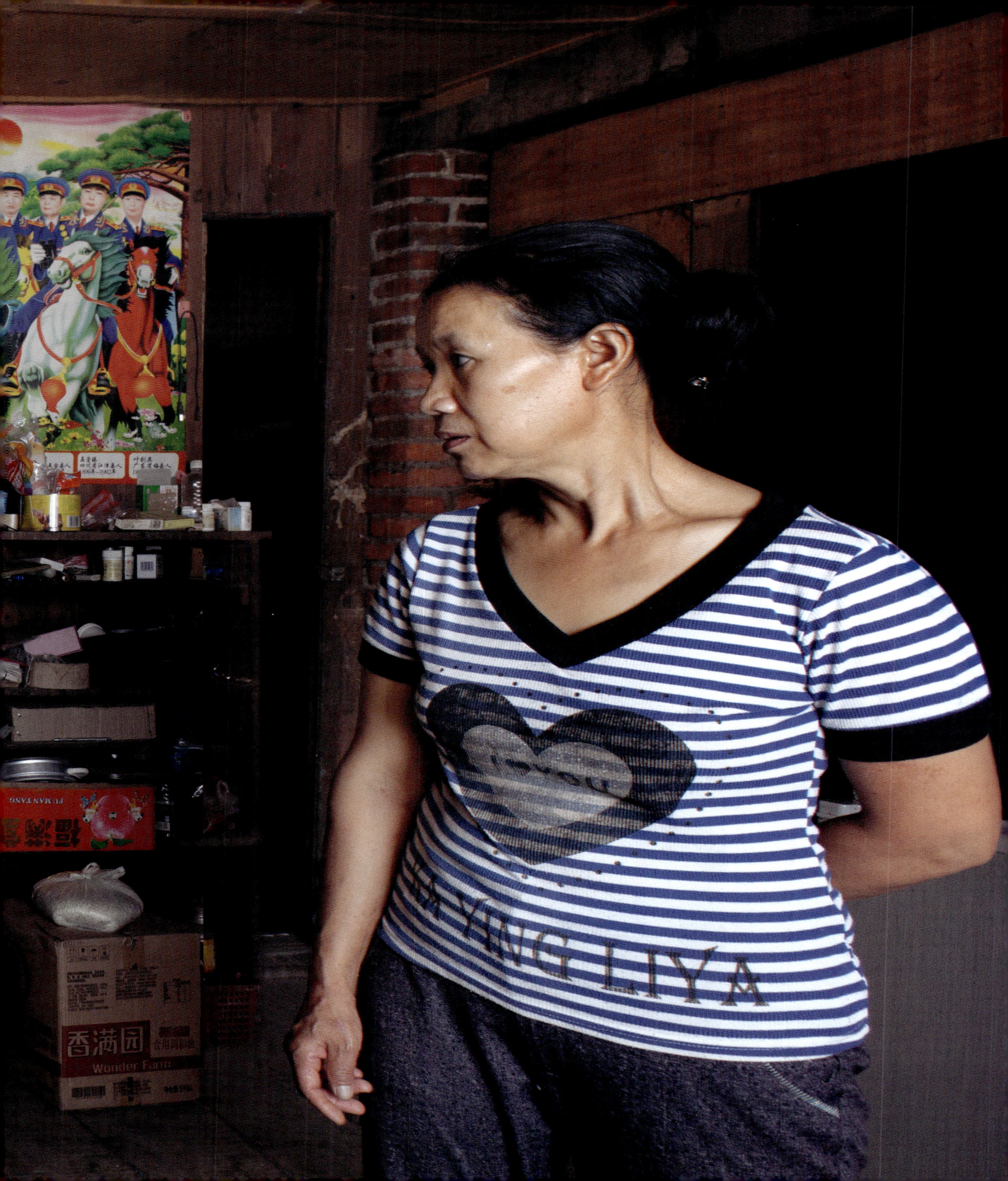

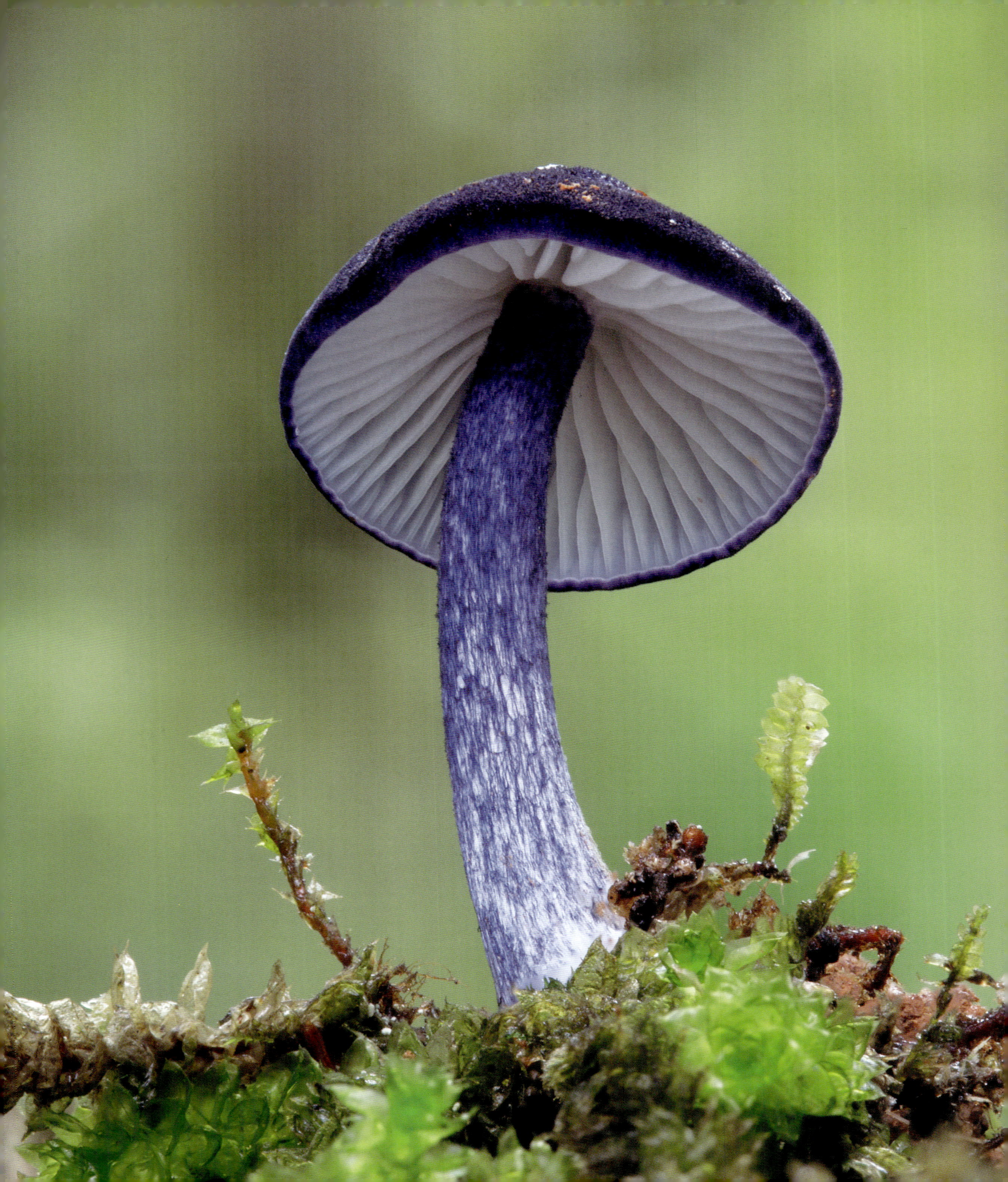

As the team from Kunming Institute of Botany arrived at Nabanhe, homes buzzed with continuous television broadcasts of Chinese World War II feature films and grand parades in the streets of Beijing. It was the lead-up to the 70th anniversary of China's World War II V-Day, and it was the first time communist China had celebrated its victory over Japan.

Stephen, a keen student of history, was fascinated to discover that Yunnan played a pivotal role in World War II, contributing to the Allied victory in Asia.[13]

By 1942, this border area of China, Myanmar and Laos provided the only remaining road or air corridor for the Allies to get supplies from India to the troops fighting the Japanese in China.

When we arrived in 2015, Yunnan was transforming into a new battleground – the epicentre of China's fight against rural poverty.[14,15,16]

'Everywhere we went, we witnessed the birth of new villages, the construction of new roads, all part of a myriad of incentives provided for families to remain in the countryside.'

Among the many potential sustainable industries gathering interest was the wild-harvesting and cultivation of mushrooms.[17]

On this fungi safari, the team hoped to unearth as many edible, medicinal and poisonous species as they could find, along with species that may benefit humanity.

Entoloma eugenei
Nabanhe National Nature Reserve, Yunnan, China
Distribution: a few localities in East Asia

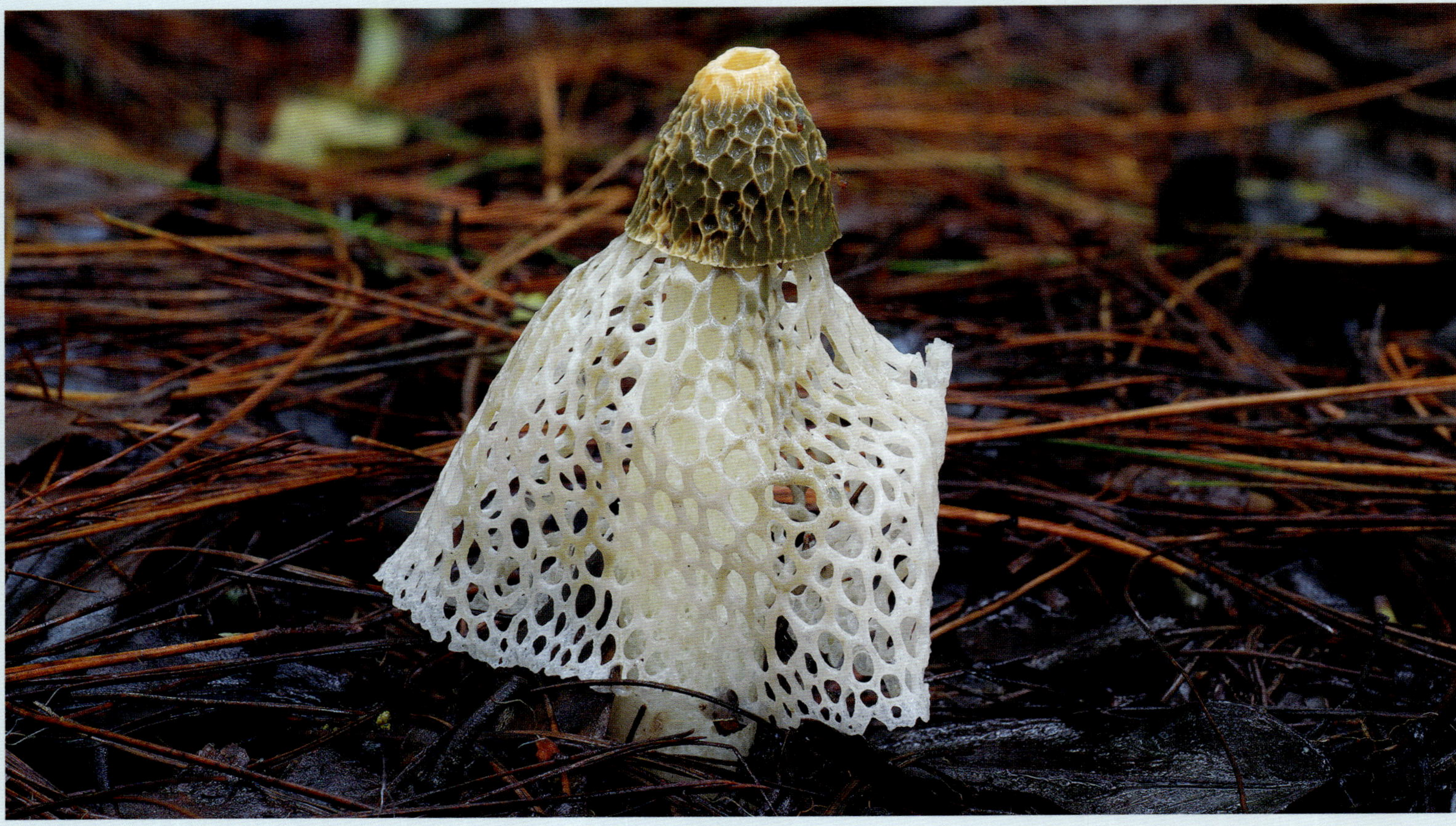

ABOVE

Phallus indusiatus, bridal veil stinkhorn
Honghe Amoshan Nature Reserve, Yunnan, China
Distribution: worldwide

Grown commercially in China as an edible and medicinal mushroom.

OPPOSITE TOP

Villagers wild-harvest *Russula griseocarnosa* and bring it to a central location where they are paid by weight. The mushrooms are then dried in a wood fired oven and delivered to an international mushroom trader in the city.
Nabanhe National Nature Reserve, Yunnan, China

OPPOSITE BOTTOM

Market streets in Yunnan's cities are lined with villagers selling wild mushrooms harvested in pine and subtropical forests. This stall was selling six delicious edible species: *Russula virescens, Lactarius deliciosus, Baorangia bicolor, Sutorius brunneissimus, Phaeoclavulina flaccida* and *Ramaria hemirubella*.

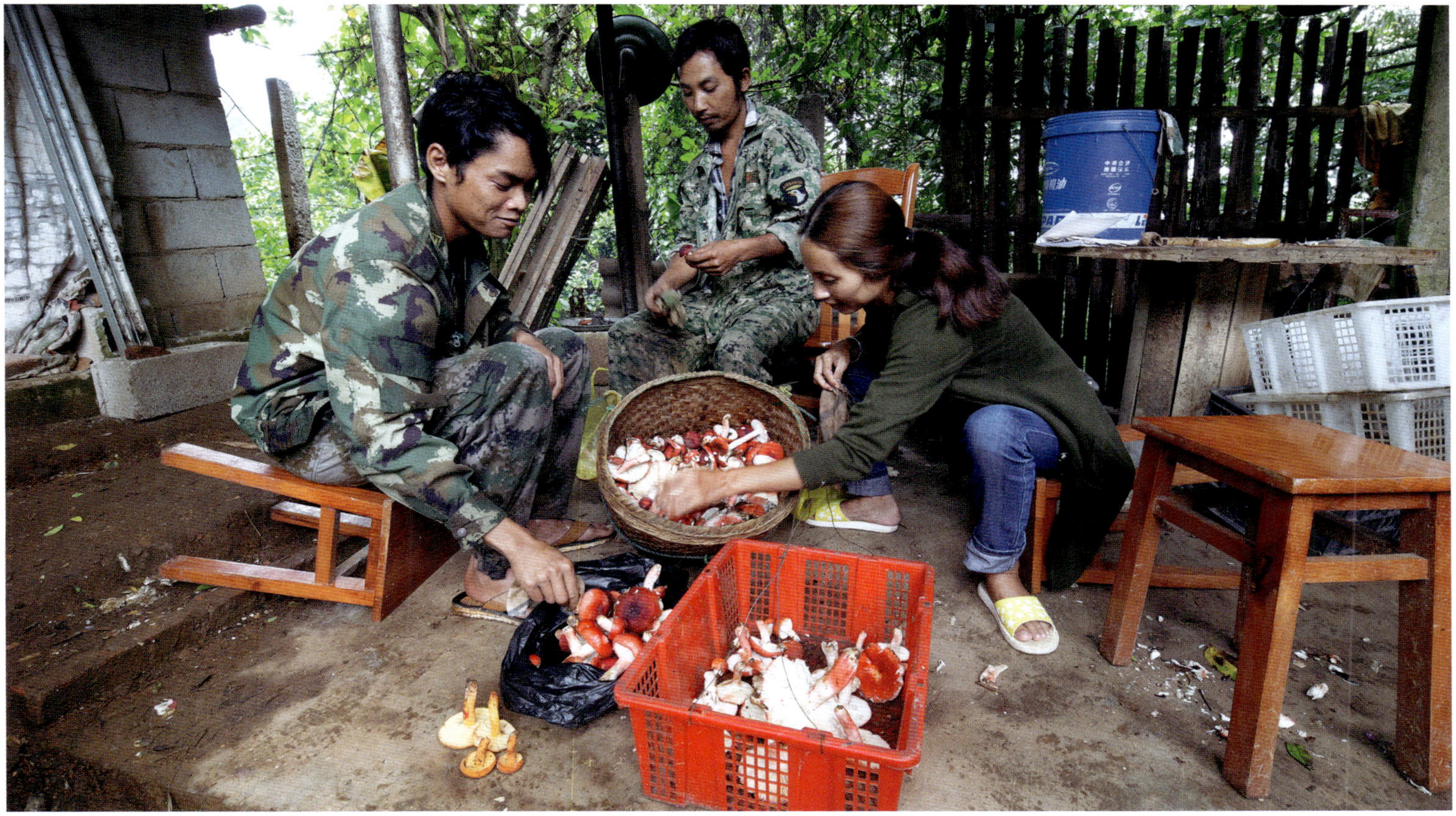

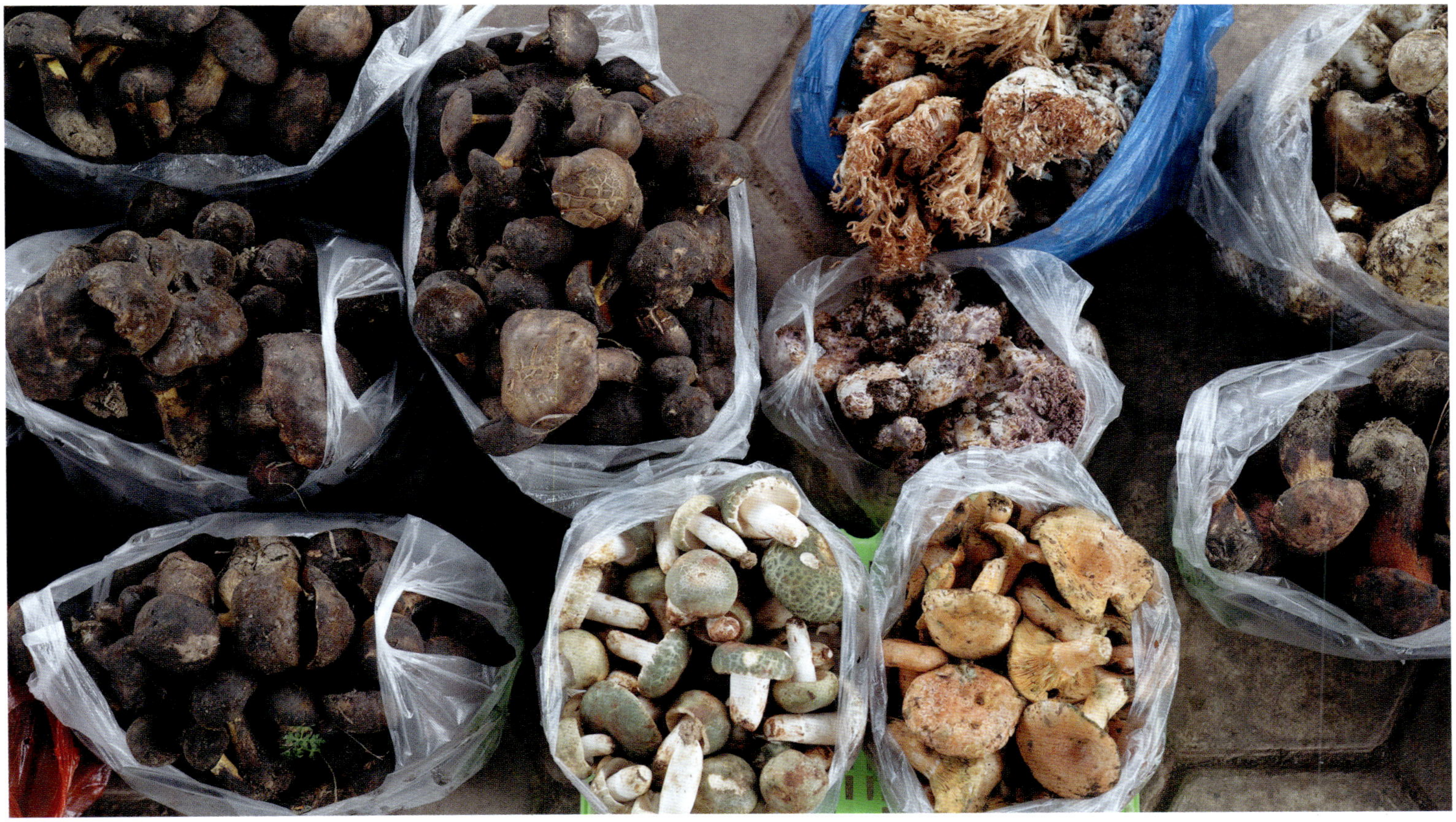

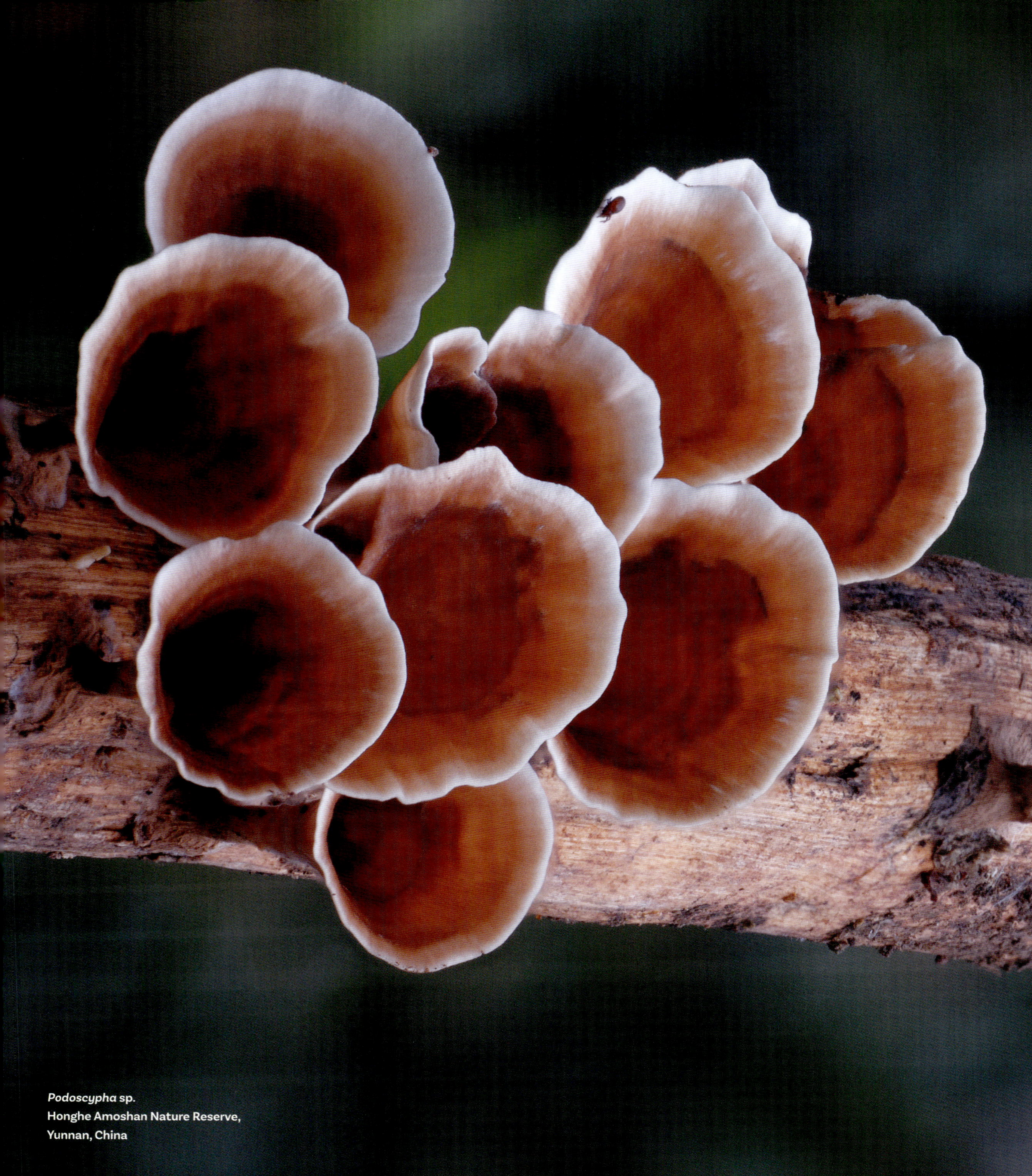

Podoscypha sp.
Honghe Amoshan Nature Reserve,
Yunnan, China

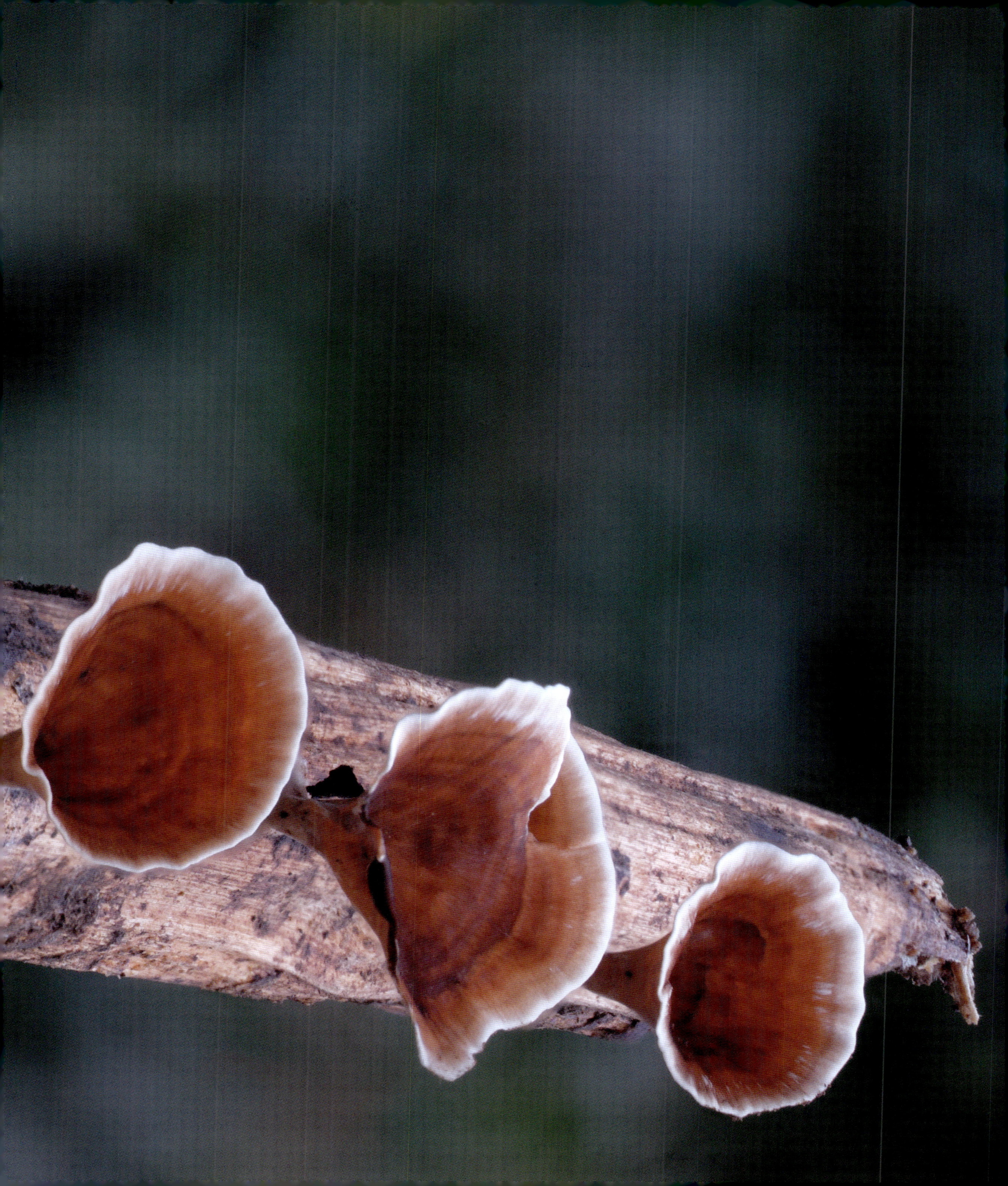

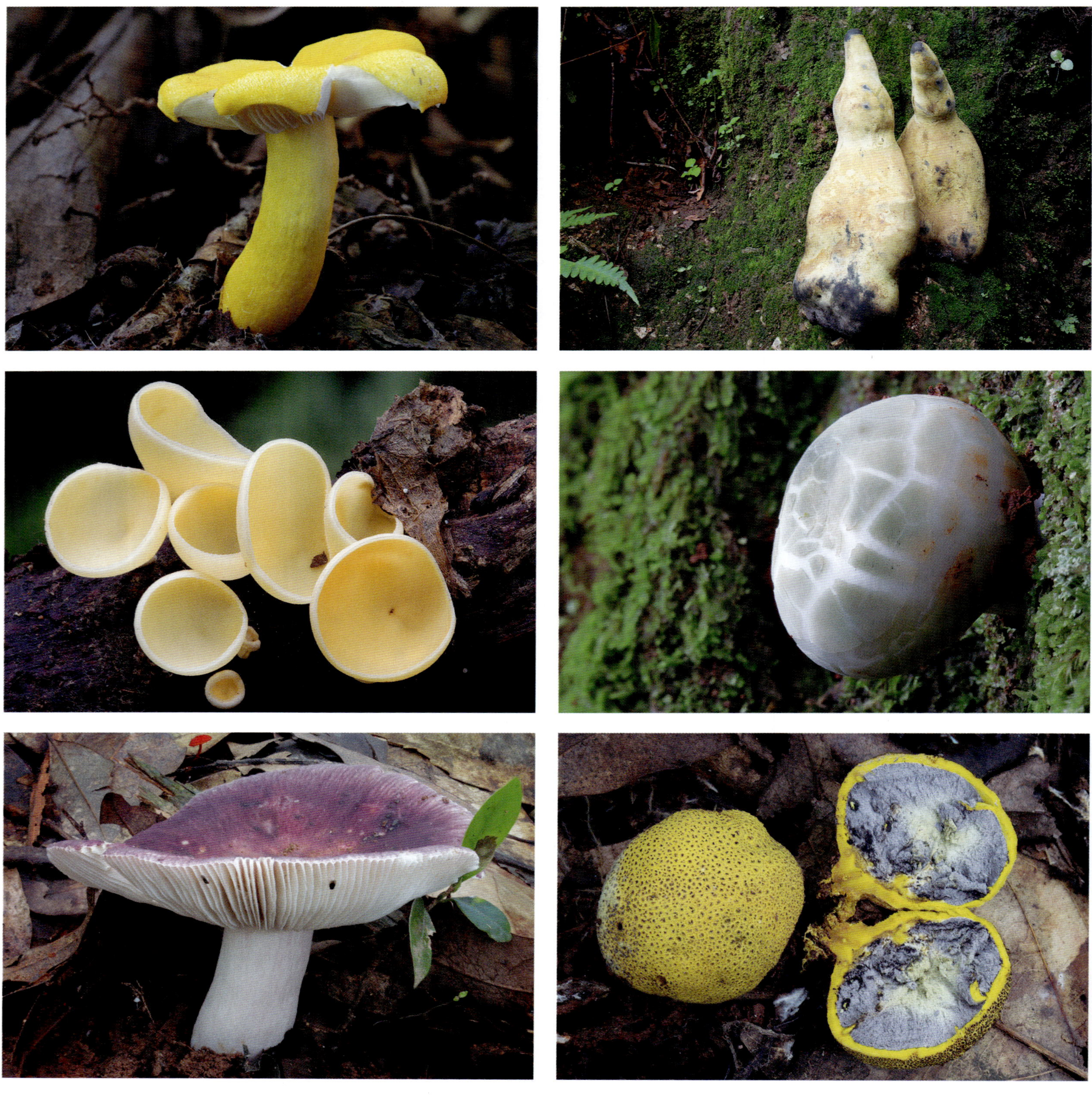

Stephen's photos would be used to compile a fungal field guide, a scientific reference for academics and local villagers alike.

With the zeal of a fungi hunter, Professor Mortimer told us that one in five species collected during the fieldwork is likely to be new to science.

Surprised and excited, we braced ourselves for a fungal adventure unlike any we had encountered so far.

The team joined forces with local rangers who unearthed species that were novel even to the seasoned eyes of Stephen's scientific collaborators.

'A visual banquet unfolded before my lens.

'Each day, I found myself documenting around 100 specimens.

'Never before had I witnessed such diversity in one place, and within one genus.

'Take, for instance, the ectomycorrhizal genus Russula. *It presented a kaleidoscope of colours – purples, reds, greens, yellows and white.*

'Most of these are edible and are wild-harvested by the villagers, dried in a simple kiln and then delivered to markets within China and internationally.'

OPPOSITE TOP LEFT
Russula sp.
Honghe Amoshan Nature Reserve, Yunnan, China

OPPOSITE TOP RIGHT
Xylaria sp.
Nabanhe National Nature Reserve, Yunnan, China

OPPOSITE MIDDLE LEFT
Cookeina indica
Nabanhe National Nature Reserve, Yunnan, China
Distribution: Asia

OPPOSITE MIDDLE RIGHT
Russula virescens, green russula
Nabanhe National Nature Reserve, Yunnan, China
Distribution: widespread in Northern Hemisphere

OPPOSITE BOTTOM LEFT
Russula sp.
Nabanhe National Nature Reserve, Yunnan, China

OPPOSITE BOTTOM RIGHT
Scleroderma flavidum
Nabanhe National Nature Reserve, Yunnan, China
Distribution: worldwide

Many mushrooms the forest revealed resembled elaborate art installations, like the very poisonous *Amanita sculpta* with its Baroque-like towers, or the surreal structure of *Cookeina tricholoma*, a cup fungus with a delicate ornamental fringe.

The purpose of the bristly hairs on *C. tricholoma* remains a mystery. They might serve as a defence mechanism against predators or assist in spreading spores.

Interestingly, the culinary appeal of this fungus varies across the globe. While the fungi-loving people of China do not favour it, it is widely consumed in Guyana, Mexico and West Africa.[18]

Amanita sculpta, sculpted volva lepidella
Nabanhe National Nature Reserve, Yunnan, China
Distribution: Asia

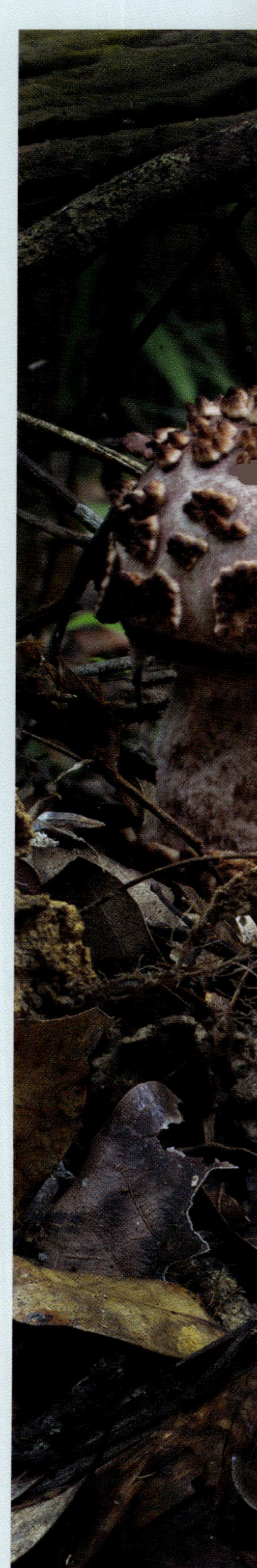

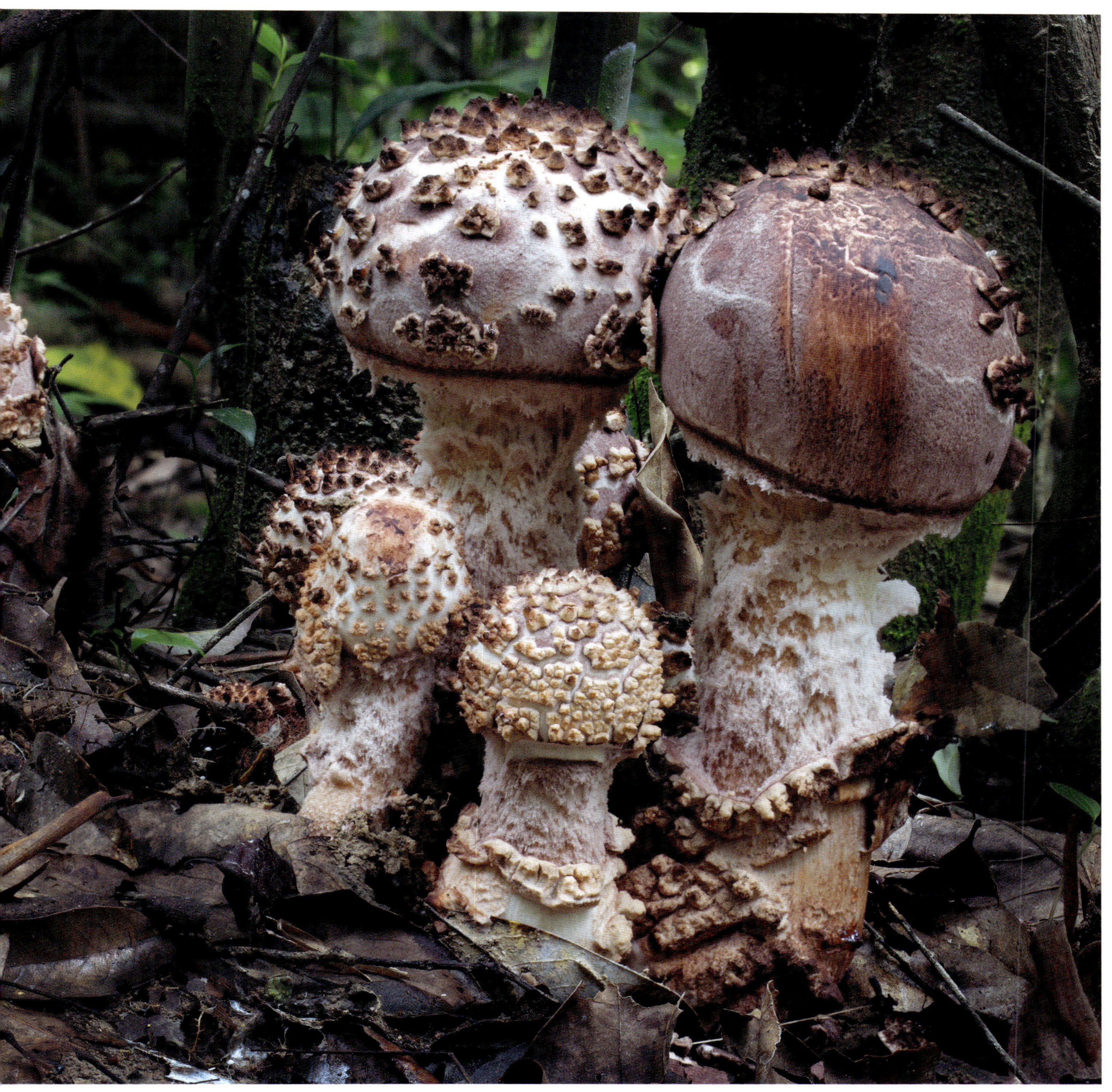

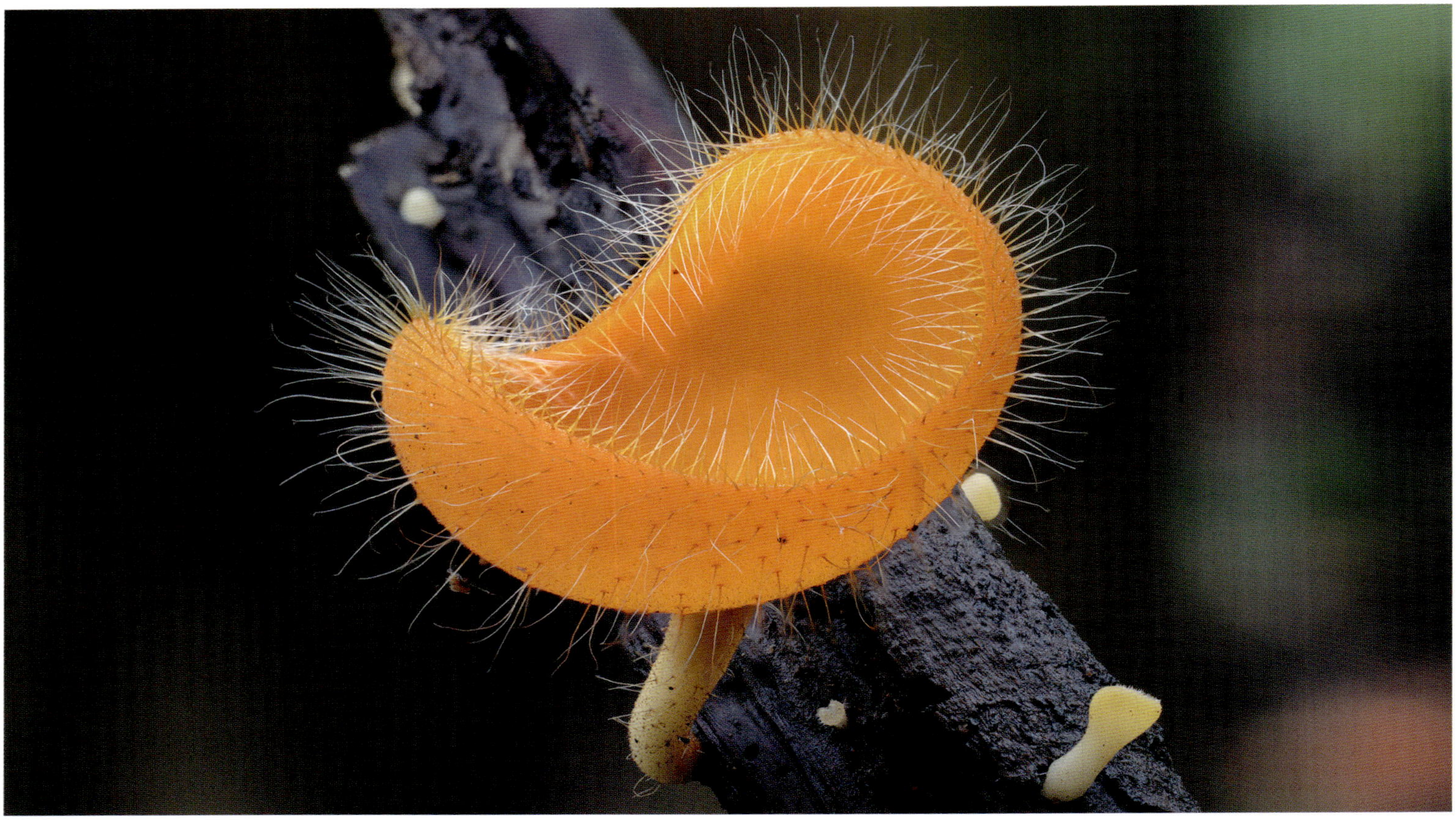

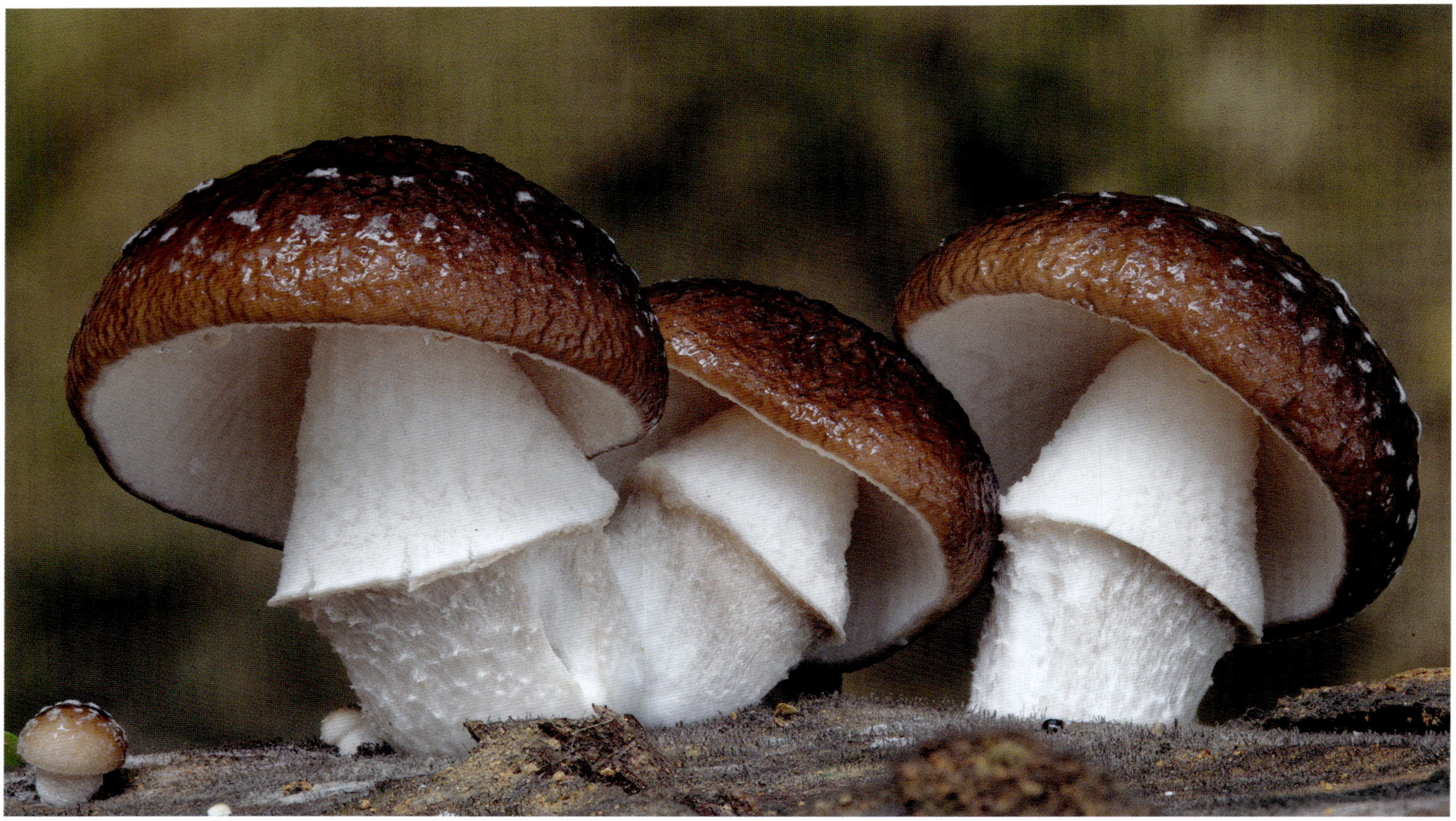

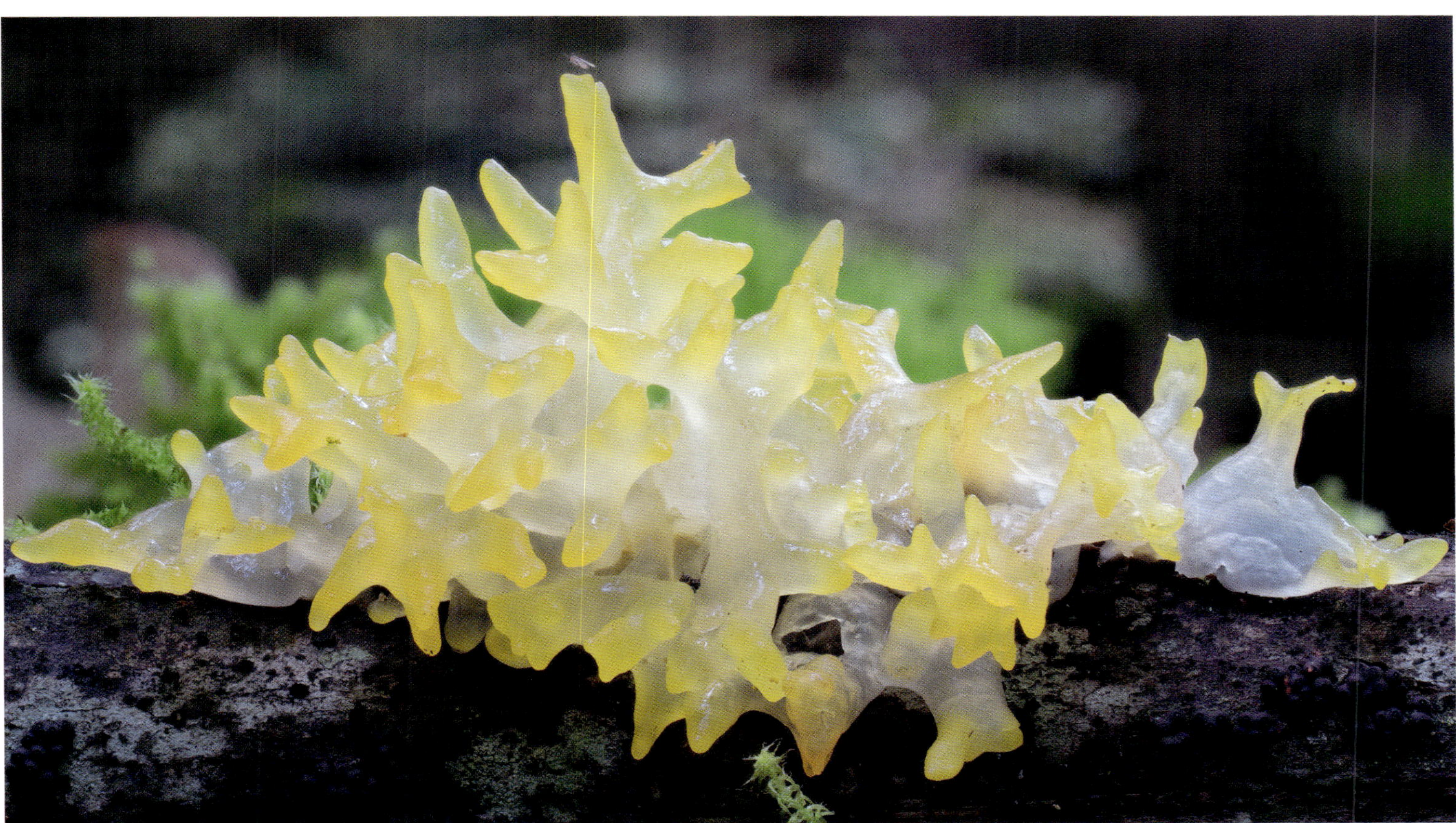

ABOVE

Tremella mesenterica, witch's butter
Nabanhe National Nature Reserve, Yunnan, China
Distribution: worldwide

OPPOSITE TOP

Cookeina tricholoma, bristly tropical cup
Nabanhe National Nature Reserve, Yunnan, China
Distribution: worldwide in the tropics

OPPOSITE BOTTOM

Oudemansiella canarii
Nabanhe National Nature Reserve, Yunnan, China
Distribution: tropical America and south-east Asia.

Among the diverse species in this region, the ectomycorrhizal *Amanita hemibapha* stands out as a fungal masterpiece.

It emerges from its egg-like pure white volva, a protective layer that encases the young mushroom. The stipe, laced with gold, provides a pedestal for the capsicum-red conical cap.

As the mushroom matures, the cap unfurls into a bowl shape, with pale gills directing the eye to the intricate skirt-like membrane that once concealed the gills.

Amanita hemibapha and its cousins *A. jacksonii* in the Americas and *A. caesarea* in Europe, are among the few edible species of *Amanita*. This genus includes some of the deadliest mushrooms in the world.

Species of *Amanita* are responsible for ~90 per cent of the deaths caused by mushroom poisoning. The toxin most responsible is α-Amanitin, a lethal poison that works by slowly shutting down the liver and kidneys.[19,20]

'With nicknames like destroying angel and the death cap, nearly all the mycologists we know caution against eating any Amanita, *unless you are absolutely sure it is an edible variety.'*

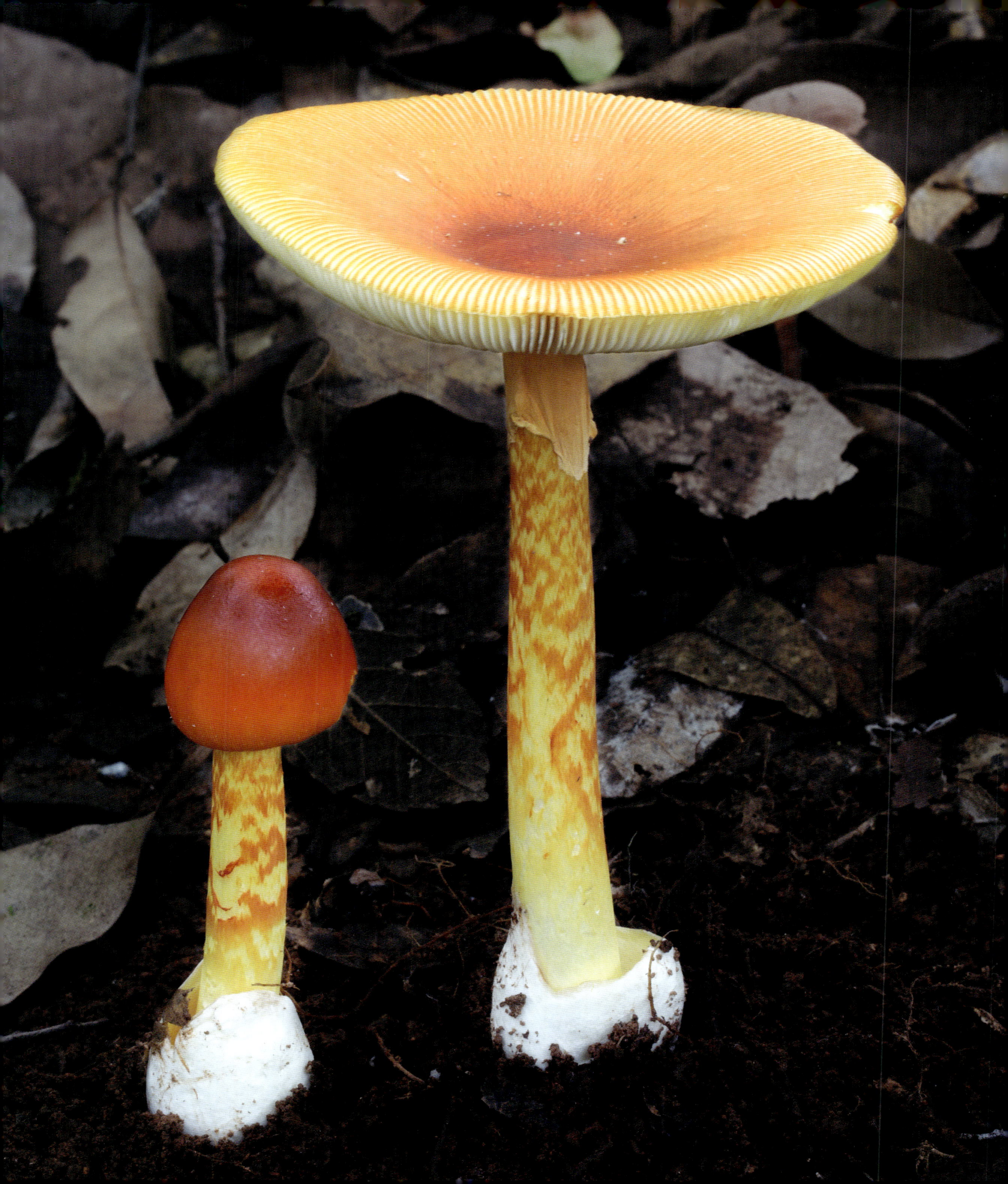

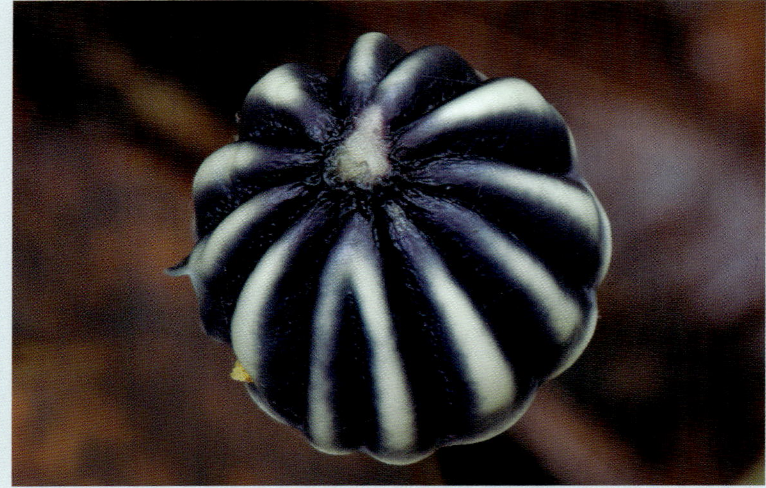

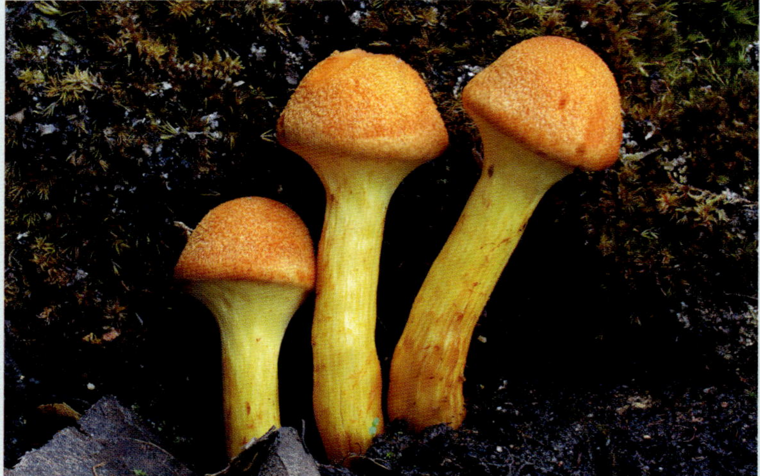

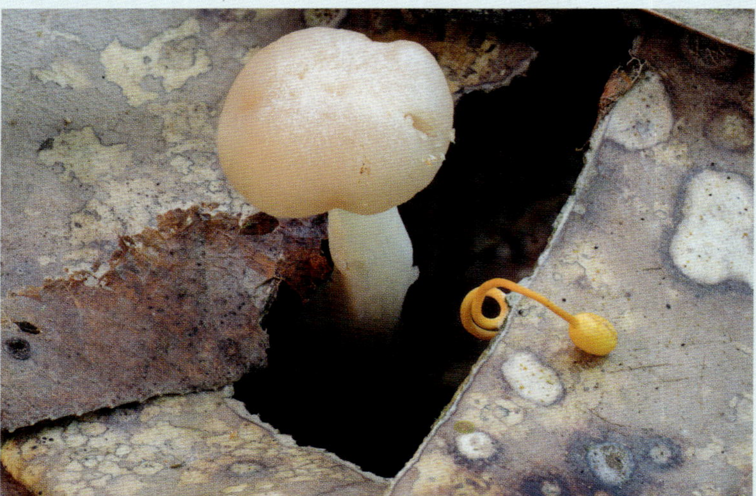

One of the most elegant specimens Stephen photographed is the unusual bold striped parasol of the saprotrophic fungus *Marasmius bekolacongoli*.

'It didn't look real, a design more fitting for a football jersey than a mushroom. I found myself pondering, could there be a purpose behind this artistry, or is it merely incidental? Another fungal mystery to be solved.'

TOP
Marasmius bekolacongoli
Nabanhe National Nature Reserve, Yunnan, China
Distribution: southern Africa and China

MIDDLE
Gymnopilus sp.
Ailao Mountain Nature Reserve, Yunnan, China

BOTTOM
Ophiocordyceps myrmecophila (right) with a small
unidentified mushroom (left)
Nabanhe National Nature Reserve, Yunnan, China
Distribution: Europe and east Asia

OPPOSITE
Gyrodontium sacchari
Nabanhe National Nature Reserve, Yunnan, China
Distribution: worldwide

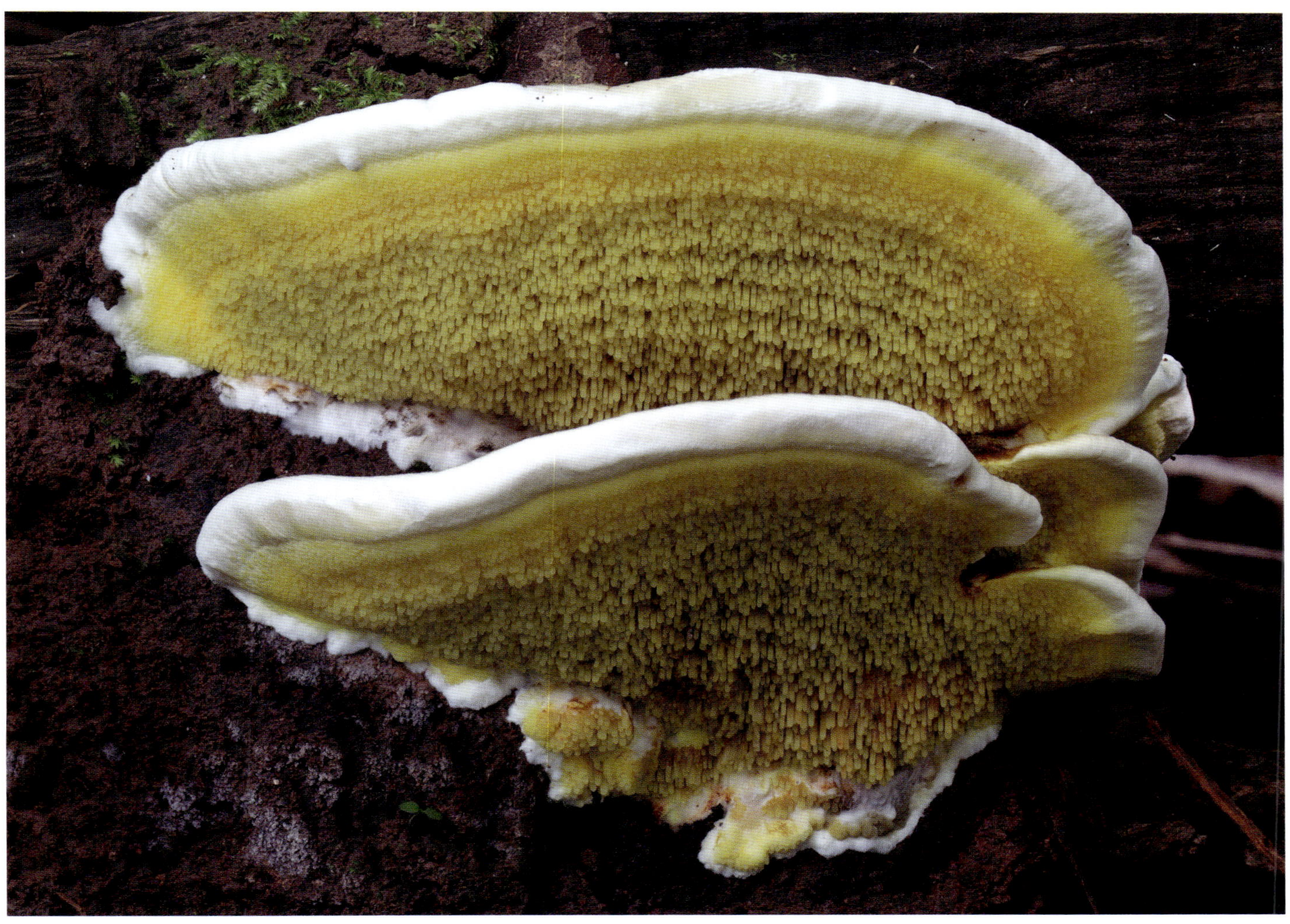

Another unexpected find was the rarely observed shelf fungus, *Gyrodontium sacchari*. It belongs to a relatively small genus, with only three known species.[21,22]

It has been found on both living and dead trees but seems to be a wood-decay fungus, causing a brown rot.

Many shelf fungi have pores on the underside but here there is a cascade of descending tooth-like pipes.

'The world of fungi is full of surprises.'

'We often glimpsed our insect and fungi expert, Ye Le, sprawled across the leaf litter on the hunt for the elusive and fascinating flesh-eating fungi.

'The colourful spikes of these fungi are challenging to find, and whenever one was discovered, the excitement was contagious.

'The diversity of this family of fungal organisms in this forest was mind-boggling.'

Ye Le found these *Ophiocordyceps* fungi inhabiting insect structures once beetles, flies, wasps, spiders, ants and cicadas.

Their bodies now merely the host for these opportunistic fungi to feed and reproduce.

Among the specimens Stephen photographed was one of Asia's most prized fungal organisms, *Cordyceps militaris.* This fungus has been used in traditional Chinese medicine for centuries. It is claimed to have curative properties for a wide range of ailments, from heart disease and kidney malfunction to fertility issues.[23]

However, wild-harvested *C. militaris* are prohibitively expensive and over-harvesting poses a significant threat to the species.[24]

Fortunately, over the past decade, the cultivation of this fungus has been highly successful, with China alone producing nearly 2 billion US dollars' worth of *C. militaris* annually.[24]

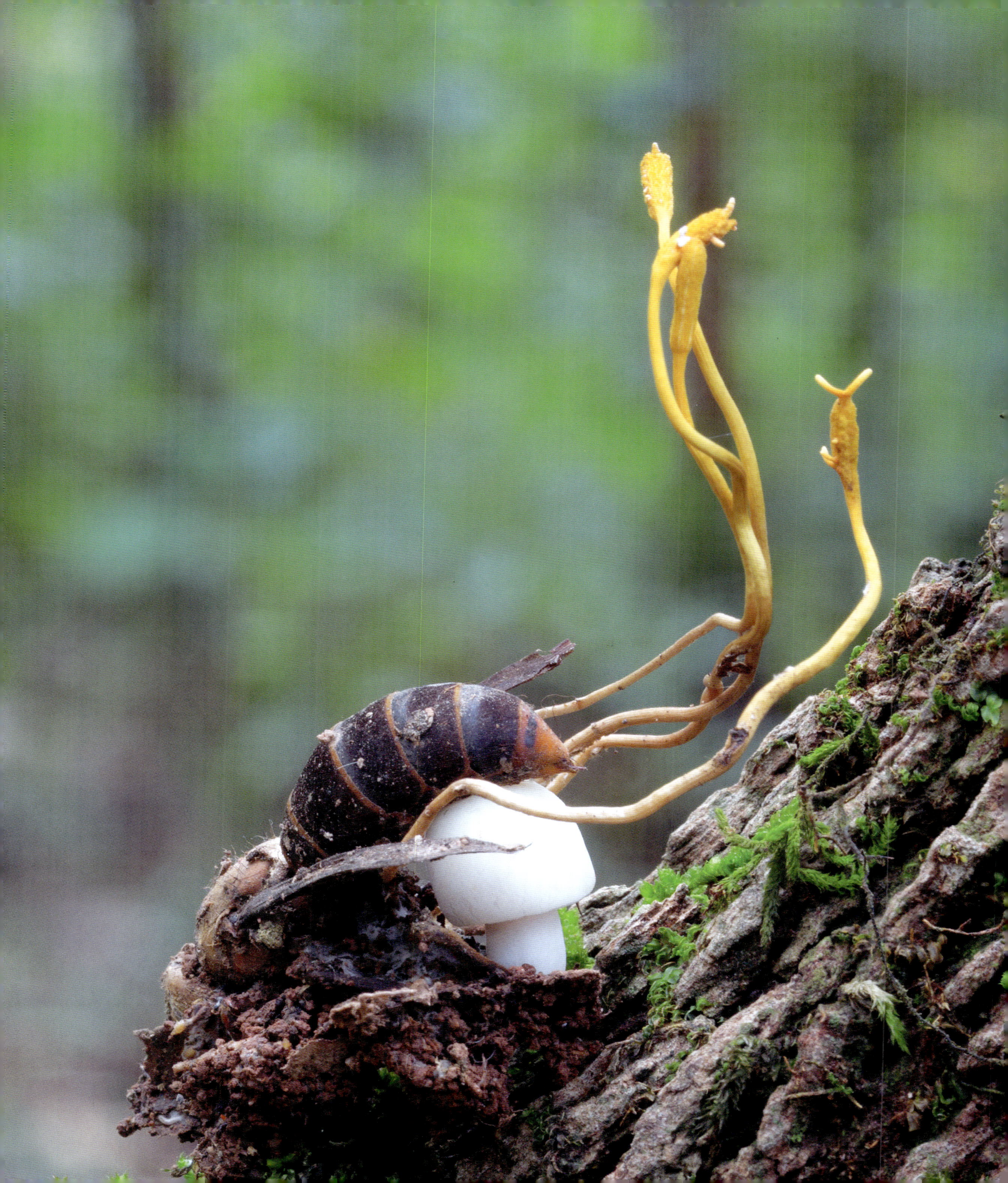

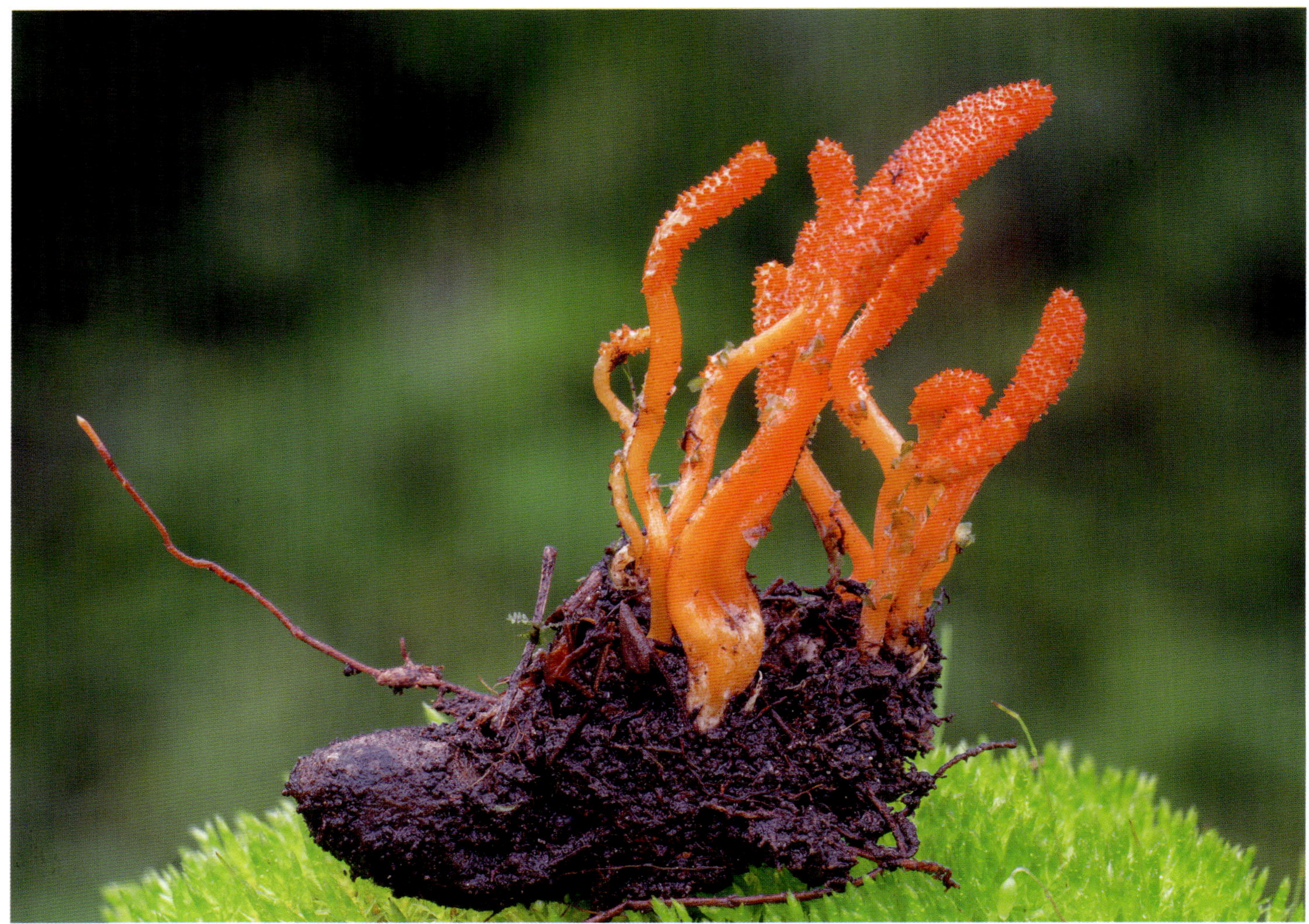

Recent scientific research has also shown that the complex compounds of *C. militaris* could indeed deliver on many of its traditional promises.[23,24]

At the end of each day, the scientists gathered, catalogued and took detailed notes on their discoveries. This process often takes longer than the actual collection of specimens in the field.

During our three-week expedition, 118 species were photographed, analysed, documented and carefully dried in a food dehydrator, preserving them for further study in the laboratory.

As a lover of nature, I was particularly intrigued by the way humans interact with the natural world. During the expedition, I filmed some of the vast ethnomycological wisdom of our local guides, as well as the fascinating scientific research conducted by our team.

TOP
Ramariopsis sp.
Honghe Amoshan Nature Reserve, Yunnan, China

MIDDLE
Calostoma sp.
Honghe Amoshan Nature Reserve, Yunnan, China

BOTTOM
Fistulina hepatica, beefsteak polypore
Honghe Amoshan Nature Reserve, Yunnan, China
Distribution: Northern Hemisphere

OPPOSITE
Cordyceps militaris, scarlet caterpillar club
Honghe Amoshan Nature Reserve, Yunnan, China
Distribution: worldwide

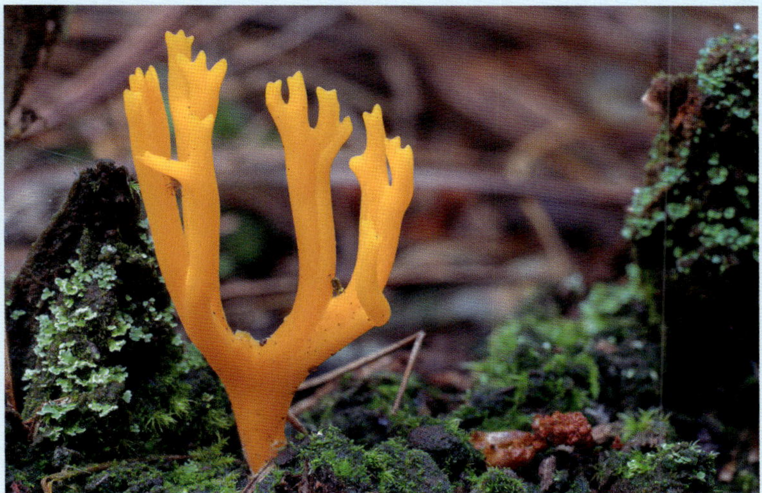

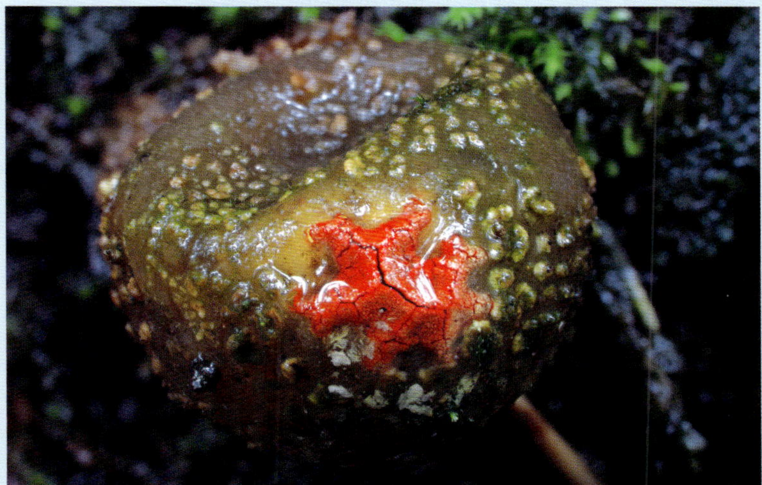

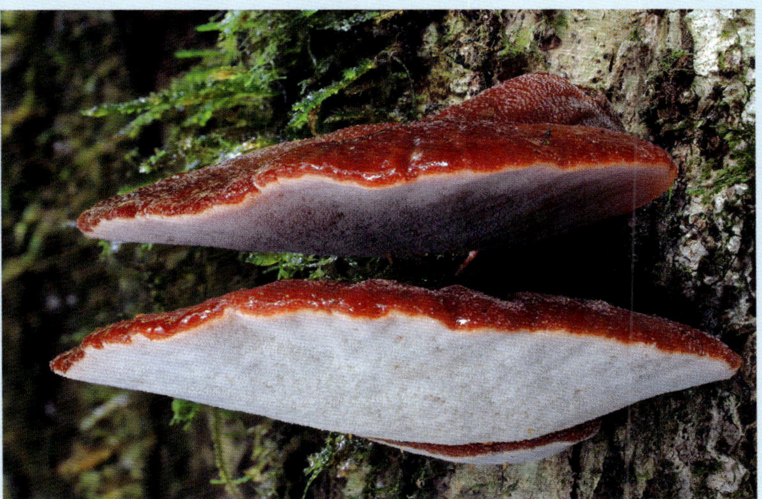

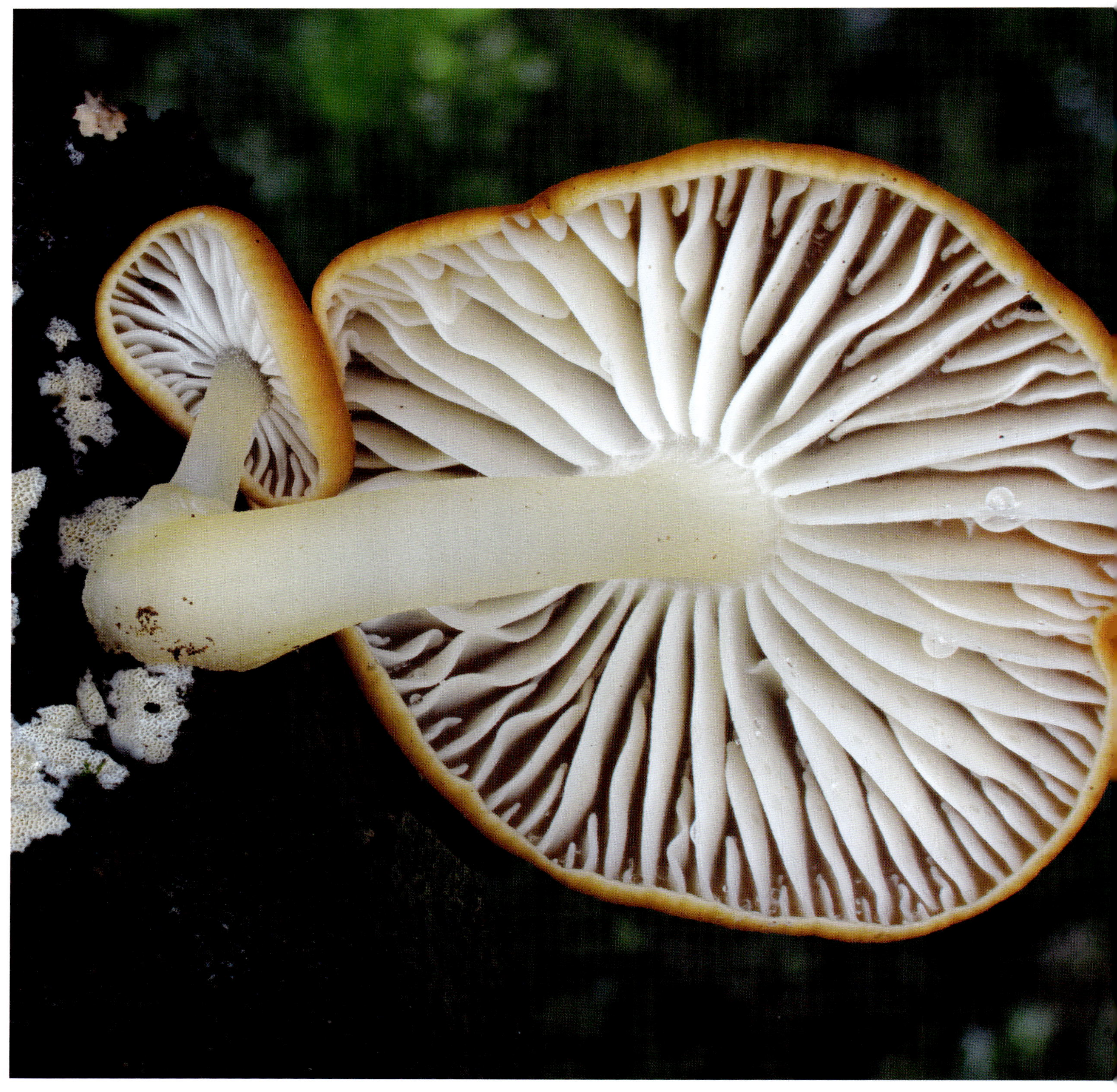

To our surprise, the educational videos we created turned out to be incredibly helpful.

The scientists used them at conferences and in funding applications, extending the reach of their project.

The field trip was a massive success. Funding was procured, a field guide was published and the seeds of collaboration were sown.[25]

This partnership, rooted in a shared fascination with fungi, blossomed into a friendship that would see us journey back to Yunnan in 2016, 2017 and 2019.

In three field trips totalling 8 weeks, we documented an additional 398 species, many of which were new records for China and some new to science.

OPPOSITE
Hydropus sp.
Ailao Mountain Nature Reserve, Yunnan, China

NEXT PAGE
Unidentified mushroom
Ailao Mountain Nature Reserve, Yunnan, China

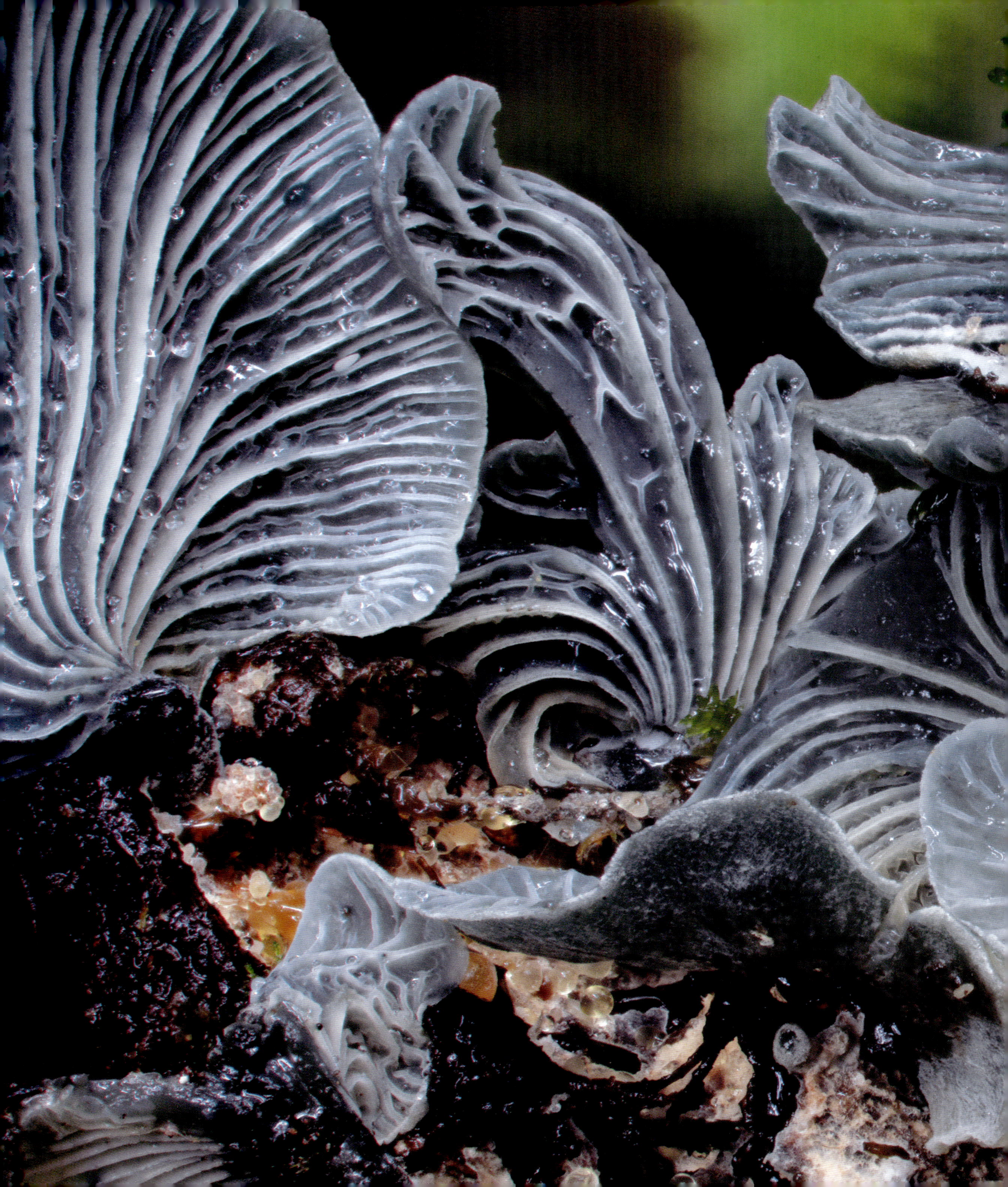

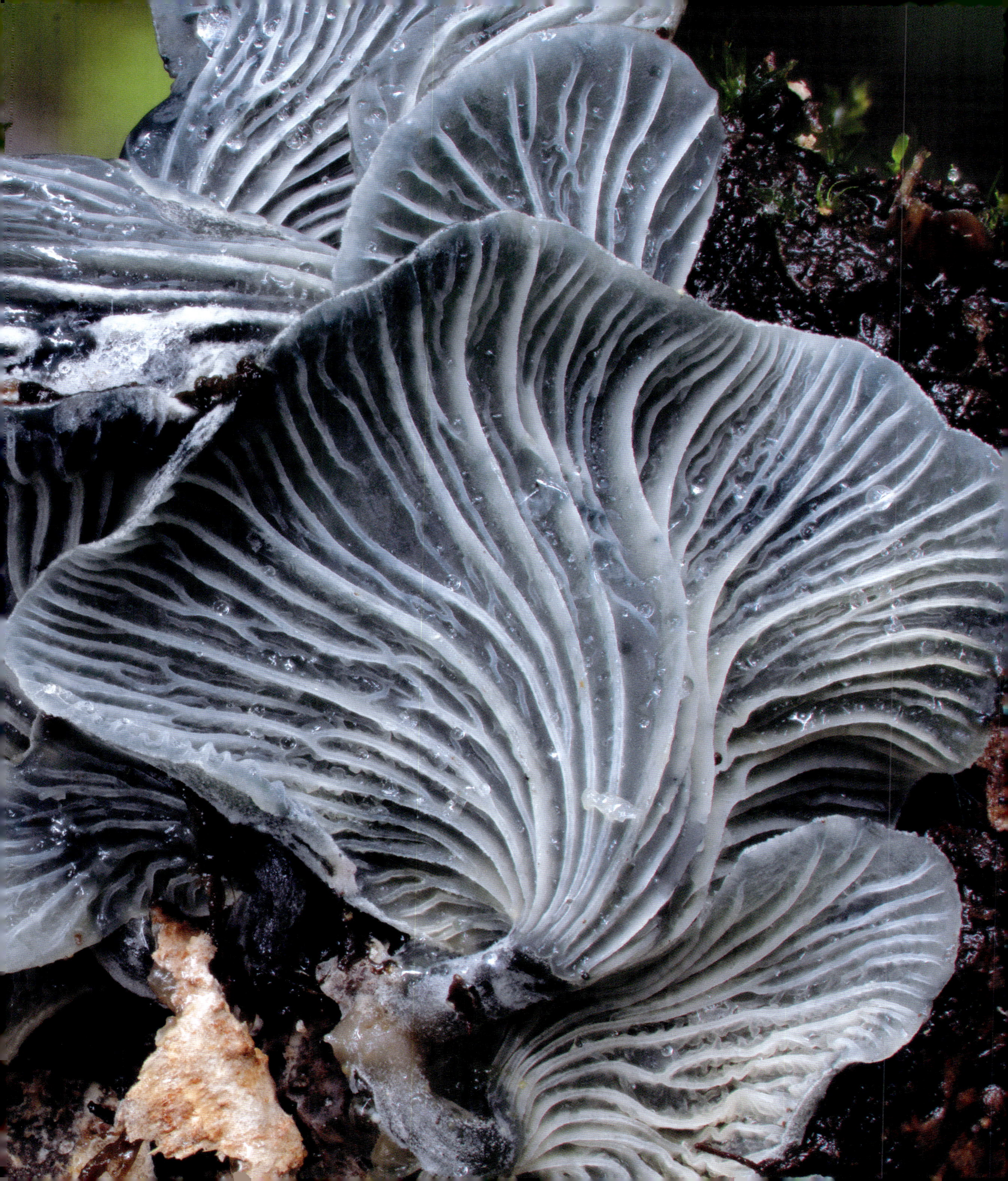

Lichen love

Imagine a world where the most enduring partnership defies all human notions of living together.

On a misty autumn morning in 2004, Stephen was trekking up to the Camels Hump, a craggy dome of lava atop Mount Macedon in Victoria, Australia.

There, clinging tenaciously to the volcanic rock formations, were lichens, their green hues deepening with the morning dew.

'Below the path on a steep, angled slab, there was a particularly lush lichen carpet.

'I carefully scrambled down to set up my camera and tripod, and that's when I noticed the most incredible structures. Miniature lichen trumpets angled upward toward the heavens.

'While I knew they were lichens, I had no idea back then that a lichen is classified as a fungus and that the true nature of these organisms has baffled scientists for centuries.'

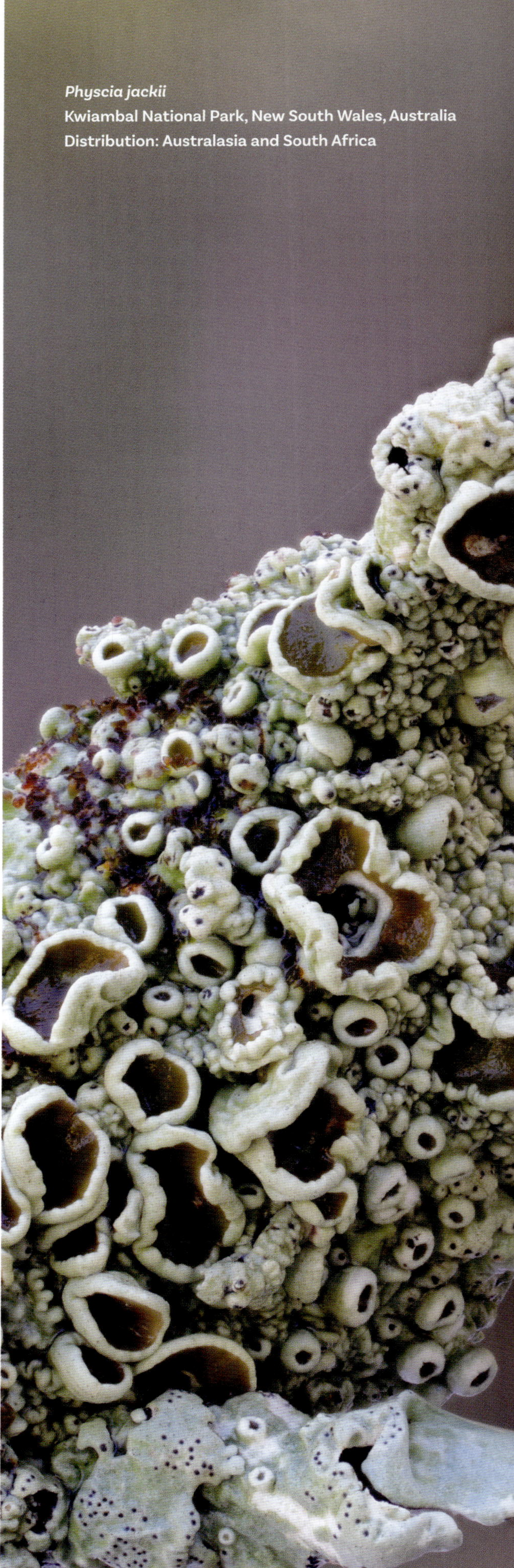

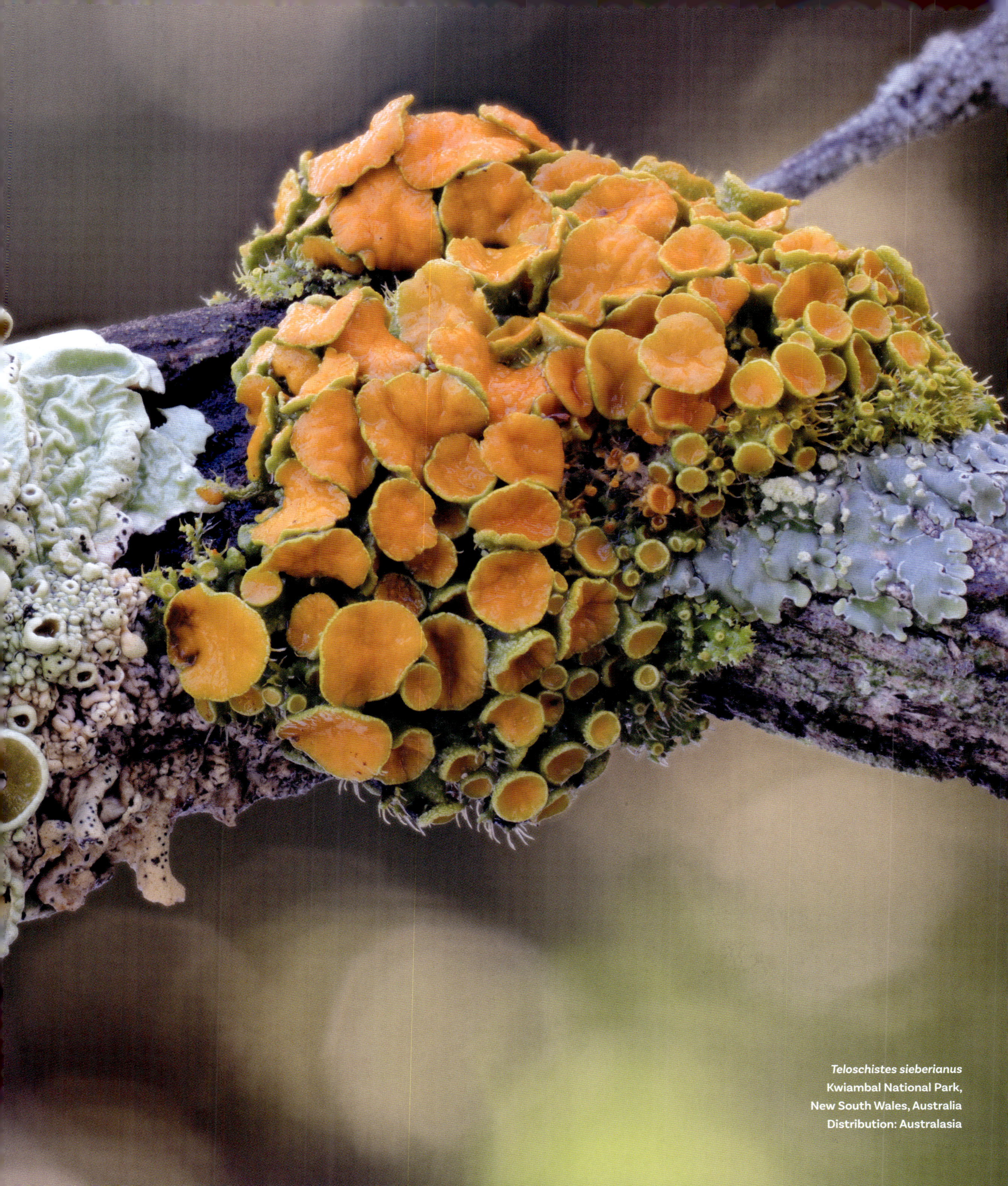

Teloschistes sieberianus
Kwiambal National Park,
New South Wales, Australia
Distribution: Australasia

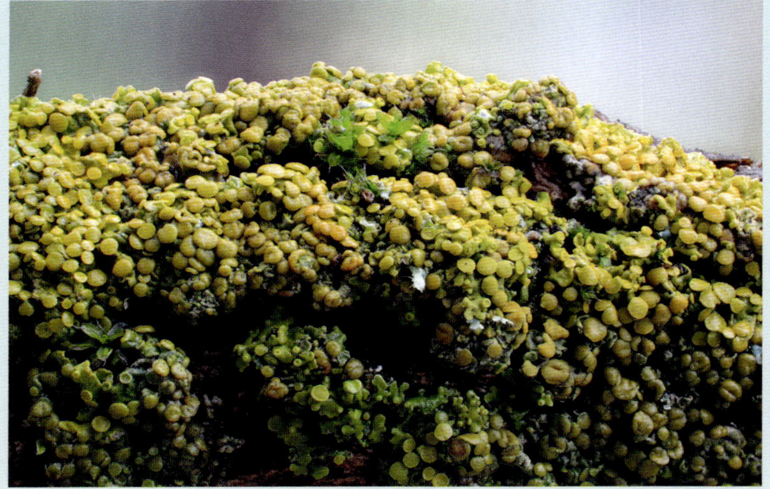

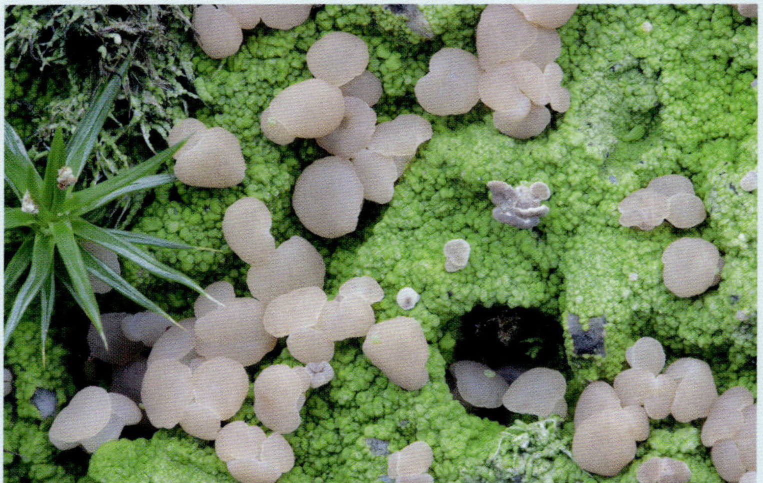

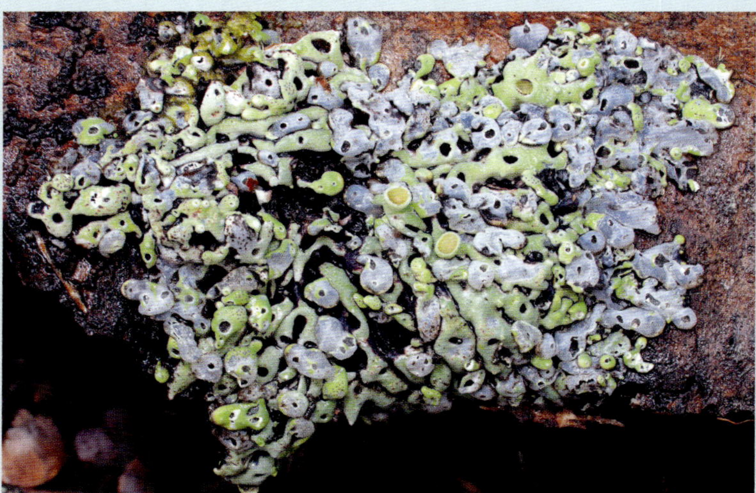

Like a jewelled mantle, lichens drape up to 8 per cent of our Earth's surface in a living mosaic.

We see them as tiny masterpieces garnishing the rainforest's towering trees, embracing boulders on alpine summits, encasing ancient tombs, cladding fences and even colonising discarded human rubbish.

Resilient pioneers of the most forbidding terrains, they flourish amid the icy wilderness of Antarctica as well as the blistering heat of deserts.

TOP
Unidentified lichen
Warren Gorge, Flinders Ranges, South Australia

MIDDLE
Baeomyces heteromorphus
Bunyip State Park, Victoria, Australia
Distribution: Australasia and southern South America

BOTTOM
Menegazzia pertransita
Lake St Clair, Tasmania, Australia
Distribution: Australasia

OPPOSITE
***Cladonia* sp.**
Boonoo Boonoo National Park, New South Wales, Australia

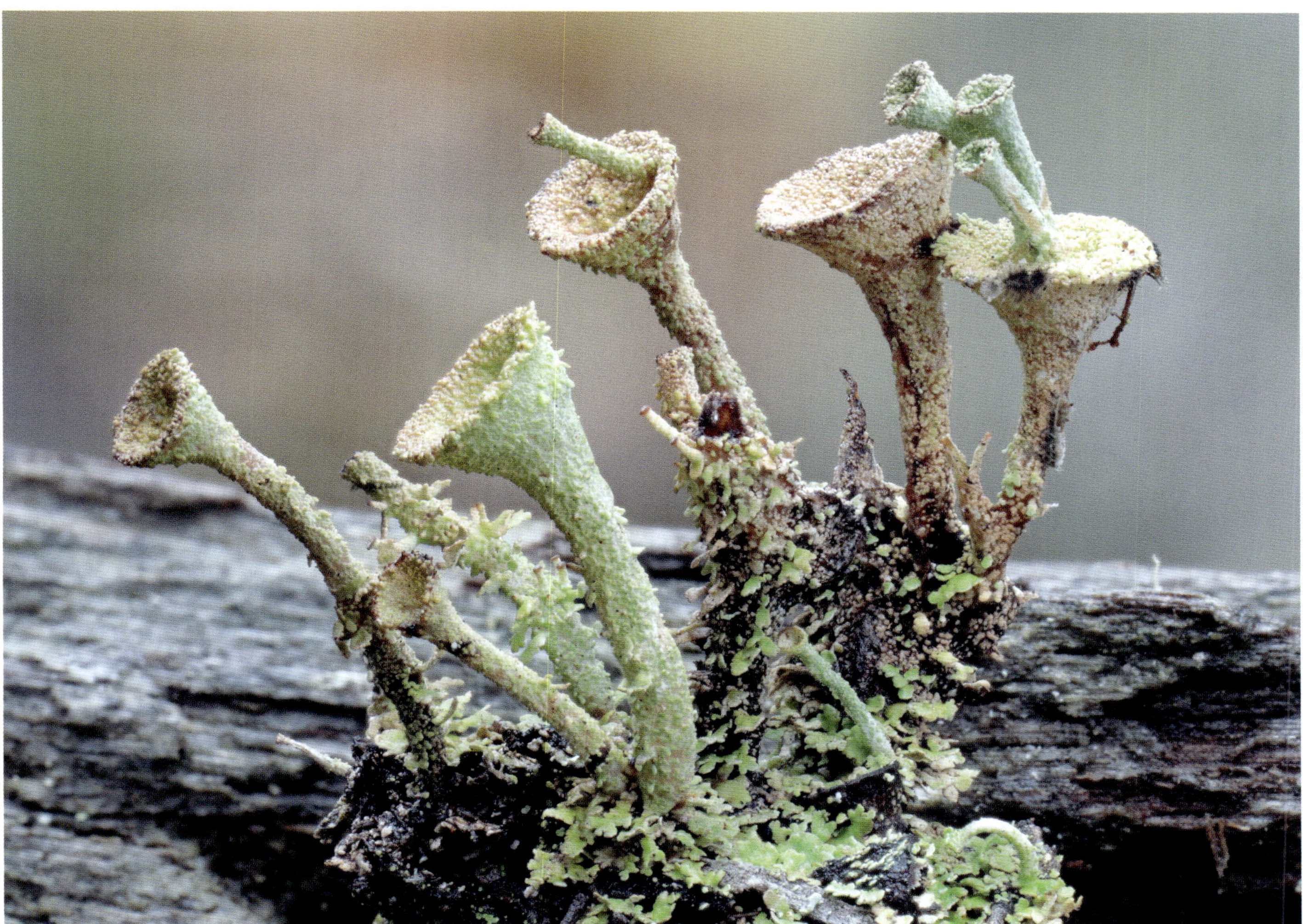

Their structures are as varied as each of the ecological niches they inhabit, with a tally of over 19,000 species described to date.[1]

Yet, for all their ubiquity, the essence of lichens has for centuries been a tale of mistaken identity.

The first lichens are thought to have evolved 400 million years ago, and until the mid-19th century, scientists believed that they were 'autonomous plants'.[2]

It was Simon Schwendener, a Swiss botanist armed with a microscope and a passion for lichenology, who detected that the anatomy of lichens didn't fit the plant theory.[3]

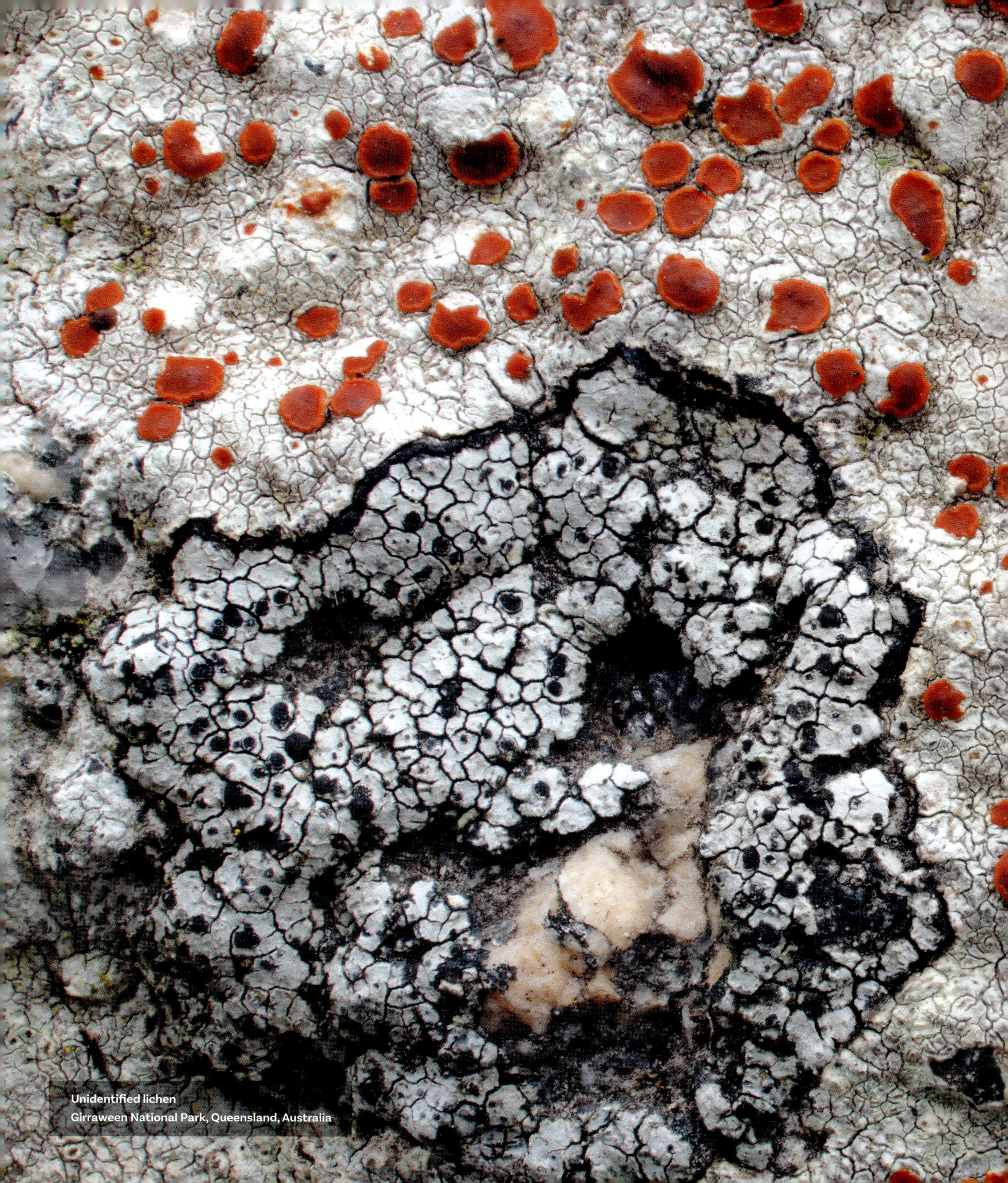

Unidentified lichen
Girraween National Park, Queensland, Australia

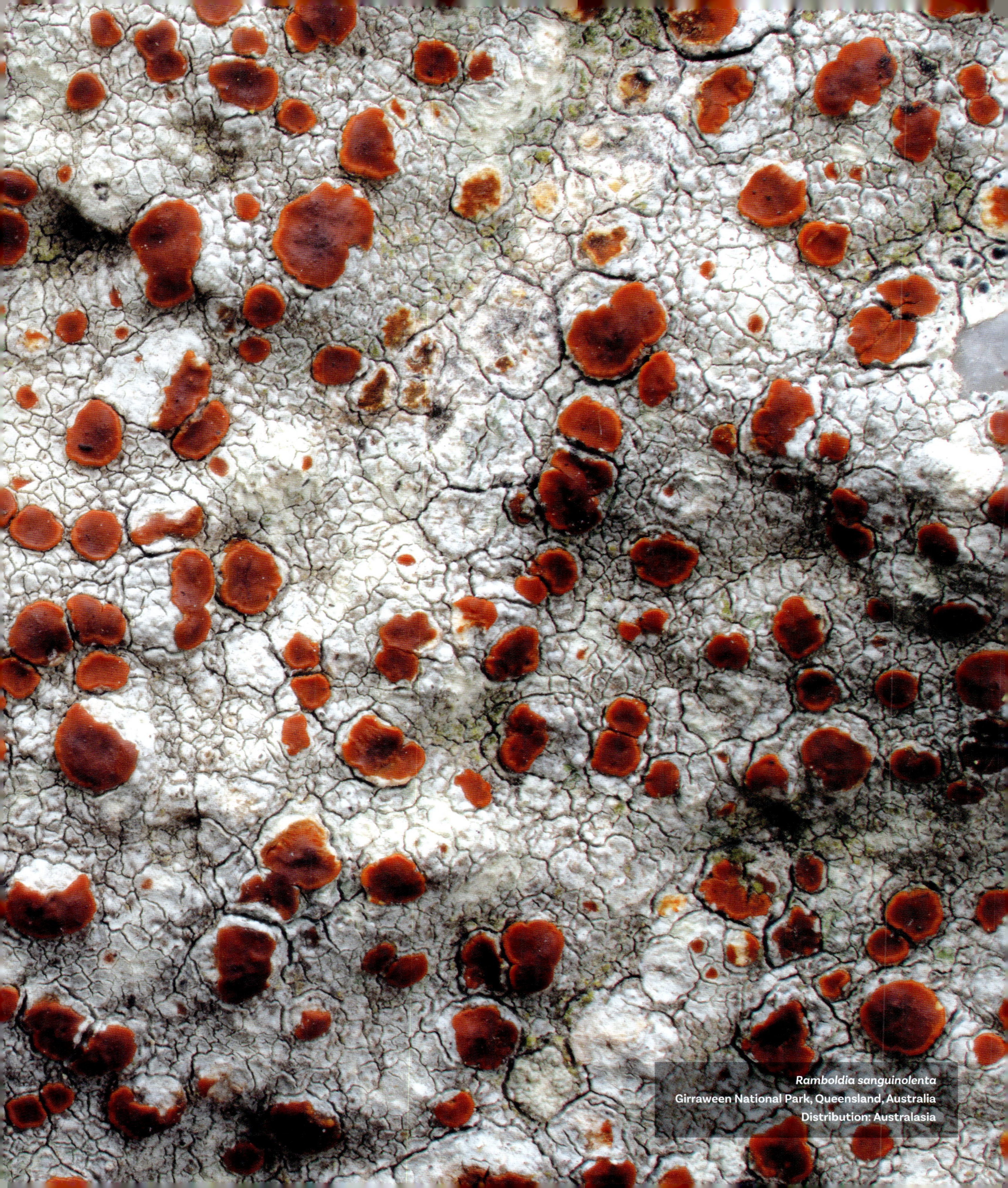

Ramboldia sanguinolenta
Girraween National Park, Queensland, Australia
Distribution: Australasia

Schwendener proposed a radical idea: lichens were not lone entities but each was a partnership between two very different organisms, a fungus and an alga.

Presented before the Swiss Natural History Society in 1867, his hypothesis of duality initially met with scepticism as the leading lichenologists of the time clung to established doctrines.[2]

Meanwhile, a growing contingent of botanists and biologists began to embrace Schwendener's vision.

It was an era ripe with scientific fervour, and the tenets of Darwinism and the survival of the fittest were hotly debated topics.

In 1866, Ernst Haeckel, inspired by Darwin, introduced the word 'ecology' to encapsulate how every living thing interacts with its environment.[4]

At the same time, the botanist Albert Frank, delving into the relationship between plants and mycorrhizal fungi, found resonance with Schwendener's ideas on lichens. He coined the term 'symbiotismus' to describe the teamwork between fungi and algae.[5]

Heinrich de Bary, a botanist and mycologist, recognised that some growths on crops were not part of the plant but parasitic fungi, dependent on their hosts for survival.

In 1879, de Bary broadened the concept of the term 'symbiosis' to describe the living together of unlike organisms across nature, whether reciprocally beneficial (mutualism), useful to one but not the other (commensal) or harmful to one but beneficial to the other (parasitic).[6,7]

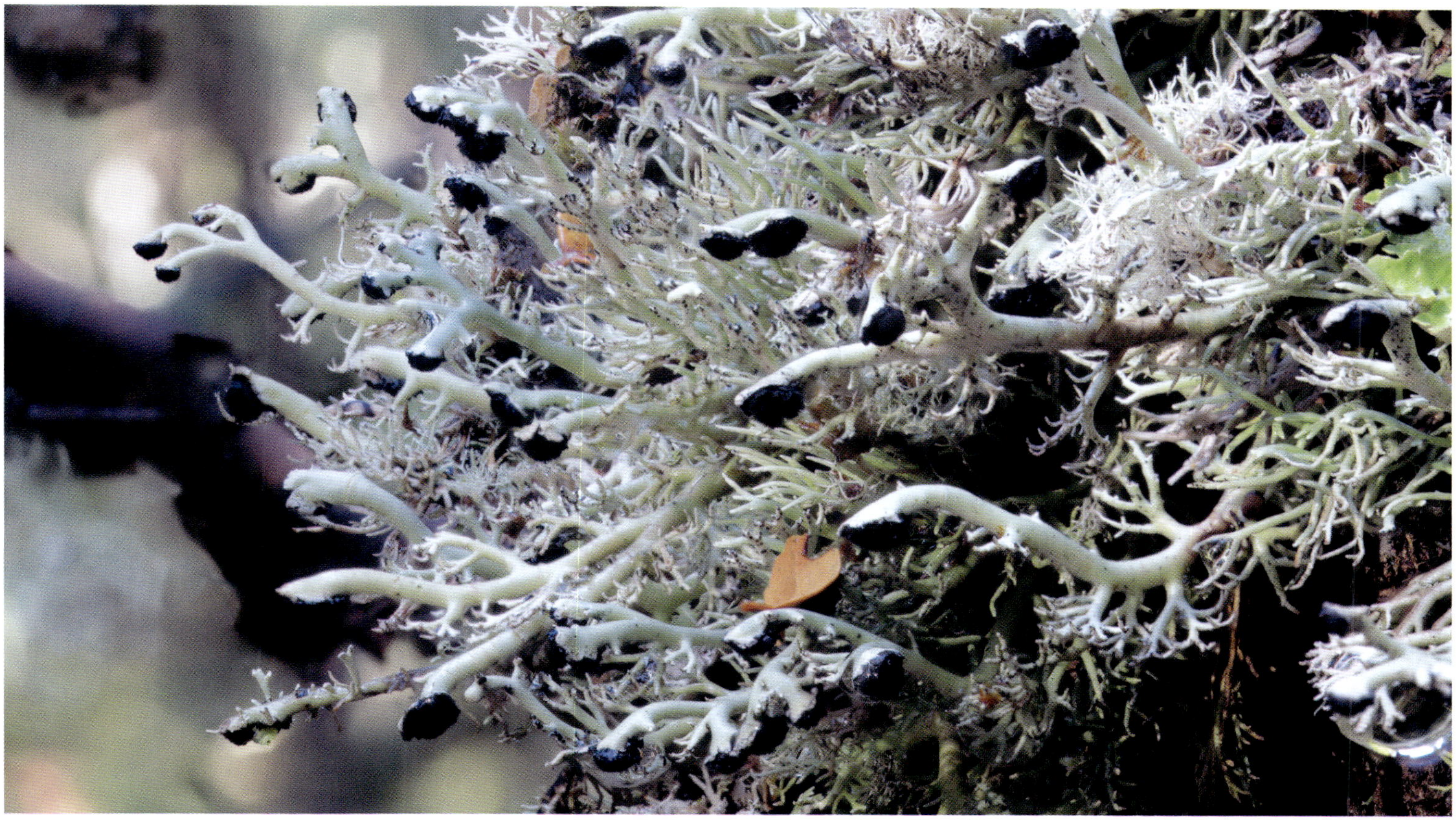

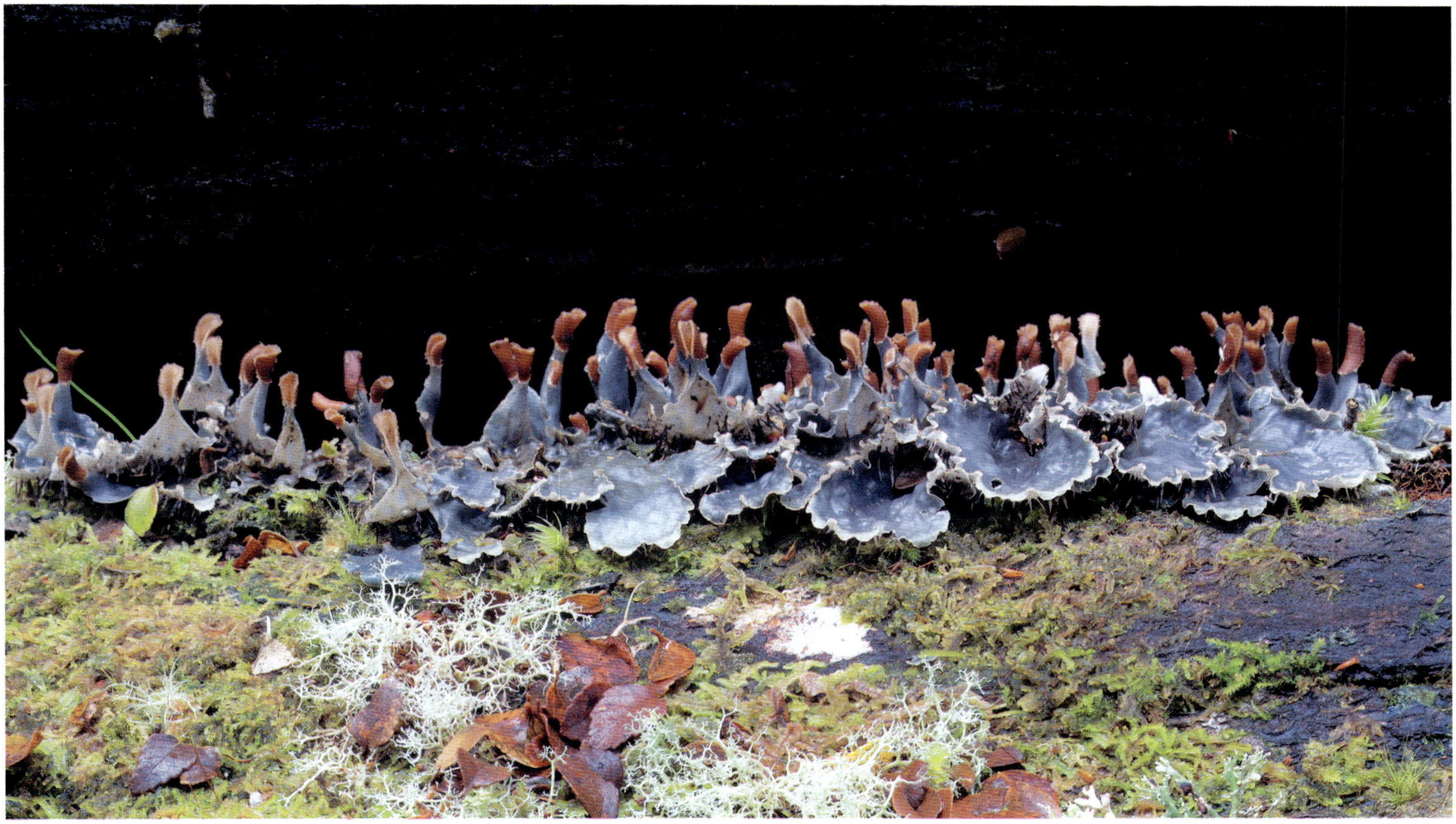

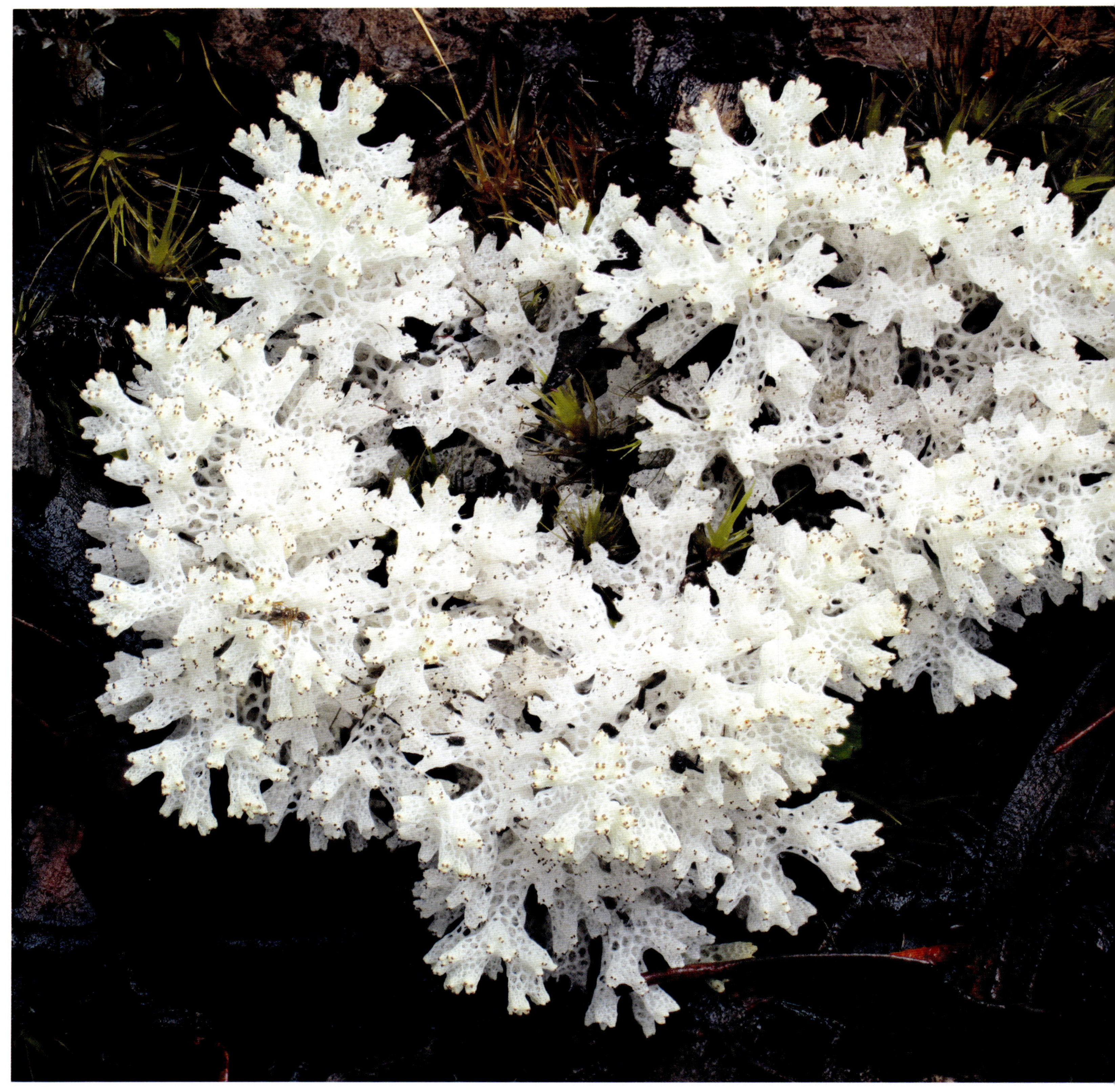

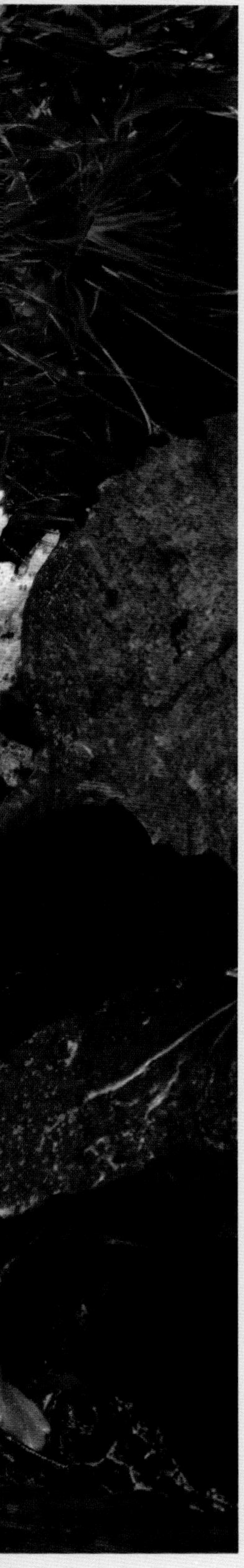

It took nearly a century of lively debates but scientists finally agreed. Lichens are not plants but a combination of a fungus (the mycobiont) and an alga or a cyanobacterium (the photobiont).[8]

The fungus and photobiont communicate through chemical signals, with the fungus wrapping hyphae around and incorporating the photobiont into the lichen's overall body, the thallus.[9]

The photobiont's job is crucial. It harnesses the power of sunlight and creates sugars through photosynthesis – sustaining both partners.

In return, the fungus becomes a master architect, building protective structures that keep the photobiont safe from predators and environmental extremes.

The fungus also gathers nutrients and moisture and anchors their mutual abode to the environment.[1]

Each lichen is a unique masterpiece, housing a distinct species of fungus. The photobiont, however, is the social butterfly of this duo and can partner up with many different lichens.

When it comes to classifying this relationship, it is the fungus that claims the spotlight, with the lichen named after its fungal architect.

Pulchrocladia retipora, coral lichen
Lake St Clair, Tasmania, Australia
Distribution: Australasia

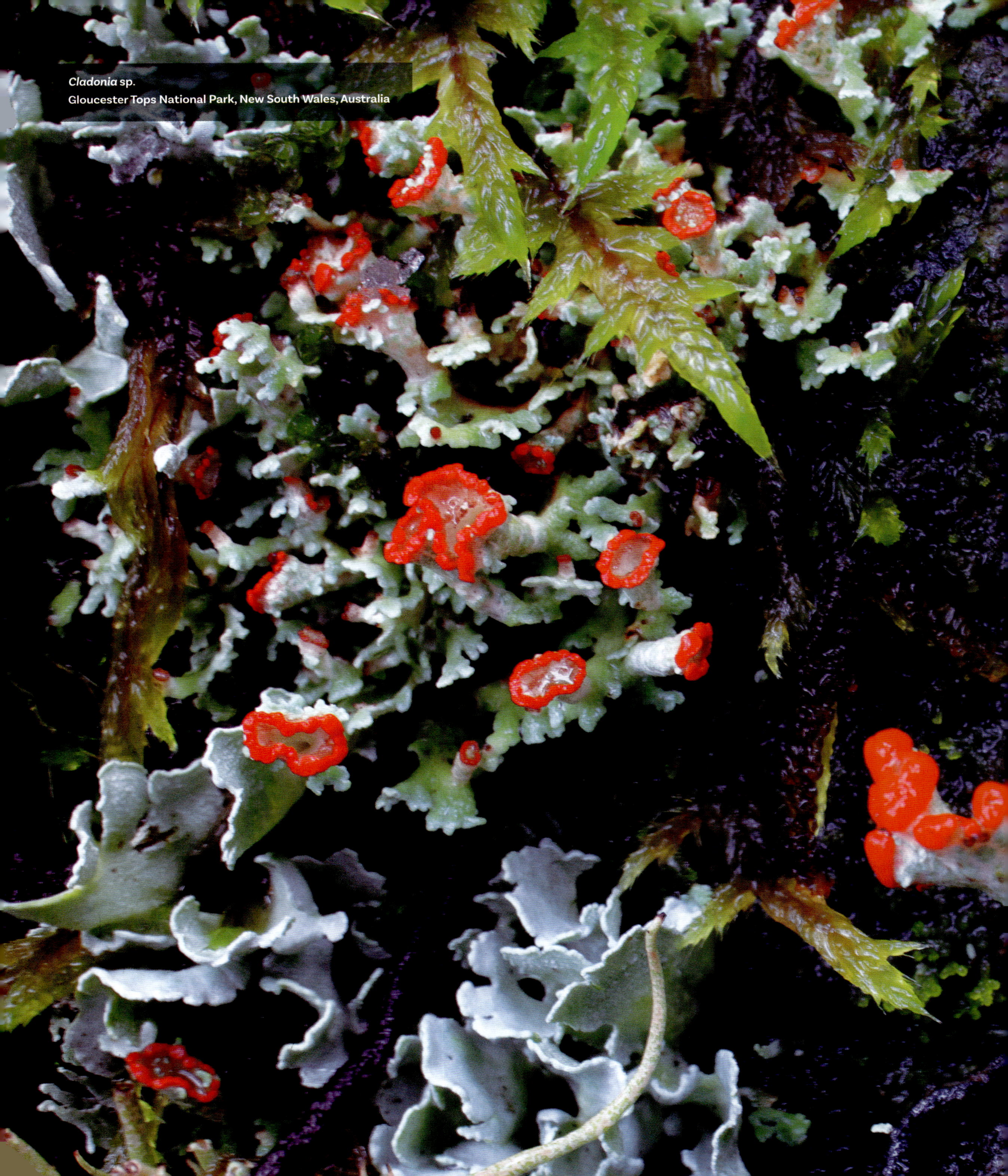

Cladonia sp.
Gloucester Tops National Park, New South Wales, Australia

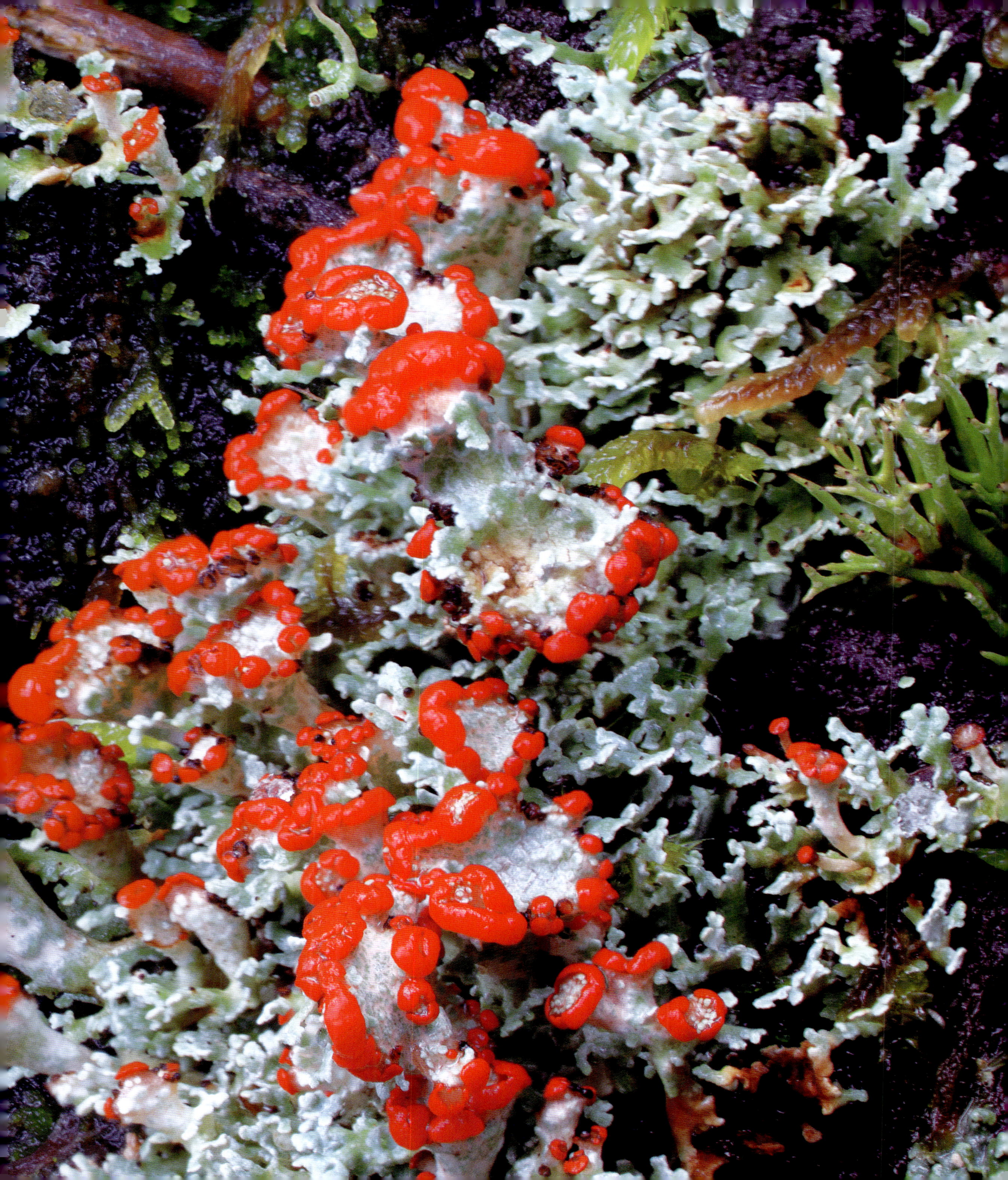

Despite these breakthroughs, the enigma of lichens persisted.

Scientists pondered over the nature of their relationship – was it mutualistic, where both partners benefit, or was the fungus the boss, keeping the photobiont on a leash? It was even suggested that fungi had discovered sustainable agriculture, a farmer tending its photobiont crop.[2,10]

The quest for answers led scientists to the laboratory, where they played matchmaker, attempting to recreate these natural unions.

They found it was possible to grow some mycobionts and photobionts separately, but they looked quite different to their combined form in the lichen thallus.[11]

In 2016, US mycologist Toby Spribille and his research colleagues, on a mission to crack the lichen code, did a genetic deep dive.

They were intrigued by horsehair lichens – two seemingly identical species. One is edible and the other is poisonous, yet they are known to consist of the same fungus and photobiont.[11]

Their plan was to use sophisticated DNA sequencing on not just a few genes but thousands to unravel the mystery.

The initial findings were baffling; the genetic script was identical, leaving the cause of toxicity unexplained.

Then came what Spribille calls his 'Eureka! moment'.

OPPOSITE TOP
Usnea scabrida
Girraween National Park, Queensland, Australia
Distribution: likely to be endemic to Australia

OPPOSITE BOTTOM
Sticta **sp.**
Yumthang Valley, Sikkim, India

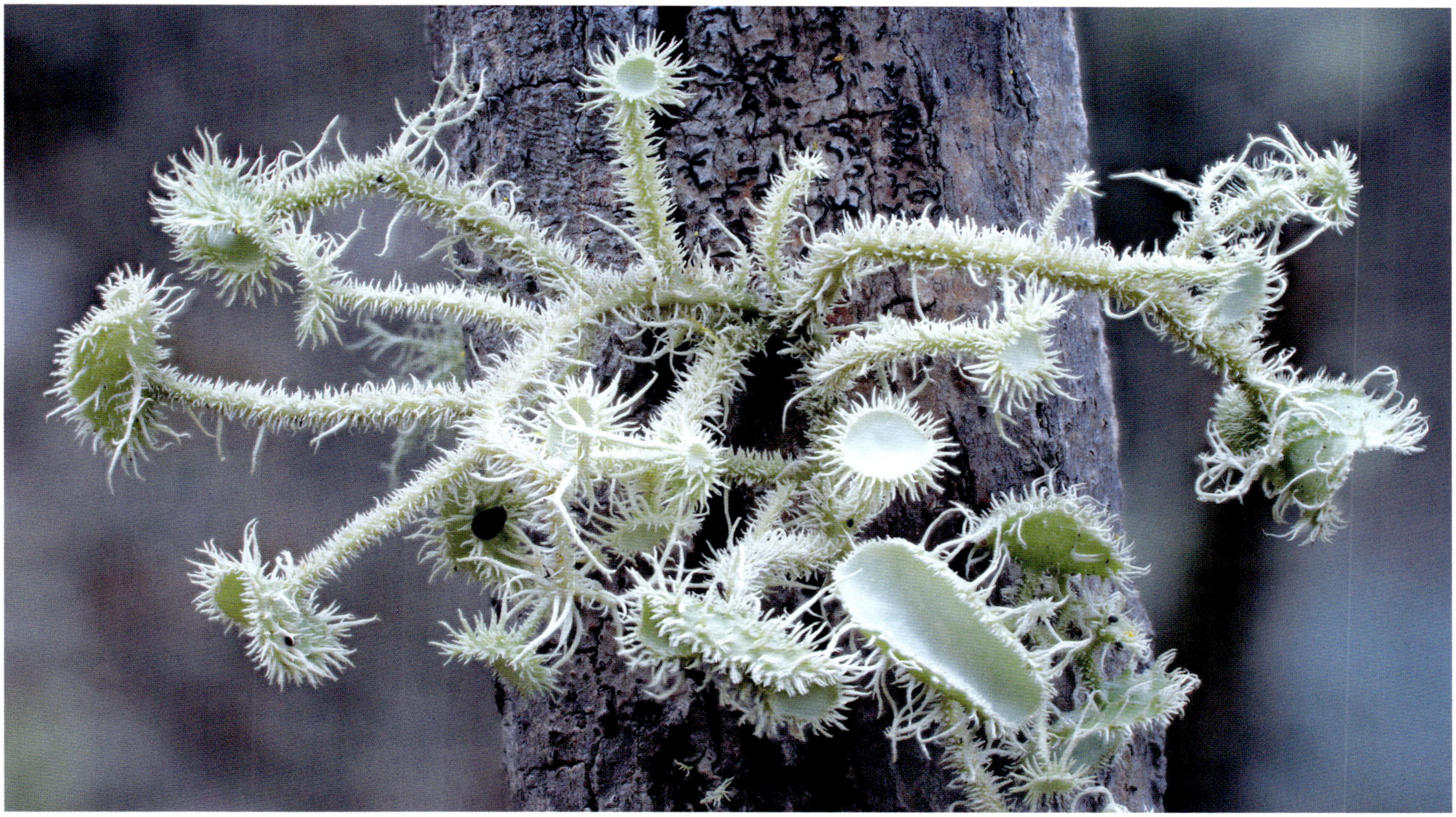

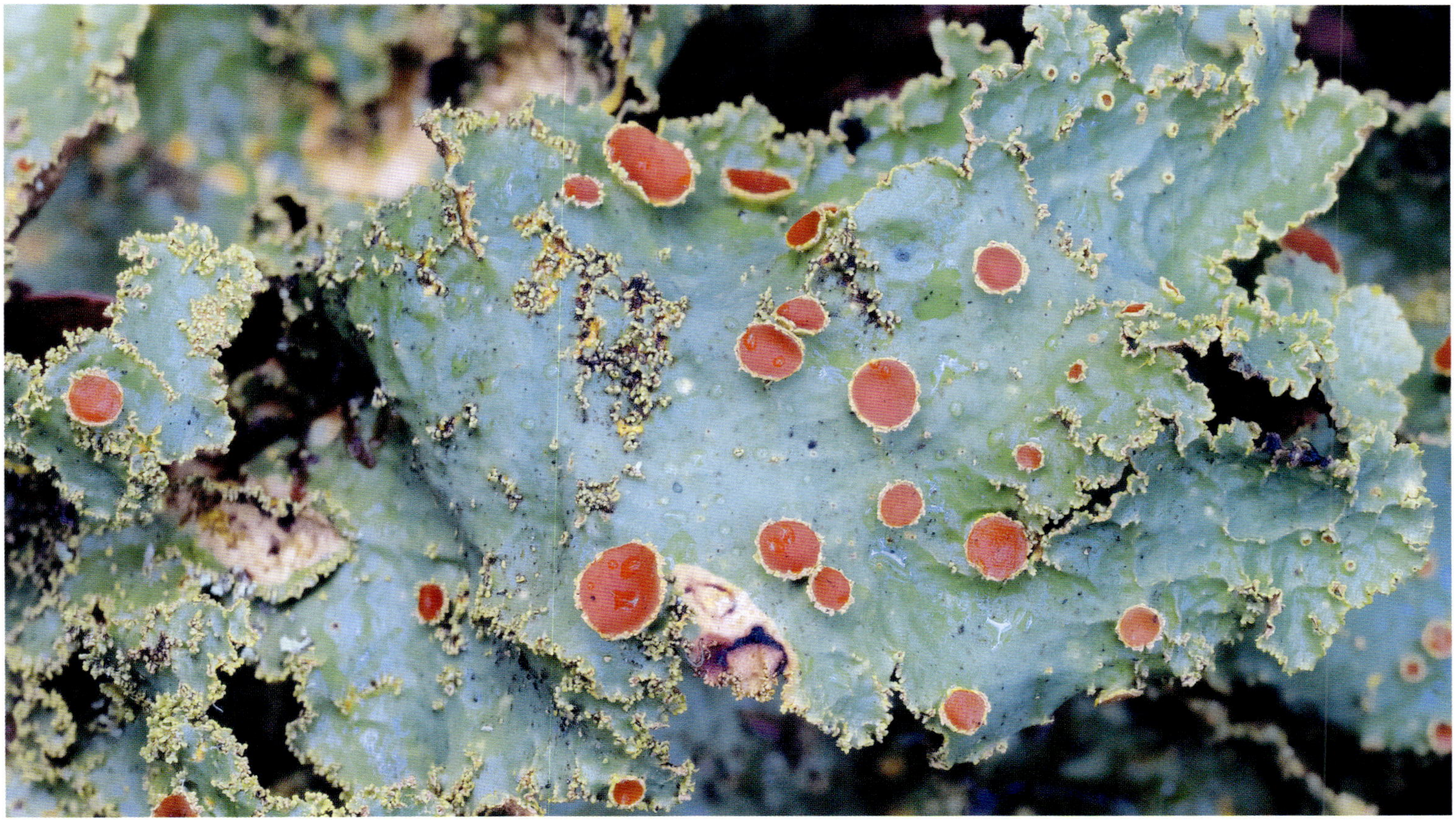

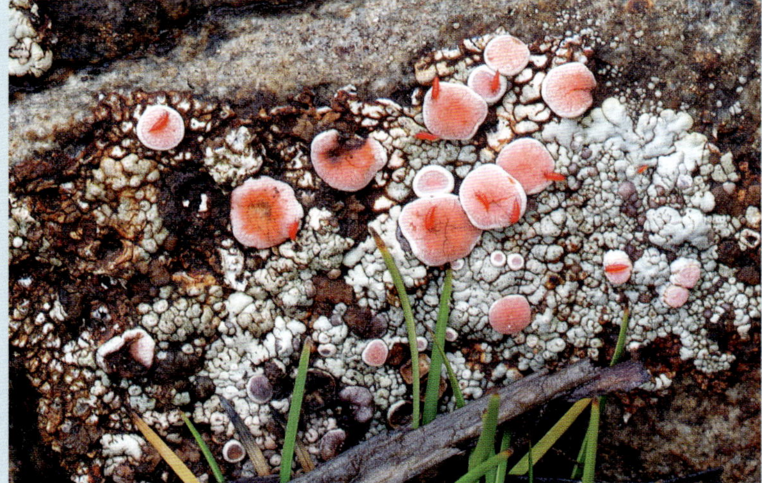

Spribille recalled that in his initial analysis he found a persistent yeast, which he dismissed as a 'rogue fungus', already on his specimens when attached to their trees. Could this yeast be the missing piece of the puzzle?[11]

Revisiting their investigation with 'fresh eyes', he and the research team found that the yeast was not only the key to the amount of toxin present but also an integral part of the lichen's structure.[11]

Then, they tested thousands of lichen specimens. Their results suggested that a trio, not a duo, is the norm.[11]

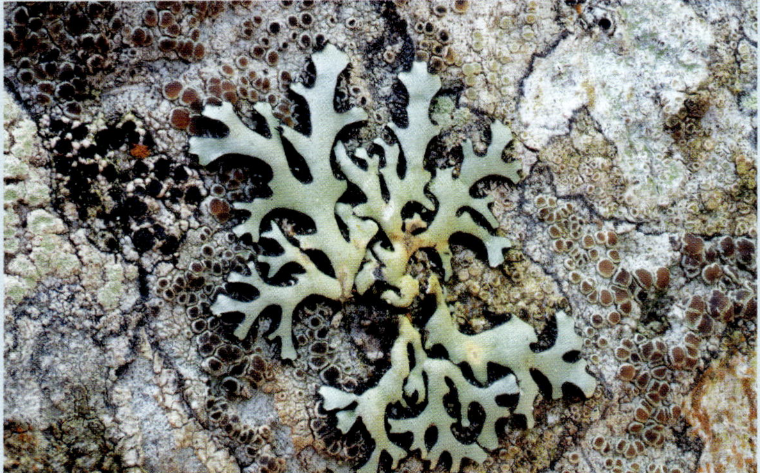

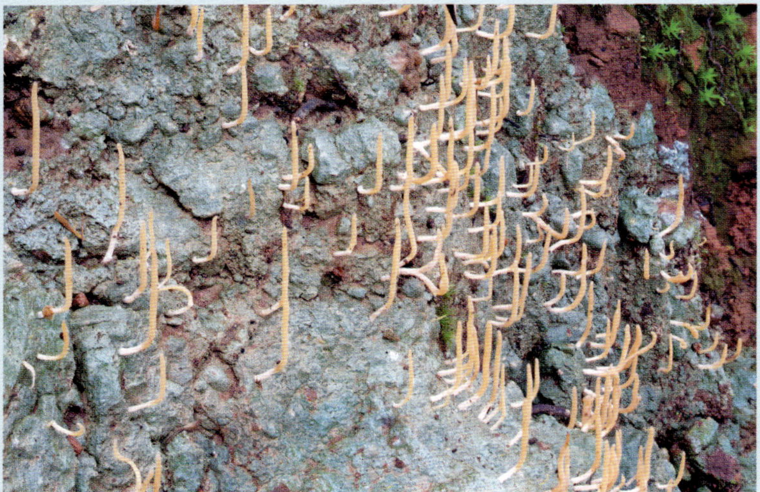

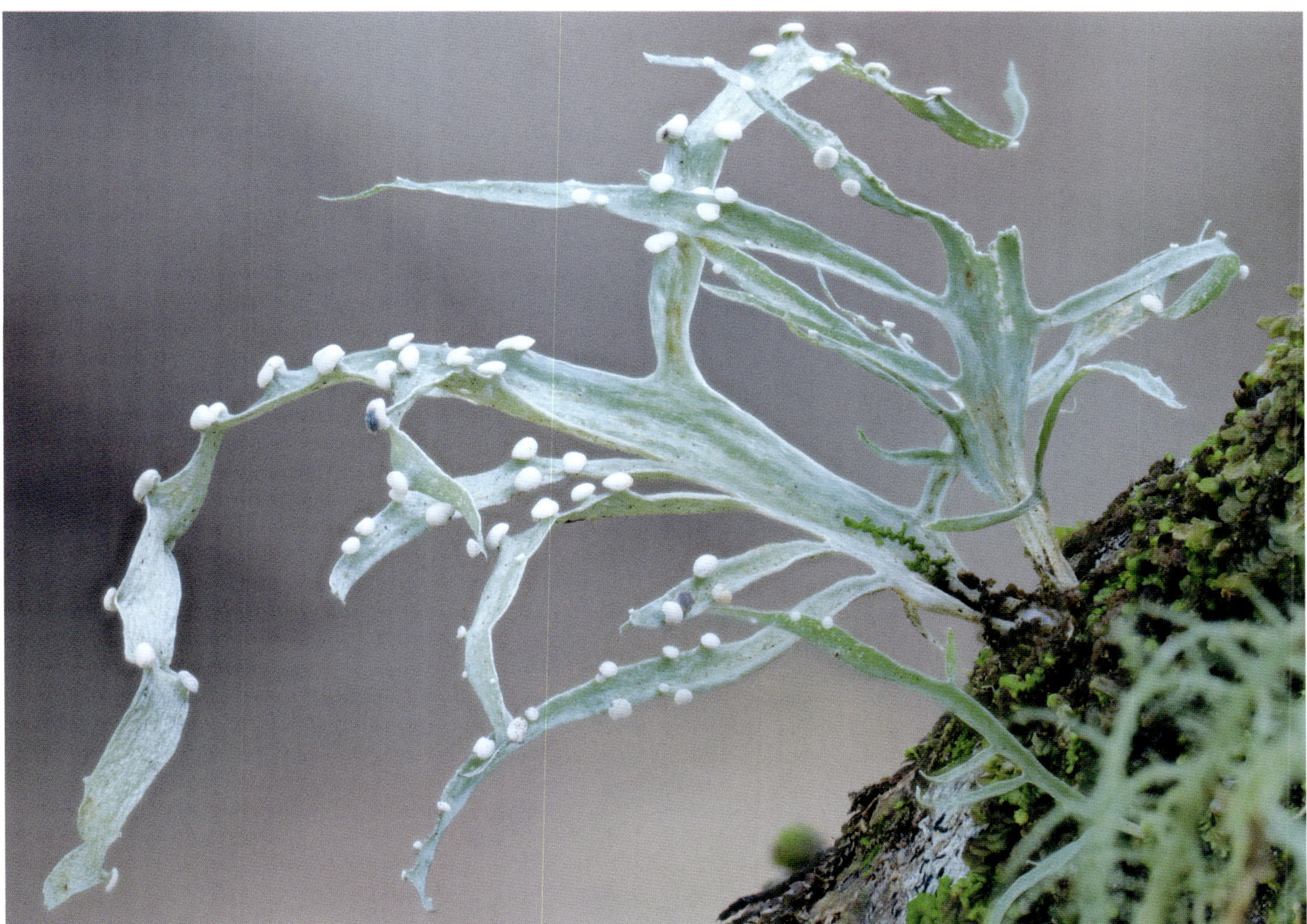

Yet, the narrative of lichens continues to be one of scientific debate. Subsequent studies have peeled back more layers, revealing not just a trio but a bustling microcosm, a community of yeasts, algae, bacteria and protists, cohabiting with a dominant fungus.[12,13]

Imagine a miniature universe humming with life, where each inhabitant plays a crucial role in their shared existence.

The revelation of lichen's complex nature has only deepened the mystery.

How do these tiny teams work together to engineer life in the most inhospitable corners of the world, crafting intricate structures of bewildering diversity?

Stephen and I marvel at the places we find lichens thriving as the ultimate survivors.

On a chilly, sleeting winter's morning, we spent hours documenting lichens coated in ice in a remnant Alpine snow gum forest rising above the flat plains of western New South Wales, Australia.

'It was magical.

'Adversity was transformed into beauty.

'Crystal ice fingers clasped pastel green branches of lichen.

'Orange and rusty coloured sporing bodies became sparkling miniature jewels, frozen art in an icy wonderland.

'My fingers were numb with the biting cold, but I couldn't resist the compulsion to capture this fleeting moment of frozen splendour.

'We've since learned we may have witnessed another of lichen's cool tricks.'

Many lichens can make ice even when it's not freezing cold in a process called ice nucleation. This is thought to be a survival strategy, a way of harvesting water in dry landscapes, and a way of slowing down the freezing process and potential frost damage.[14,15]

Hypogymnia **sp. and** *Usnea* **sp. coated in ice**
Mount Kaputar National Park, New South Wales, Australia

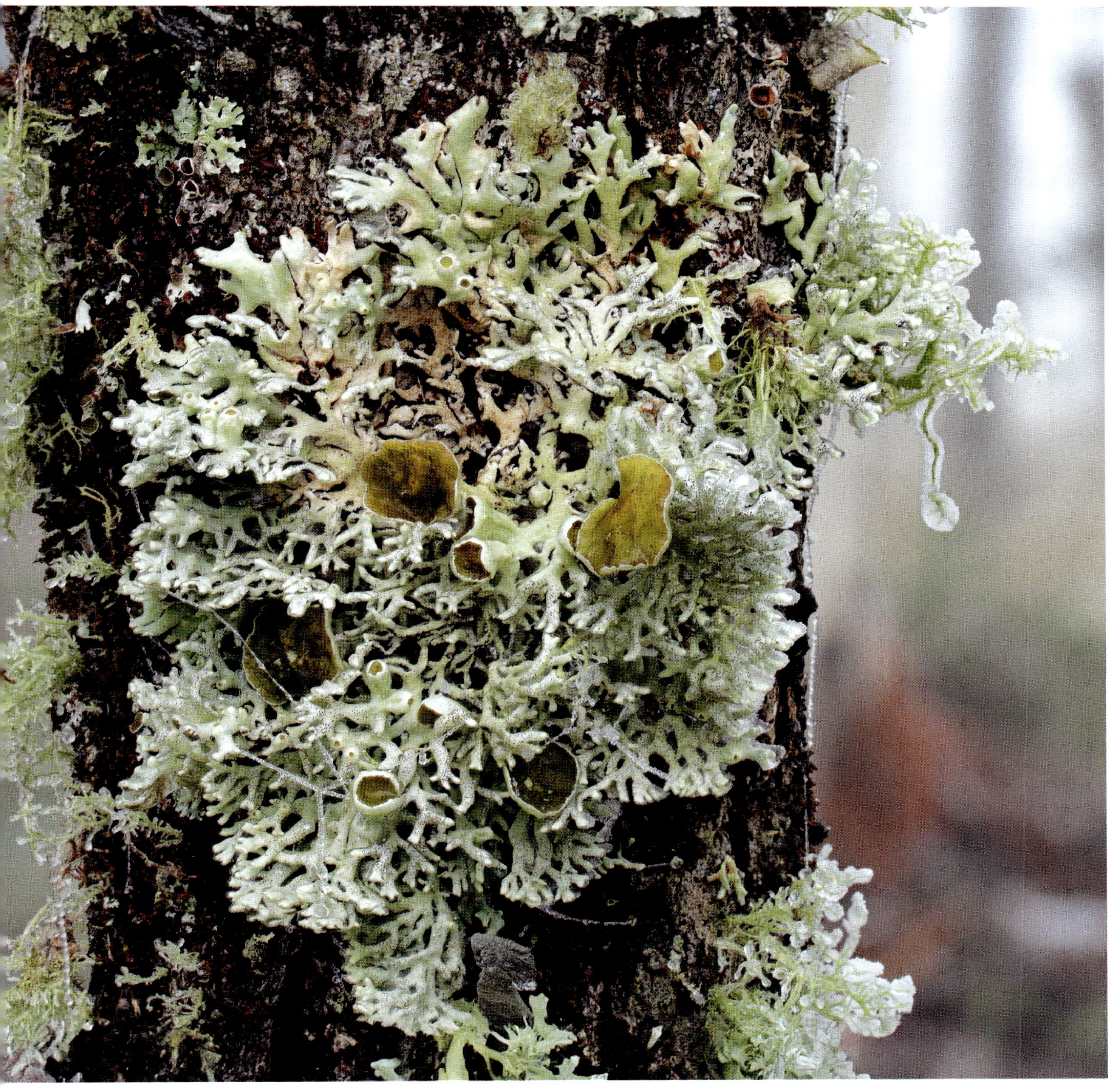

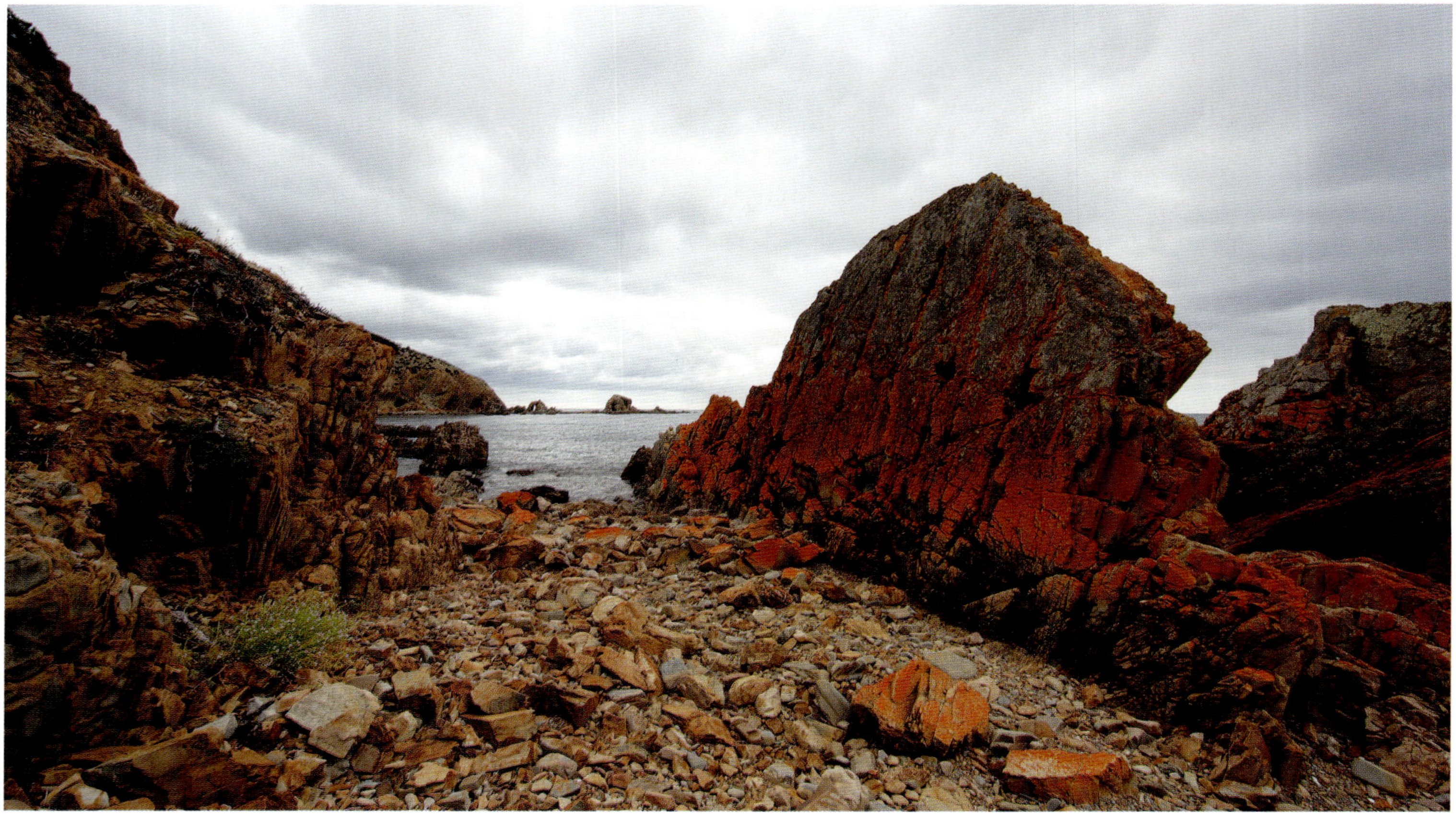

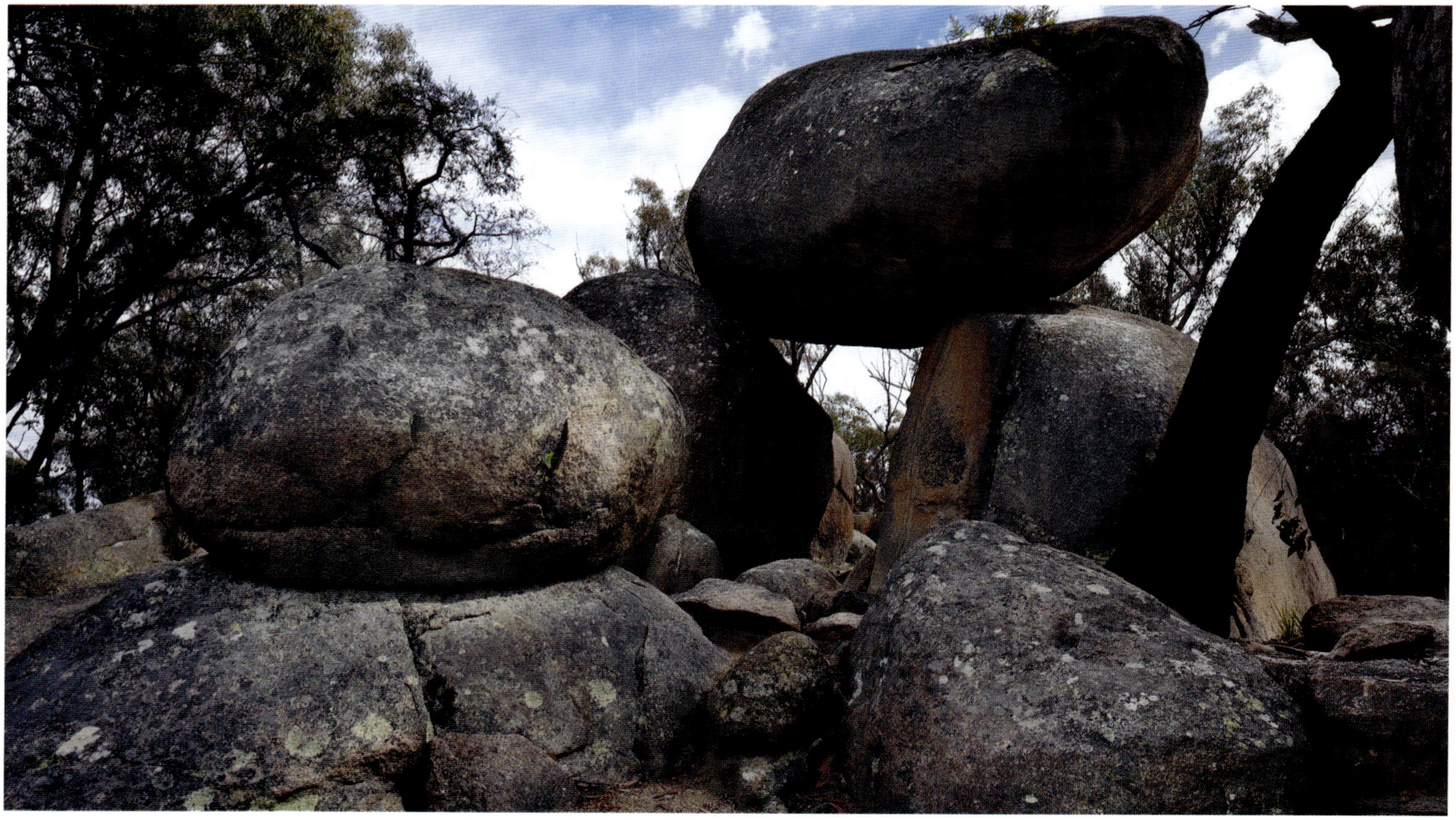

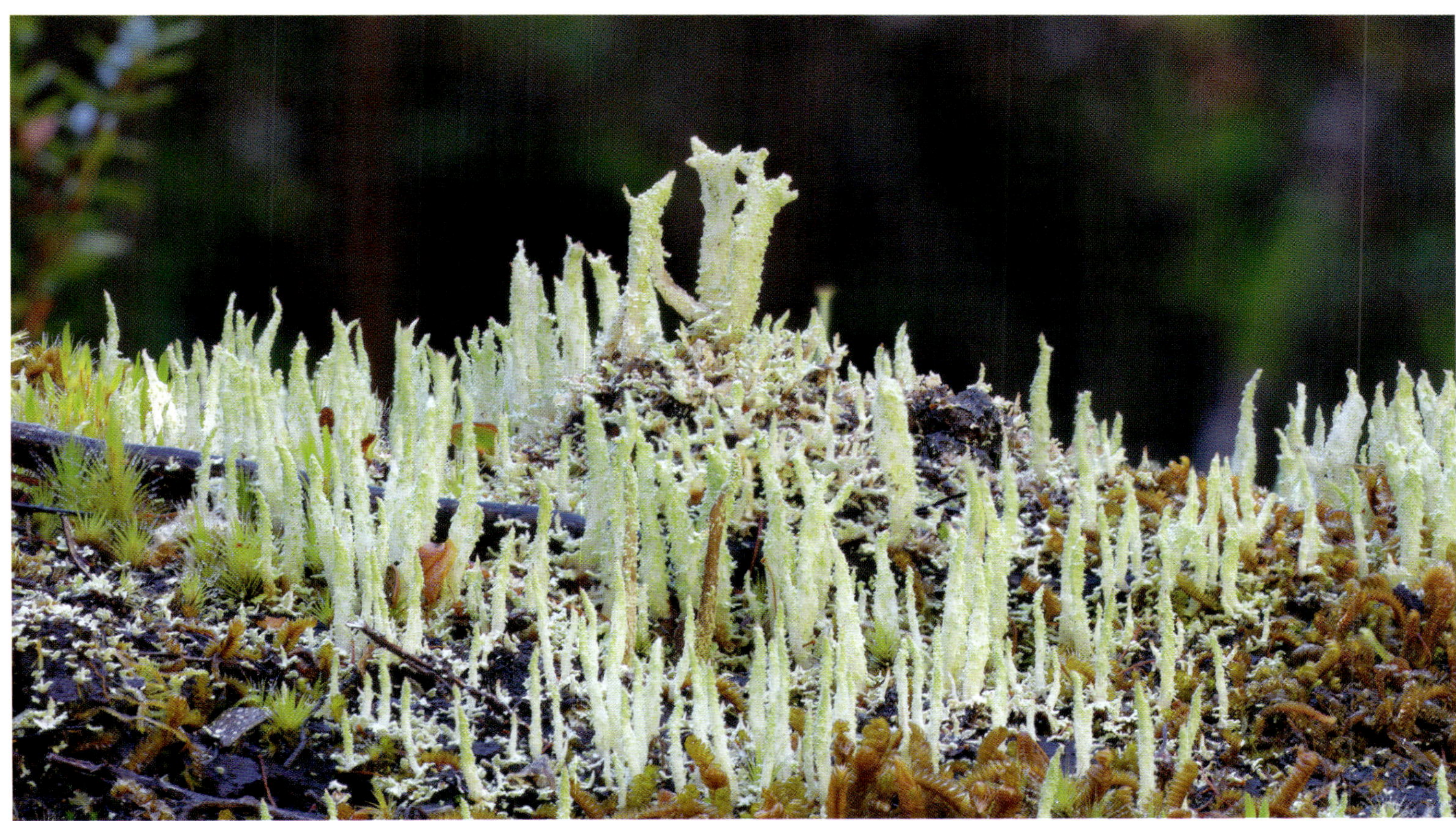

Stephen has also documented lichens clinging on the rocky shores of the Southern Ocean's islands, where neither fungus nor its photobiont could endure alone. Yet, in unison, they defy the relentless ocean sprays, the searing sun and the bitter howling winds. Their unique exchange of nutrients and water allows them to oscillate between desiccation and rejuvenation.

ABOVE
Cladonia sp.
Tantauco National Park, Chiloe, Chile

OPPOSITE TOP
Caloplaca gallowayi
King George Beach, South Australia, Australia
Distribution: Southern Australia

OPPOSITE BOTTOM
Over millions of years a combination of wind, water
and lichens have sculpted the Granite Arch.
Girraween National Park, Queensland, Australia

We have witnessed lichens' capacity to claim the cracks caused by cooling and heating and then use an arsenal of chemicals to help sculpt monumental arches from granite monoliths. It is a process that creates soil in places where lichens are often the first living things to grow.

In an abandoned mountain retreat, we stumbled upon a mechanical necropolis – a fleet of cars and buses disappearing under layers of lichens and mosses that were collecting moisture and nurturing corrosion. We wondered how long it would take for nature's demolition team to return metal to minerals and back to the earth.[16,17]

Tombstones in graveyards tell us tales of time. Stones that are two centuries old are adorned with a rich tapestry of lichen species, while newer markers only have a sparse scattering of fresh growth.

The crusty lichens that dot the Antarctic landscape are known for their incredibly slow development. In this extreme environment, some lichens may grow only 1 centimetre every hundred years, while others might take a thousand years to achieve the same feat.[18]

Lichens are the ultimate testaments to longevity, often outliving entire civilisations. In the frosty heart of Lapland, there is a humble crusty lichen – *Rhizocarpon geographicum*. This little survivor is ancient, around 9,000 years old.[19]

While humans were starting to dabble in agriculture, this lichen was already growing, and it is still going strong today. It is a living piece of history, quietly marking the passage of millennia on the rocks it calls home.

Sadly, we have also seen lichens disappear from a landscape.

OPPOSITE TOP
**Crusty and leafy lichens make their home
on a gravestone dated 1826.
Tenterfield Cemetery, New South Wales, Australia**

OPPOSITE BOTTOM
**Lichens and mosses colonise an old bus.
Salisbury, New South Wales, Australia**

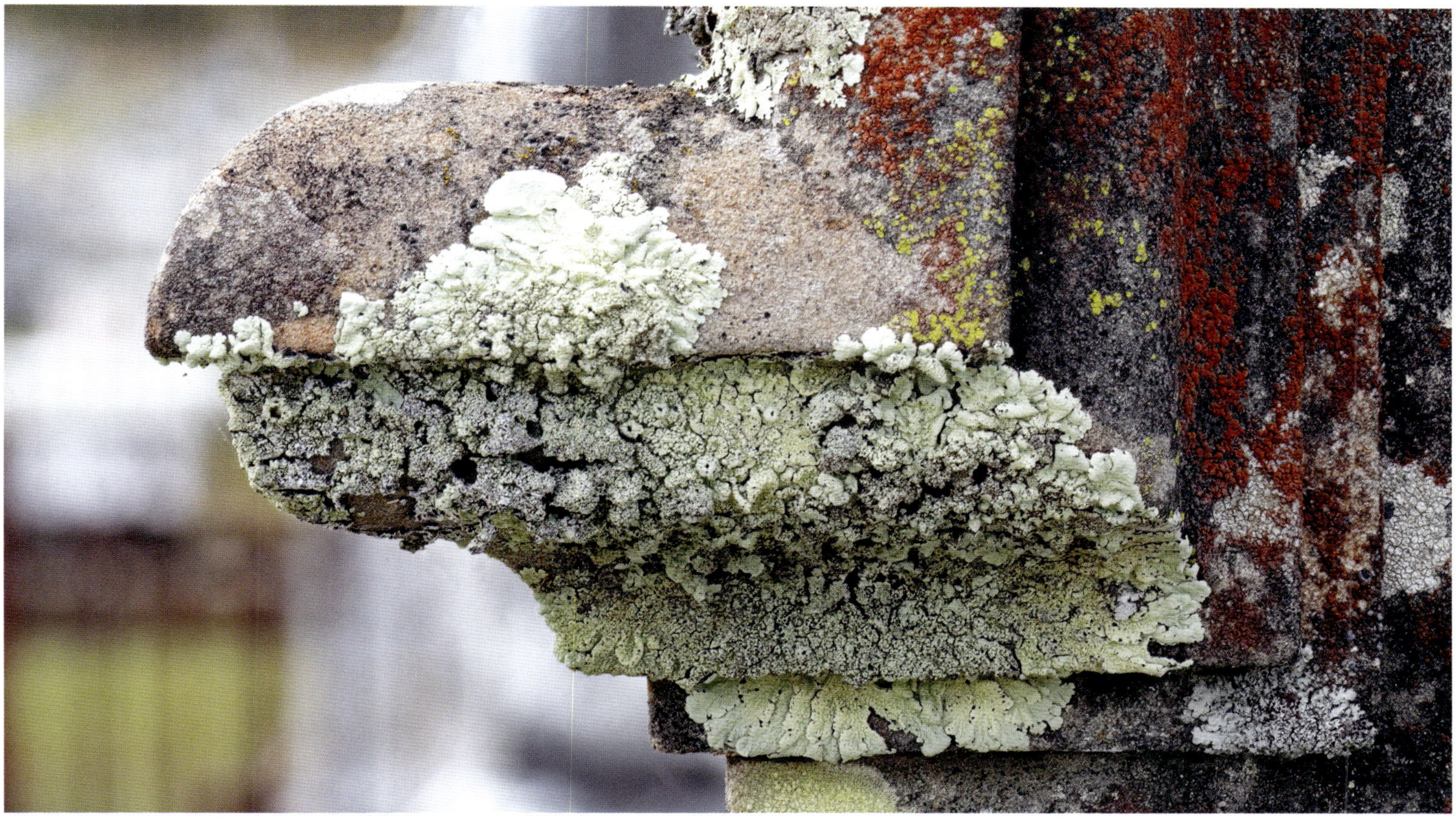

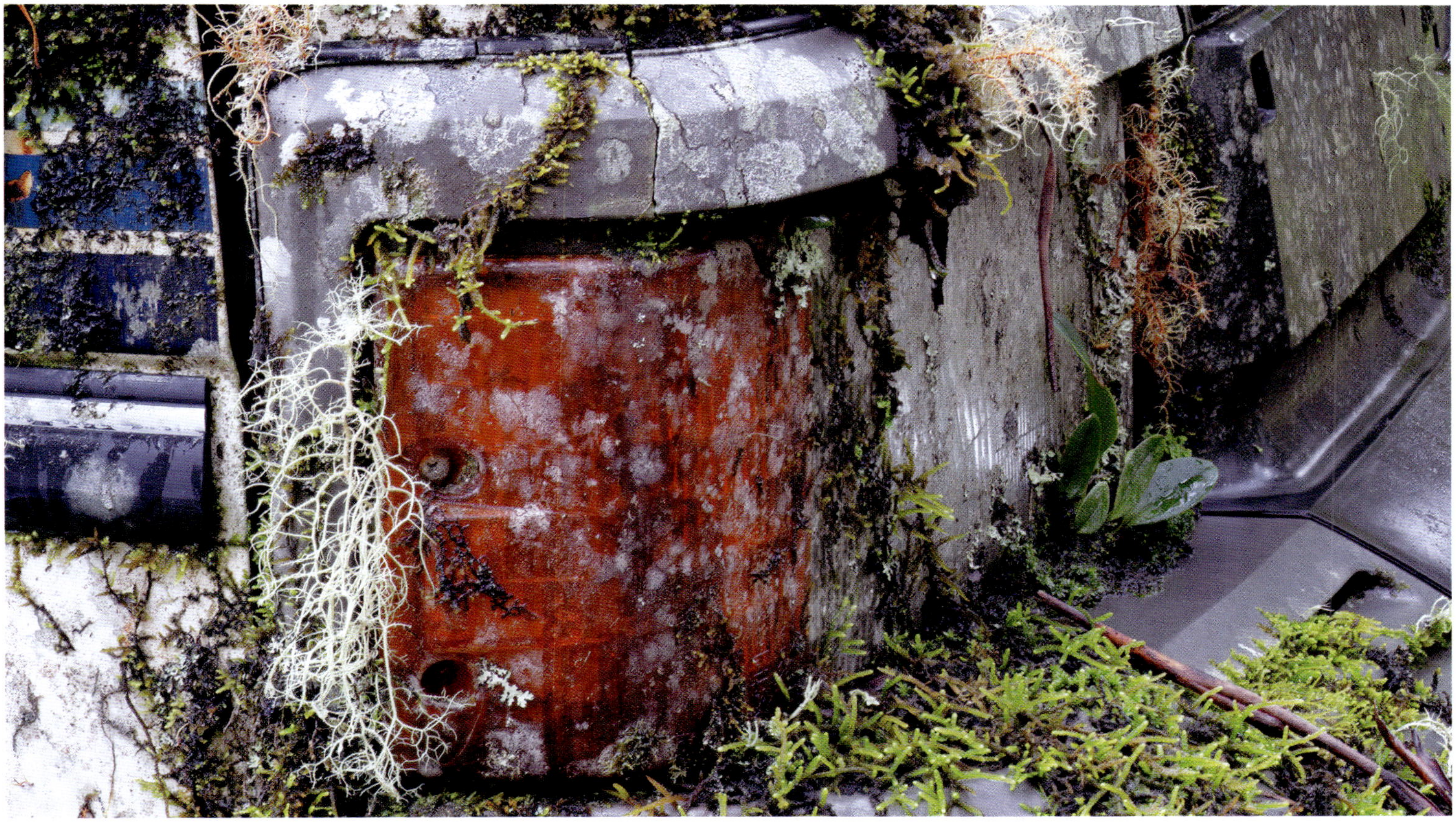

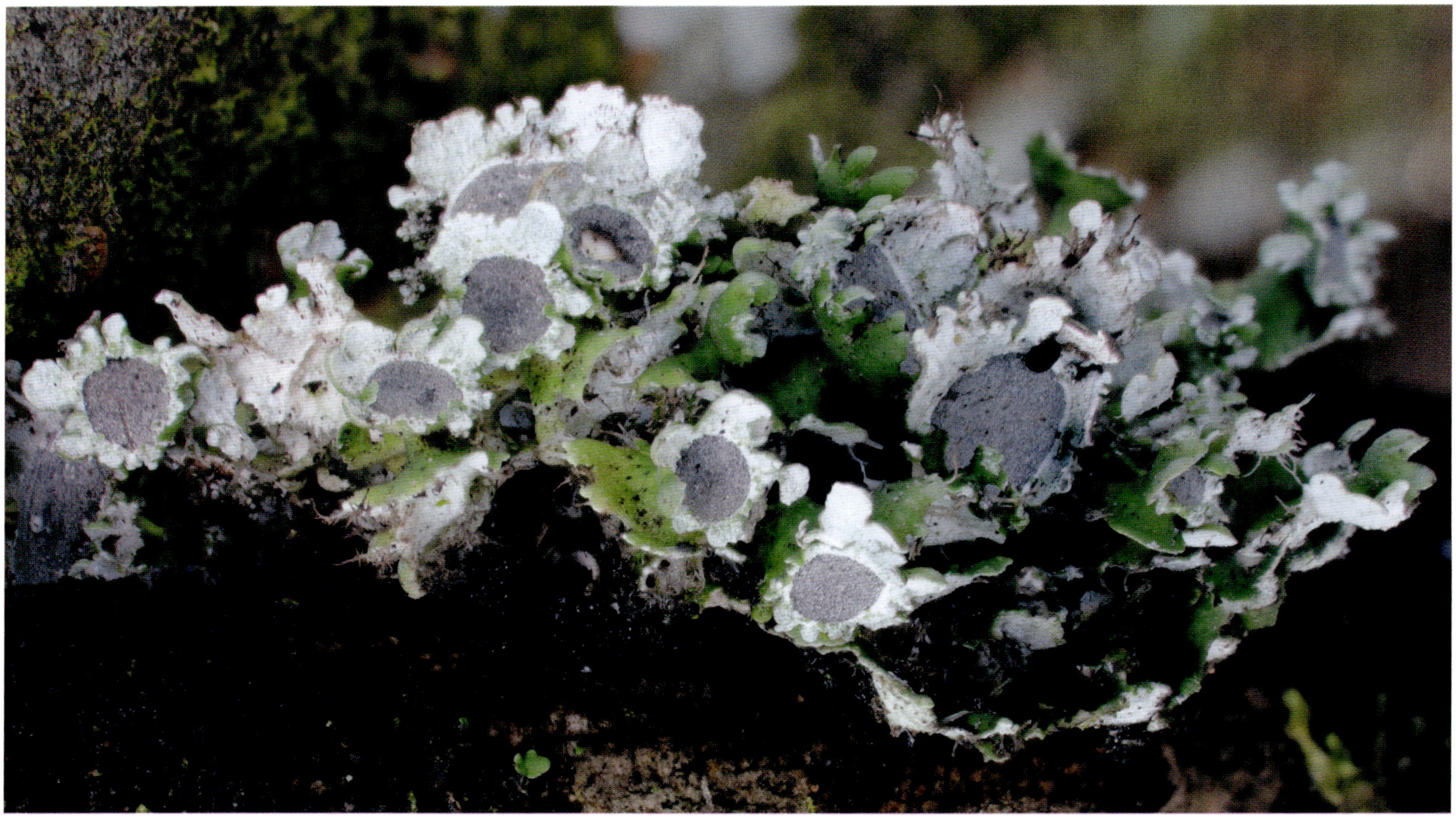

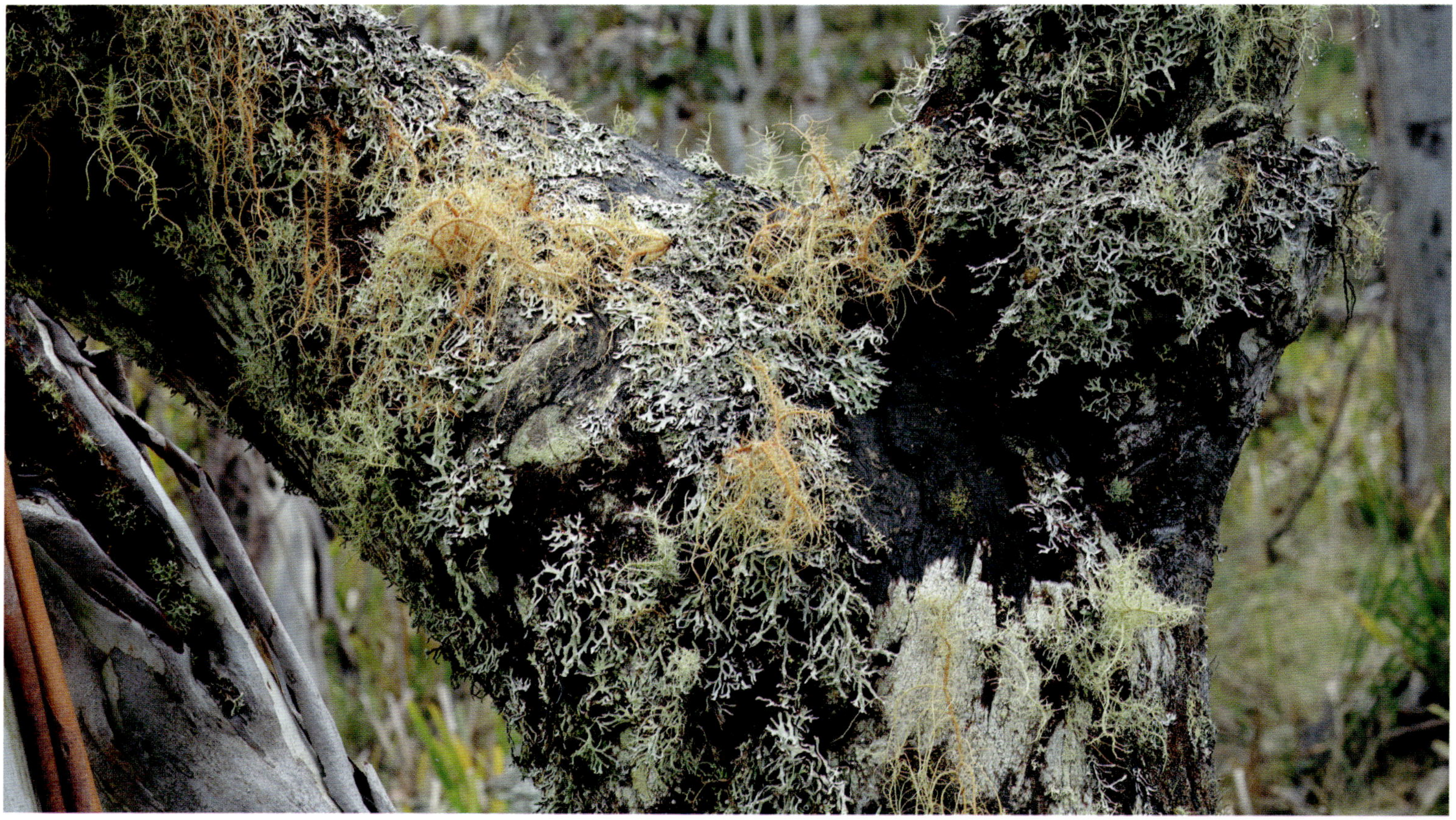

After the 2019 bushfires in eastern Australia, we have returned every year to a forest where slow-growing banksias and cypress pines were once festooned with bearded and leafy lichens. Like many of their kind, they seemed to prefer specific host trees, possibly finding the perfect home in the deep bark crevices that only occur on old trees.[20,21]

The fires decimated the cypress and banksias and their lichens, leaving only ash. Five years later, small seedlings are appearing, but the many beautiful lichens that used their parents as hosts are not present.

We hope they will recover eventually, but the increasing frequency of fires in this region does not bode well for this lichen community.

It warns us that an organism that is so resilient to extremes can also be fragile.

Lichens love clean air.

With no root system to draw up nutrients and water, they are incredibly adept at absorbing substances directly from the atmosphere.

When poisonous gases and heavy metal particles pollute the air, lichens respond. Indeed, since the 1960s, scientists have been using lichens as natural detectives in the quest to monitor air quality.

OPPOSITE TOP
Heterodermia barbifera
Ialong, Meghalaya, India
Distribution: tropical areas, mainly in Central and South America and Asia

OPPOSITE BOTTOM
Usnea rubicunda, red beard lichen, in a garden of lichens on an old
banksia that was severely burnt in the 2019 bushfires.
Boonoo Boonoo National Park, New South Wales, Australia

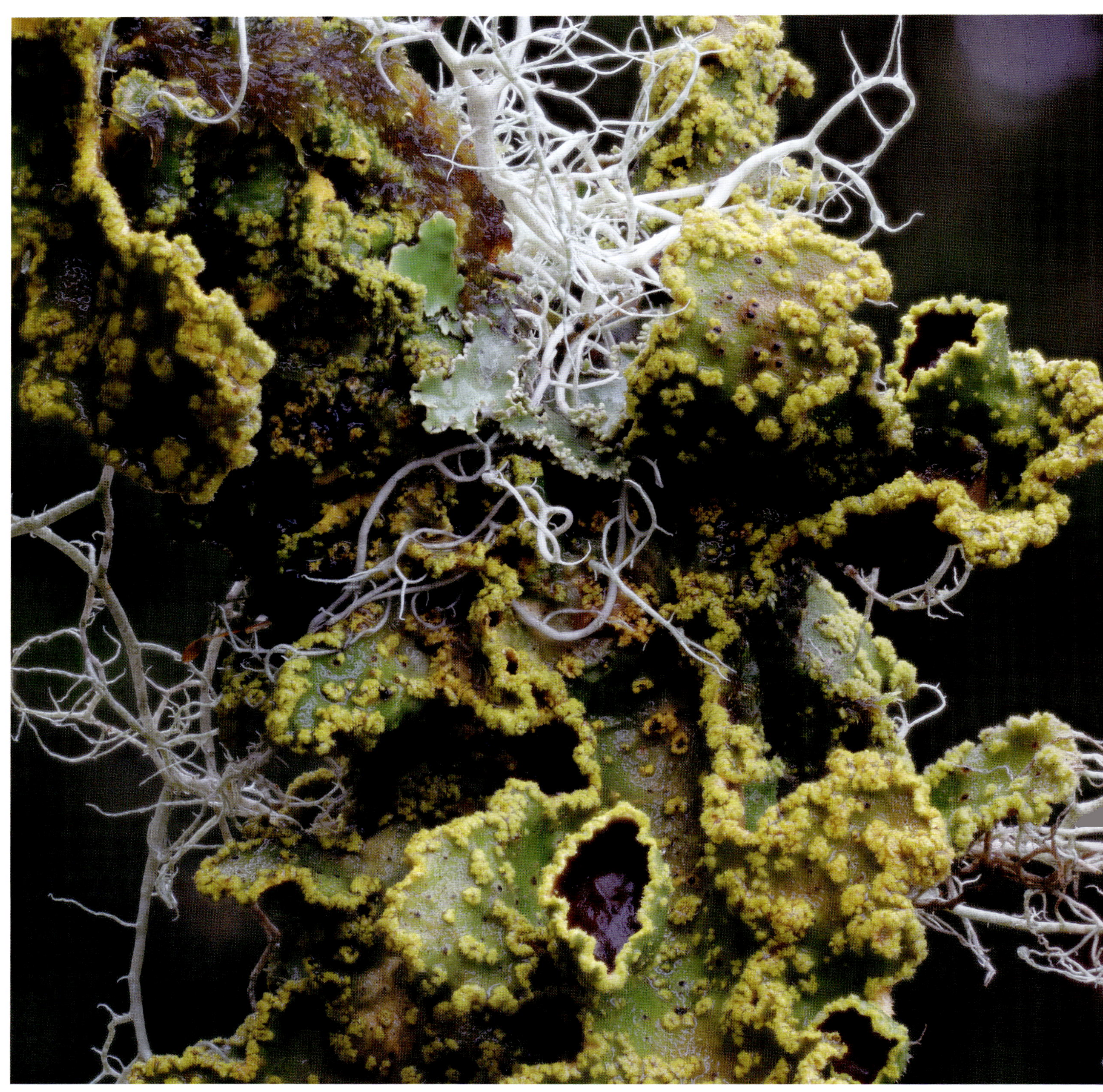

Lichens are like the canary in the mine; their health depends on the air's composition. Some perish in the presence of these contaminants, while others thrive.

Air monitoring records show lichens disappearing not only due to high pollution levels but also changing climates. For instance, in the United Kingdom, Europe and the Himalayas, researchers are finding that high alpine species that enjoy cool, wet weather are vanishing.[22,23,24]

'My curiosity about lichens began with my love affair with their diversity and the spectacle of their beauty.

'As the story of lichens has taken shape over the past couple of decades, I have become just as enamoured with how they live their lives.

'They're redefining our understanding of symbiosis, transcending simple ideas of predator and prey or mutual agreements between two partners, revealing more complex and cooperative interactions than we ever envisioned.

'Lichens challenge us to expand our scientific horizons and to safeguard these organisms that so elegantly teach us the art of coexistence.'

Crocodia rubella
Weindorfers Forest Walk, Tasmania, Australia
Distribution: Australasia

Safari to the end of the world

The China experience gave us a taste for a unique form of ecological, purposeful travel focused on the kingdom of fungi.

Once kindled, this newfound passion refused to be extinguished, and in 2016, we decided to attempt a similar adventure at 'the end of the world' in the rugged landscapes of Patagonian Chile.

We asked our now-expanding international fungal network, 'Is there anyone in Chile who would like to connect with us for a fungi foray?'

All fungal pathways led to one person – Giuliana Furci.

Giuliana is a self-taught mycologist whose fascination with fungi began in 1999. While studying aquaculture at the Universidad de Los Lagos, she and colleague Carolina Magnasco embarked on a journey across Chile.

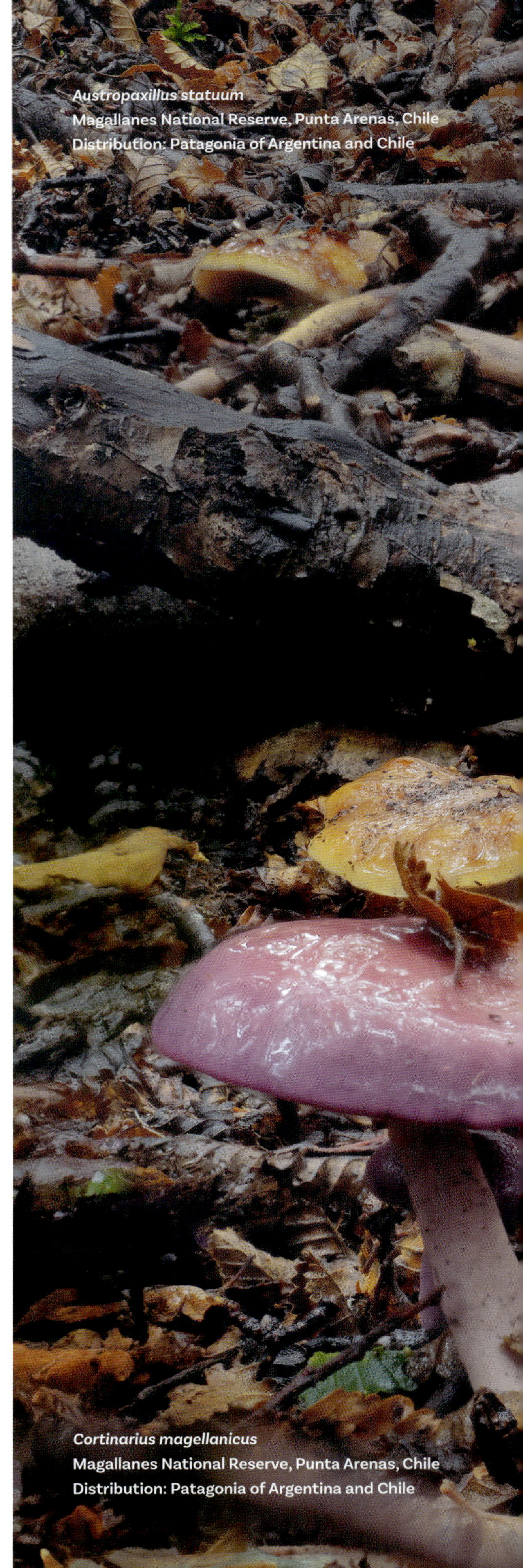

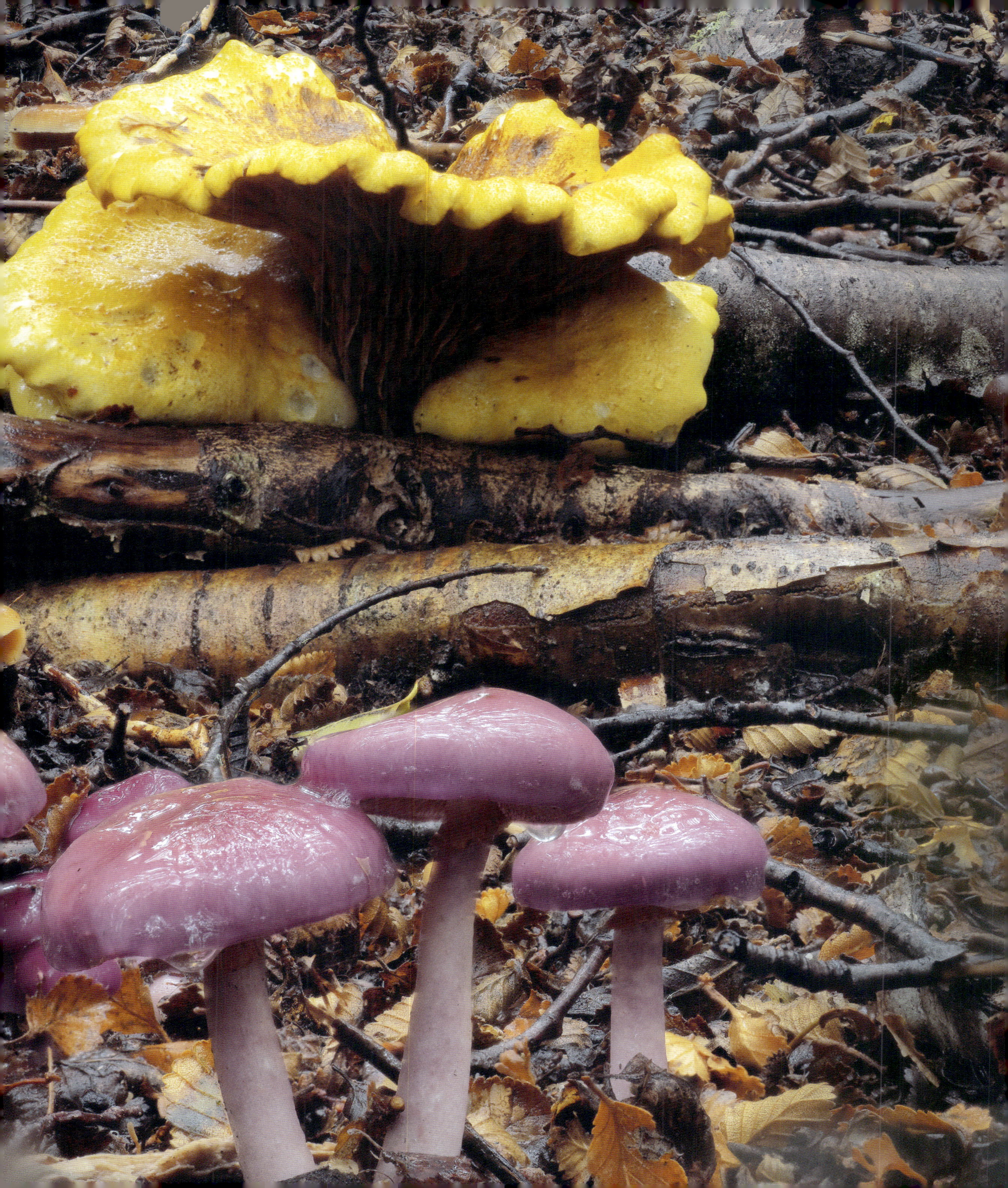

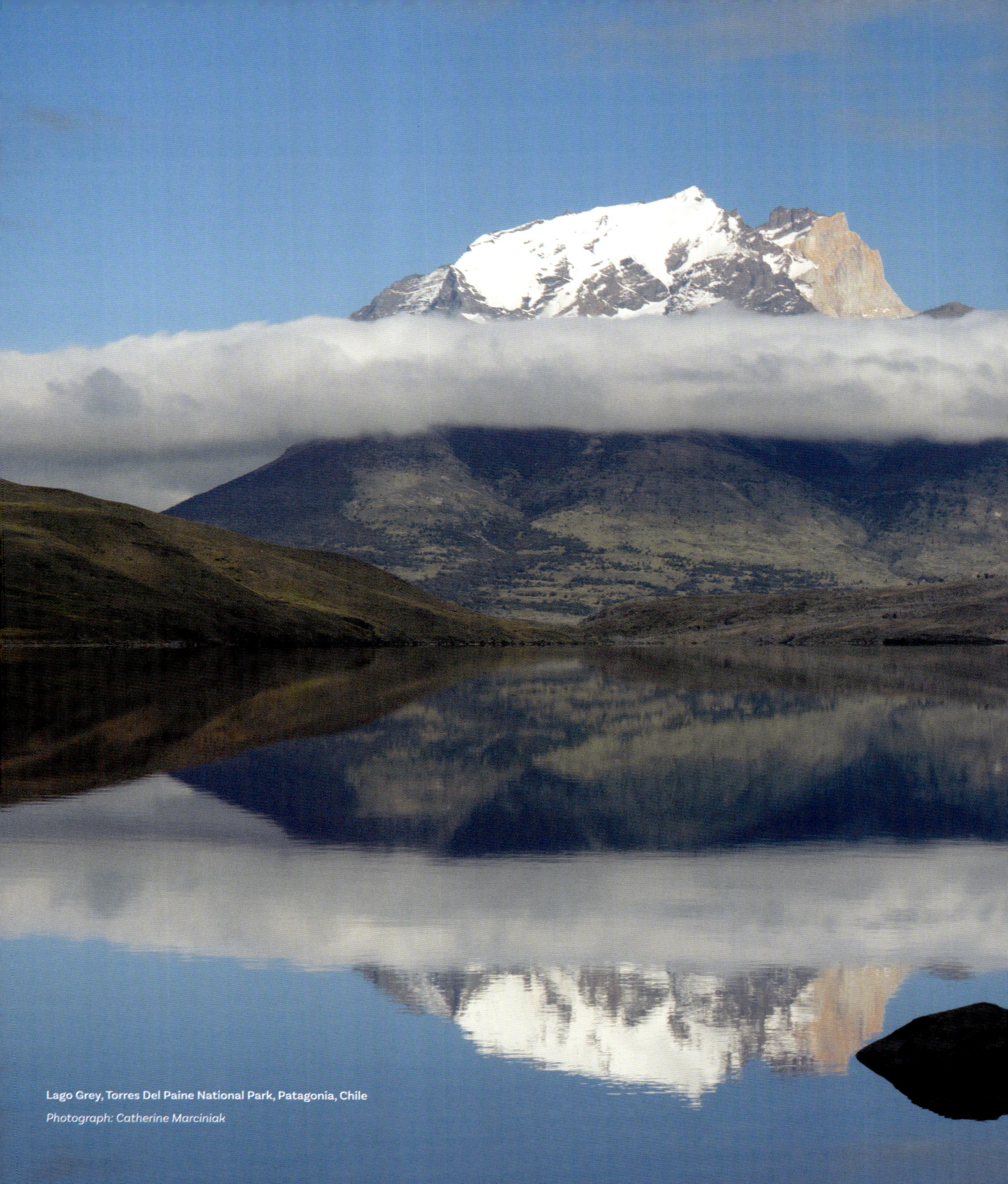

Lago Grey, Torres Del Paine National Park, Patagonia, Chile

Photograph: Catherine Marciniak

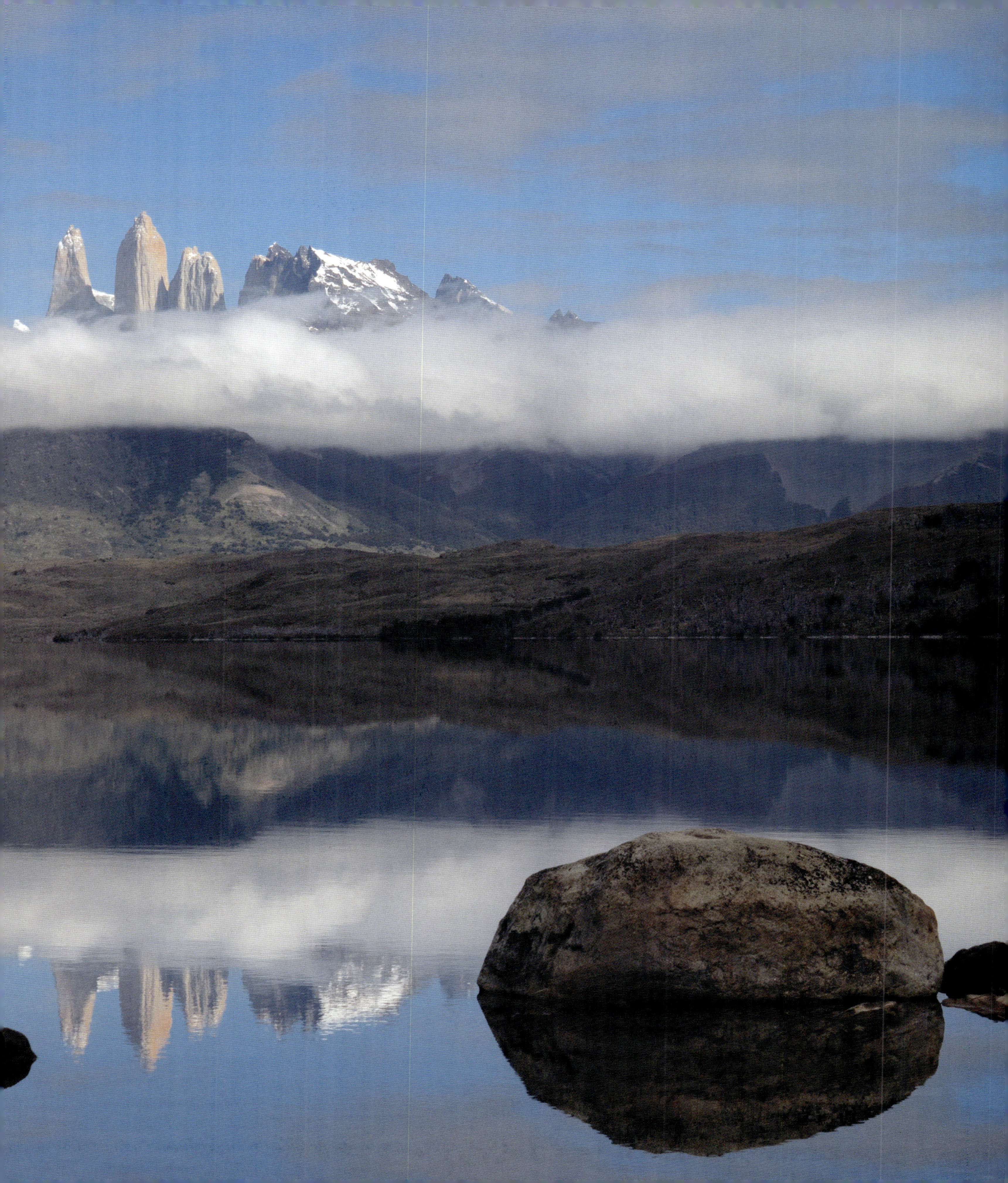

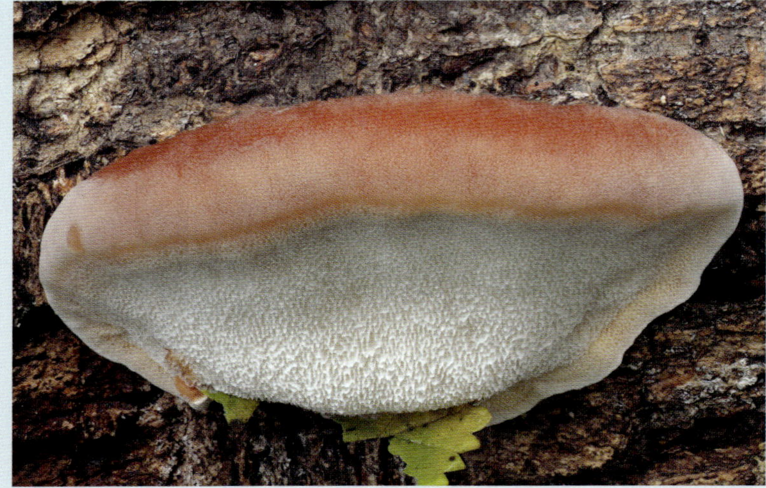

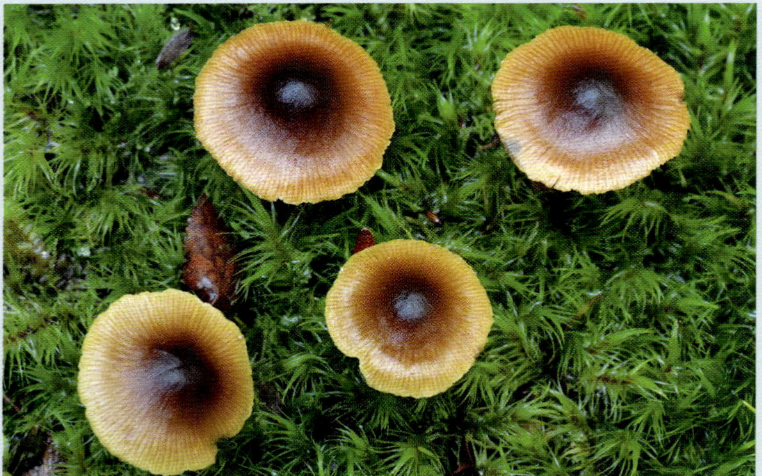

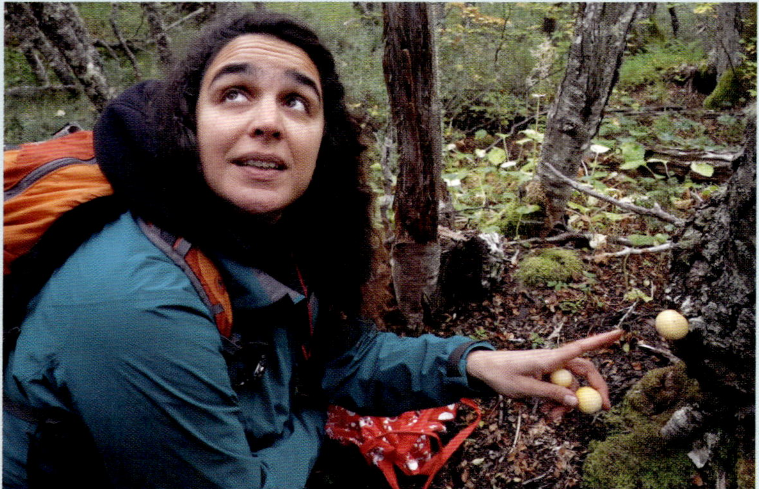

They stumbled upon a cornucopia of fungi. Their 6,000 photographs and hundreds of collections of fungal specimens birthed an all-consuming passion and a new project.

In 2007, Giuliana published Chile's first fungi guide with field photography, a remarkable achievement.

Her passion for fungi led her to establish the Fungi Foundation, a not-for-profit organisation that records and shares fungal knowledge and inspires the conservation of the fungal kingdom internationally.

Her work was instrumental in triggering the inclusion of fungi in Chilean environmental legislation in 2010, making Chile the first country to legally protect fungi.[1,2,3]

TOP
Postia sp.
Magallanes National Reserve, Punta Arenas, Chile

MIDDLE
Cortinarius sp.
Tantauco National Park, Chiloe, Chile

BOTTOM
Giuliana Furci, Fungi Foundation
Co-leader of the Patagonian fungi field trip

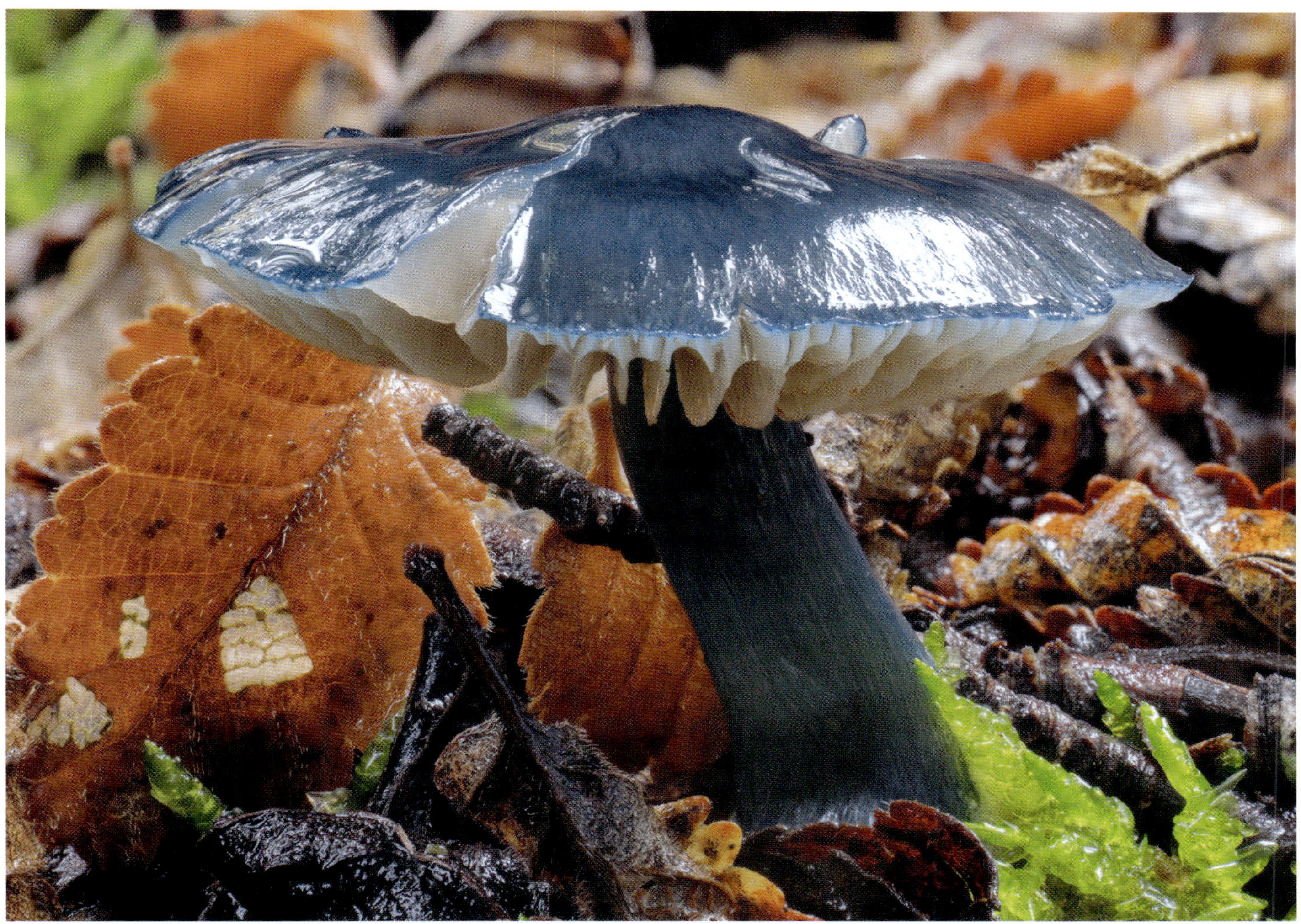

'*What that means is that Chile considers the impact on all living beings, including fungi, before undertaking any large projects like building a highway, dam, or housing development.*'
Giuliana Furci[3]

Today, Giuliana is one of the most influential advocates of fungi globally, and her enthusiastic passion for this under-recognised kingdom is infectious.

Entoloma sp.
Magallanes National Reserve, Punta Arenas, Chile

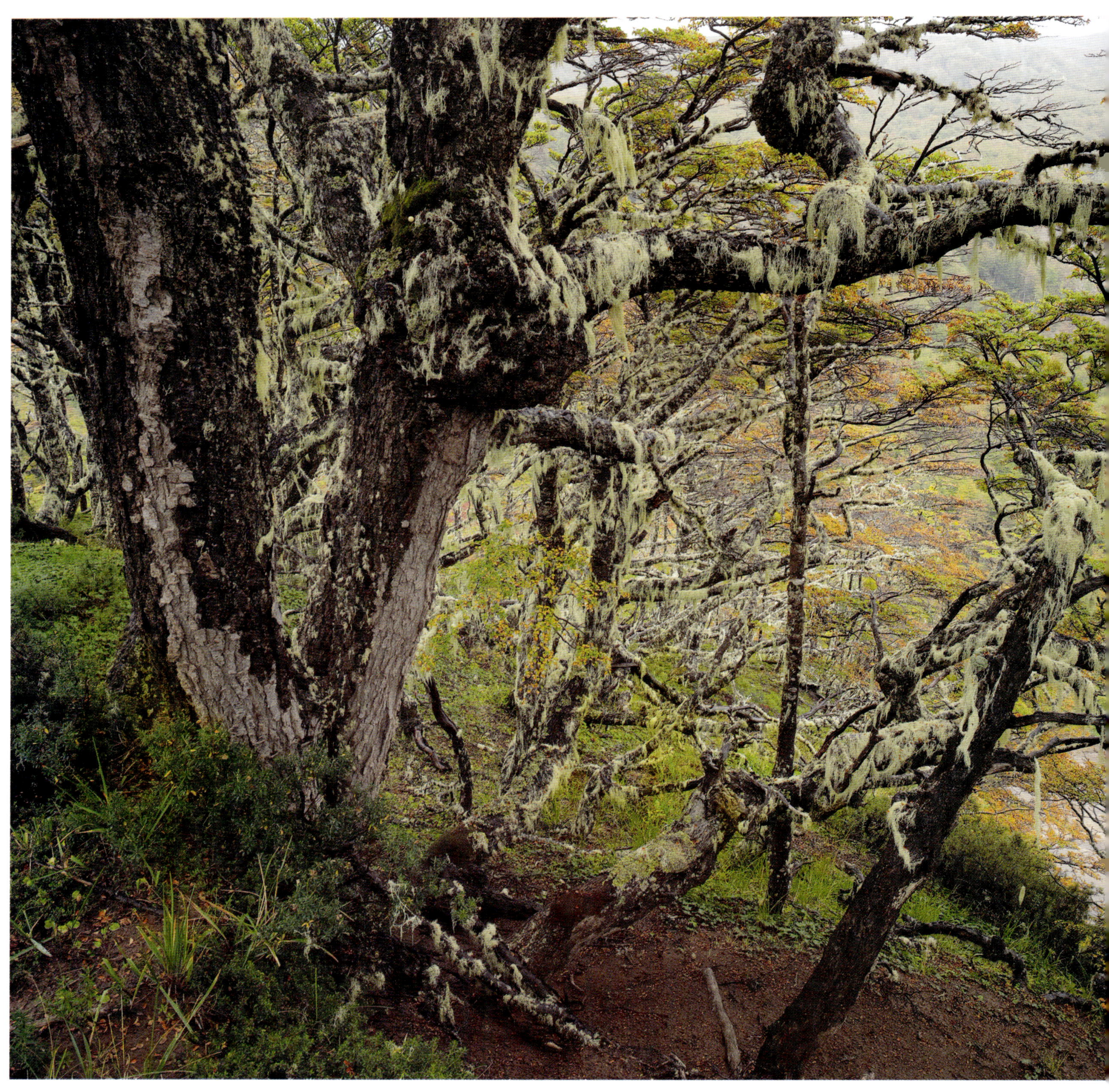

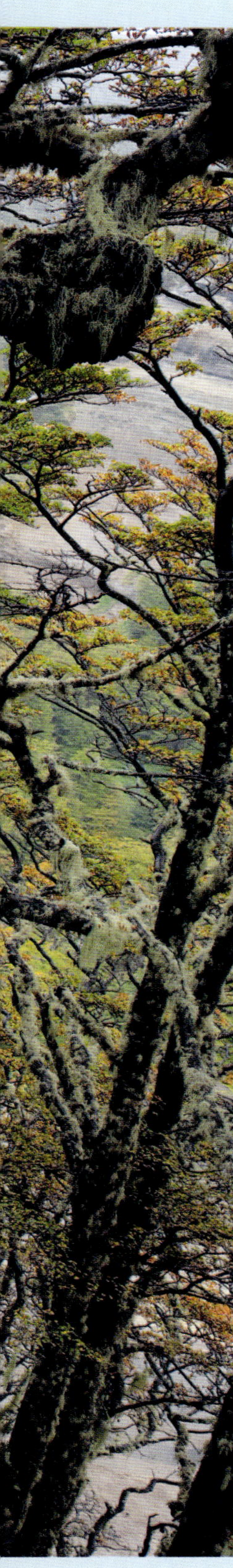

In 2016, Giuliana orchestrated an extraordinary opportunity for both Stephen and myself – an invitation to join a fungi field survey in the breathtaking landscapes of southern Chile.

This collaborative endeavour supported by funding from the US National Science Foundation, brought together scientists from Argentina, North America, and Chile.

Their shared mission was to meticulously document the rich biodiversity and evolutionary history of the ectomycorrhizal fungi entwined with the majestic *Nothofagus* trees in Patagonia.

The biogeography of this region bears the indelible marks of the breakup of the supercontinent Gondwana.

About 50–35 million years ago, Australia and South America were connected via Antarctica, allowing the ancestral Gondwanan biota to spread and diversify on each separate continent, laying the foundation for the unique ecosystems we see today.[4]

In the vast temperate rainforest of Patagonia, the *Nothofagus* trees reign supreme; their cousins are also found in Argentina, Australasia and New Caledonia.[4,5]

These forest relics, largely unexplored for fungi, present a tantalising new frontier for the fungi hunter.

Here mushrooms adorn a canvas of damp forest floors and moss-covered trees – a treasure trove awaiting discovery.[6]

**Autumn in the *Nothofagus* forests
of Magallanes National Reserve,
Punta Arenas, Chile**

When introduced to the field team, we found ourselves in the company of some of the world's most respected mycologists.

Professor Matthew Smith, who co-led the team with Giuliana, is the Curator of the Fungarium at the University of Florida.

Don Pfister, the Asa Gray Research Professor of Systematic Botany and Curator Emeritus of the Harvard University Herbarium, is considered a legend in the world of mycology.

Mycologist Tuula Niskanen, now the Senior Curator at the Finnish Museum of Natural History, specialises in studying the ectomycorrhizal genus *Cortinarius*, one of the most dominant fungal groups in the *Nothofagus* forests.[7]

Postdoctoral Researcher Alija Mujic, now an Assistant Professor at the California State University Fresno, planned to spend his time with knees pressed against the forest floor, meticulously scouring beneath the surface in search of the enigmatic and understudied truffles.[3]

These truffles were not the rare and coveted European culinary delicacies but rather elusive fungi that sustain the forest's creatures and remain largely unknown to science.

Ectomycorrhizal in nature, these fungi attached to roots form symbiotic relationships with trees and manifest in many colours and textures – from the rough and warty to the soft and squishy.

OPPOSITE
Cortinarius sp.
Magallanes National Reserve, Punta Arenas, Chile

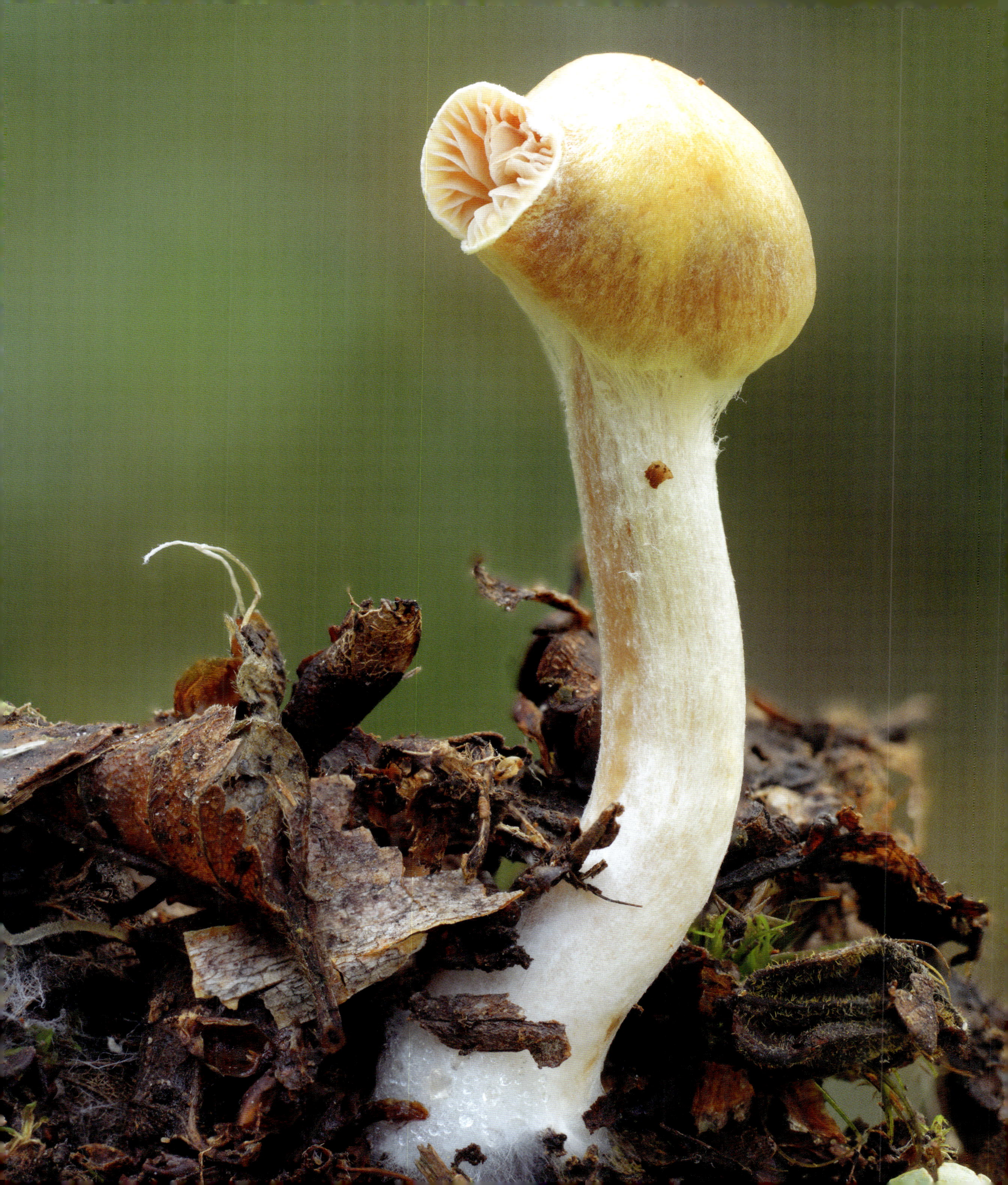

We would spend five days with the survey team, exploring and learning about the fungal wonders of southern Patagonia and the Lake District.

This remote corner of the world and its untamed wilderness proved incredibly biodiverse for fungi.

In the autumnal landscape, the *Nothofagus* trees were shedding their leaves, painting the forest with shades of golden yellow and fiery red.

The moist and chilly weather created the perfect conditions for fungi to flourish.

We saw mushrooms of different shapes, sizes and colours everywhere.

There were familiar faces, akin to their relatives in the Australian landscapes.

Stephen spotted a distinctive blue *Mycena cyanocephala*, a cousin of the stunning *M.interrupta* that grows in the southern parts of Australia.

OPPOSITE TOP
Polyporus gayanus
Parque Nacional Villarrica, Pucón, Chile
Distribution: Australia and South America

OPPOSITE BOTTOM LEFT
Mycena cyanocephala
Parque Nacional Villarrica, Pucón, Chile
Distribution: Central and South America

OPPOSITE BOTTOM RIGHT
Mycena interrupta, **pixie's parasol**
Dip Falls, Tasmania, Australia
Distribution: Australasia

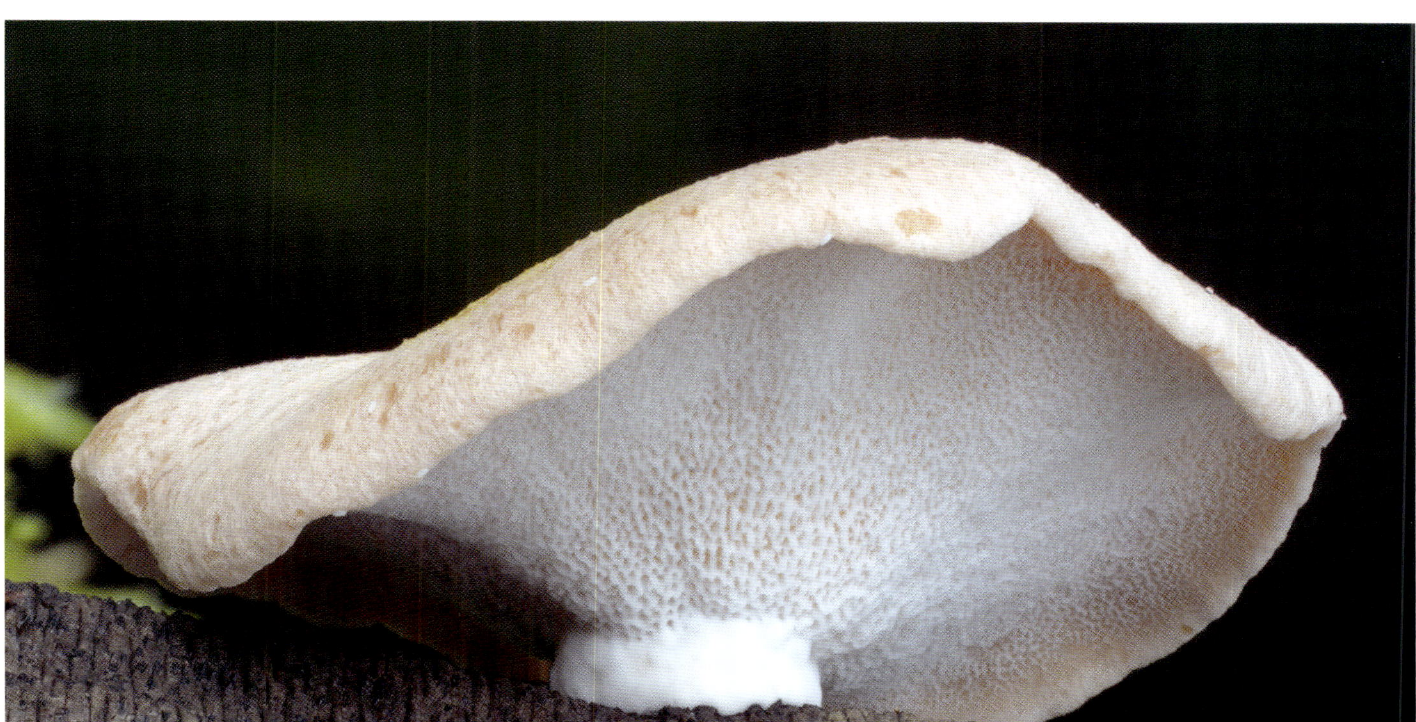

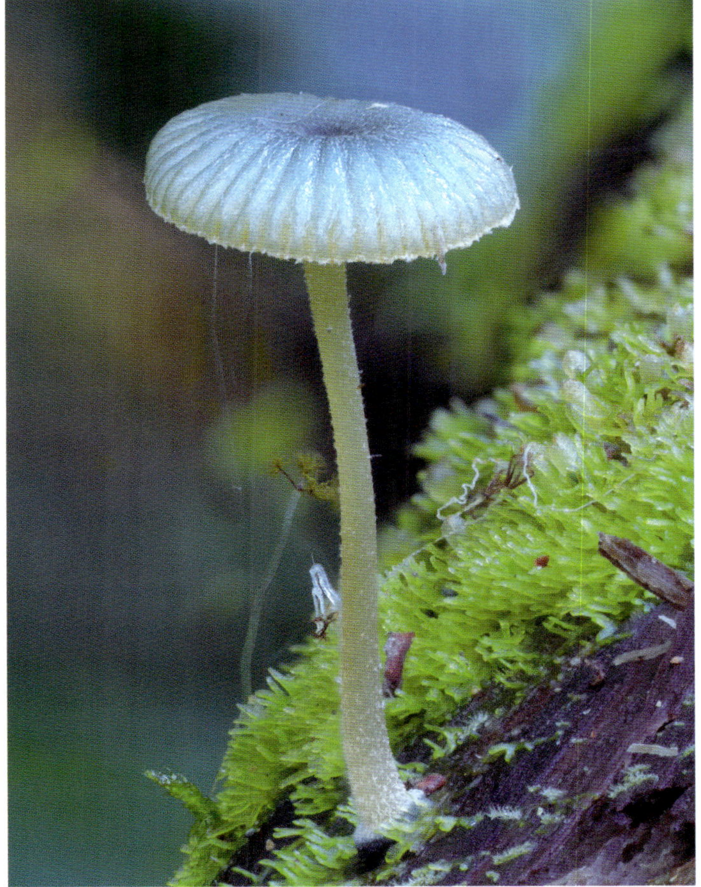

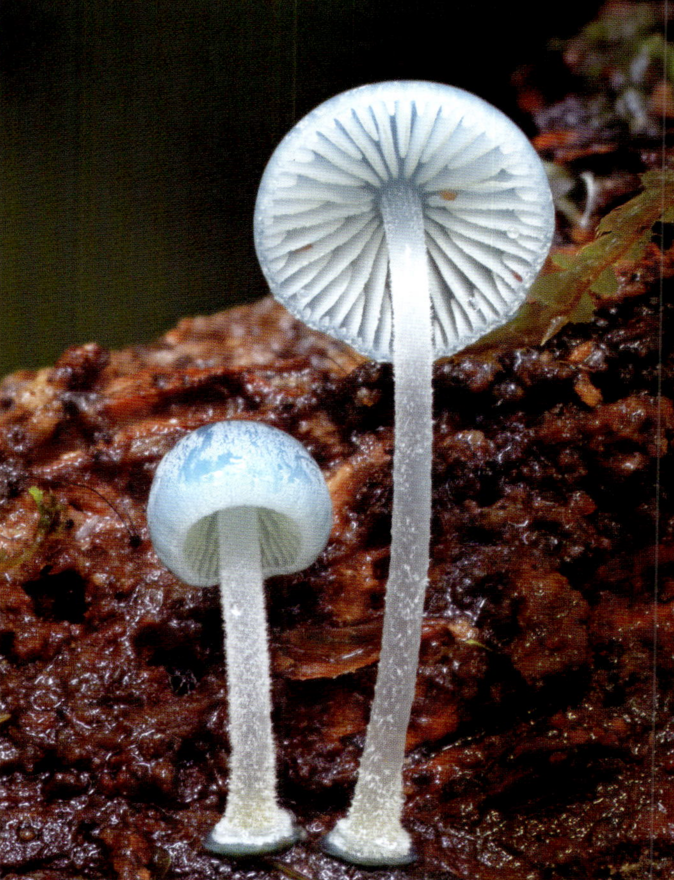

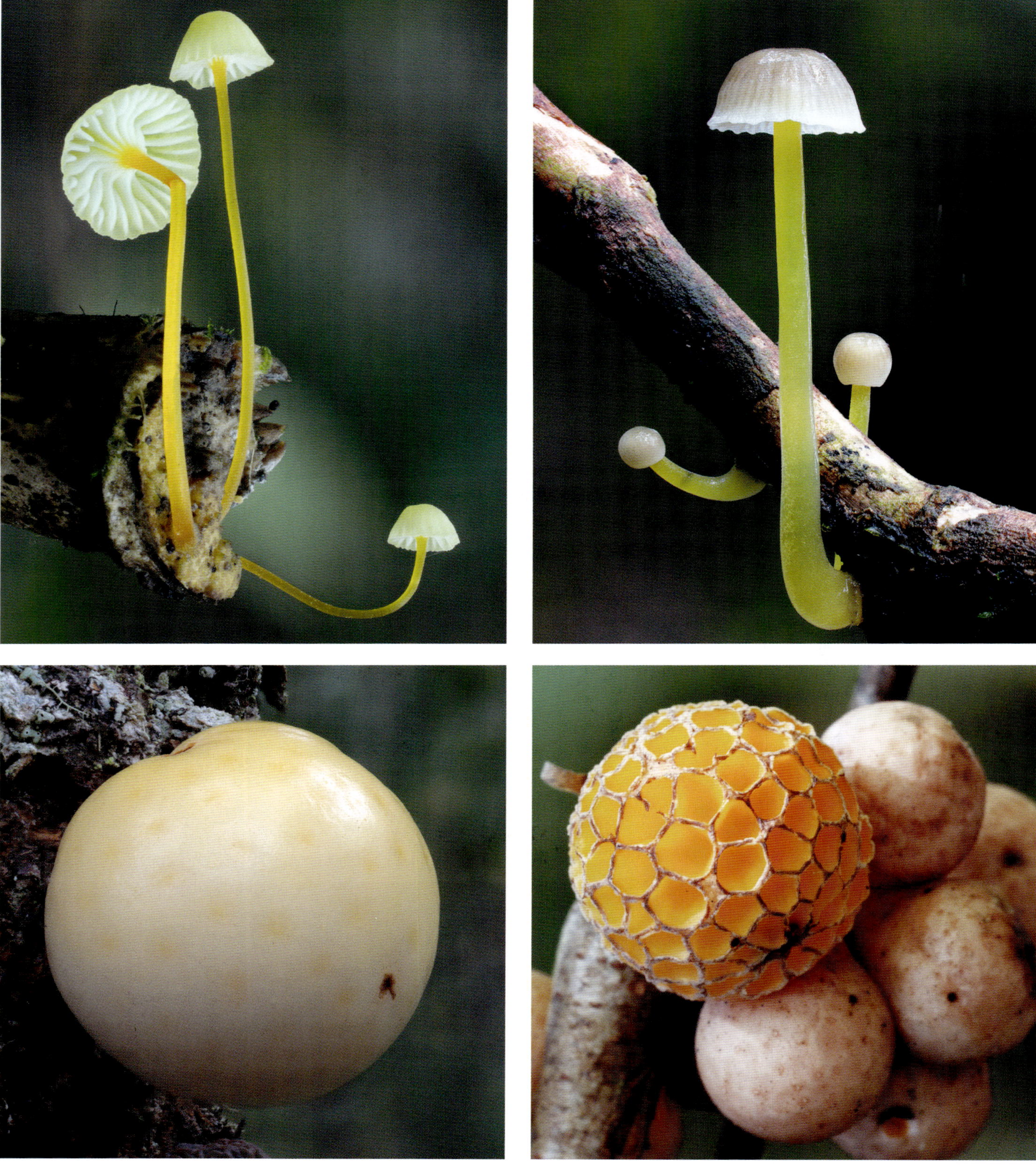

He also recognised the delicate, translucent *Mycena chusqueophila*, reminiscent of the Australasian *M. subviscosa*.

Among the most intriguing fungal organisms we encountered were two species of *Cyttaria*, a genus of parasitic fungi that grows exclusively on living *Nothofagus* trees.

Cyttaria does not harm its host tree but can induce the growth of large galls upon its branches.

The galls become nurturing shelters, providing both nutrients and protection to the fungus within. From these peculiar galls emerge spherical fruiting bodies, resembling vibrant yellow or orange golf balls.

The Patagonian *Cyttaria darwinii* strongly resembles its cousin *C. gunnii*, which grows on *Nothofagus* trees in Australia.

The dried specimen of *C. darwinii* housed in the fungarium at Royal Botanic Gardens, Kew, London, was collected by none other than Charles Darwin himself. It is one of the earliest fungal specimens in the collection.[8]

OPPOSITE TOP LEFT
Mycena chusqueophila
Parque Nacional Villarrica, Pucón, Chile
Distribution: southern South America

OPPOSITE TOP RIGHT
Mycena subviscosa
Queenstown, Tasmania, Australia
Distribution: Australasia

OPPOSITE BOTTOM LEFT
Cyttaria darwinii, Darwin's fungus
Magallanes National Reserve, Punta Arenas, Chile
Distribution: southern South America

OPPOSITE BOTTOM RIGHT
Cyttaria gunnii, beech orange
Philosopher Falls, Tasmania, Australia
Distribution: Australasia

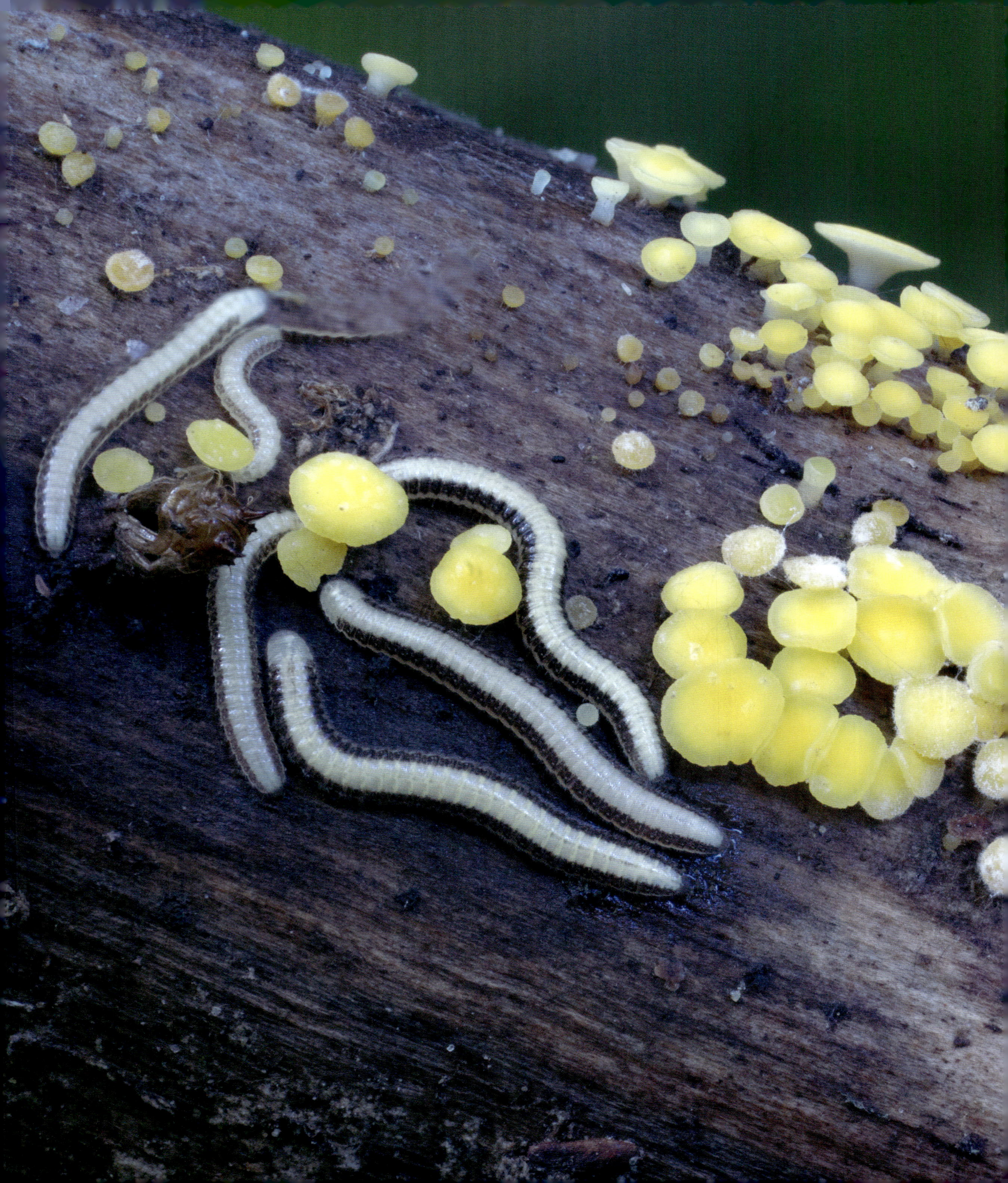

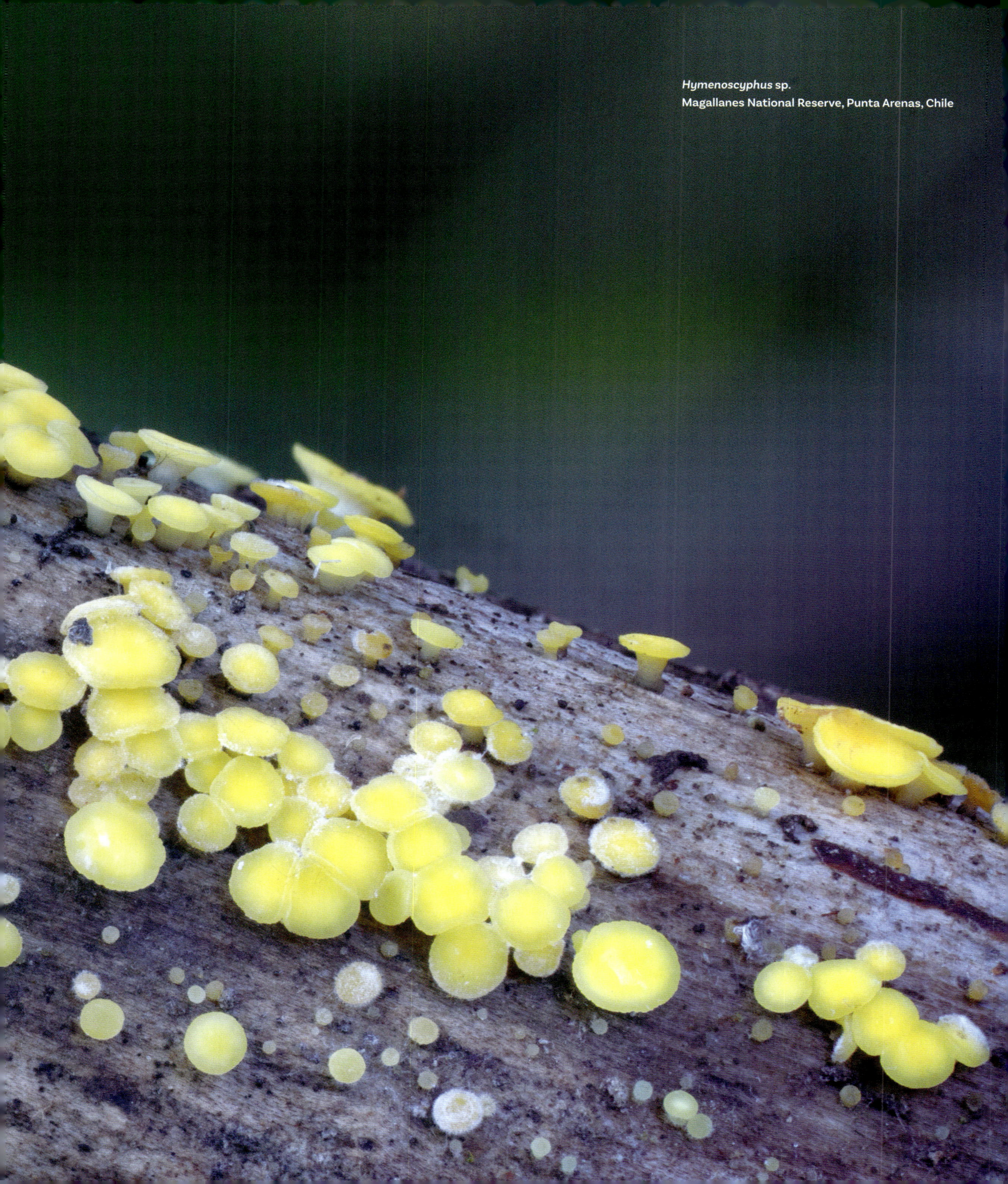

Hymenoscyphus sp.
Magallanes National Reserve, Punta Arenas, Chile

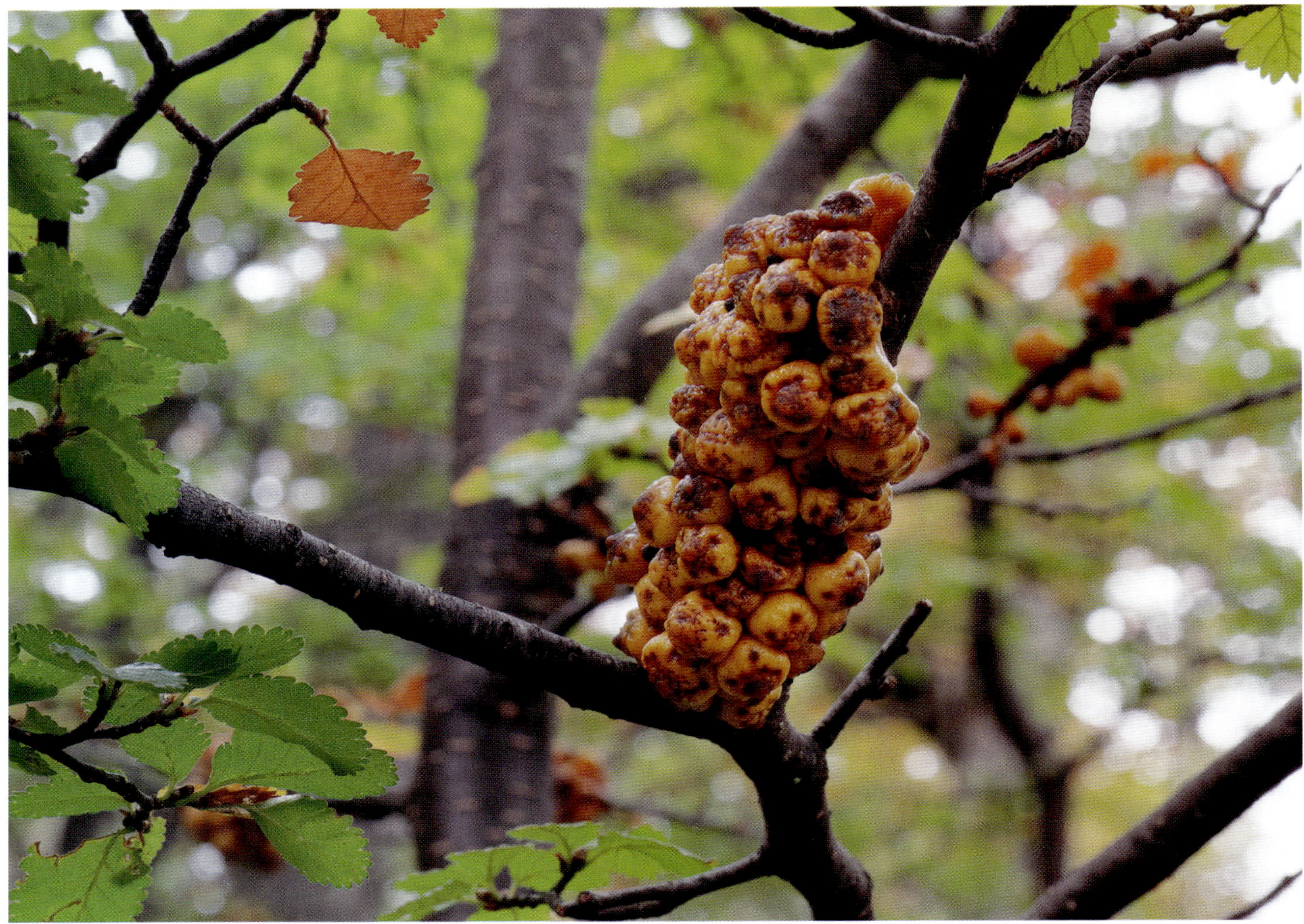

In December 1831, the recently graduated naturalist Charles Darwin set sail on H.M.S. *Beagle* on what is lauded as one of the most significant voyages of exploration in maritime history.

When this survey expedition landed in Tierra del Fuego, Darwin observed *Cyttaria* growing in abundance.

In his journal, he wrote that the Indigenous Yagan and Mapuche people 'appear to think these excrescences … an estimable dainty'.[9,10,11]

Cyttaria johowii
Magallanes National Reserve, Punta Arenas, Chile
Distribution: southern South America

'Giuliana told me the taste of Cyttaria darwinii *is reported to be like a sweet jelly, but the one I sampled in Patagonia would need more sugar and spice to satisfy my sweet tooth.'*

Many mushrooms were new to us, such as the tiny orange prongs of a *Pseudomitrula* and the 'earth tooth' *Geomorium singeri*, found deep within the leaf litter.[12,13]

TOP
Pseudomitrula **sp.**
Magallanes National Reserve, Punta Arenas, Chile

MIDDLE
Geomorium singeri, **earth tooth**
Magallanes National Reserve, Punta Arenas, Chile
Distribution: southern South America

BOTTOM
Cortinarius **sp.**
Magallanes National Reserve, Punta Arenas, Chile

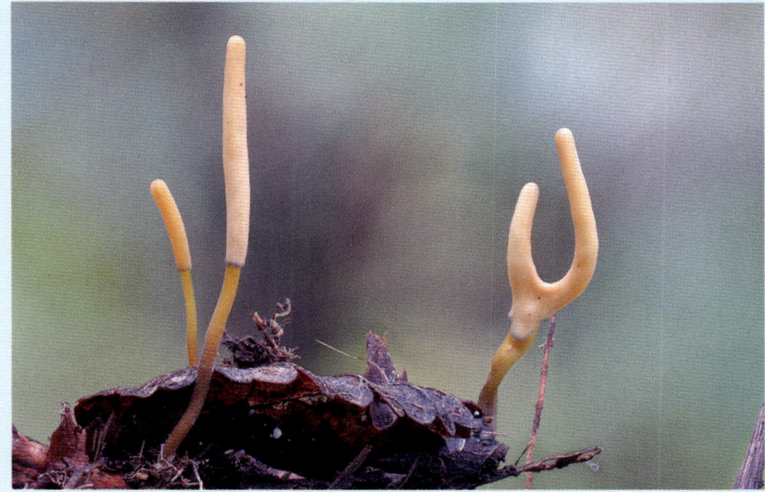

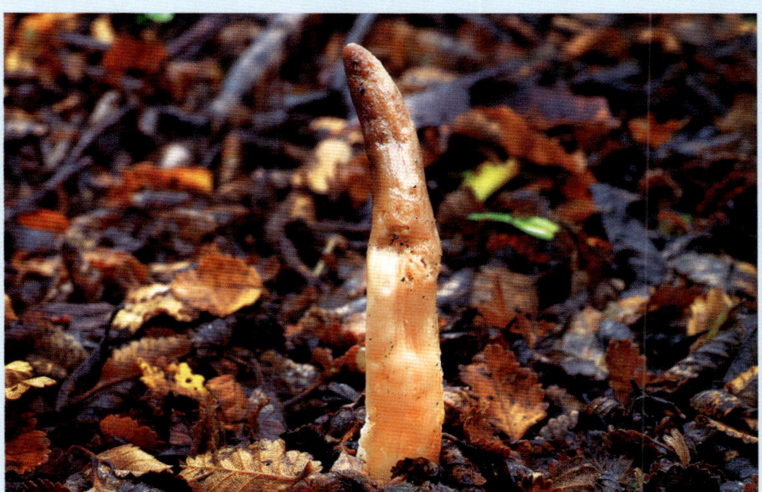

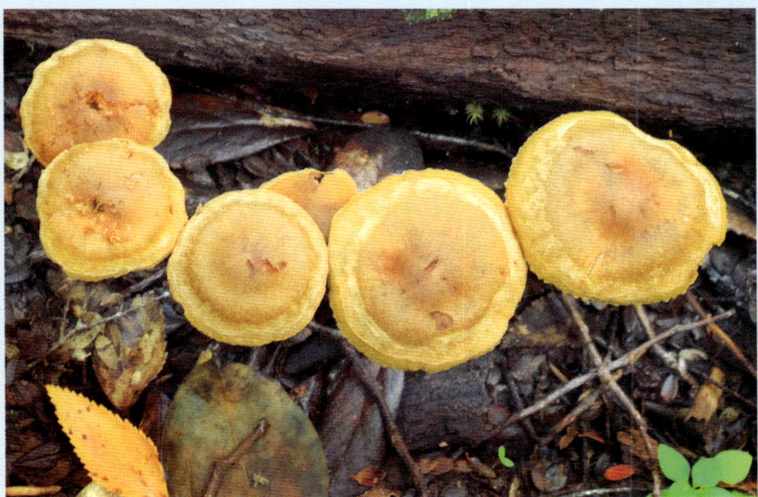

We were also astonished at the vast number of *Cortinarius* species emerging through the carpet of *Nothofagus* leaves.

Stephen became enamoured with the vibrant purples of *Cortinarius magellanicus*.

Cortinarius mushrooms have some distinctive traits that run in the family.

When they are young, they have a striking fibrous veil that covers their developing gills. As the mushroom matures, this veil falls, leaving behind a glimpse of its remains tethered onto the stipe. The remnants are ephemeral and can quickly disappear. These unique characteristics make *Cortinarius* mushrooms a beautifully layered character for the fungi photographer.

Tuula Niskanen predicted that between 200 and 300 species of the *Cortinarius* genus could be hidden throughout the vast southern *Nothofagaceae* forests.[3,14]

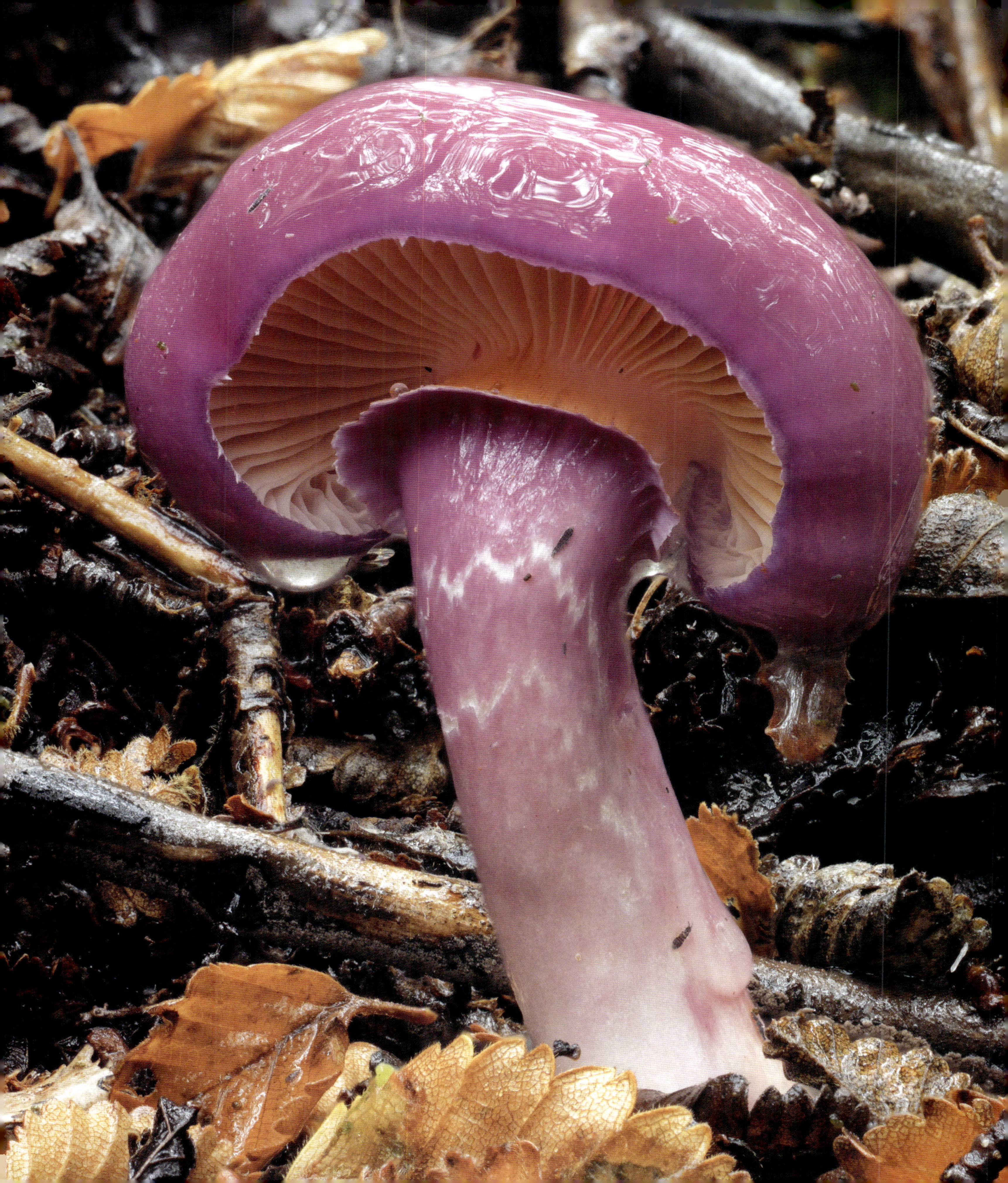

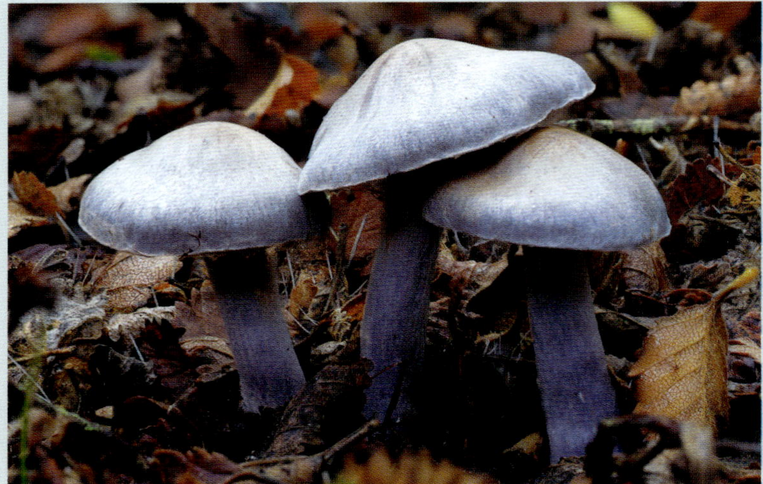

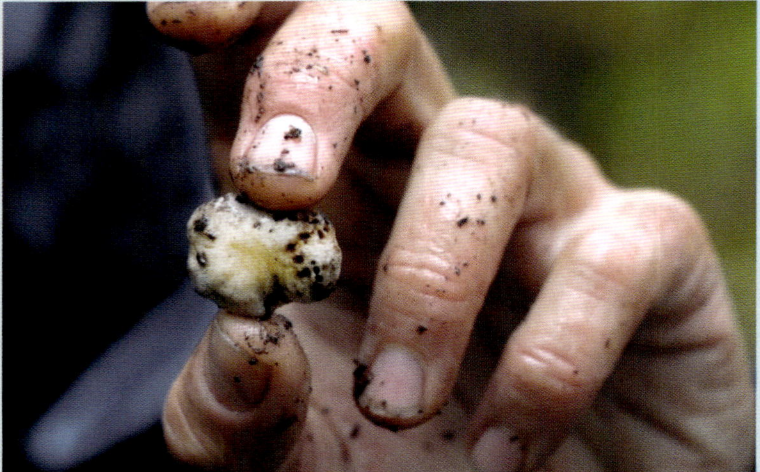

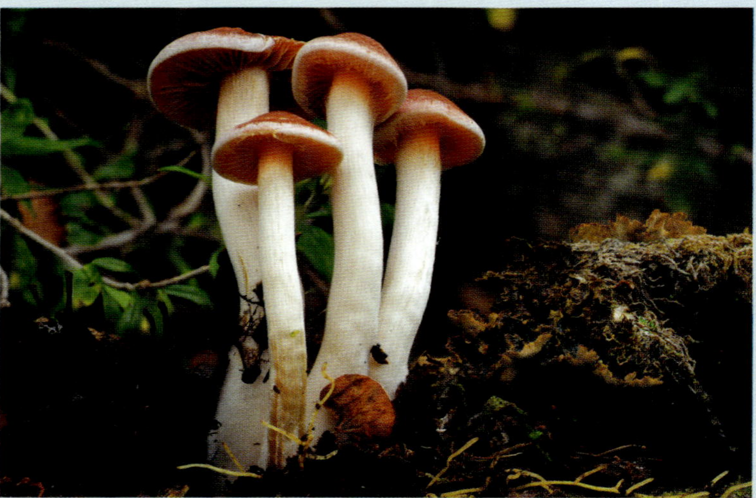

When people think of *Cortinarius*, they usually envision an above-ground structure with a cap and stipe.

However, during the field trip, Alija Mujic discovered that 16 of the truffle-like fungal species he found in the top layers of the soil also belonged to the genus *Cortinarius*.

Two have been identified as new to science, and four are still under investigation.[7,14]

TOP
Cortinarius sp.
Magallanes National Reserve, Punta Arenas, Chile

MIDDLE
Cortinarius sp.
Magallanes National Reserve, Punta Arenas, Chile

BOTTOM
Cortinarius sp.
Magallanes National Reserve, Punta Arenas, Chile

OPPOSITE
Cortinarius sp.
Magallanes National Reserve, Punta Arenas, Chile

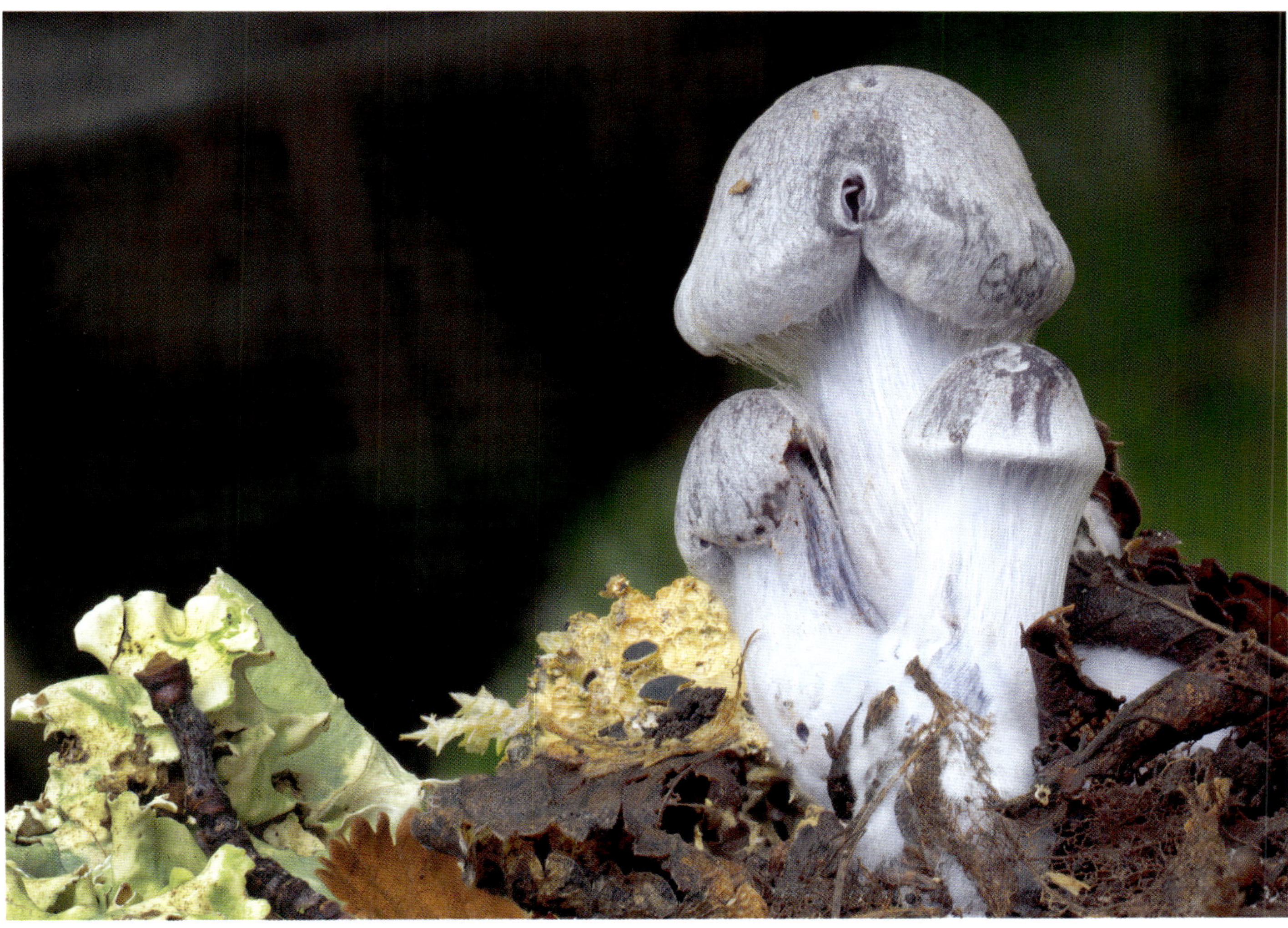

Another fascinating finding was that these truffles had formed a special partnership with the local bird population.

In the hidden world of the truffle, reproduction unfolds with a touch of collaboration, an alliance with creatures from the animal kingdom. Encased within the mushroom body, the truffle's spores cannot be spread by wind or a passing insect. Instead, the truffle produces a perfume irresistible to animals – perhaps a mouse, a squirrel, a bear or a bandicoot.

As these creatures eat the truffle, they ingest most of the flesh, but the spores pass through unharmed in their faeces.

The spores germinate in the soil and the life cycle repeats.

In Patagonia, Matthew and Alija noticed local birds scratching around in the leaf litter.

Back in the laboratory at University of Florida, Chilean doctoral student Marcos Caiafa, now a postdoctoral associate, examined the bird droppings for fungal spores and found 'copious viable spores from truffles and other mycorrhizal fungi'. This groundbreaking study confirmed that our feathered friends, who have long been recognised as crucial agents in the dispersal of plant seeds, also play a pivotal role in the widespread dispersal of fungal spores.[15]

Between 2015 and 2019, scientific teams led by Matthew Smith and Giuliana Furci collected and analysed more than 1,000 specimens in several Patagonian field trips. Their investigations resulted in 15 previously unknown species of fungi, 20 peer-reviewed papers, and a new field guide by Giuliana dedicated to the fungi of southern South America.

Matthew is excited about the discoveries. He also reflects that the incredible biodiversity they documented highlights a 'Northern Hemisphere bias' in mycology, which has resulted in the unique ecosystems of the Southern Hemisphere being understudied.

The brief safari was a transformative experience for us. It led to a deep friendship with Giuliana, which all parties cherish. It also increased our ability to identify species and understand their connectivity in the forest.

Little did we know that our next fungi safari in the foothills of the Himalayas, another understudied region of the world, would demand that we draw deep on every bit of this expanding mycological knowledge.

OPPOSITE
Bondarzewia guaitecasensis
Huilo Huilo, Neltume, Chile
Distribution: southern South America

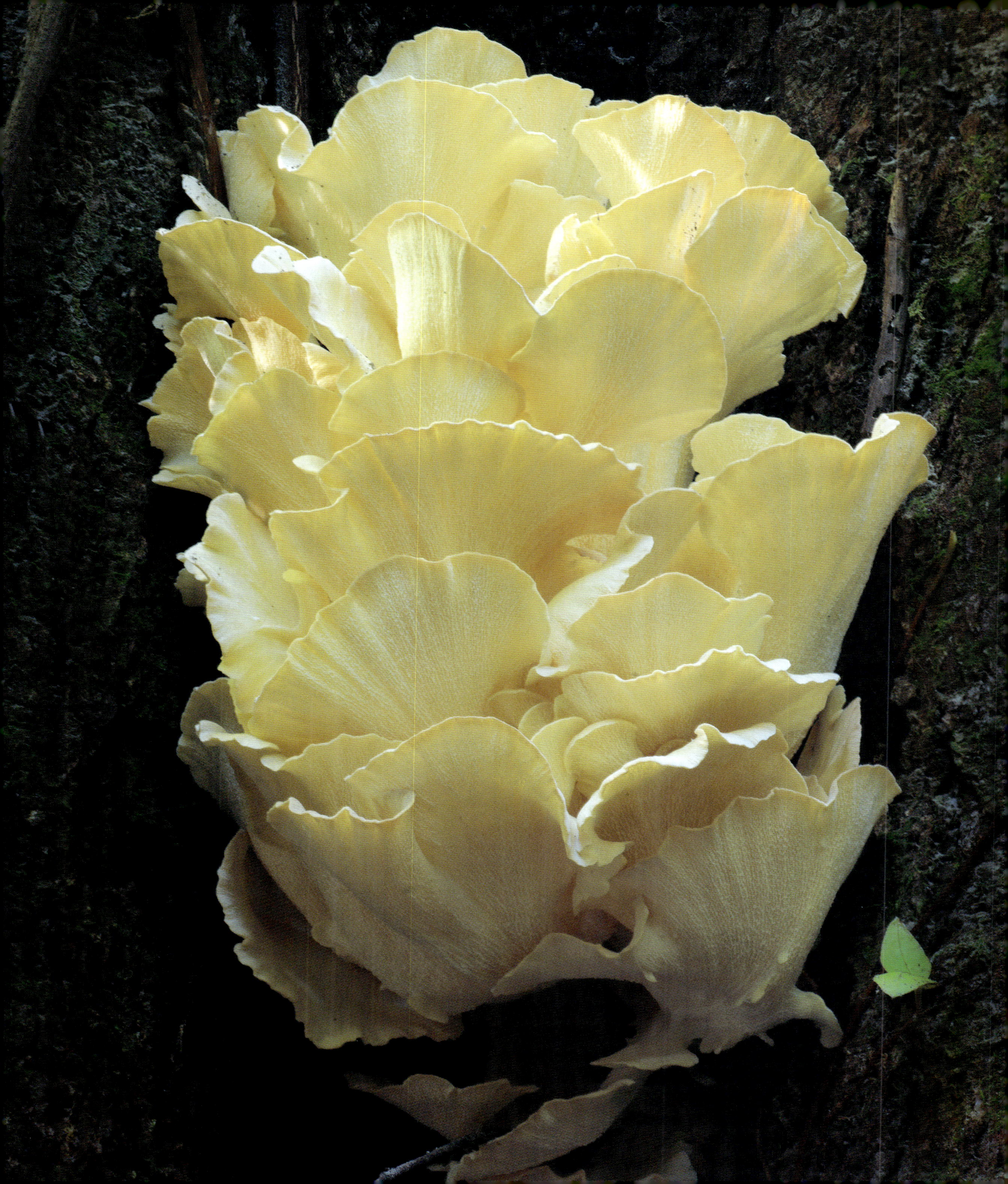

Glowing in the dark

Under the cloak of darkness, fungi paint the night with celestial-like constellations hidden in plain sight.

It was a moonless night following three days of rain. The foliage in Stephen's small rainforest sanctuary hung heavy with moisture.

The heat and humidity were intense.

Having read of the tropics' reputation as the realm of mushrooms that glow in the dark, Stephen was on a quest to witness these luminescent characters firsthand.

Mycena chlorophos, affectionately known as 'green pepe', is our local species in the rainforests of eastern Australia. It is one of more than 125 known bioluminescent fungi worldwide.

This subtropical variety grows on dead twigs, branches and logs.

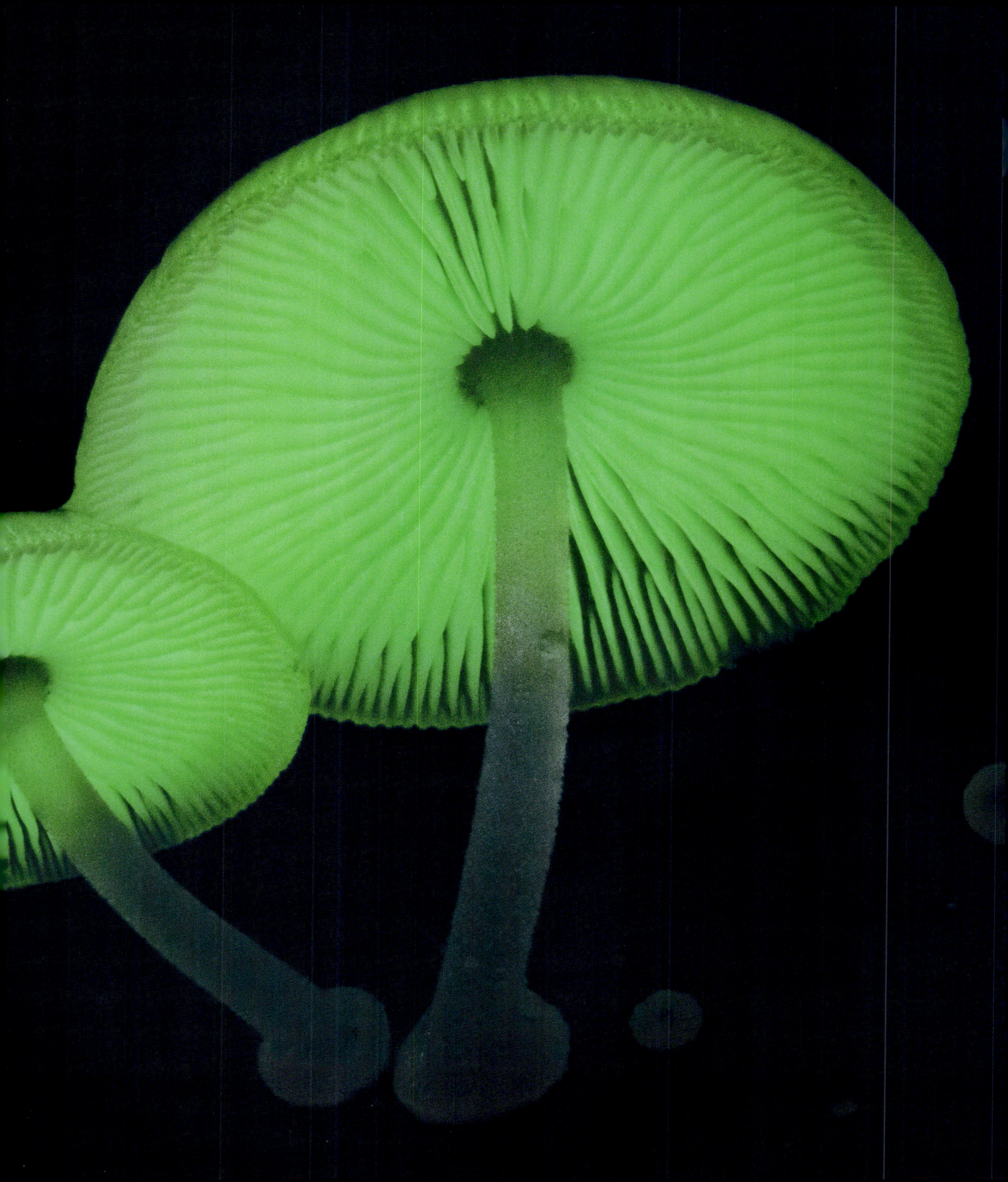

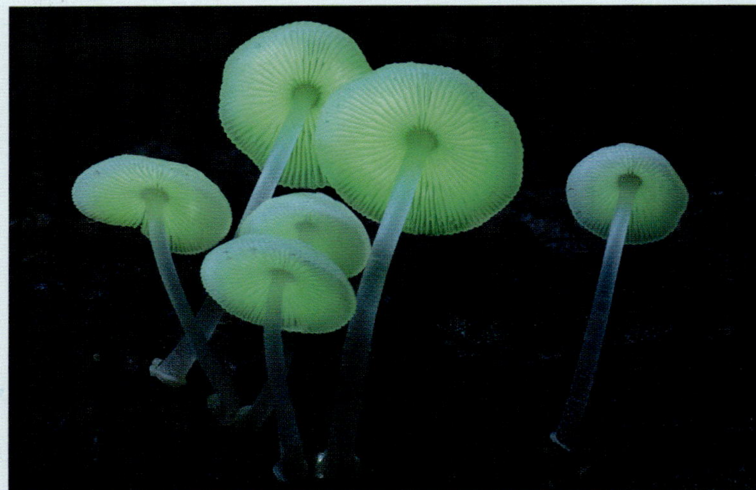

'Switching off the houselights, I ventured into the densest patch of forest I could find and extinguished the torch.

'I expected it would take several minutes for my eyes to adapt to the dark because I knew bioluminescent fungi could be quite dim.

'But as soon as the torchlight disappeared, there they were, tiny beacons of light sprinkled through the undergrowth, glowing an ethereal green.

'They cast their spell on me.'

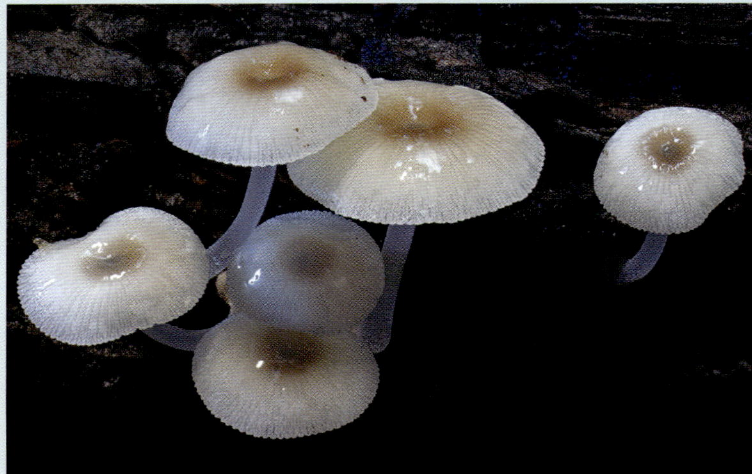

TOP

To photograph *Mycena chlorophos* in the field requires planning. Stephen uses a tripod, remote shutter control and a touch of torchlight to give a sense of the substrate and surrounds. His exposure settings on the Sony a7rV 60-megapixel camera are ISO 800, an aperture of F11 combined with a shutter speed that varies between 2–15 seconds, depending on the luminosity of the specimen.

MIDDLE

Mycena chlorophos in daylight is a pretty but slimy white mushroom. A glutinous membrane covers the cap and stipe.

BOTTOM

The same *M. chlorophos* mushroom in the dark. The degree of brightness varies from specimen to specimen.

OPPOSITE

One of Stephen's early photographs of *M. chlorophos* after a summer shower. The mushrooms start as non-luminous buttons; their glow develops and builds in intensity as they mature.

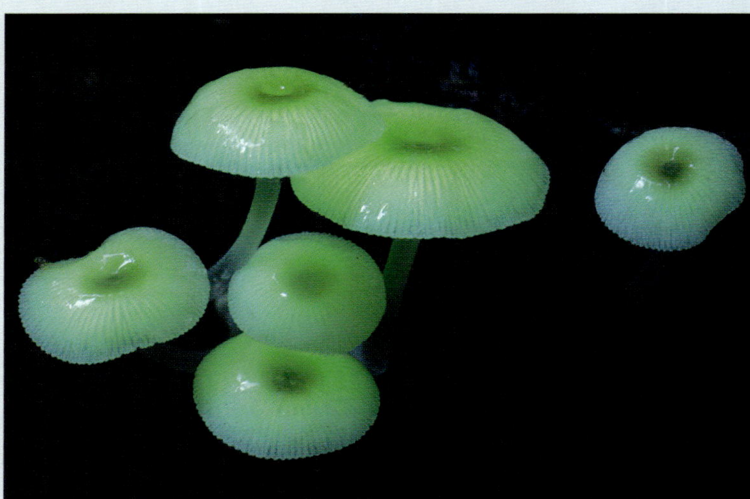

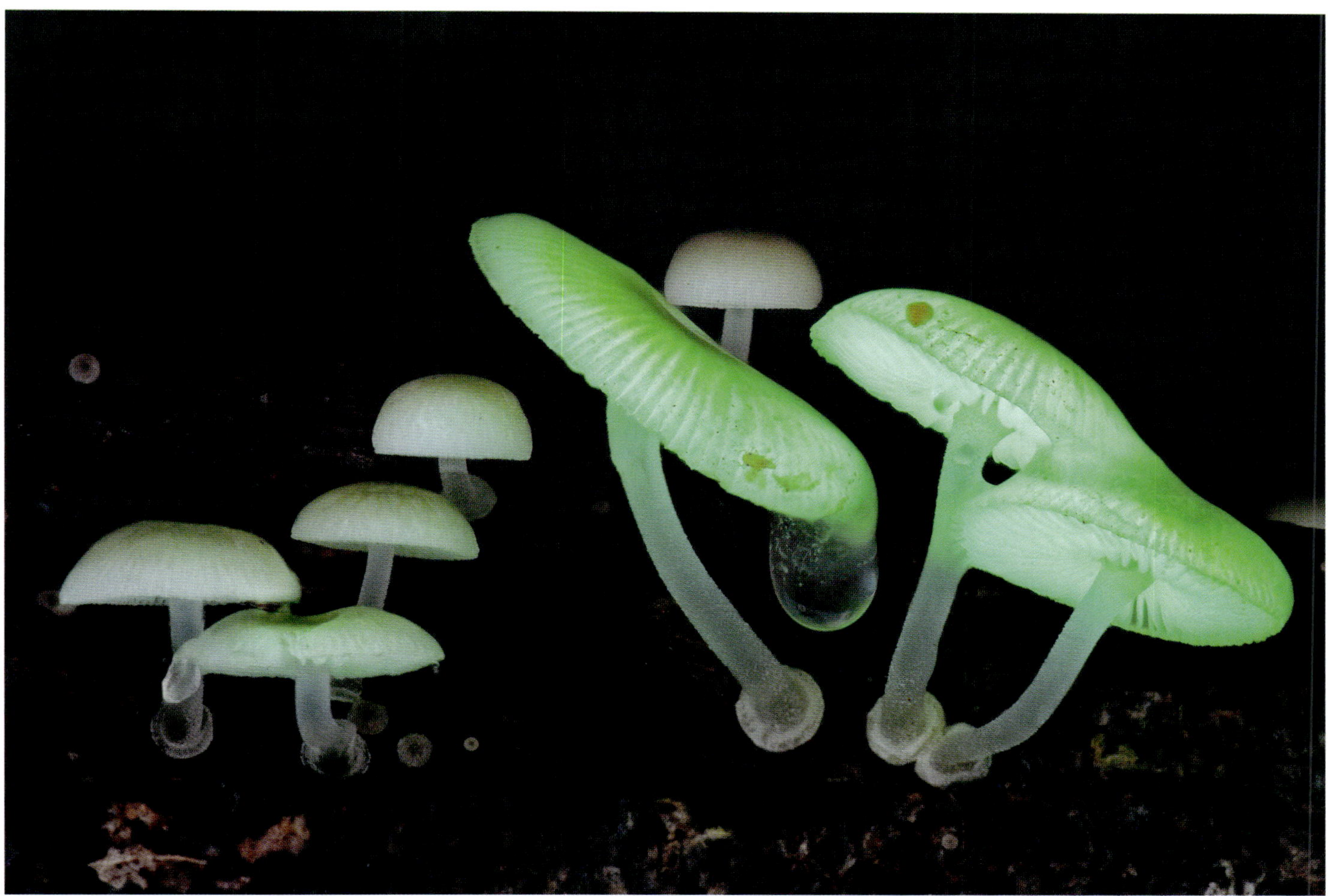

The sporing bodies of *M. chlorophos* by day are unremarkable, a white to brown mushroom, with a thick layer of glutinous jelly on the cap and stipe, but in the dark they are nature's own radiant jewels.

Their glow comes from within, a biochemical generator using the enzyme luciferase. In a complex reaction of acids and oxygen, luciferase activates a molecule called luciferin, which in Latin means the 'light bearer'. The resulting energy is released as light without heat, creating the mesmerising bioluminescent display in these glowing mushrooms.[1]

On countless balmy summer evenings, I would bid farewell to Stephen as he vanished for hours. With camera in hand, he braved the mosquitos and the rustling of unseen critters in the damp dark, eager to capture the nocturnal spectacle of *M. chlorophos*.

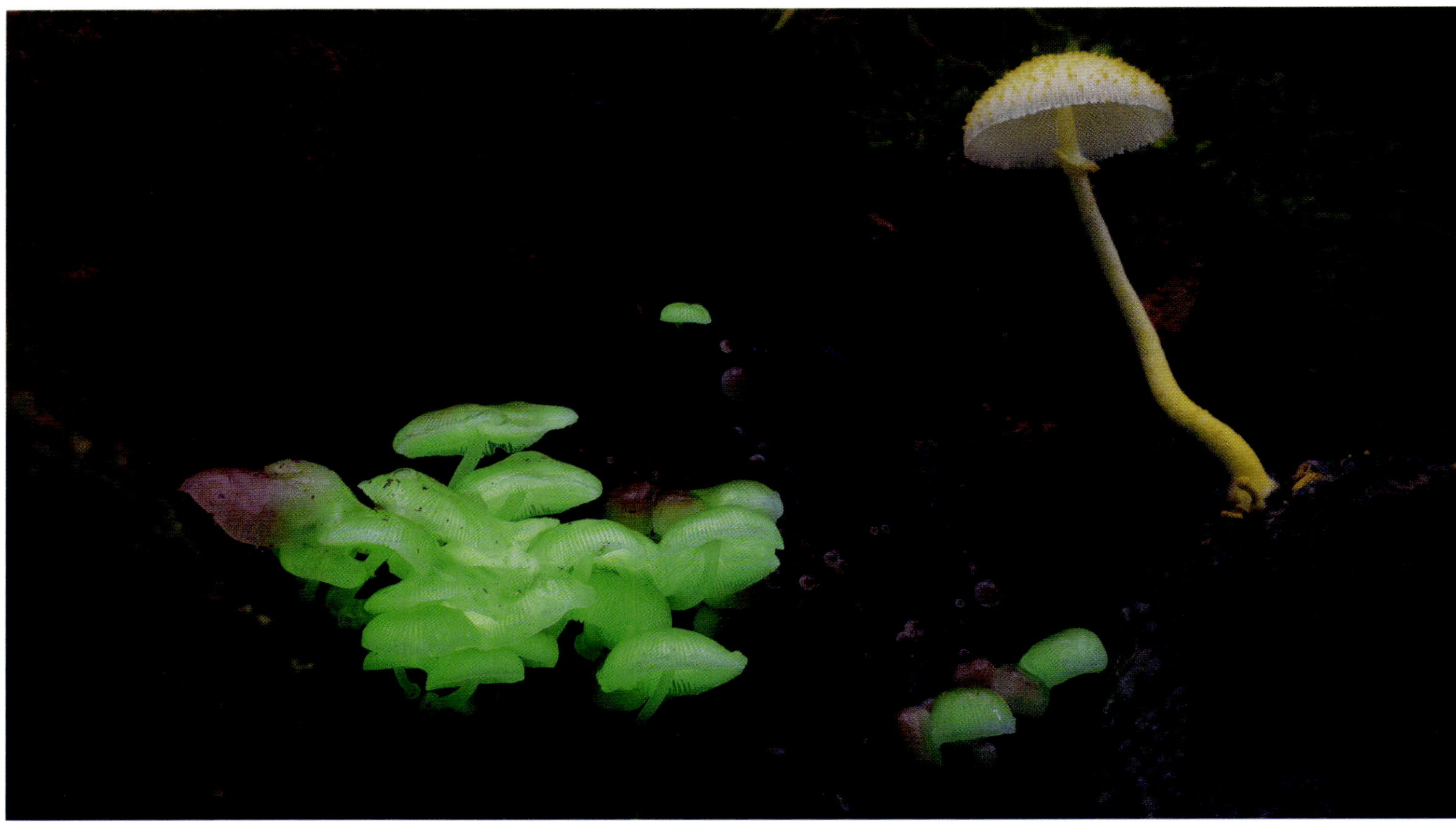

'Bioluminescence, for me, has always held a fascination.

'From the spectacle of a firefly dance to the shimmering canvas of the ocean turned blue–green by minuscule plankton, to the pinpricks of light of glow-worms amid their snares, mimicking star constellations on the ceiling of a cave.

'When night falls, it's as if a veil of magic and mystery descends upon the world.'

ABOVE

This frame is from a time-lapse where *M. clorophos* and its miniature set were mounted on a turntable. As the growing *M. chlorophos* were revealed the yellow *Leucocoprinus birnbaumii* fungi-bombed the rotating scene. It turned out to be a graceful addition to the action-packed sequence featured in the forests episode of the BBC documentary series *Planet Earth II*.

OPPOSITE TOP

During warm spring nights in the Australian subtropics, fireflies light up the evening sky with a moving spectacle of bioluminescence. They emit green and yellow light using luciferase and other compounds. There are over 2,000 firefly species known to science, with each species having its own unique flashing light patterns to attract mates.

Photograph: Catherine Marciniak

OPPOSITE BOTTOM

Despite their name, glow worms are not worms but are the larvae of various species of fungus gnats and beetles. Their glow creates the illusion of a starry night sky on the ceiling of a cave or earth grotto, attracting small insects, which they then capture using sticky beaded threads.

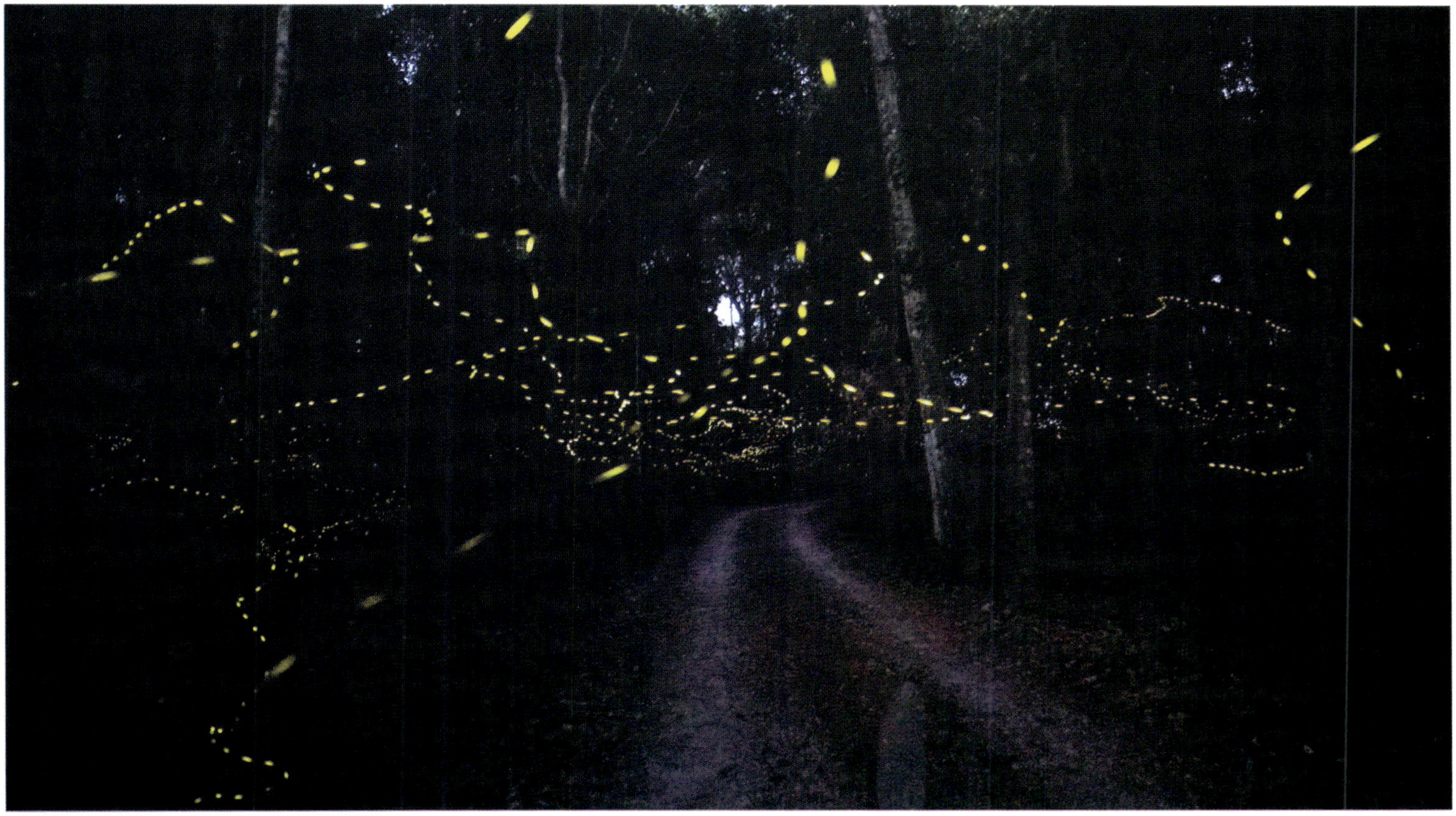

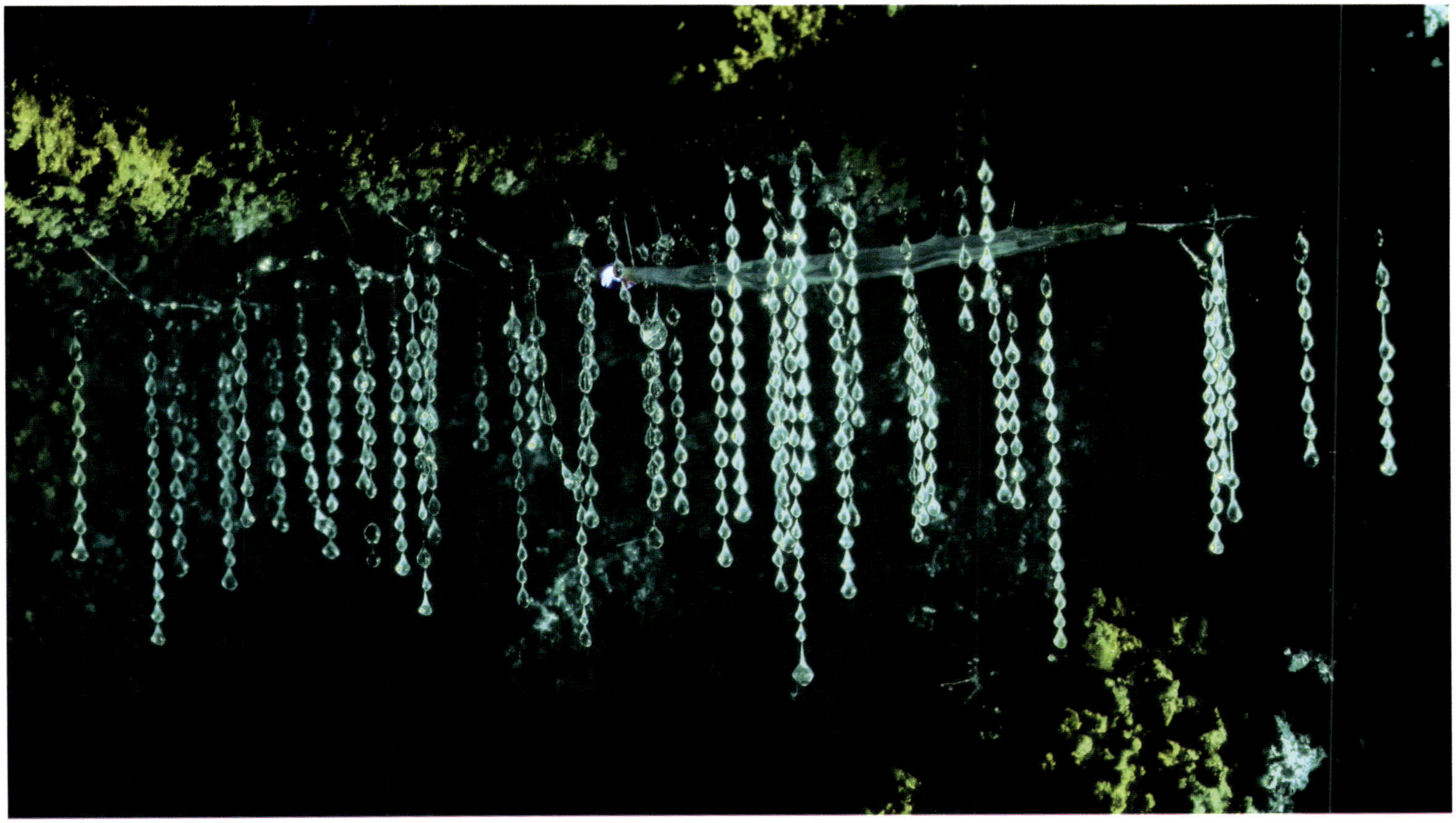

Stephen's summer explorations in the dark were just the prologue to an extraordinary journey.

When he uploaded his photographic portraits of *M. chlorophos* to the internet, the world noticed.

The BBC, renowned for its groundbreaking nature documentaries, stumbled upon Stephen's virtual galleries and contacted him in 2015 as part of their research for their new series *Planet Earth II*.

It was to be presented by Sir David Attenborough and was set to be the biggest nature documentary series for a decade. The BBC wanted to include a major storyline on luminous fungi.

'They asked me if I knew where they could find luminous fungi growing.

'"It is tricky," I explained. Its growth hinges on temperature and humidity.

'If they were to send a crew out, we could assist, but they would need to be opportunistic, triggering a shoot when conditions are just right.

'Then, as an aside, I told them I'd done a few experiments with time-lapsing fungi, including M. chlorophos, *and asked if they were interested in seeing the clips.*

'The producers were quick to respond with a resounding "yes".

'Time-lapses of our local subtropical fungi danced across living room screens all over the world, in not only Planet Earth II *but many subsequent nature documentaries.'*

The creators of the 'Jungle' episode in *Planet Earth II* included 75 seconds of Stephen's fungi time-lapses, which doesn't sound like much but in television land, that is significant screen time.

OPPOSITE
A rotting branch covered with **M. chlorophos** can be just bright enough to use as a torch to light a path through the night forest.

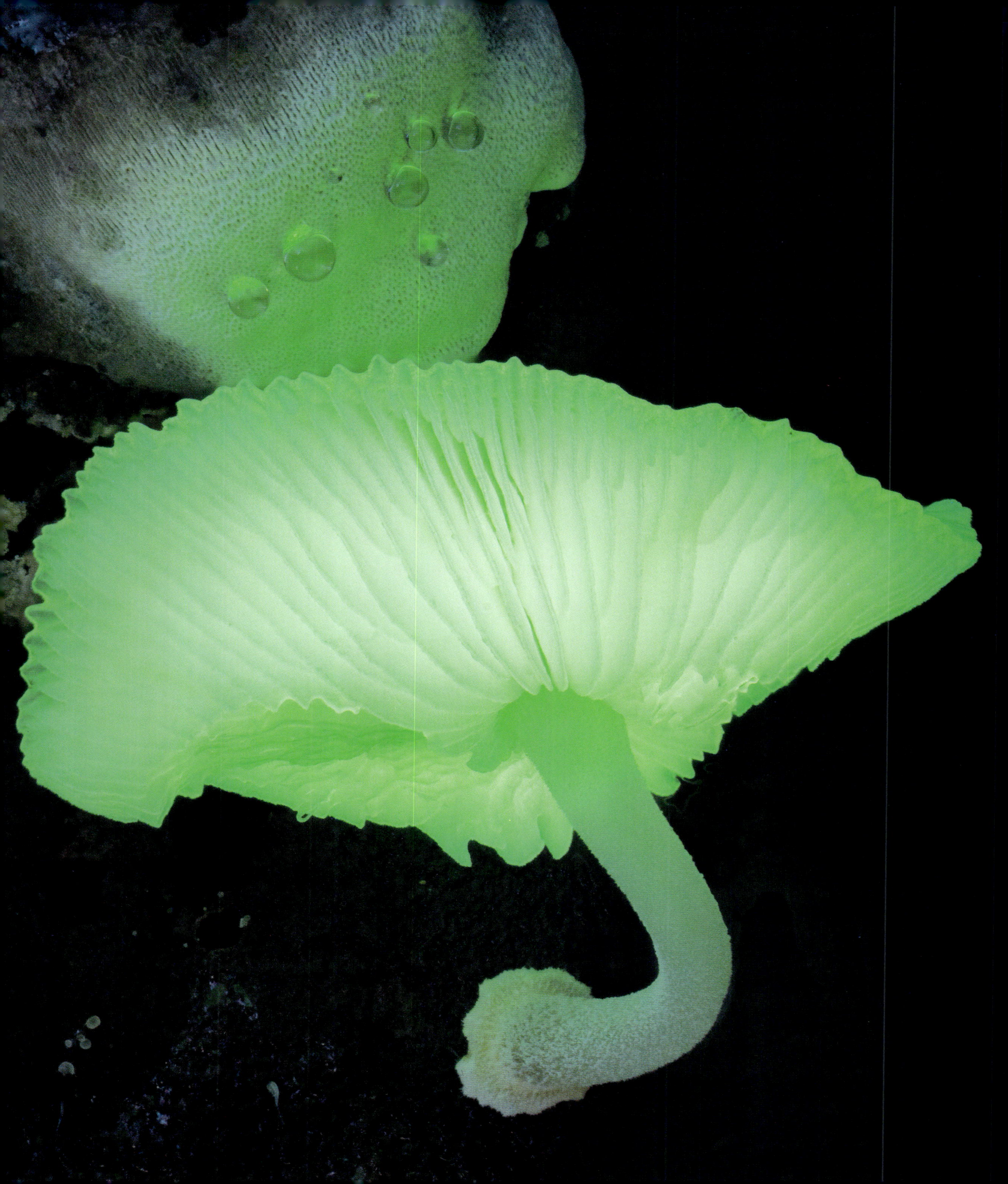

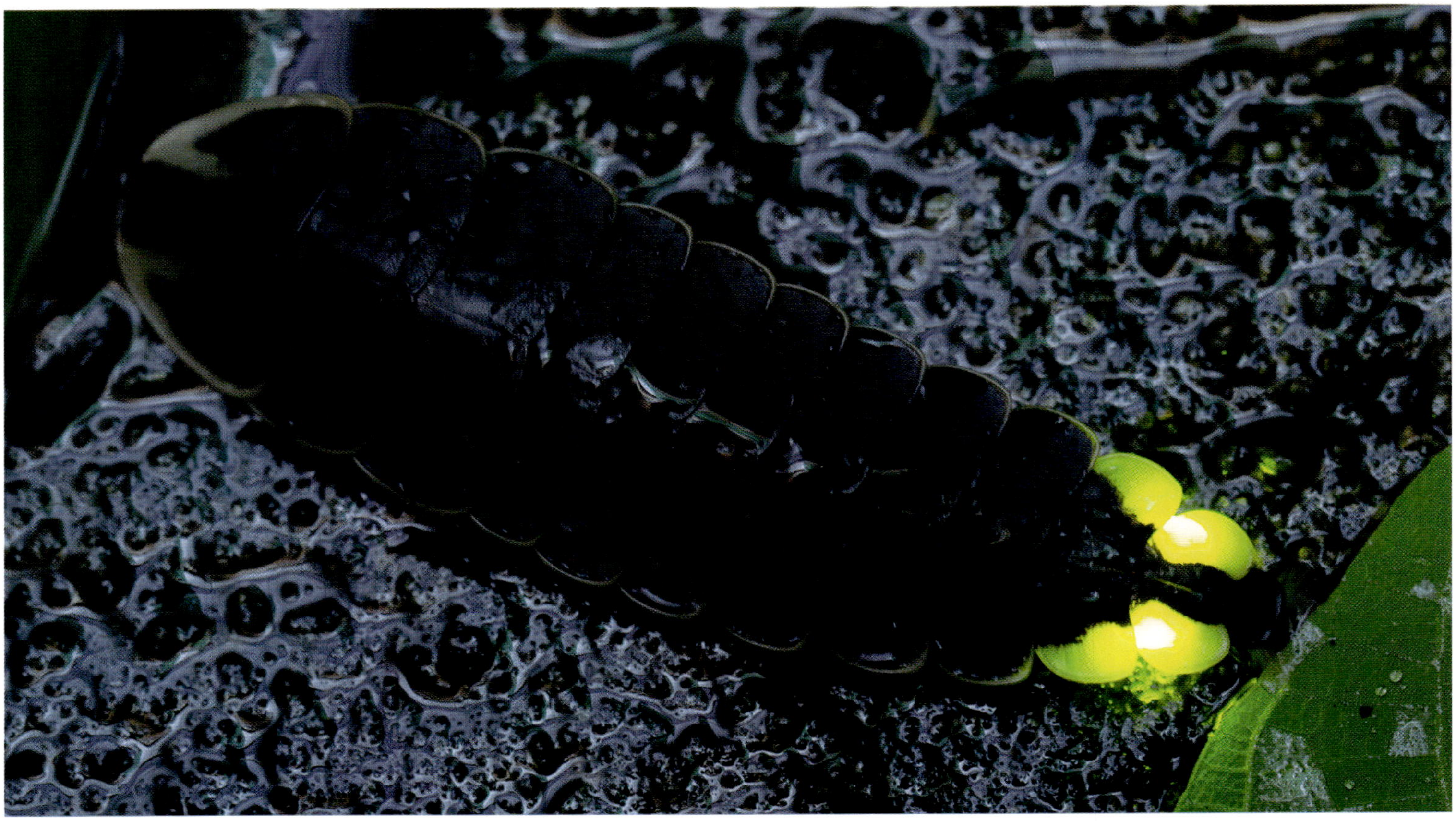

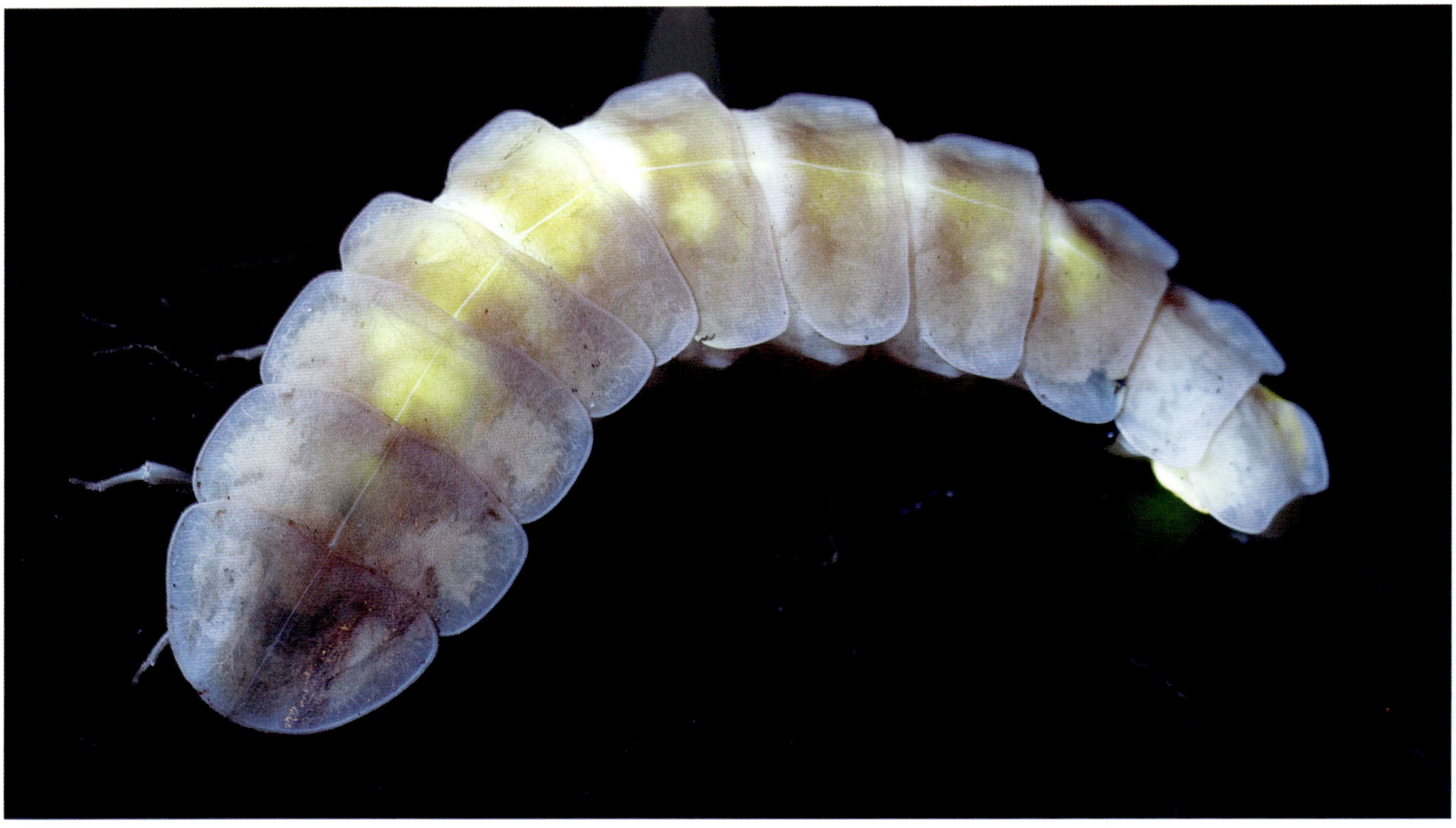

The sequence starts with a moonlit forest, and then our subtropical fungi take centre stage, characters in a story about why mushrooms glow.[2]

The science was based on 2015 findings that some bioluminescent species have circadian rhythms, an internal clock that dims their glow by day and then turns it up at night. The research team, led by Cassius Stevani from the São Paulo University in Brazil, was on a quest to discover why.[3]

They used plastic mushrooms covered in sticky adhesive. Half were controls with no light. The other half were built with internal LEDs that mimicked the wavelength of the light of a local fungus, *Neonothopanus gardneri*.[3]

These luminous mushroom imposters were beacons for all kinds of flying insects that collected in the glue, suggesting that the fungus used its chemistry to attract night bugs, co-opting them to spread its spores.[3]

In *Planet Earth II*, a male click beetle with bioluminescent lights on its head and abdomen is given a cameo.

Documented by a BBC crew, we observe this humble creature bewitched by the distant light of a mushroom he mistakes for a female click beetle. His search among glowing gills for his mate is in vain, but the fungus has created an unwitting courier as fungal spores cling to his body. Continuing his nocturnal search, the click beetle disperses the spores wherever he goes.

This story of bioluminescence is a drama of courtship and trickery. Other organisms use their light to communicate with their own kind, as camouflage, to defend themselves from predators, or as a weapon to prey on others.[1,4]

OPPOSITE TOP
The tiny eggs of fireflies hatch into larvae with elongated bodies, soft exoskeletons and bioluminescent organs that communicate and scare off predators. Stephen photographed this firefly larva in northern Thailand.

OPPOSITE BOTTOM
Using the light of an ultraviolet torch on a firefly larva in Sri Lanka Stephen realised these insects are not only bioluminescent, they are also biofluorescent.

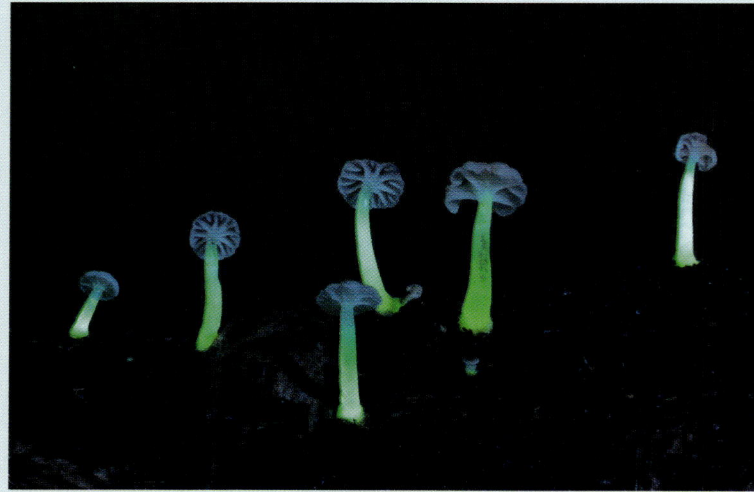

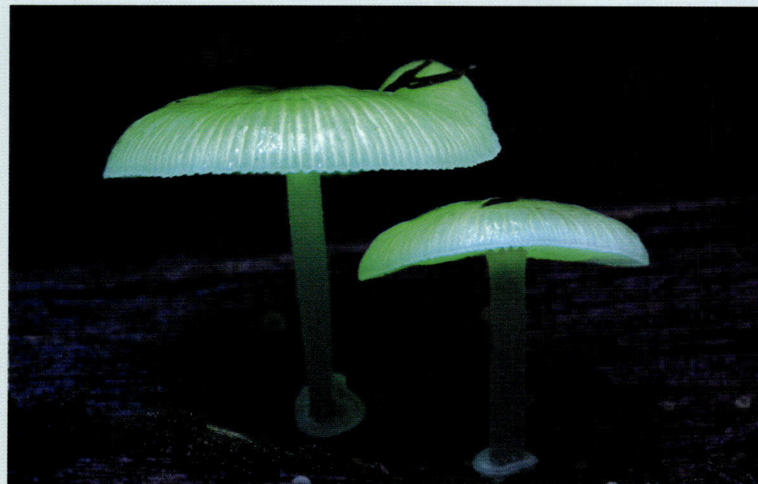

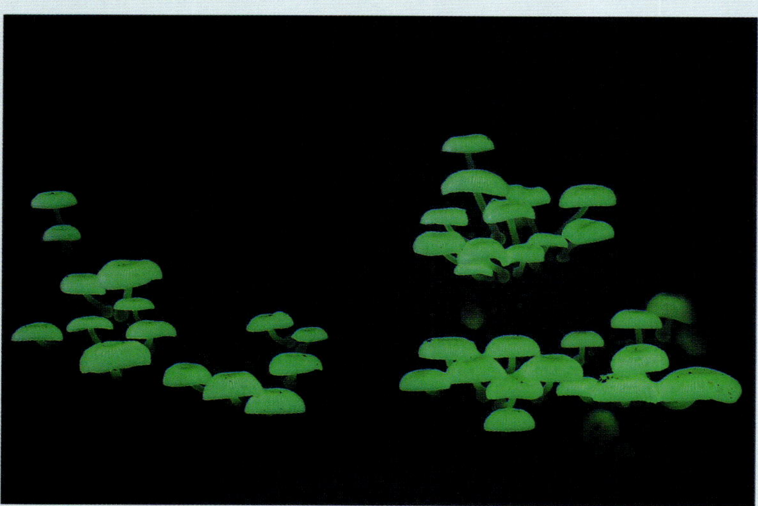

Stephen's time-lapses of bioluminescent fungi allow us to observe how some creatures interact with glowing fungi.

We have seen hundreds of wood dwellers, tiny beetles and grubs, tirelessly scuttling over the wood like speedy vacuum cleaners under the glowing parasols.

Some nibble at the mycelium while others feast on the fallen spores.

But the critters who emerge from the shadows of our fungarium and devour whole mushrooms in a single gulp are slugs and snails, which then spread the spores in their slimy poo.

We never tire of how these time-lapses allow us to peek into the usually unseen world of mushrooms.

TOP

In 2018, Stephen documented a new to science, bioluminescent species growing on bamboo in Meghalaya, India. This species, named *Roridomyces phyllostachydis*, exhibits an intriguing trait: its stipe glows brightly, while the caps remain non-luminous, a striking contrast to *M. chlorophos*.

MIDDLE

At an optimal temperature of 27°C, *M. chlorophos* is a vibrant neon green, with its fresh sporing body emitting a continuous wavelength of ~530 nanometres.

BOTTOM

Captured in a time-lapse sequence, this image of *M. chlorophos* was recorded on rotting logs in a plastic crate placed in Stephen's home bathroom shower. The mushrooms transitioned from button size to full maturity in just three days.

OPPOSITE

Omphalotus nidiformis, the ghost fungus
Border Ranges National Park, New South Wales, Australia
Distribution: Australasia

On the Sony AR72, the exposure was ISO 1600, aperture F8 shutter speed of 2 minutes.

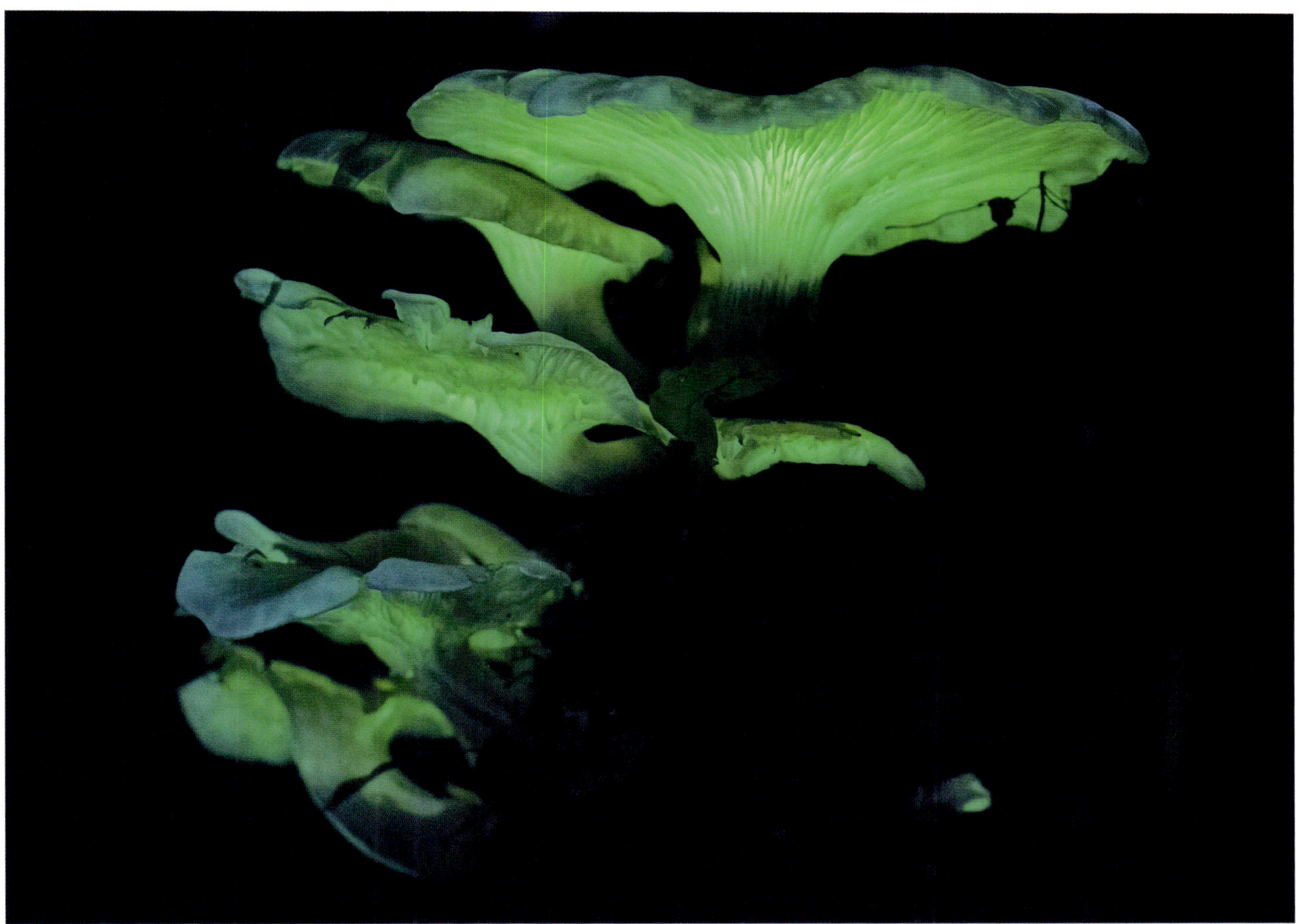

As the cooler temperatures of autumn move across southern Australia, hundreds of fungi enthusiasts head out into the dark to spend the night in damp and chilly forests.

They are hunting for another bioluminescent species, *Omphalotus nidiformis*, known as the ghost fungus.

This species is much larger than *M. chlorophos*. The sporing body is like a fan and it is found growing on stumps and standing dead trees.

The most surprising place we've documented it is in a commercial pine forest in South Australia, where it has cleverly adapted to growing in clusters on cut stumps of a North American tree species.

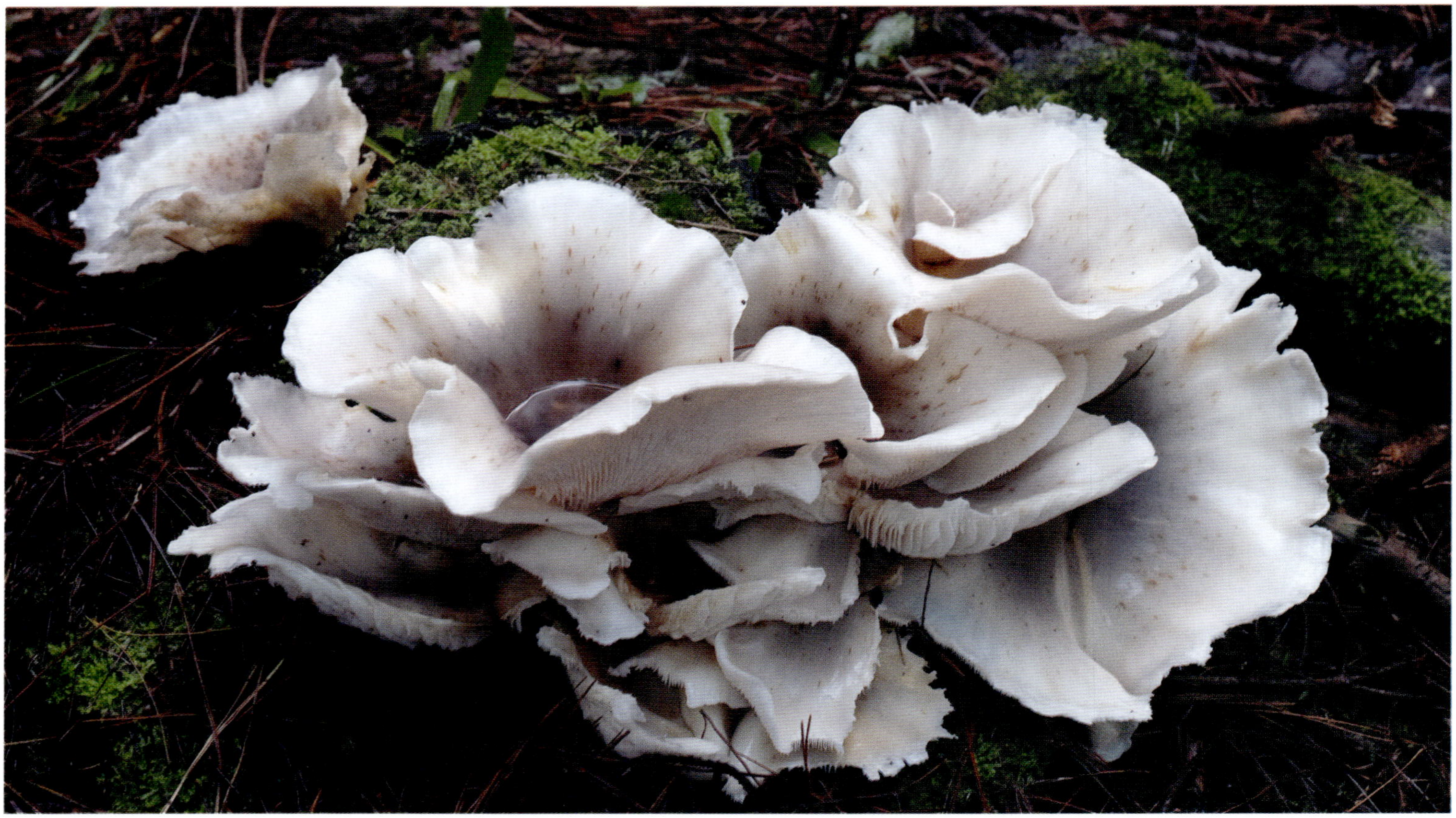

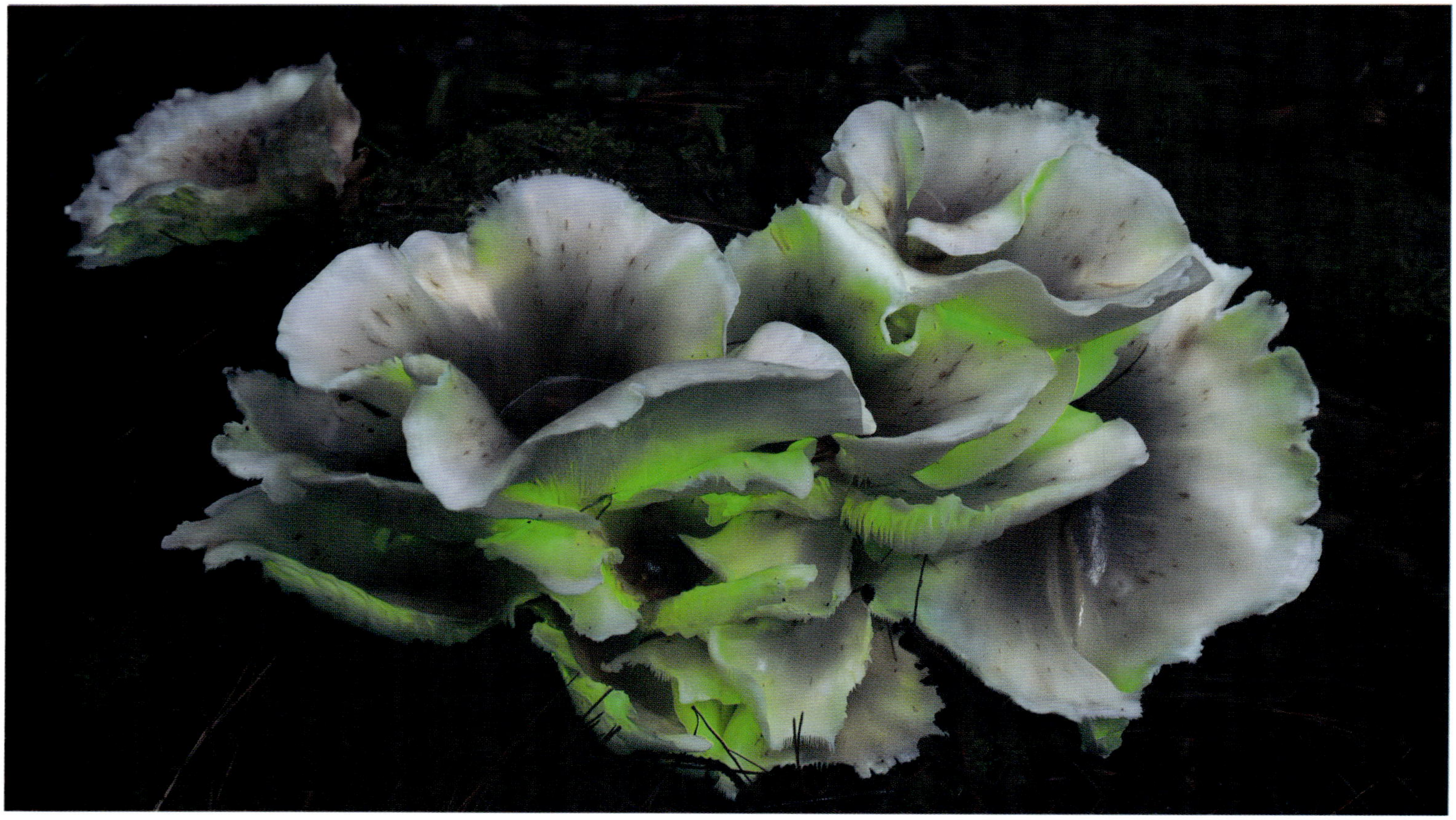

After a couple of days of autumn showers, there can be so many sporing bodies of ghost fungus in this plantation that it is occasionally turned into a tourist destination with the fun moniker Ghost Mushroom Lane.

Our visit to this forest, 2,000 kilometres from home, was a rare opportunity for us to photograph *O. nidiformis* and attempt to time-lapse it in the field.

We knew better than to venture out into the darkness without a plan, so we reconnoitered by daylight. We were lucky enough to spot two clusters of enormous white mushrooms ~ 30 centimetres across.

'For this dance of light and darkness, I took some time to plan the ideal settings for the photograph. The forest at night is no place for fumbling.

'Using a torch to help the camera focus on the near edge of the mushroom, I then manually focus bracketed.

'Taking the time-lapse shots was a whole other beast.

'The moon was almost full, casting a strong silvery light that could drown the mushrooms' glow.

'Our window was narrow – one night, one chance. Every 30 seconds, the shutter clicked for three hours – just enough time for hopefully 15 seconds of magic.

'And the result – we were lucky. In those fleeting frames, we'd managed to capture in pixels the supernatural beauty of O. nidiformis and its glow pulsing in the night.'

OPPOSITE

Omphalotus nidiformis at dusk in the forest bordering Ghost Mushroom Lane, South Australia. The top image is the first frame of a 15-second time-lapse shot in the field. The bottom image is frame number 41. For the second image the Sony ar7V the exposure was ISO 1600, aperture F8 and shutter speed 144 seconds.

NEXT PAGE

Under powerful ultraviolet lights, lichens on the trunk of a Bangalow palm, *Archontophoenix cunninghamiana*, reveal their biofluorescent charm, showcasing vivid colours and intricate patterns. The 365-nanometre UV lights, equipped with a high pass filter, bring out this hidden beauty without creating the purple tint that can be caused by visible light interference.

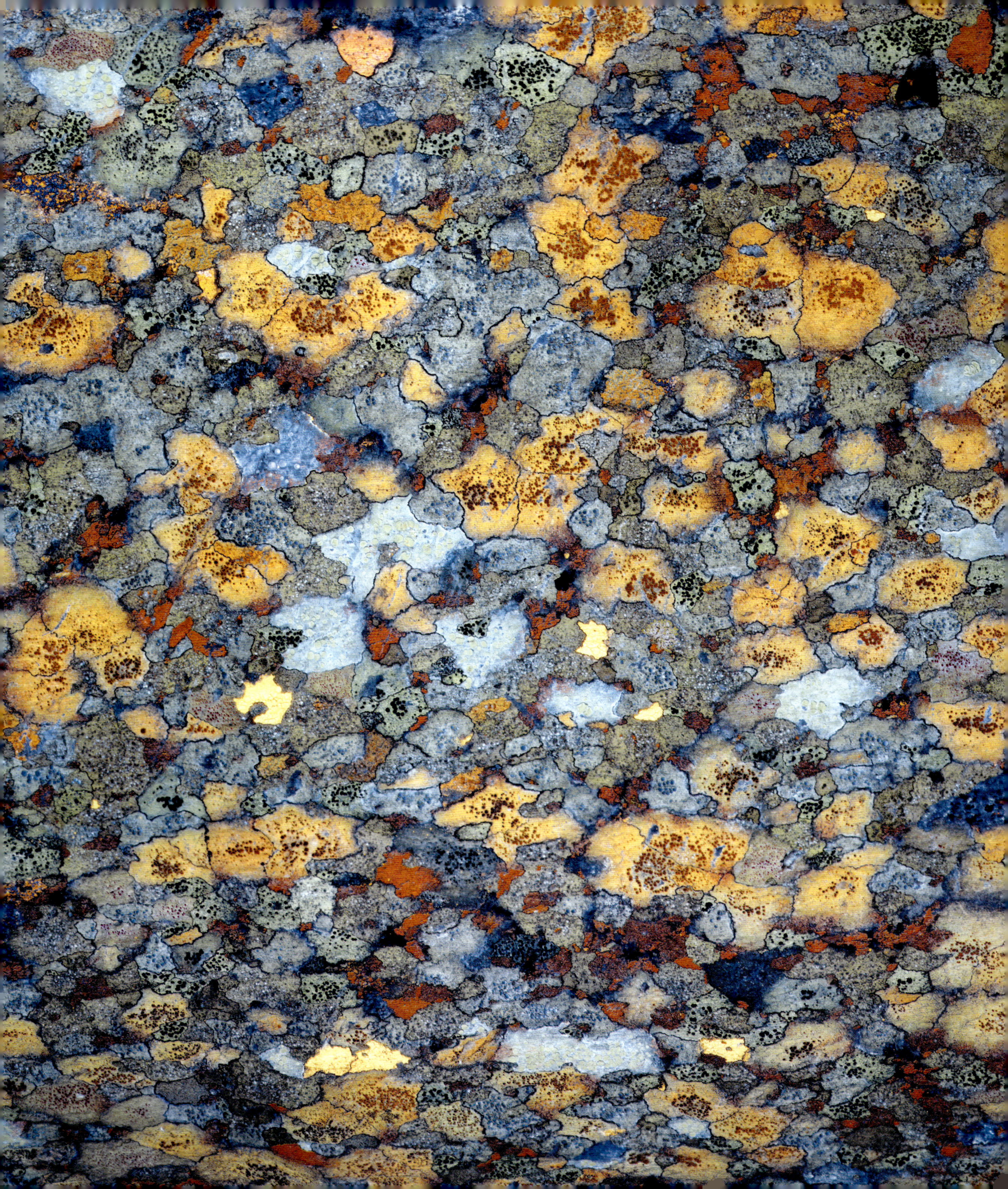

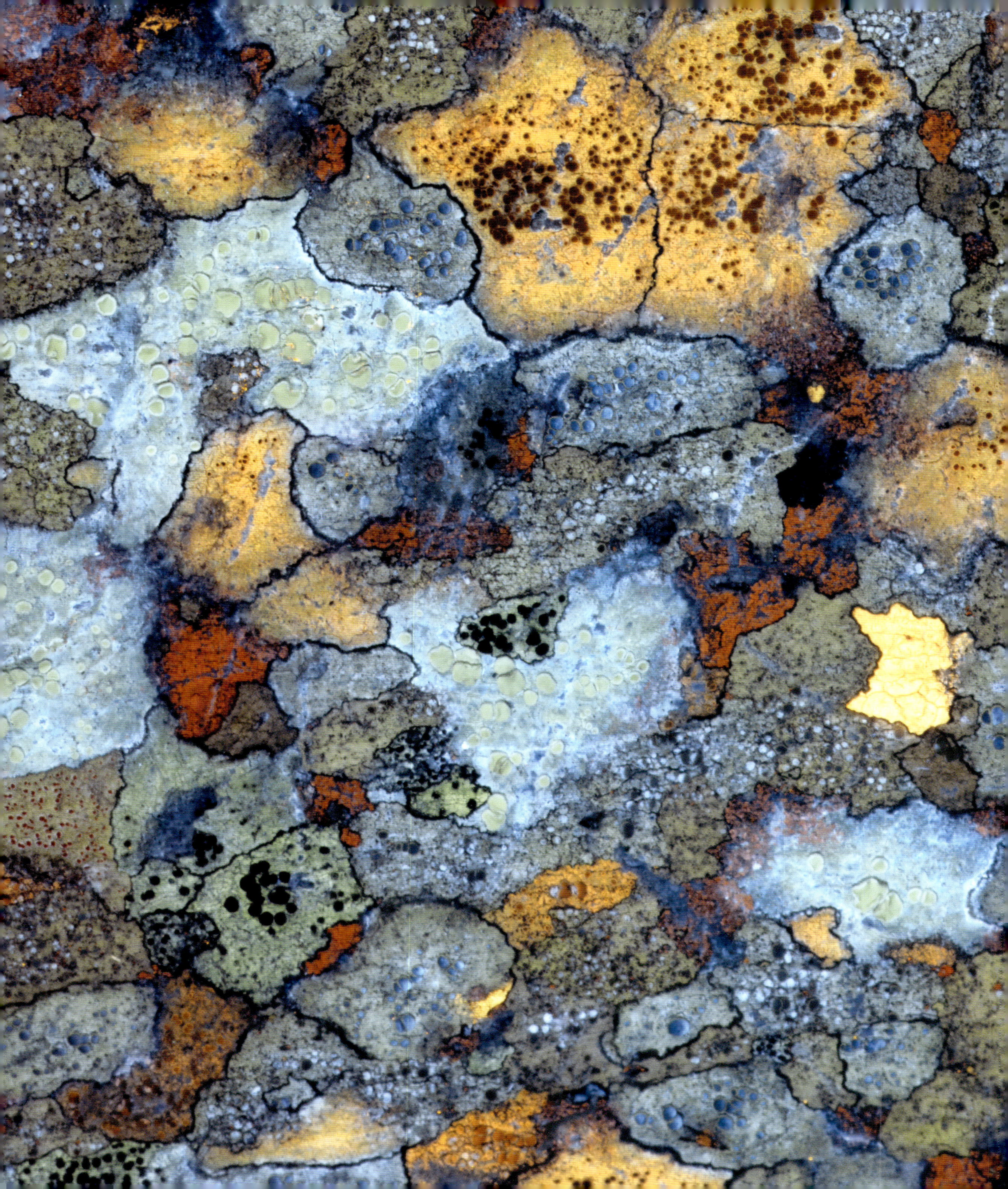

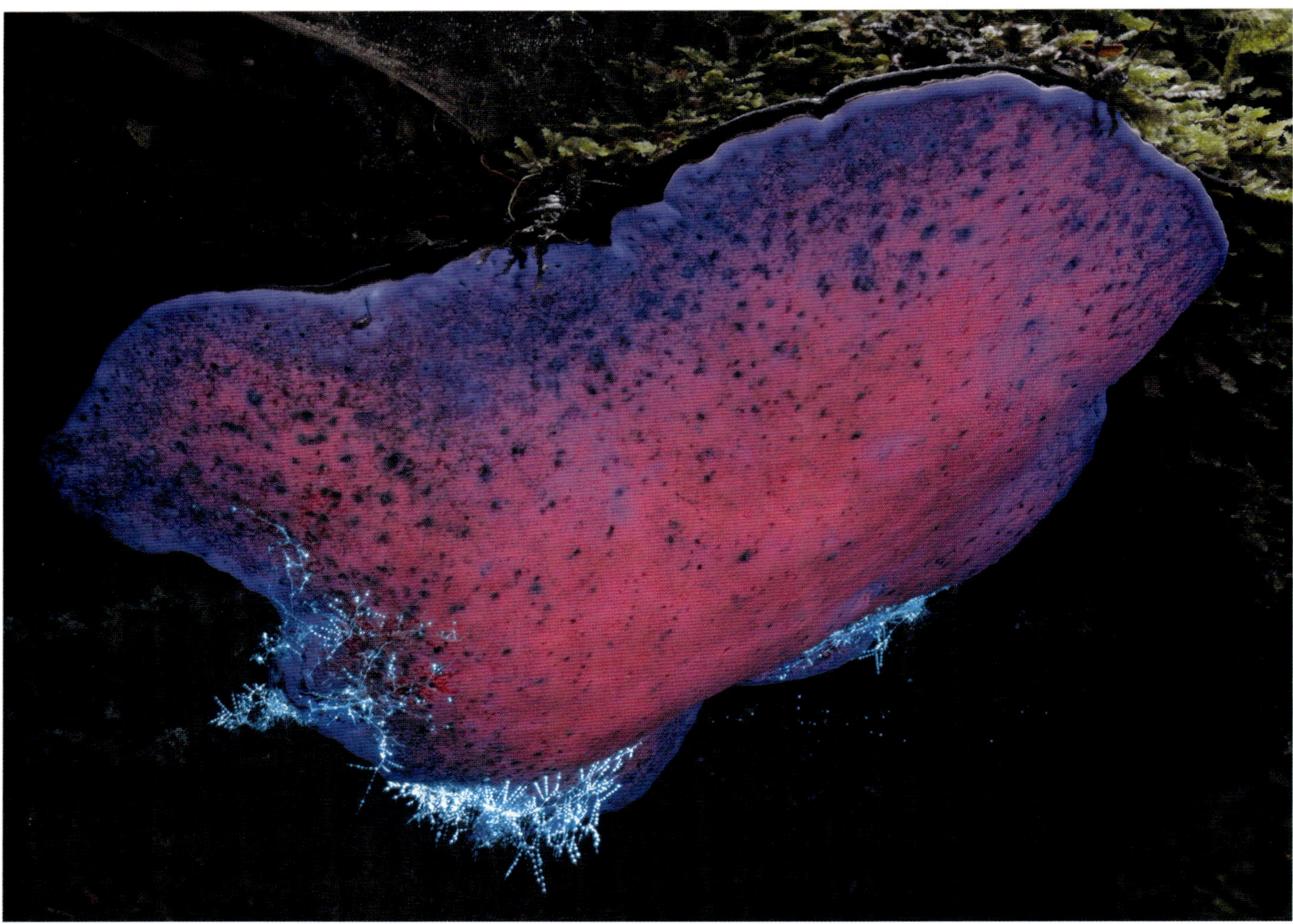

Stephen's recent obsession is shining a light on another hidden luminous world.

Adding an ultraviolet (UV) torch to his camera kit, he is learning that we are surrounded by organisms, including fungi, that absorb UV light and convert it to visible light, a hidden flamboyant palette of unexpected colours.

ABOVE

Illuminating the underside of a *Ganoderma* in Tasmania with an ultraviolet torch reveals not just brilliant blues and pinks but also a fascinating discovery: a fungus gnat intricately weaving its net using the mushroom as a loom. While the purpose of such webs is still under investigation, they probably serve to collect a meal of fungal spores and possibly capture insects to supplement the omnivore's diet.

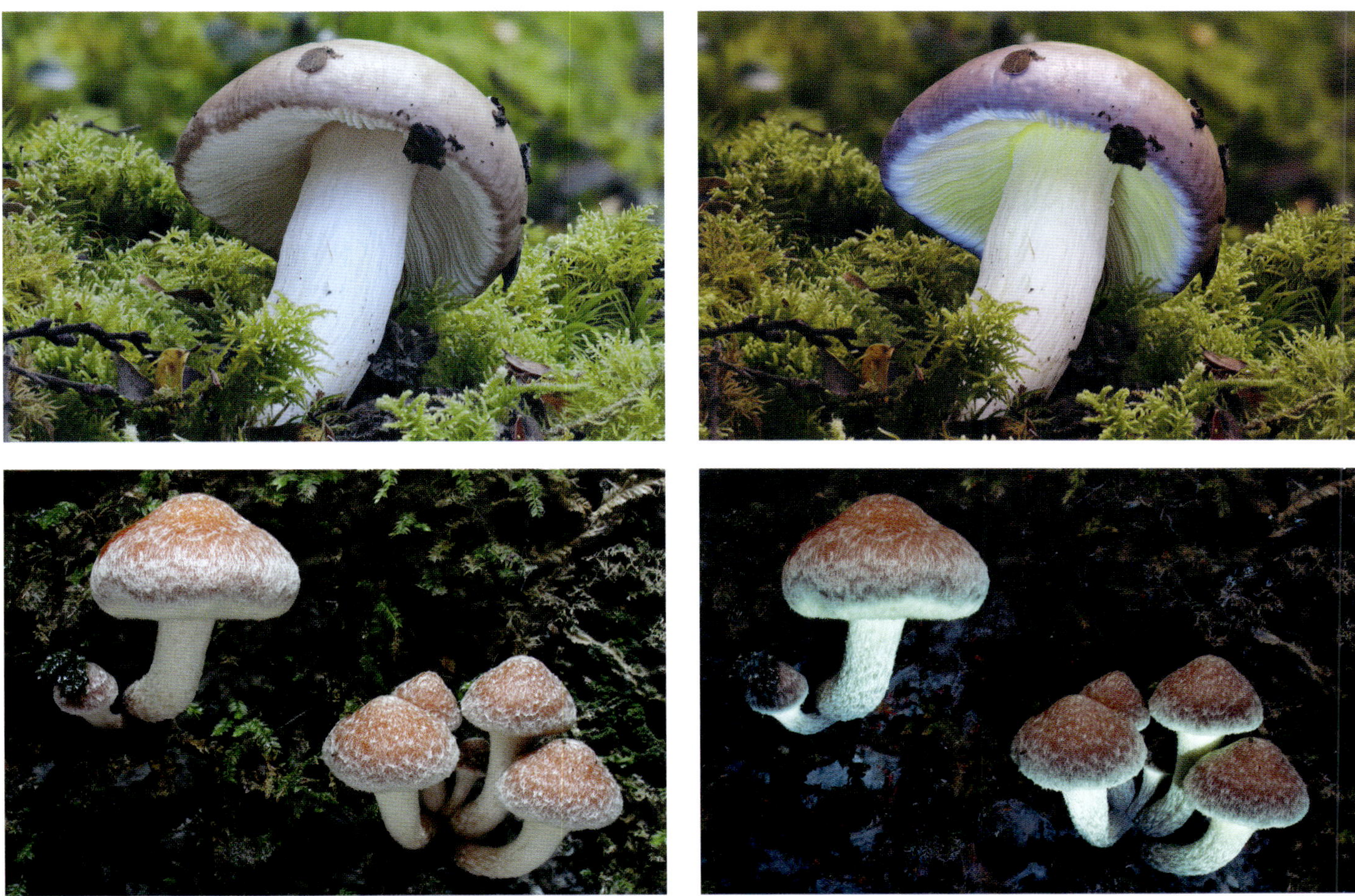

'Quite a few species of fungi fluoresce.

'Under the light of my UV torch, an entoloma becomes a brilliant indigo, hypholomas turn aqua, and the underside of a ganoderma transforms from milky cream to neon pink and blue, with sparkling necklaces, the fluorescing webs made by fungi gnats.'

ABOVE TOP
Russula sp.
Fiordland National Park, Te Anau, New Zealand
Fluoresces yellow and blue under UV light

ABOVE BOTTOM
Hypholoma sp.
Philosopher Falls, Tasmania, Australia
Fluoresces green under UV light

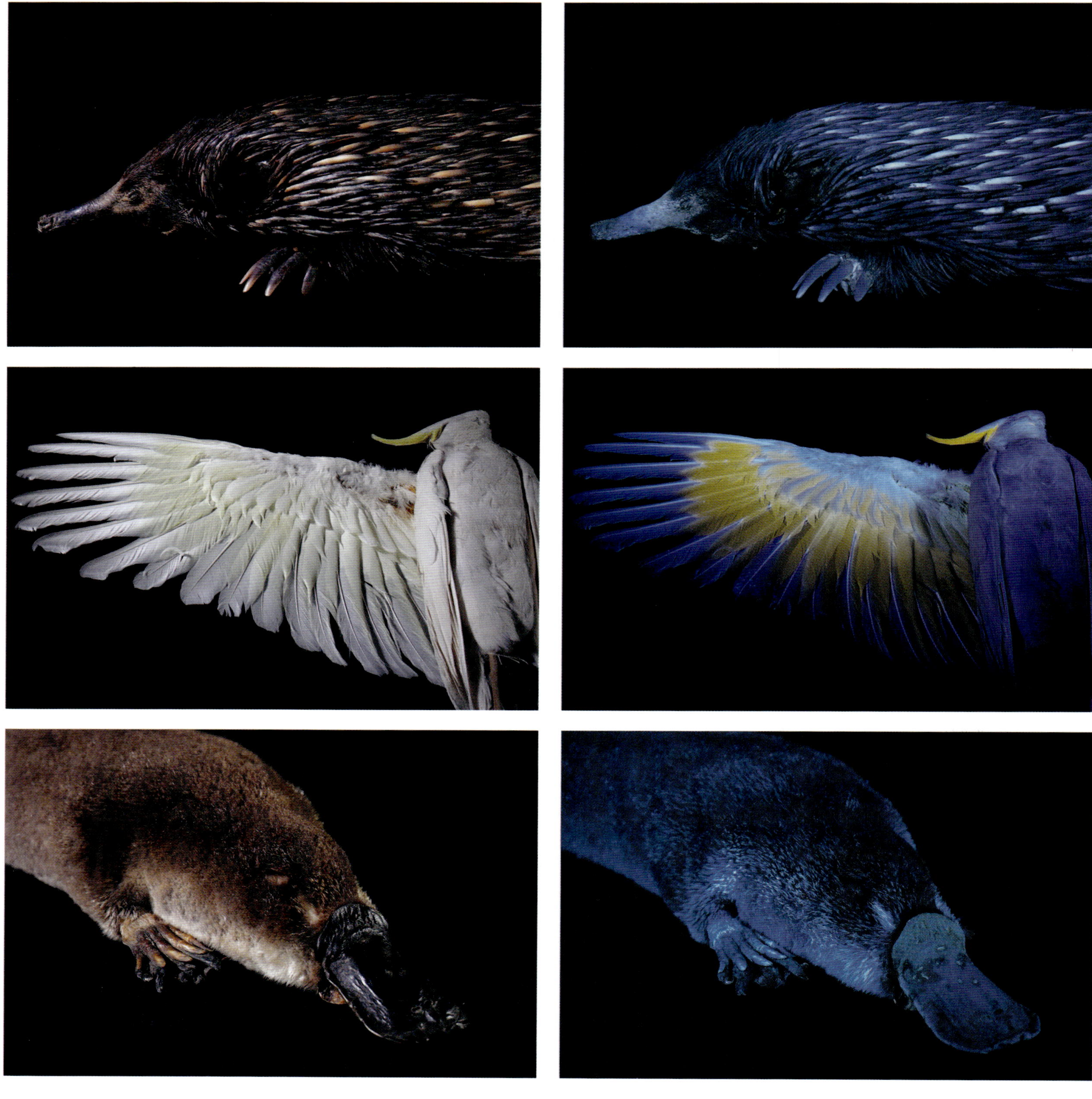

There has been a great deal of interest in biofluorescence ever since scientists discovered that some of our cutest creatures put on an astonishing light show at night.

At least 125 species of mammal fluoresce, including platypus, squirrels and echidnas, with the majority of these animals being nocturnal or living under the canopy of forests.

Many fish, frogs, birds, insects and arachnids also fluoresce.[5]

Scientists have pondered the purpose of the striking display of fluorescence exhibited by these organisms.

They speculate that like bioluminescence, these colourful reactions to UV light may allow predators to blend into their environment and ambush their prey, or act as an ancient form of camouflage, or used to identify members of one's own species, or could even be incidental. [5]

What we do know is that the vibrant hues we see under UV light in nature are the result of a complex interplay between biology, environment and evolution.

OPPOSITE TOP
Tachyglossus aculeatus, short-beaked echidna
Queensland Museum, QM J19029
Echidna spines fluoresces blue under UV light

OPPOSITE MIDDLE
Cacatua galerita, sulphur-crested cockatoo
Queensland Museum, SkinQMO.27319 Wing QMO.32696
Feathers fluoresces yellow and blue under UV light

OPPOSITE BOTTOM
Ornithorhynchus anatinus, duck-billed platypus
Queensland Museum, QM J8753
Fur fluoresces blue under UV light

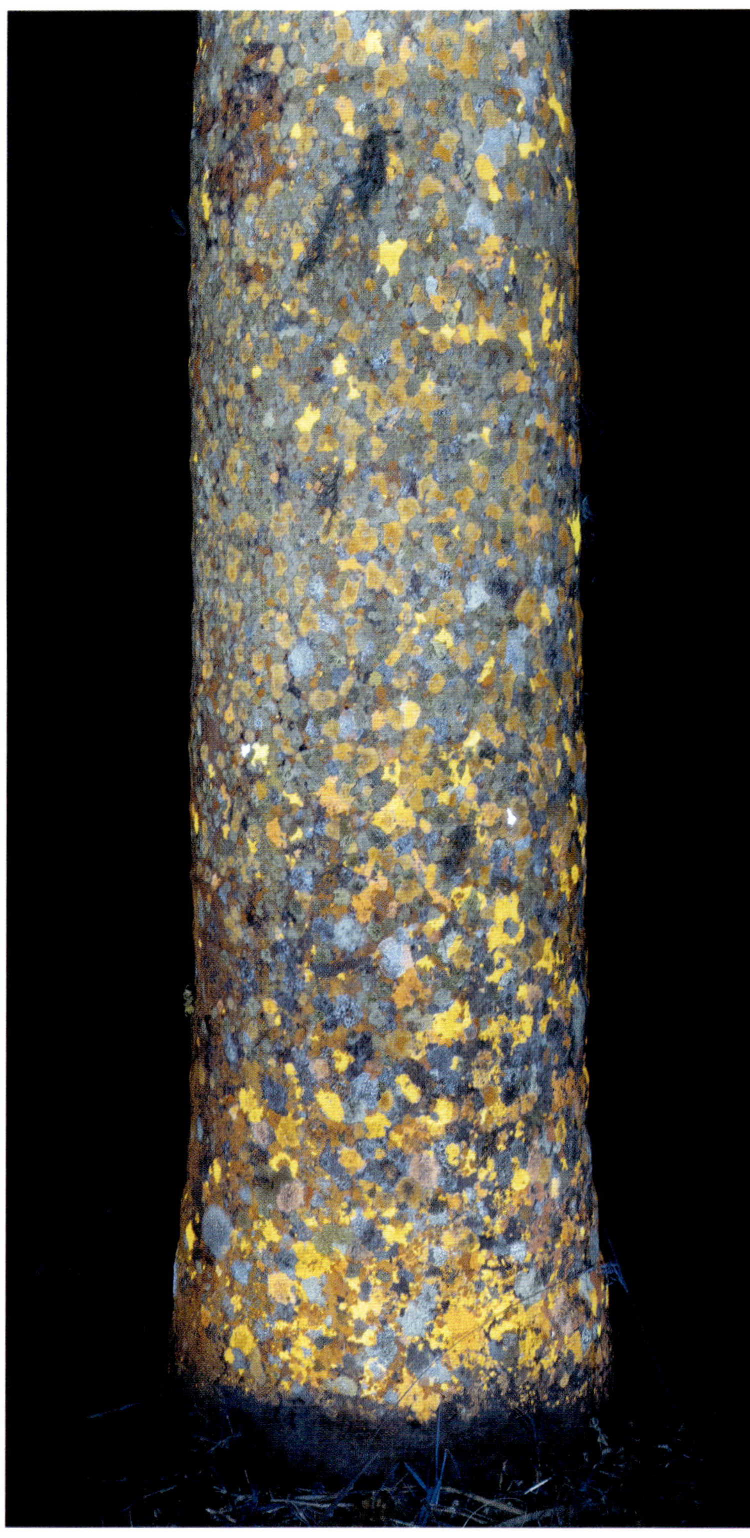

The fungal group that provides some answers to why certain organisms fluoresce are lichens.

'I've photographed these tough, beautiful fungi, on palm trees in the garden, concrete steps, and rocks.

'Boring white crusts, when unveiled under UV, reveal shimmering golds, aquas and oranges.

'I've learned that these colours are a window into how lichens can master life in the most exposed places on the planet.'

LEFT
The subtle shades of green and grey crust lichens on the trunk of a Bangalow Palm transform under UV light into a mesmerizing display of vibrant fluorescence, revealing hidden patterns and colours unseen by the naked eye.

OPPOSITE
The colour a lichen fluoresces under UV light can provide some clues about its species but is not always diagnostic. The same fluorescence colour can occur in different species, and some species may not fluoresce at all.

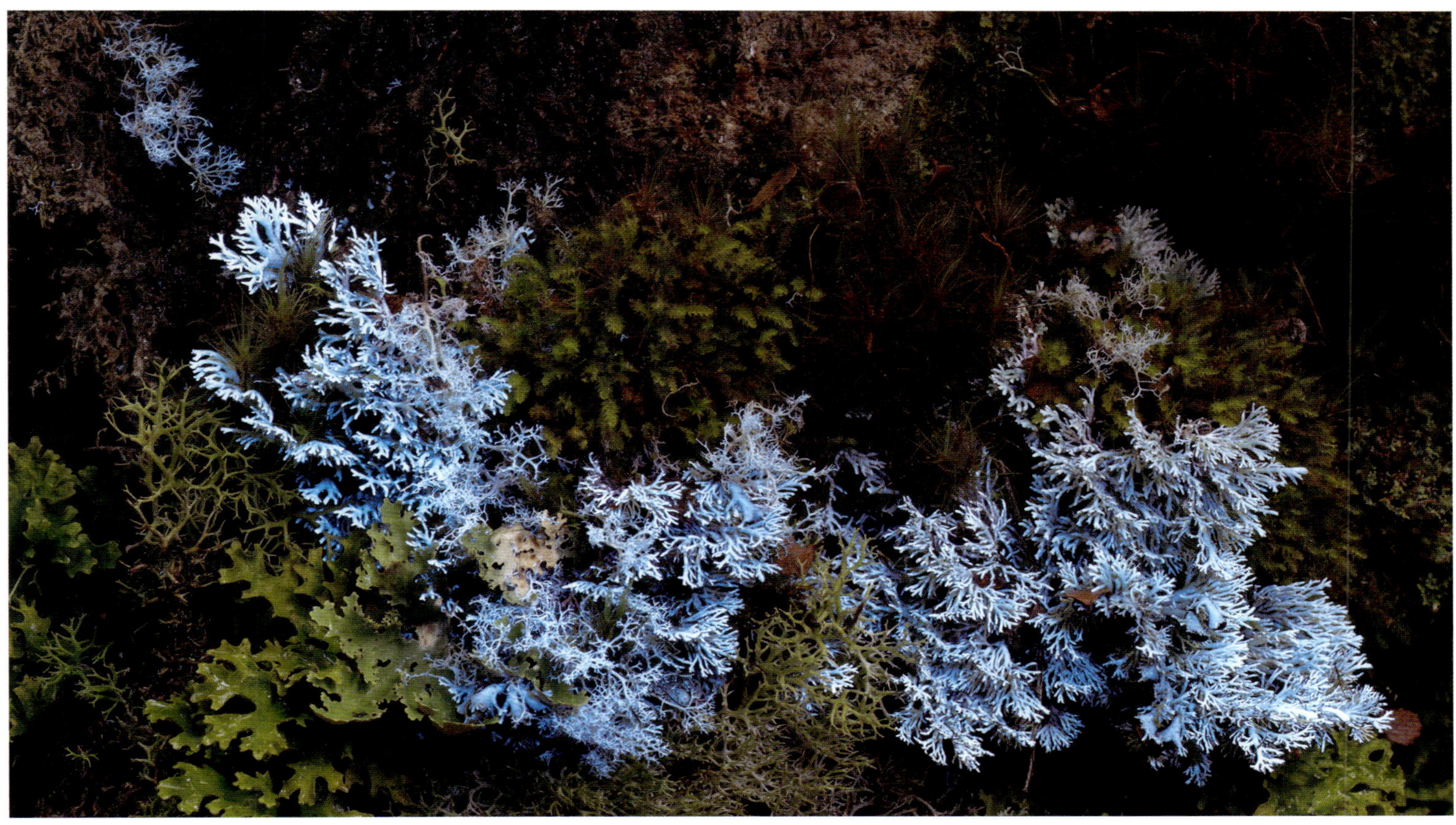

Remember that lichen is a sum of many parts, including a fungus (the mycobiont) and algae or cyanobacteria (the photobiont).

One of the jobs of the mycobiont in many lichens in extreme environments is to create a sunscreen by producing chemical compounds called anthraquinones.[6]

These crystallise on the upper layers of the lichen's thallus, becoming a lattice that absorbs and reflects UV radiation, shielding the lichen from the sun's damaging effect on its structures and DNA.[6,7]

Using this sunscreen, these lichens can control the amount of UV light they absorb as they contract and resurrect.[7]

In other lichens, the ability to fluoresce could also deter herbivores or attract organisms to spread spores, contributing to the lichens' survival.[8]

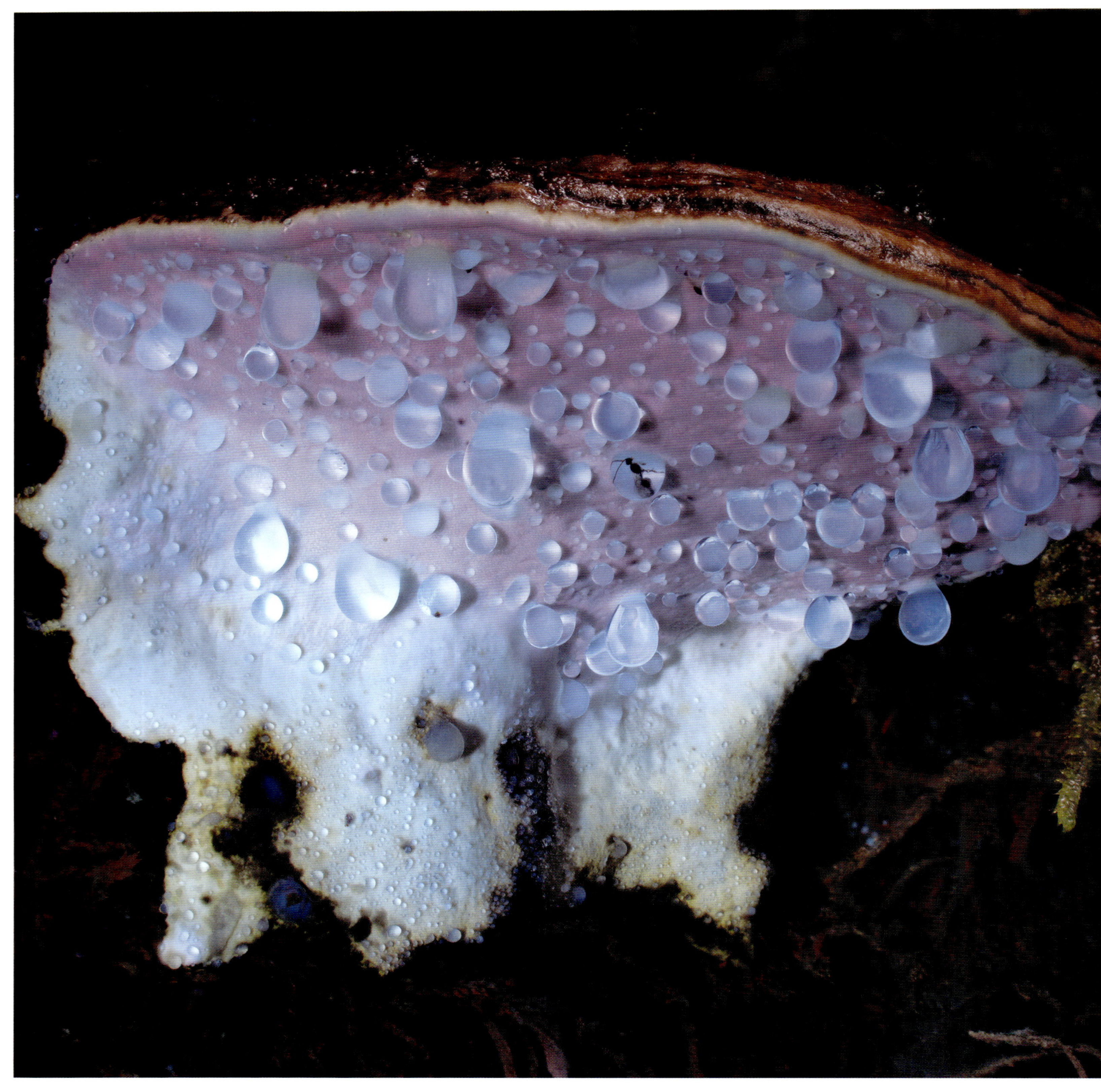

The more we delve into the intriguing world of biofluorescence, the more questions arise.

What role does the power to fluoresce play for those fungi on the forest floor, where harmful UV rays rarely penetrate?

And why do some of the most boring-looking lichens transform into a spectrum of colours under Stephen's UV torch?

Despite all the unsolved mysteries, the wonders of biofluorescence and bioluminescence are already benefitting humanity in numerous ways.

Multiple studies show that fluorescing compounds and pigments found in lichen's natural UV protection could be used to improve sunscreen for humans.[9,10,11]

The enzyme luciferase, which is responsible for bioluminescence, has already proved invaluable in gene tagging, drug discovery, hygiene control and environmental pollution analysis. It even allows us to observe biological processes in real-time within the living bodies of small mammals.[12]

But the fascination with bioluminescence doesn't stop there.

Ganoderma sp. showing guttation (formation of water droplets)
Water droplets form at the tips of the hyphae that make up this ganoderma's sporing body. Guttation is due to high internal pressure, often from rapid water uptake or humidity. Here, under UV light, the droplets glow blue, contrasting with the pink fluorescence of the underside. Within a droplet, a tiny insect is caught in the spotlight.

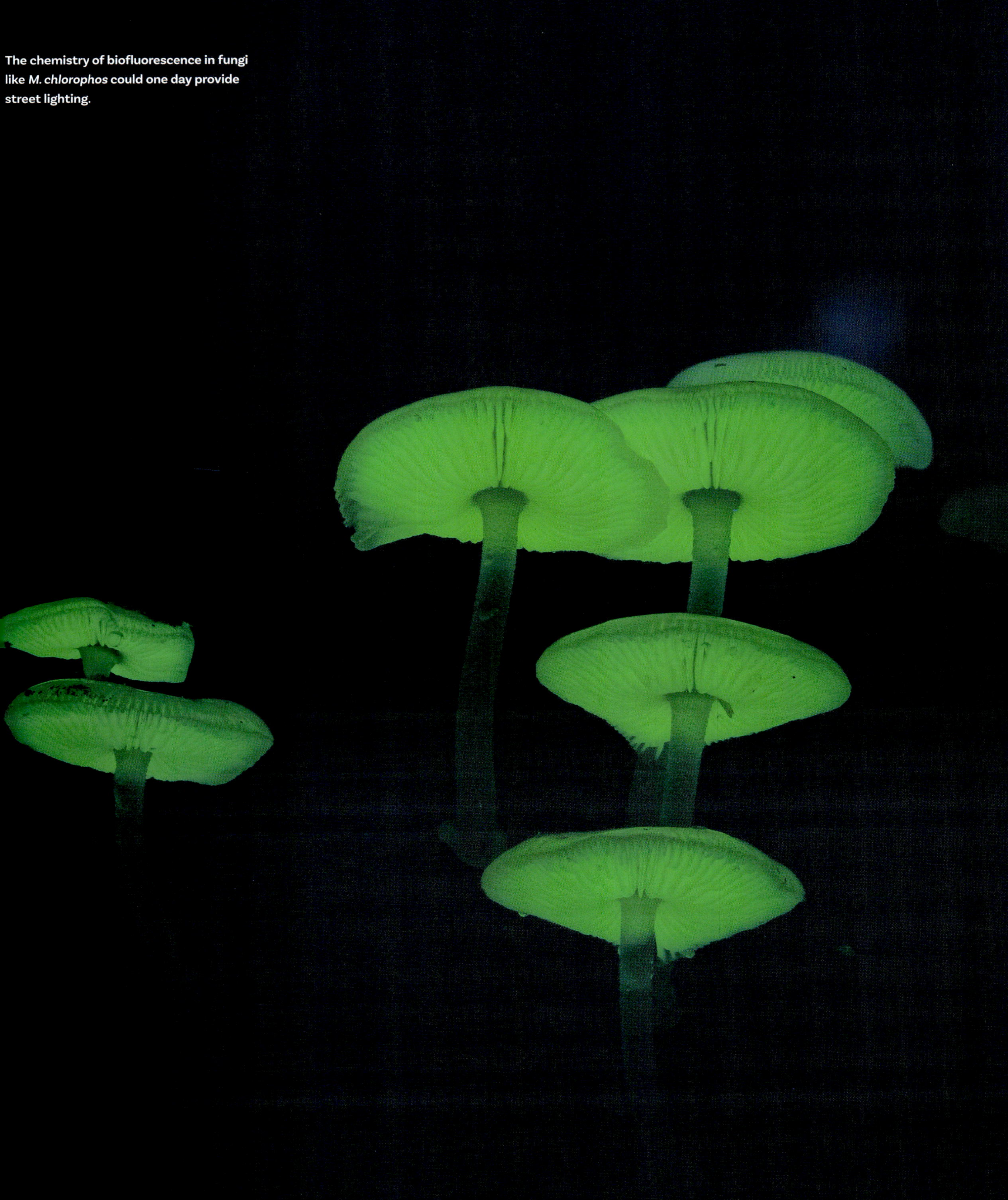

The chemistry of biofluorescence in fungi like *M. chlorophos* could one day provide street lighting.

Inspired by the chemistry of luminescence, scientists are also exploring novel projects that could transform how we light up our cities.

By creating bioluminescent plants and bacteria through genetic engineering, their vision is to replace traditional streetlights with glowing flora or lamps filled with shimmering microbes.[12]

'Imagine walking down a street lined with the soft, natural glow of bioluminescent organisms.

'That is sure to bring a touch of magic to our urban landscapes.'

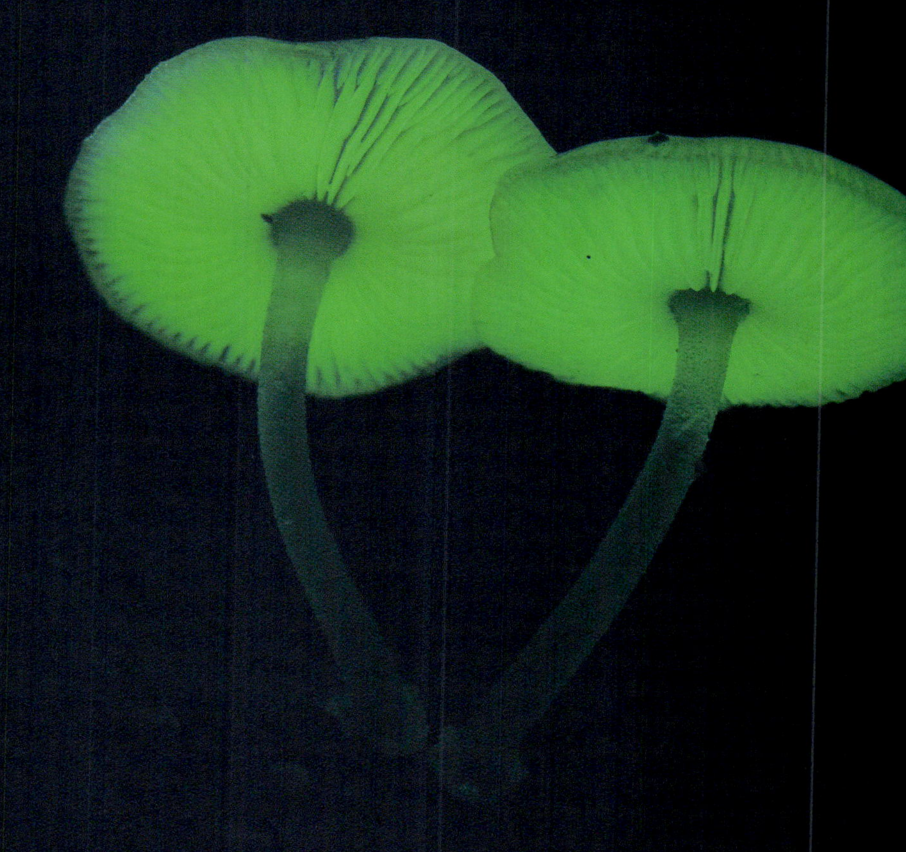

In the shadow of the Himalayas

We cannot protect biodiversity for future generations if we don't know what is there.

Beneath the towering peaks of the Eastern Himalayas there is a vast expanse of uncharted forests.

Fungal treasures in this remote and rugged terrain have never seen the curious eyes of a mycologist.

However, by the late 2010s, interest in fungi was spreading across the region, partly due to the efforts of our collaborators at the Kunming Institute of Botany in China.

Their initiative, the Centre for Mountain Futures in the Eastern Himalayas, works with local communities and governments to 'develop sustainable solutions that promote biodiversity and conservation, enhance livelihoods, and mitigate the impacts of climate change'.

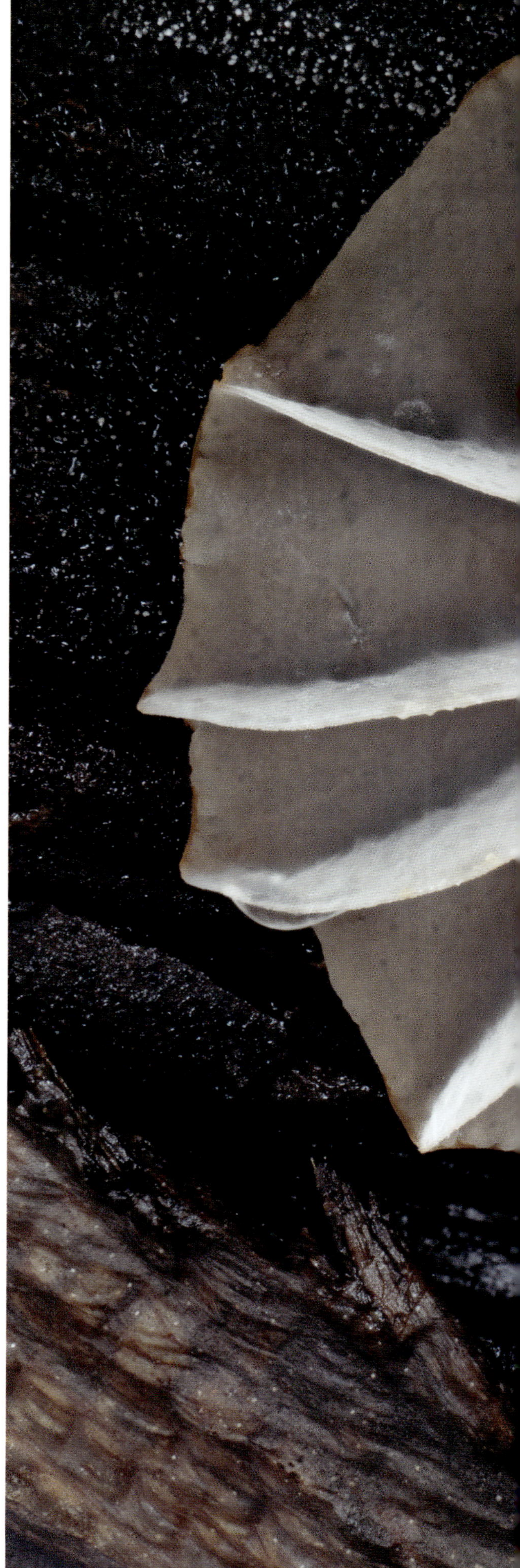

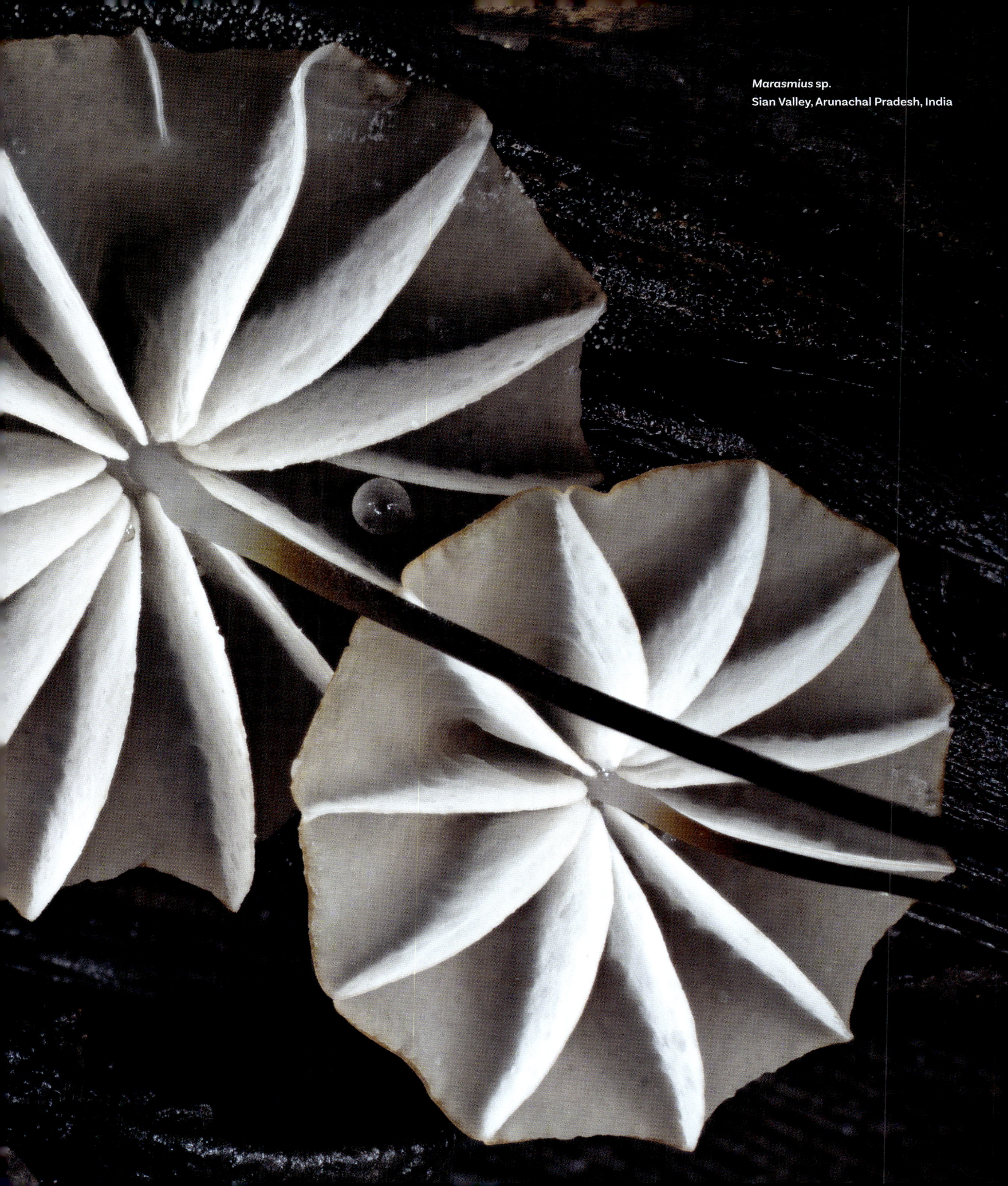

Marasmius sp.
Sian Valley, Arunachal Pradesh, India

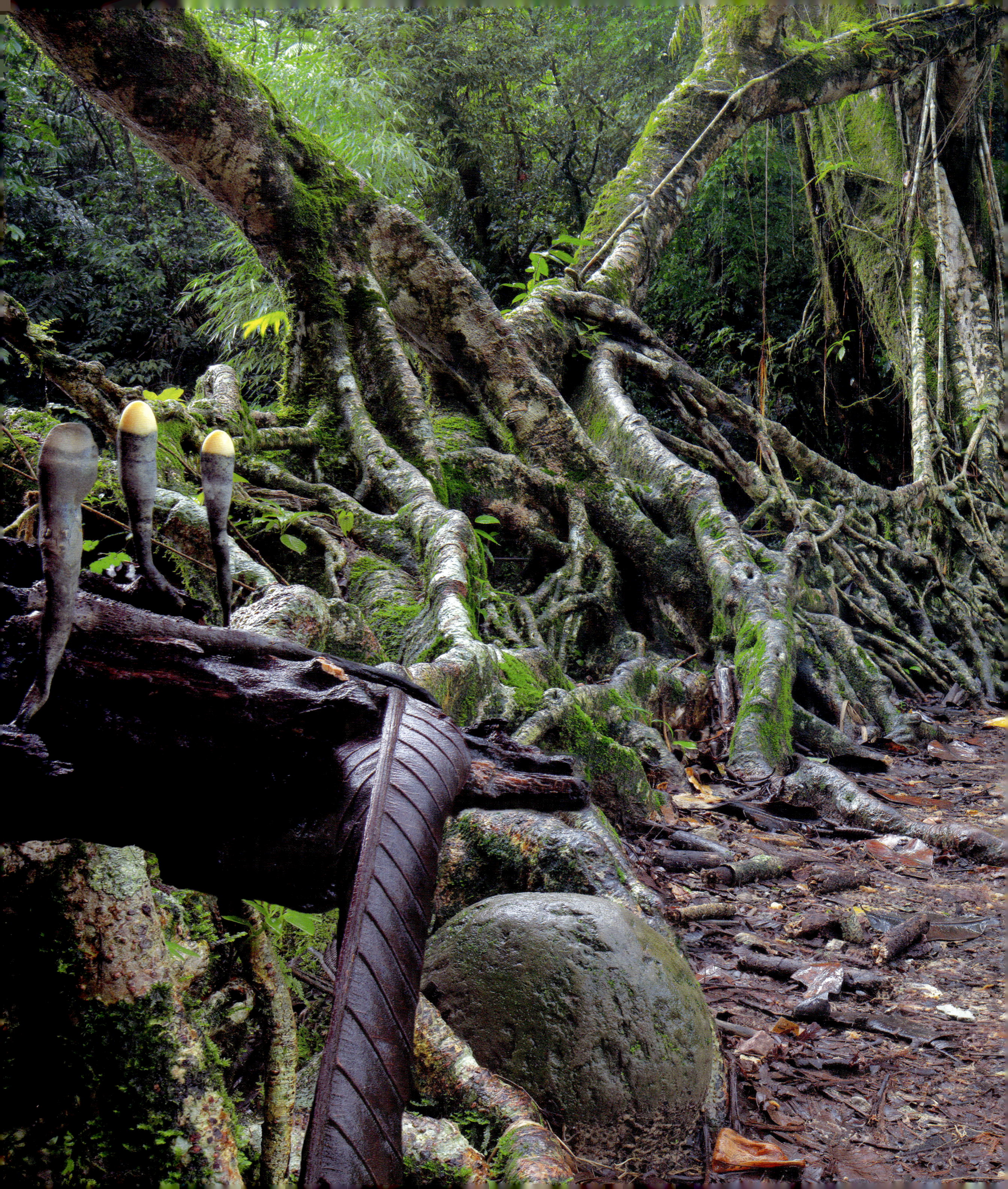

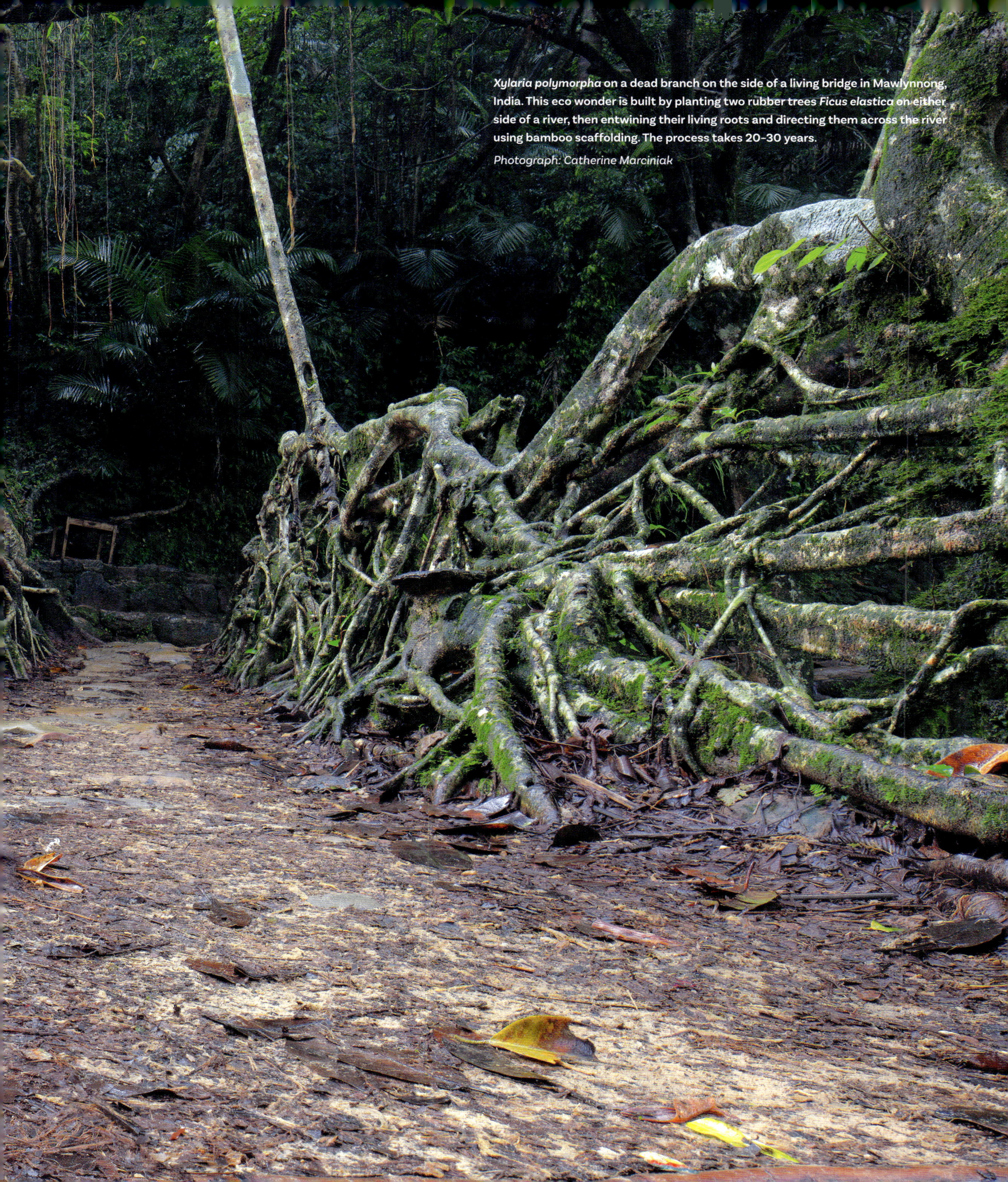

Xylaria polymorpha on a dead branch on the side of a living bridge in Mawlynnong, India. This eco wonder is built by planting two rubber trees *Ficus elastica* on either side of a river, then entwining their living roots and directing them across the river using bamboo scaffolding. The process takes 20–30 years.

Photograph: Catherine Marciniak

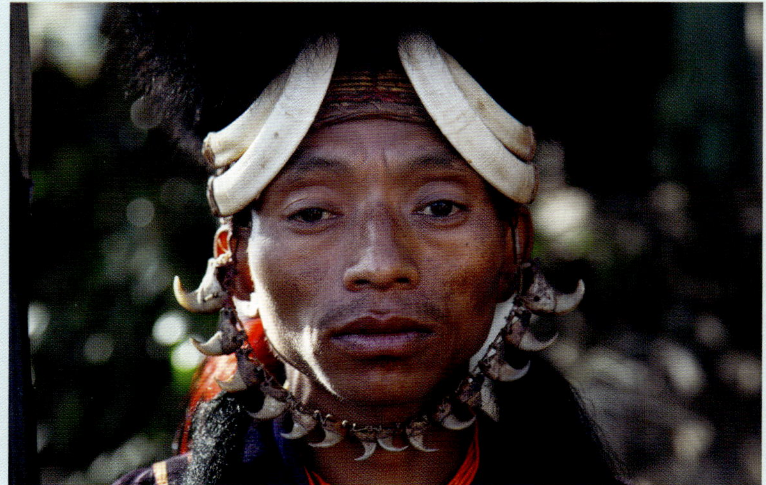

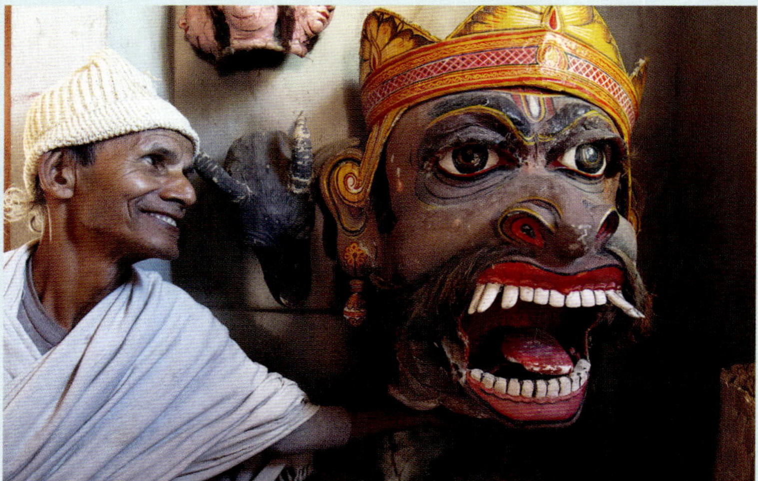

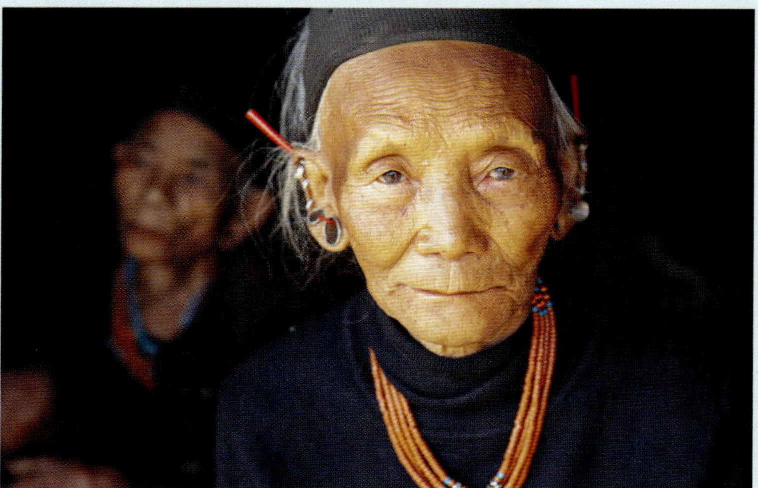

Professor Peter Mortimer urged their Mountain Futures partners to add fungi to their conservation agendas, and the Balipara Foundation in India took the call to action seriously.

They operate in Meghalaya, Assam, Manipur, Nagaland, Tripura, Mizoram, Arunachal Pradesh and Sikkim.

Each state in this enchanting region offers breathtaking vistas and unique customs and beliefs.

From the vibrant Hindu festivals of Assam to the magical allure of Meghalaya's living root bridges, and the proud legacy of Nagaland's former head-hunters, it's a corner of the world that is a haven for humanity's rich tapestry of diversity.

TOP

A Nagaland tribesman embracing tradition at the Hornbill Festival in Kohima, India.

Photograph: Catherine Marciniak

MIDDLE

Masks made of a mix of bamboo, mud and natural fibres are used by monks on Majuli Island in traditional spiritual drama and dance performances.

BOTTOM

The Konyak people in Nagaland live in a strategic trading area right on the border of India and Myanmar.

Photograph: Catherine Marciniak

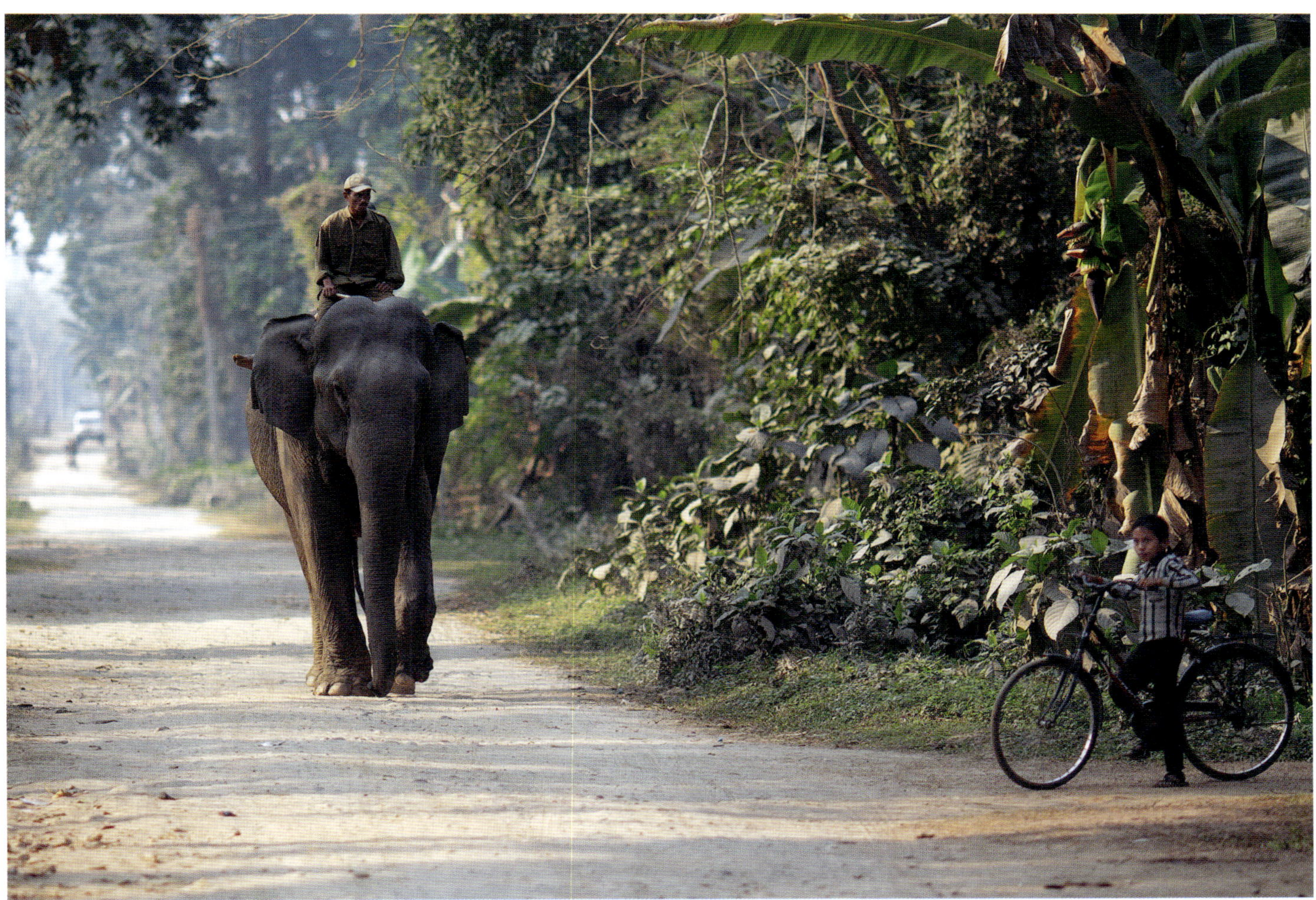

Founded in 2016 to restore forests across Assam's elephant country, the Balipara Foundation's vision quickly grew. They began restoring vast forest habitats, establishing wildlife corridors and developing eco-friendly agriculture and agroforestry.

Recognising the intrinsic bond between people and nature, they collaborate closely with rural communities, empowering them to become stewards of their own landscapes, to ensure a flourishing biodiversity for generations to come.

ABOVE
Escalating habitat loss in Assam, India, intensifies the struggle
between elephants and humans for dwindling resources, creating
a pressing conservation challenge.

It is within this context that, in 2018, the Balipara Foundation invited us to join them on a month-long field trip to document the wild fungi of this Eastern Himalayan region.

Gautam Baruah, an environmental scientist and head of Balipara Foundation's Rural Futures initiative, led the expedition.

For each of the forests we were to visit, he had lined up foraging experts from nearby villages to locate the fungi and tell us what they knew about their edibility and toxicity.

Their knowledge was invaluable but we soon realised they were relying on us to be the mycological experts.

Prior to this, our international fungi field trips included mycologists who identified species and collected specimens for DNA sequencing.

In the remote country of India's north-east, no mycologist specialising in the taxonomy of forest fungi could be found for this pioneering fungi safari.

We were not sure we were up to the task as we only had our past experiences to rely on. We hoped we had learned by osmosis.

We came up with a fallback plan to help identify what we were finding. Stephen would not only photograph the fungi finds but we would also film his conversations with the village experts in a documentary, thereby sharing our discoveries with the international science community.

Marasmius cf. luteolus
Hoollongapar Gibbon Sanctuary, Assam, India

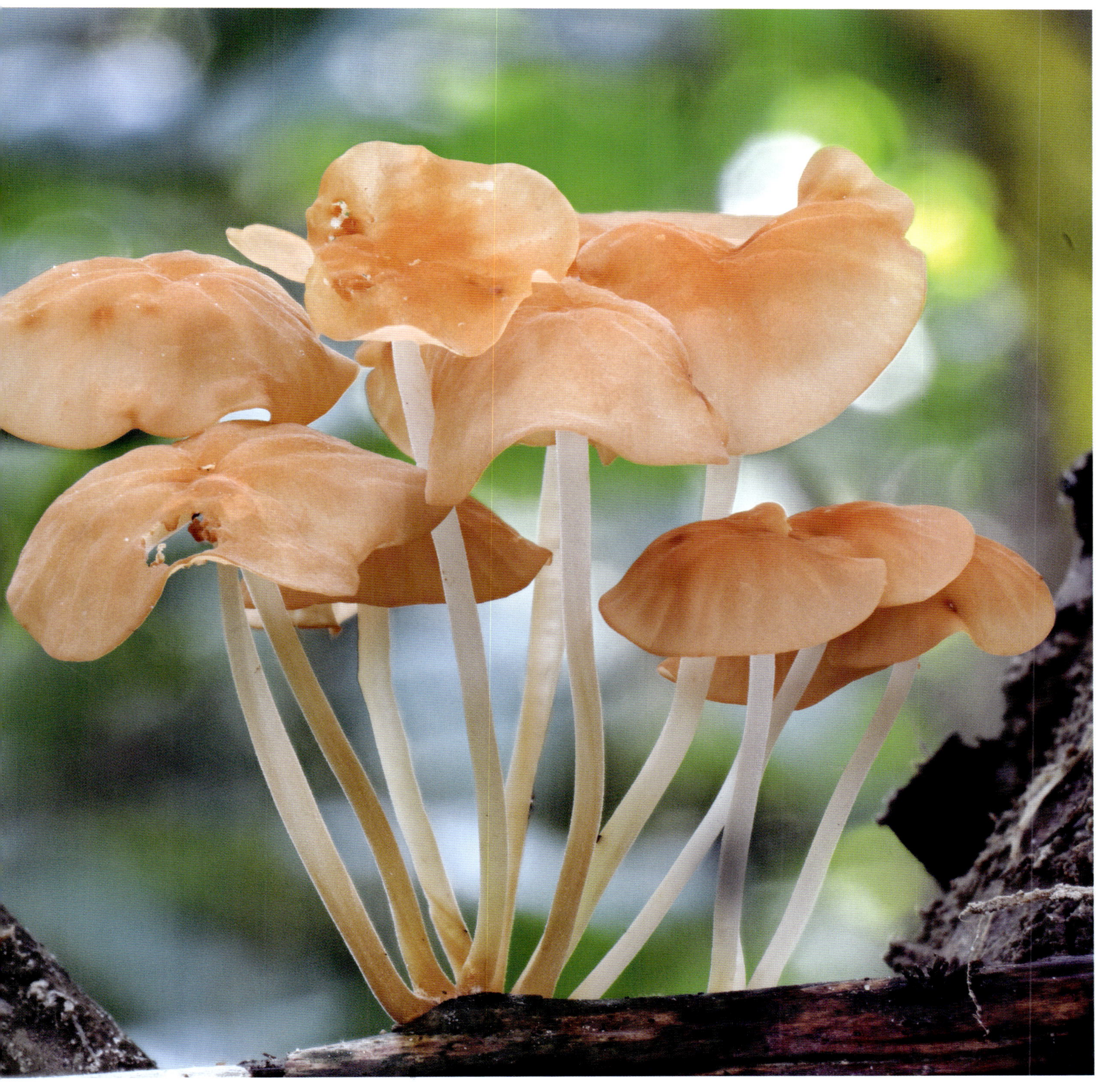

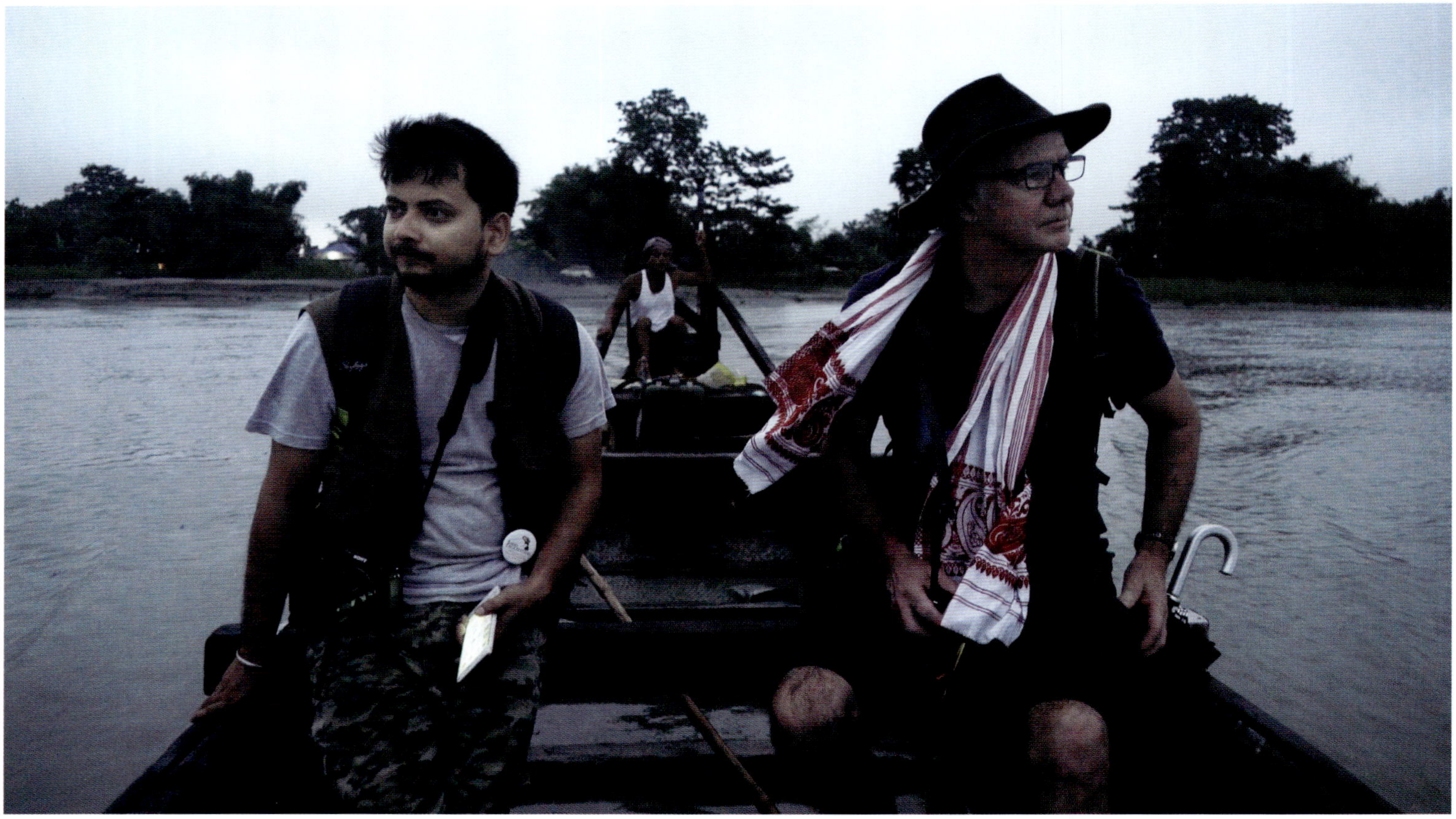

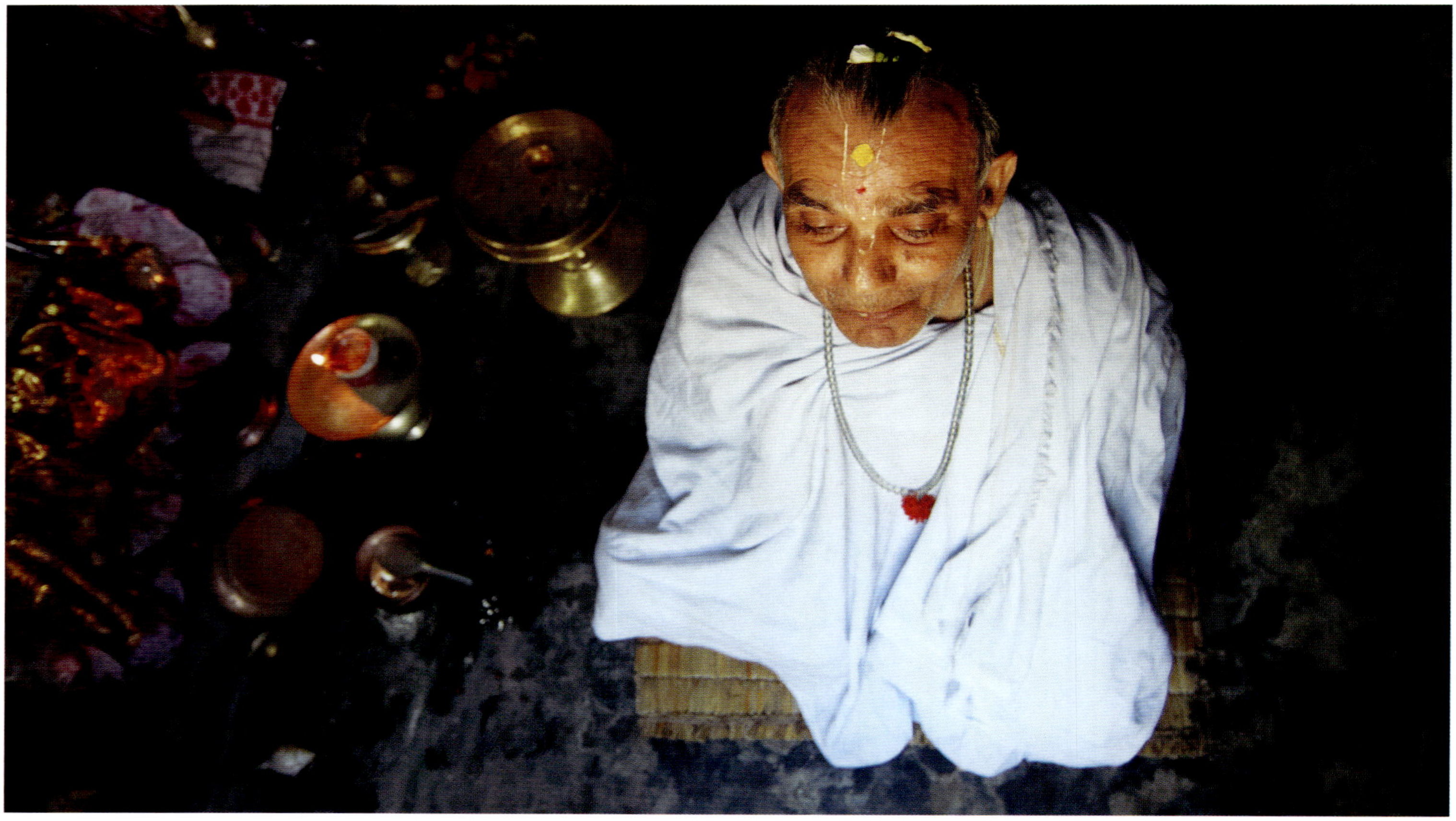

Our journey started on Majuli Island in the heart of the Brahmaputra.

This mighty river, born in the highlands of Tibet, flows southwards through Assam.

We crossed from the mainland to the Island just before sunrise with stunning dawn pastels and the spluttering percussion of an ancient outboard motor providing a backdrop to our journey.

Majuli Island is home to the Mising tribe, fisher people and farmers, whose daily lives are intertwined with the ebb and flow of the river's dry and wet seasons.

The island is also home to thousands of Hindu monks, some as young as 5 years old. Living in monasteries that date back to the 16th century, they follow ancient traditions that blend worship with the arts.

Through mask theatre, song, drumming, gymnastics and dance they pay homage to their spiritual beliefs. It is a melting pot of traditions, the convergence of ancient Hinduism and Buddhism, where the beauty of the rituals is irresistible to the eye of a photographer.

On Majuli Island we were to meet the inspiring Padma Shri Jadav Payeng, known as the Forest Man of India. He had embarked on a colossal mission to plant life back onto his island.[1]

In 1979, the swollen Brahmaputra swept away thousands of homes in Assam, leaving over a million people homeless.[2]

Dawn crossing of the Brahmaputra River in Assam, India.

Majuli Island is home to over 20 Vaishnavite monasteries, traditional prayer halls where monks worship Vishnu.

Jadav Payeng was just 16 years old at the time and was distraught by the impact on wildlife. Thousands of snakes were washed up onto his island, where land clearing meant there was no shade or tree cover. They suffered and died slowly in the scorching post-flood summer sun.

On consulting his village elders for advice, Jadav Payeng was given 50 bamboo saplings to plant in the sandbars. He then used earthen pots filled with water to create a drip irrigation system.

The bamboo grew quickly, creating and holding fertile soil in place. Seeds from the upstream Himalayan forests carried by the river washed onto the island and took hold in the bamboo.

Jadav Payeng propagated and planted 10,000 seedlings annually to supplement the regrowth.

And so, a jungle was born.

Now, after 40 years, the forest is over 500 hectares and is home to an array of wildlife, including deer, leopards, rhinos, elephants, snakes and a species of vulture that was threatened with extinction.[2]

This remarkable story is testament to the power of one person's determination and the resilience of nature when given the chance to flourish.

Marasmius cf. aurantioferrugineus
Wild Mahseer, Assam, India

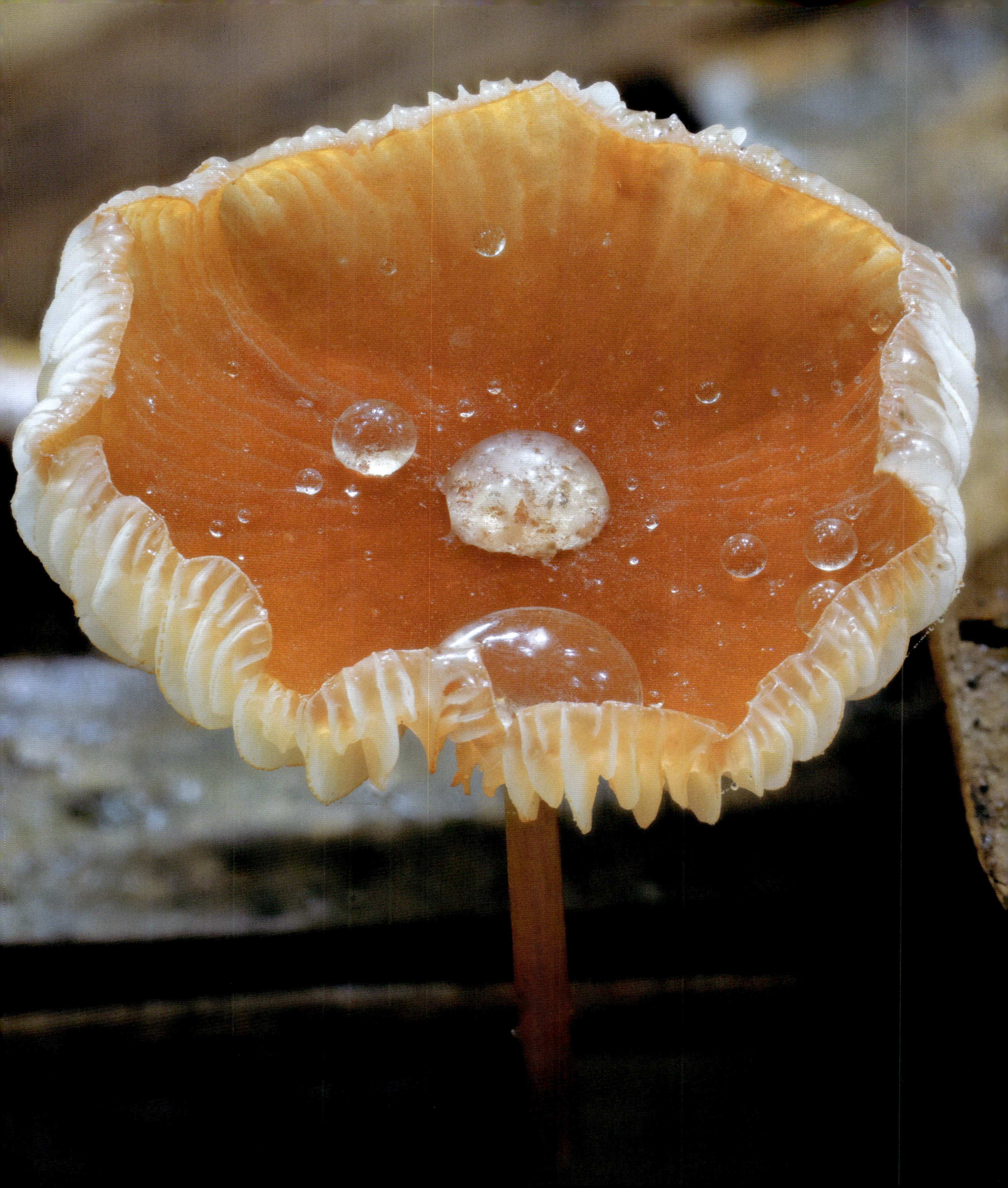

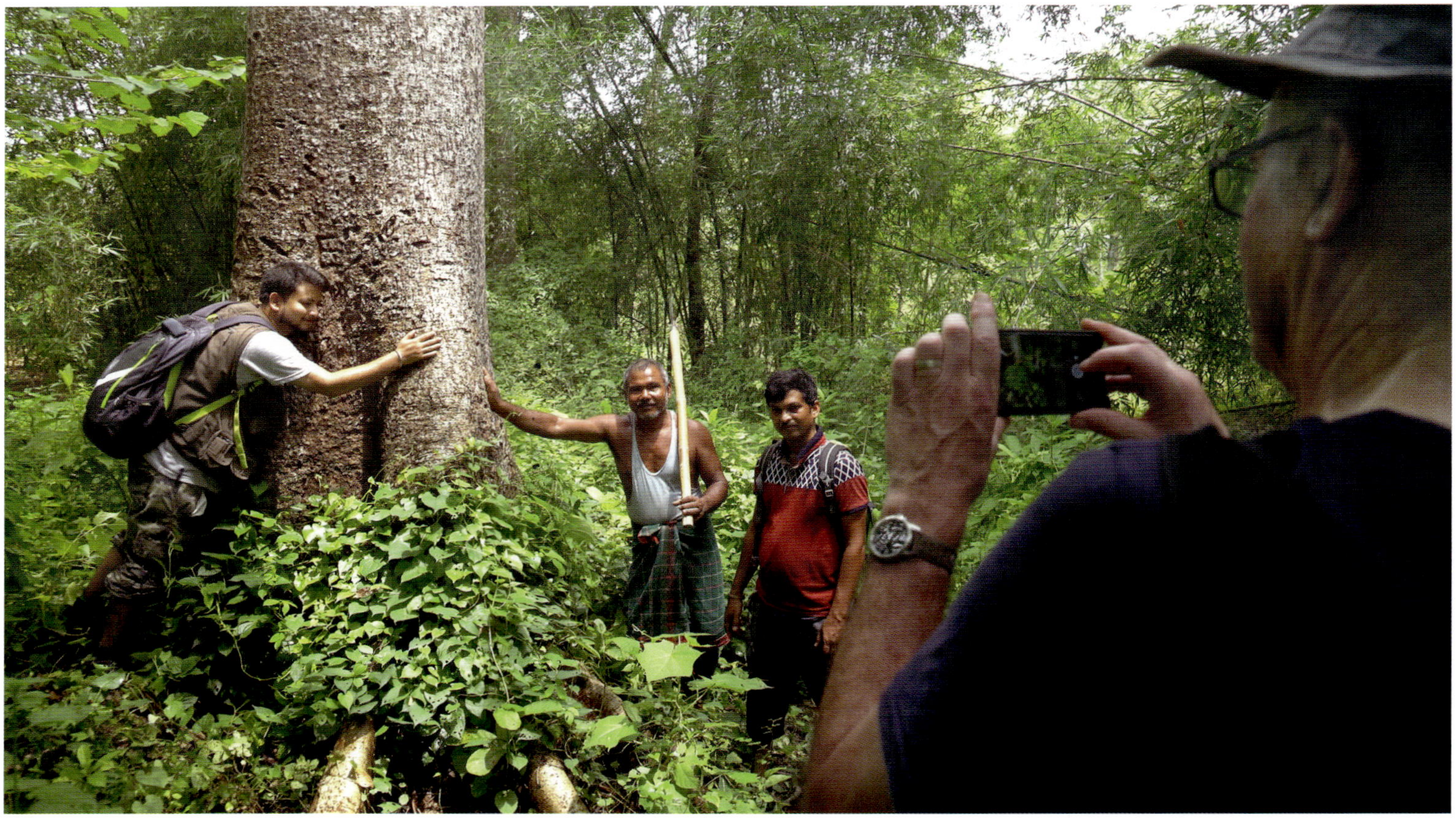

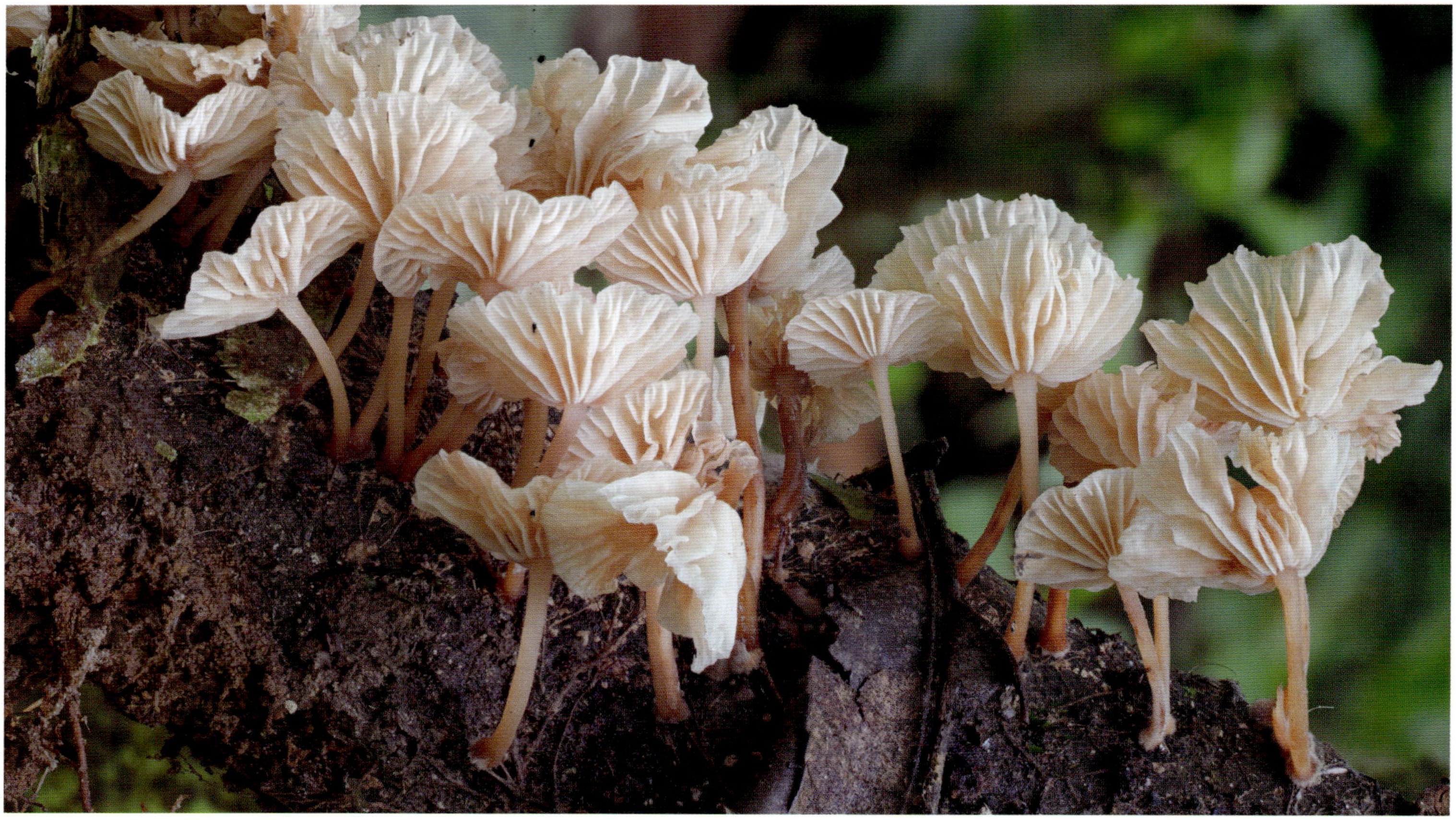

Trekking 8 kilometres to reach Jadav Payeng's forest was no easy feat. We lugged our camera gear across flooded creeks as temperatures climbed to the mid-30s with suffocating tropical humidity.

But as soon as we stepped into the shade of his towering trees and bathed in his forest, it was clear that our morning's efforts were well worth it.

Jadav Payeng had an intimate relationship with every tree, every blade of bamboo and every mushroom. He had a keen eye for observing the interactions between organisms, and shared how mushrooms attracted insects that attracted birds, that spread the seeds of the trees – the giant web of nature.

Despite this wisdom, he openly admitted his Mising tribe had lost knowledge about edible mushrooms over the past two generations, with many people having accidentally consumed poisonous mushrooms and suffered the consequences.

At the same time, Jadav Payeng was aware that people in Assam were not getting enough protein from their diets of legumes and rice, with children being particularly at risk of protein malnutrition, leading to physical stunting and poor brain development.[3,4]

Mushrooms are a high-quality source of protein, and our colleagues from the Balipara Foundation were eager to learn if there were any species in Jadav Payeng's Molai Forest that could be cultivated or wild-harvested to increase protein in diets.[5,6]

In this regrowth forest, we were greeted by a variety of fungi, thriving on the fallen wood around us.

OPPOSITE TOP
Padma Shri Jadav Payeng proudly shows us a towering *Bombax ceiba* tree, nurtured from a seedling he planted more than 30 years ago.

OPPOSITE BOTTOM
Unknown species of mushroom
Sian Valley, Arunachal Pradesh, India

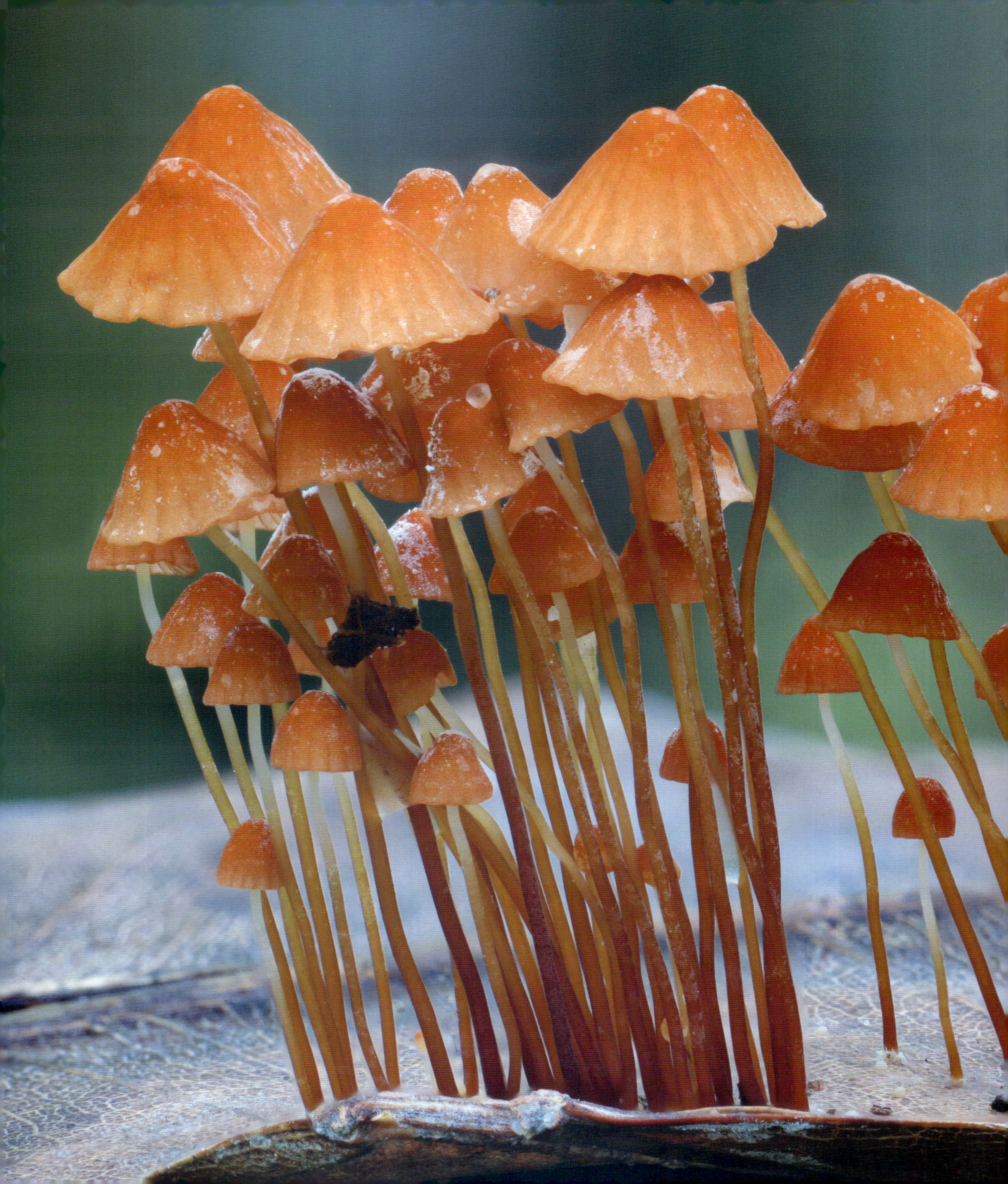

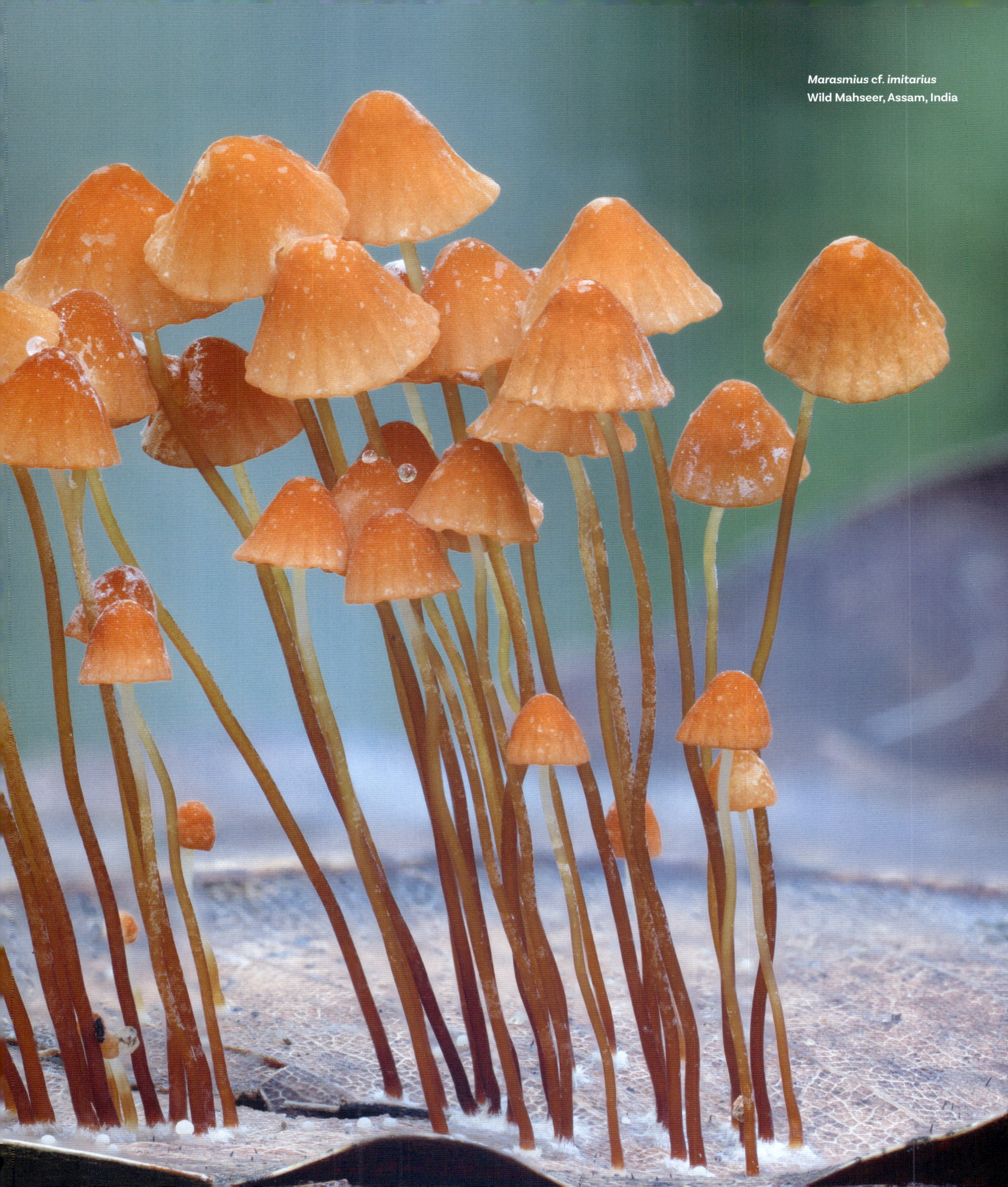
Marasmius cf. imitarius
Wild Mahseer, Assam, India

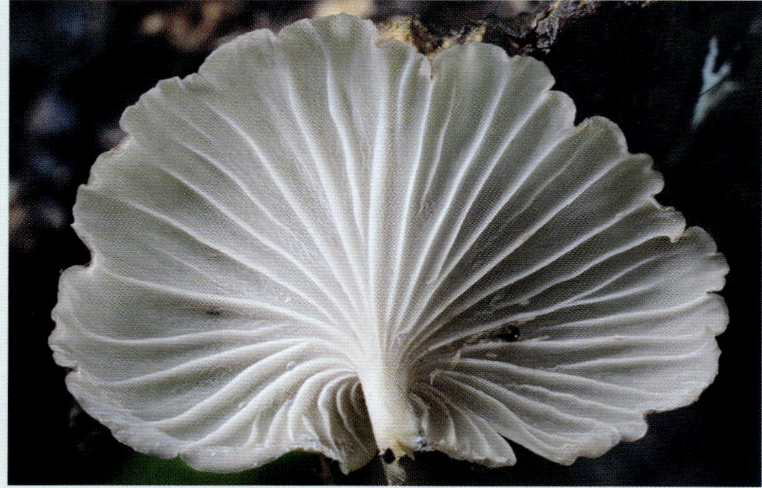

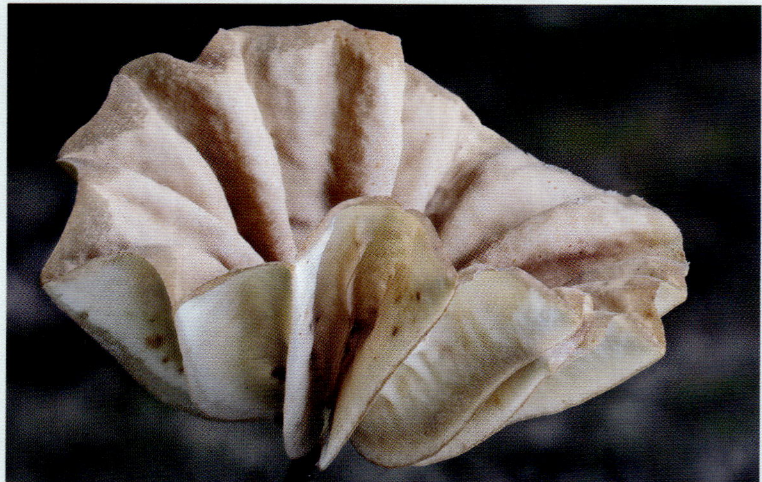

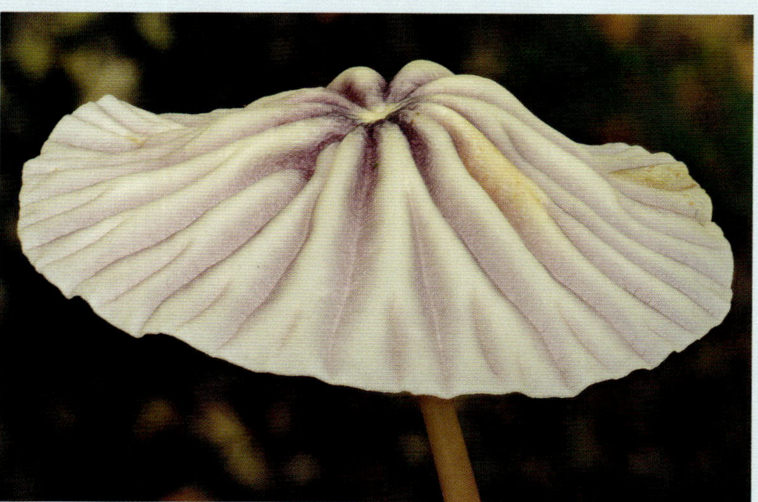

A small stick was adorned with pink-gilled mushrooms sporting frilly upturned petticoats, while others shimmered with crystal raindrops from the soft morning shower.

Stephen captured the beauty of many small mushrooms with chiselled gills emerging from dead twigs and fallen branches.

These fungi play a vital role in the ecosystem, breaking down organic matter and returning nutrients to the soil.

TOP
Trogia sp.
Sian Valley, Arunachal Pradesh, India

MIDDLE
Marasmius cf. *leveilleanus*
Hoollongapar Gibbon Sanctuary, Assam, India

BOTTOM
Marasmius cf. *bekolacongoli*
Sian Valley, Arunachal Pradesh, India

OPPOSITE TOP
Xylaria sp.
Mawphlang, Meghalaya, India

OPPOSITE BOTTOM
Scutellinia scutellata, eyelash cup
Yumthang Valley, Sikkim, India
Distribution: worldwide

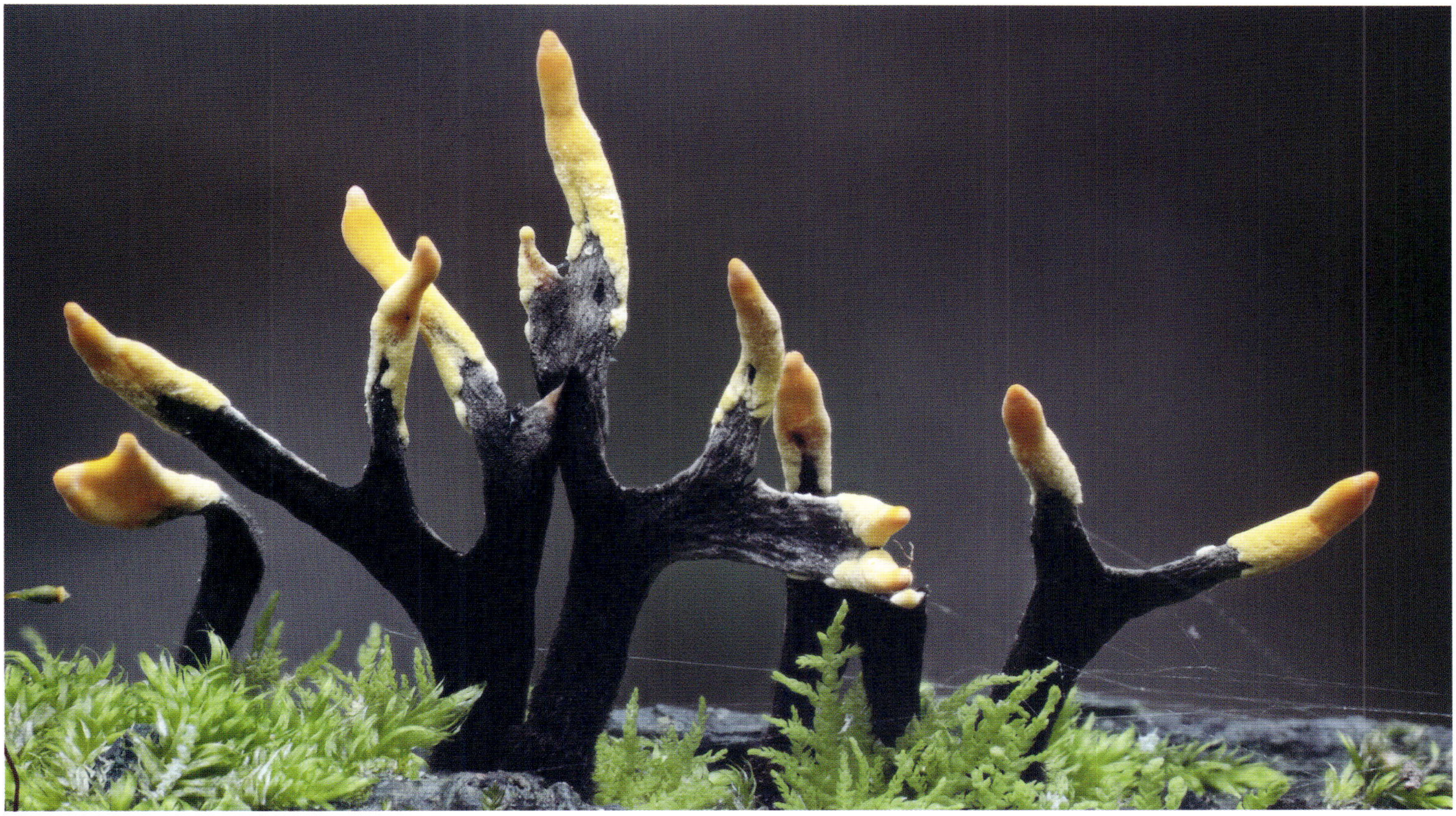

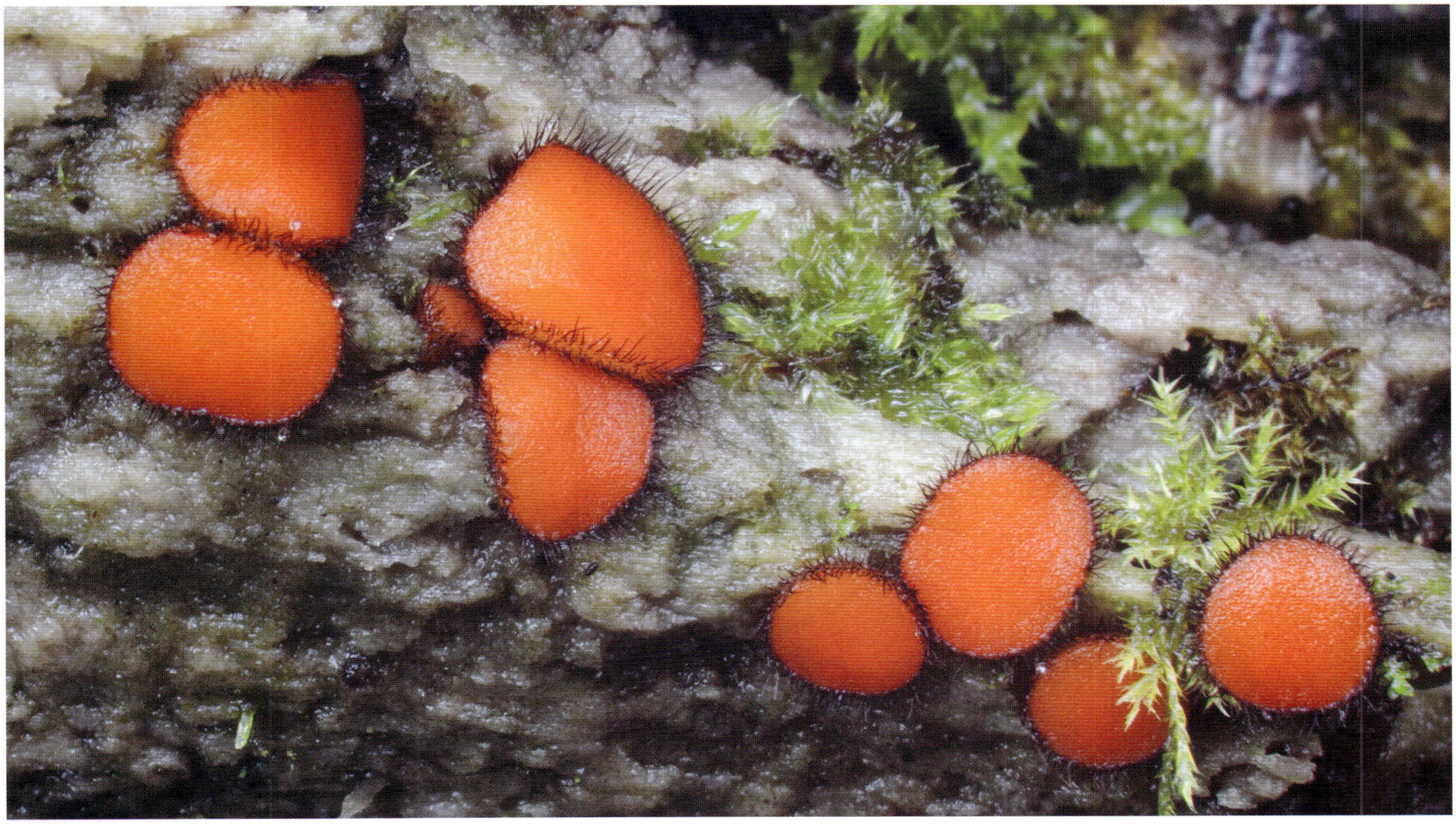

Despite their critical importance, we didn't recognise any from groups known to be edible for humans.

Then we spotted a promising candidate.

Growing from an earth mound was a cluster of grey mushrooms with hard points at the top of the cap.

These mushrooms are a species of *Termitomyces*, a fungal genus not only famous for their delicious taste but also for their incredible alliance with termites, a partnership that began around 30 million years ago.[7,8]

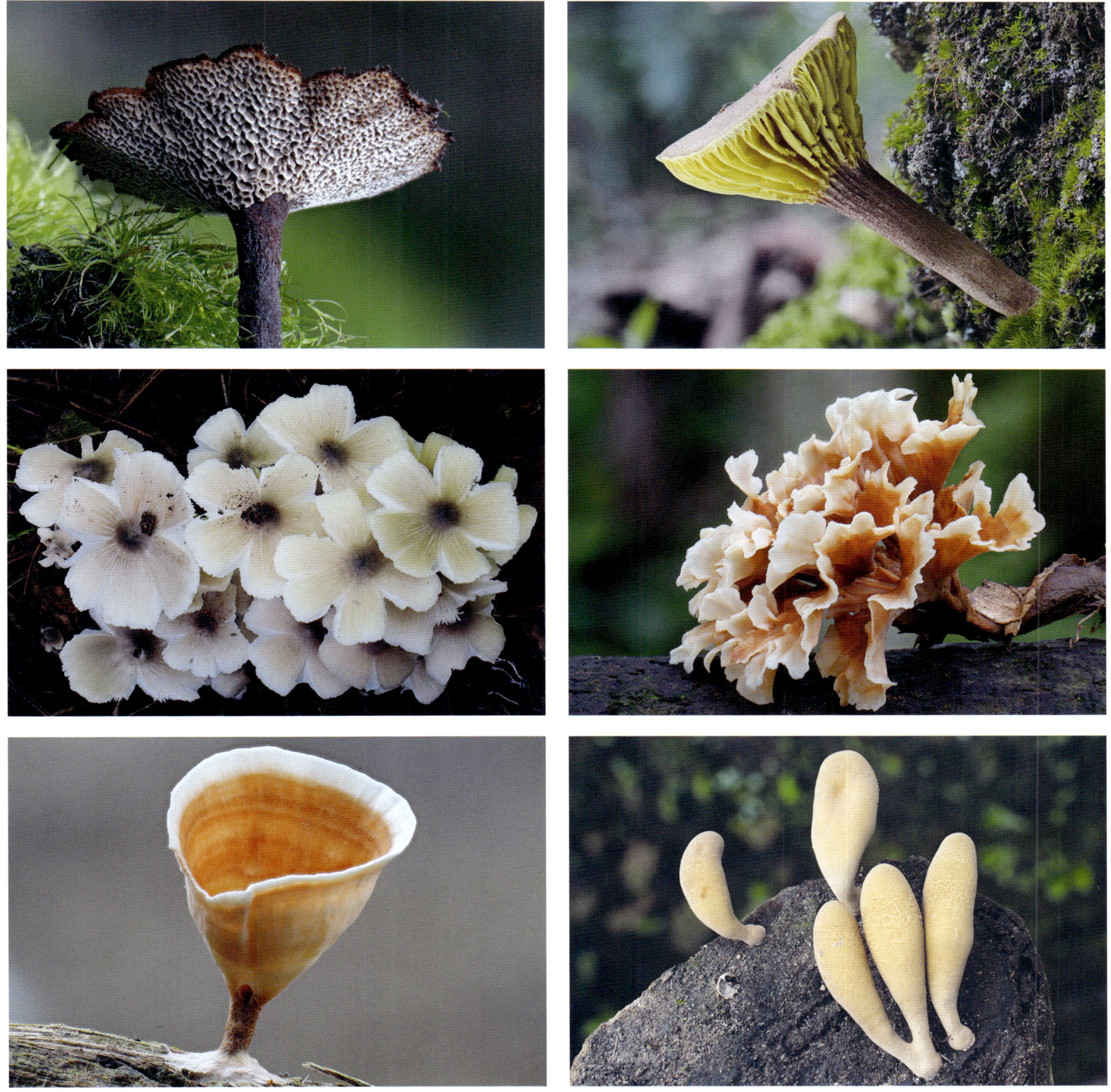

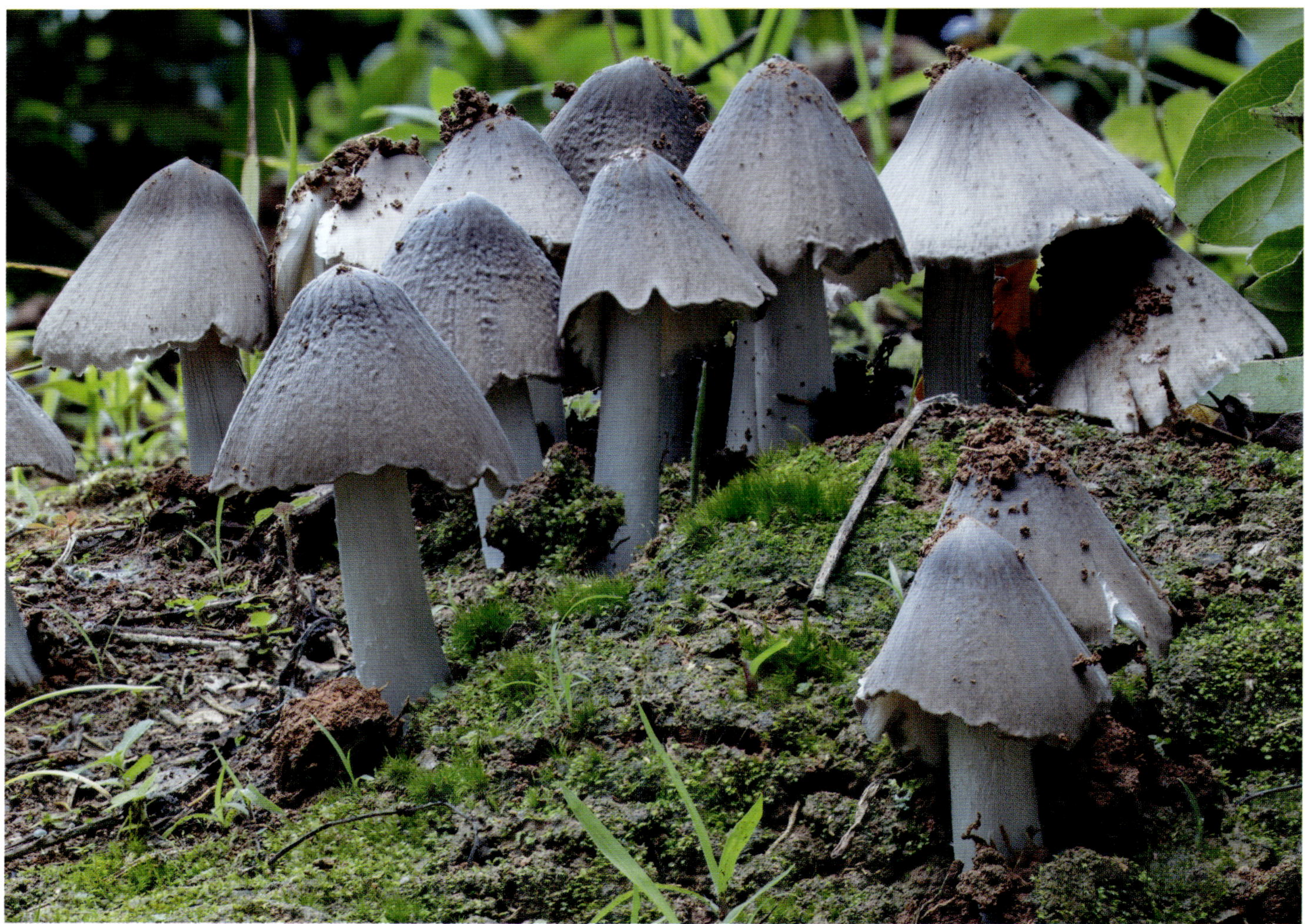

Beneath the mound is a fungus-farming termite nest.

A favourite food of these termites is decaying plant matter, but this species of termite does not have all the gut microbes they need to digest the lignin and cellulose that bind plants together. They have outsourced that job to their fungal partner.[9]

Above ground, the termites forage for decaying plants and spores they collect from *Termitomyces* mushrooms, which they mix together in their gut.[8,10]

Termitomyces intermedius
Nabanhe National Nature Reserve, Yunnan, China
Distribution: east Asia

Then, in the labyrinth of the nest, they sow any undigested lignin, cellulose and spores they excrete into specialised comb-like structures, an inoculated substrate for their fungus garden.[8,9]

Their nest provides the perfect conditions for the fungus to grow, with high humidity and a constant temperature.[8,9]

Like human farmers nurturing a crop, the termites then keep the fungus fed with a continuous supply of semi-digested plant matter.

As the fungus-enriched comb matures, it produces small fleshy fungal nodes – a tasty meal for the termites.[8,9]

TOP
Cortinarius sp.
Yumthang Valley, Sikkim, India

MIDDLE
Cookeina insititia, fringed goblet
Dibrugarh, Assam, India
Distribution: Asia and Australasia

BOTTOM
Irpex sp.
Mawphlang, Meghalaya, India

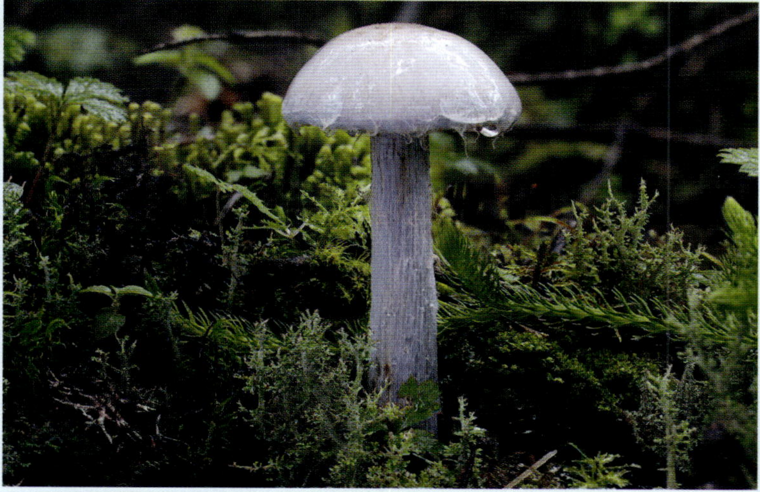

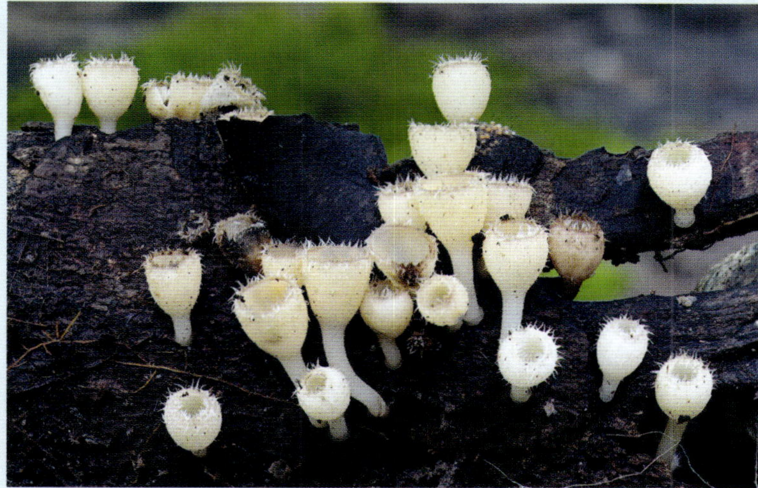

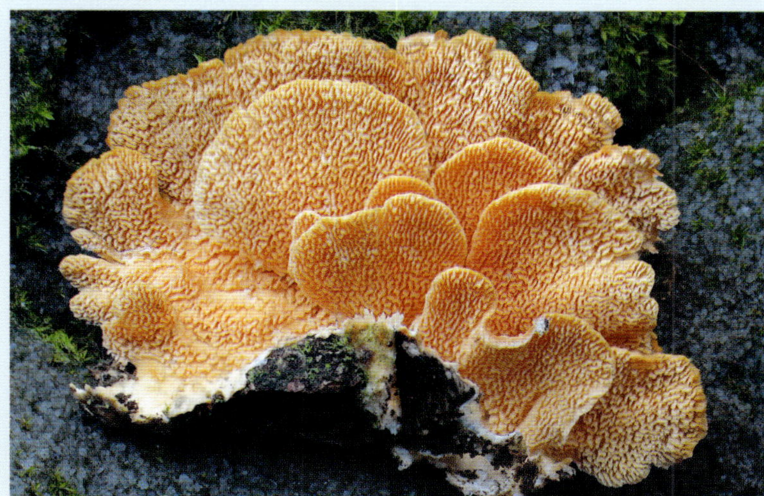

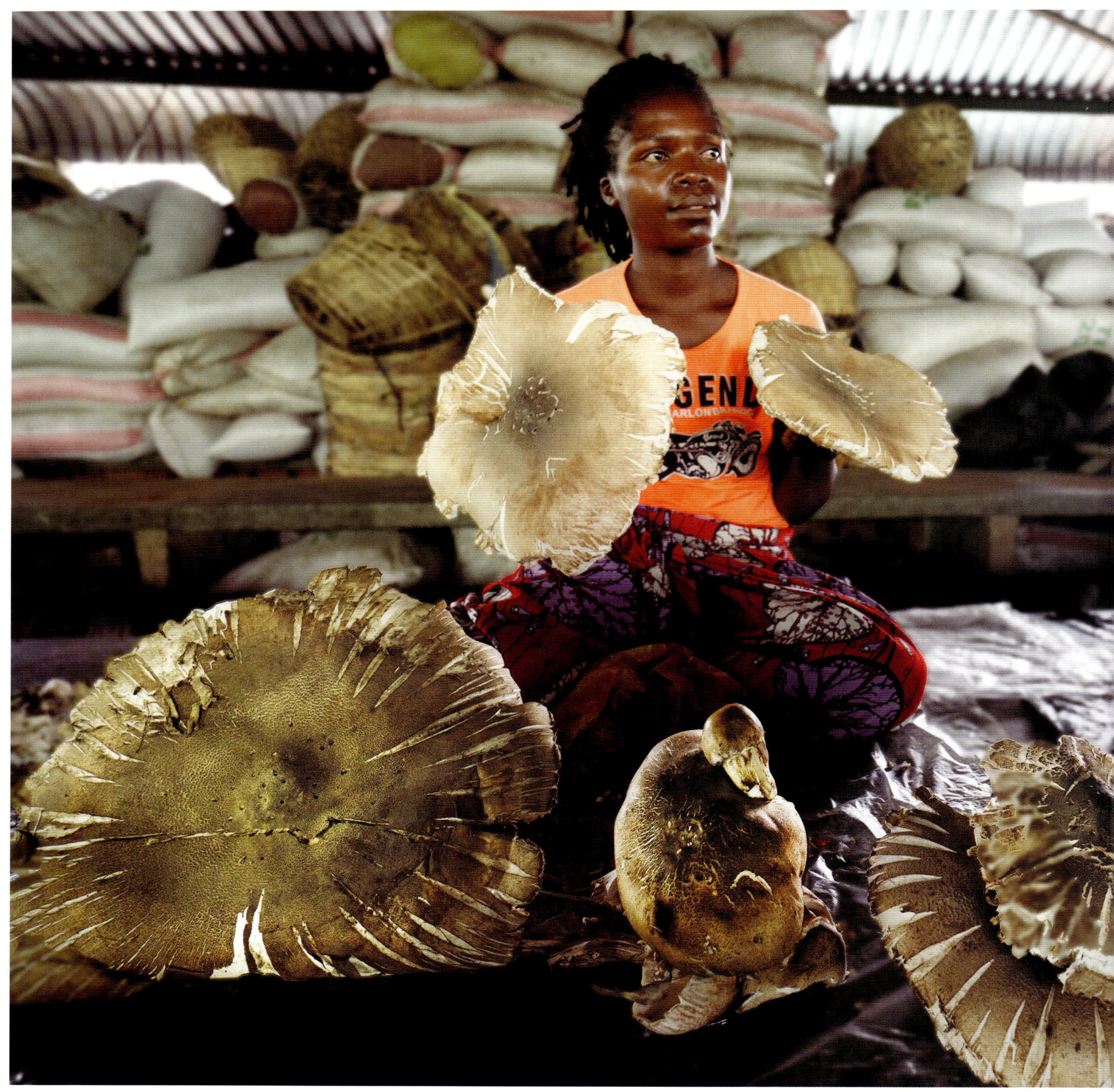

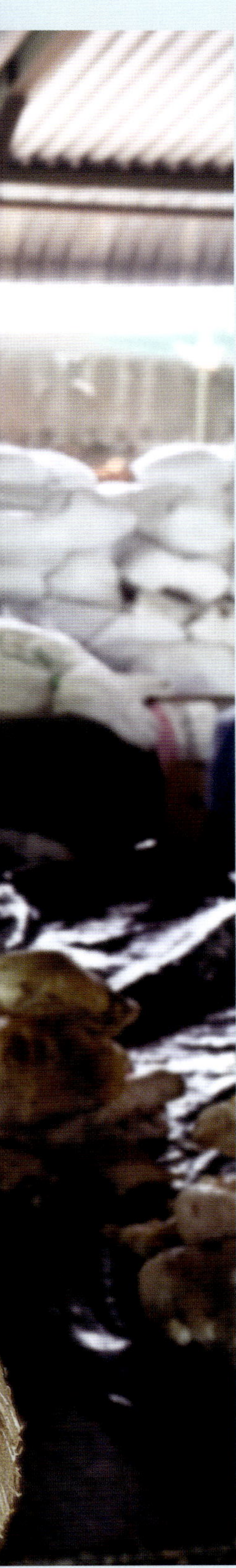

Enough of the comb substrate is left by the termites for the fungus to produce its sporing bodies, the mushrooms we see above ground.

The appearance of these distinctive umbrella-like structures, with their tough pointy tops designed to thrust a route up through the soil, coincides with the onset of the rainy season, which prompts termite swarming.

The mushrooms produce spores, and the cycle of this sustainable system of farming repeats.

Despite the best efforts of humans, cultivating these mushrooms beyond the termite mound remains elusive.

However, in nature, the alliance has been so successful that over millennia, it has resulted in the evolution of around 60 different species in sub-Saharan Africa and Asia.[7]

Among them is *Termitomyces titanicus*, an African species that can grow to over a metre in diameter, making it one of the largest mushrooms in the world.[11]

During the expedition, we were fascinated to find that while some tribes confidently wild-harvested *Termitomyces* as a nutritious delicacy, others, only a valley away, remained completely unaware of their potential as a food source.[7]

Astridah Mwansa sells wild-harvested *Termitomyces letestui*
Soweto Market, Lusaka, Zambia
Distribution: equatorial Africa

Photograph: Catherine Marciniak

It was in the mountains of Meghalaya that we met some of the most knowledgeable fungi foragers.

This region, situated on the border of India and Bangladesh, receives over 10 metres of rainfall each year, making it a perfect destination for a fungi safari.

Local guides from villages on the forest fringe led us through pine and subtropical forests, where we discovered some of the most stunning specimens of edible mushrooms we have ever seen.

We were introduced to the dainty *Laccaria amethystina*, a deep violet edible species that the locals use as a colourful addition to salads or stir-fries.

The milkcap *Lactifluus volemus* was another popular species among the locals as well as insects.

This mushroom is one of the easier species to identify, as it secretes a white, milk-like liquid when the cap is damaged or cut.[12]

In the markets of Meghalaya, it was selling for around 400 rupees or US$6.00 per kilo, making it a good source of income for the locals.

We also documented several varieties of edible *Russula*.

Most of the culinary mushrooms we encountered were ectomycorrhizal partners of the conifers and broadleaf trees in these higher-altitude forests.

OPPOSITE TOP LEFT
Laccaria amethystina, amethyst deceiver
Dzüko Valley, Nagaland, India
Distribution: Northern Hemisphere

OPPOSITE TOP RIGHT
Russula lepida, rosy russula
Yumthang Valley, Sikkim, India
Distribution: Northern Hemisphere

OPPOSITE BOTTOM
Lactifluus cf. *volemus*
Ialong, Meghalaya, India

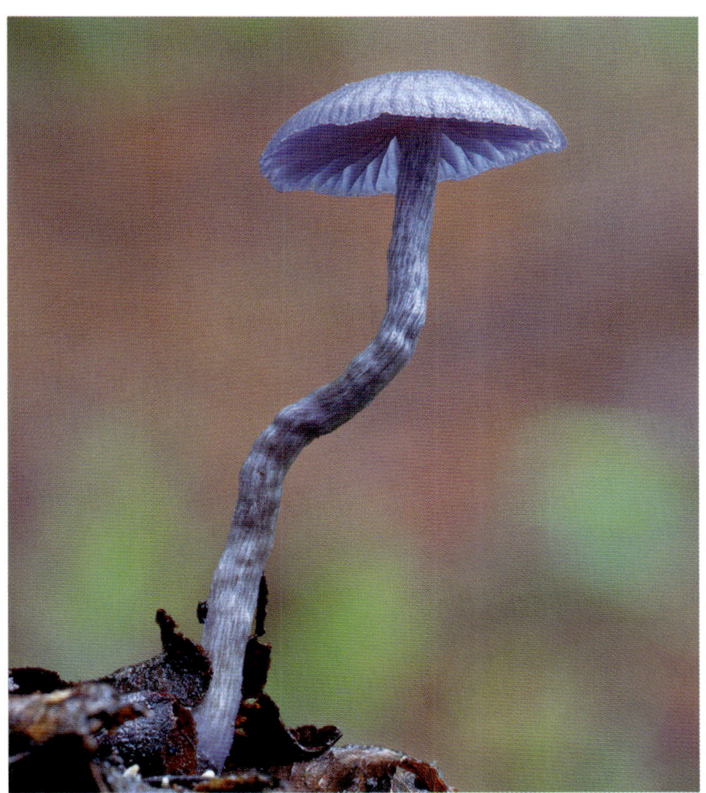

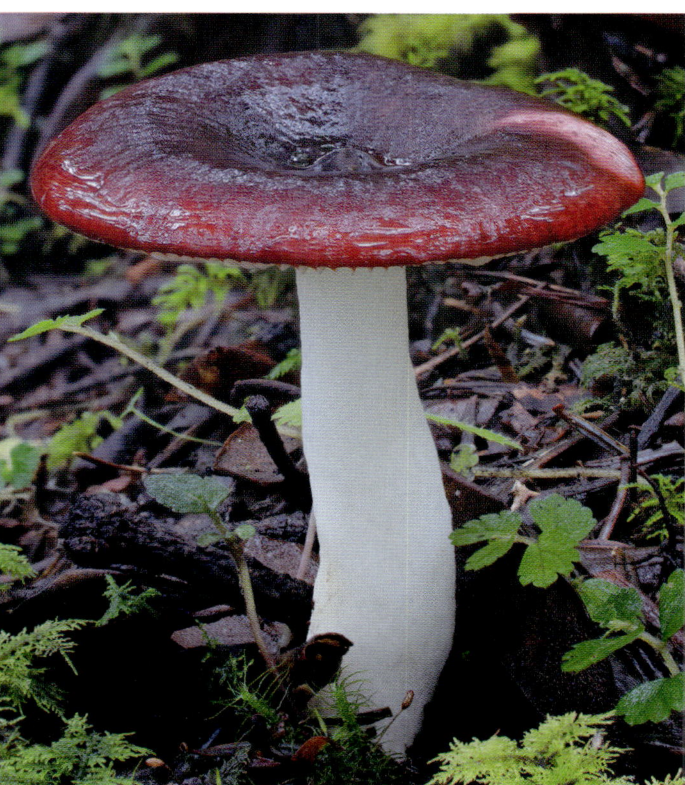

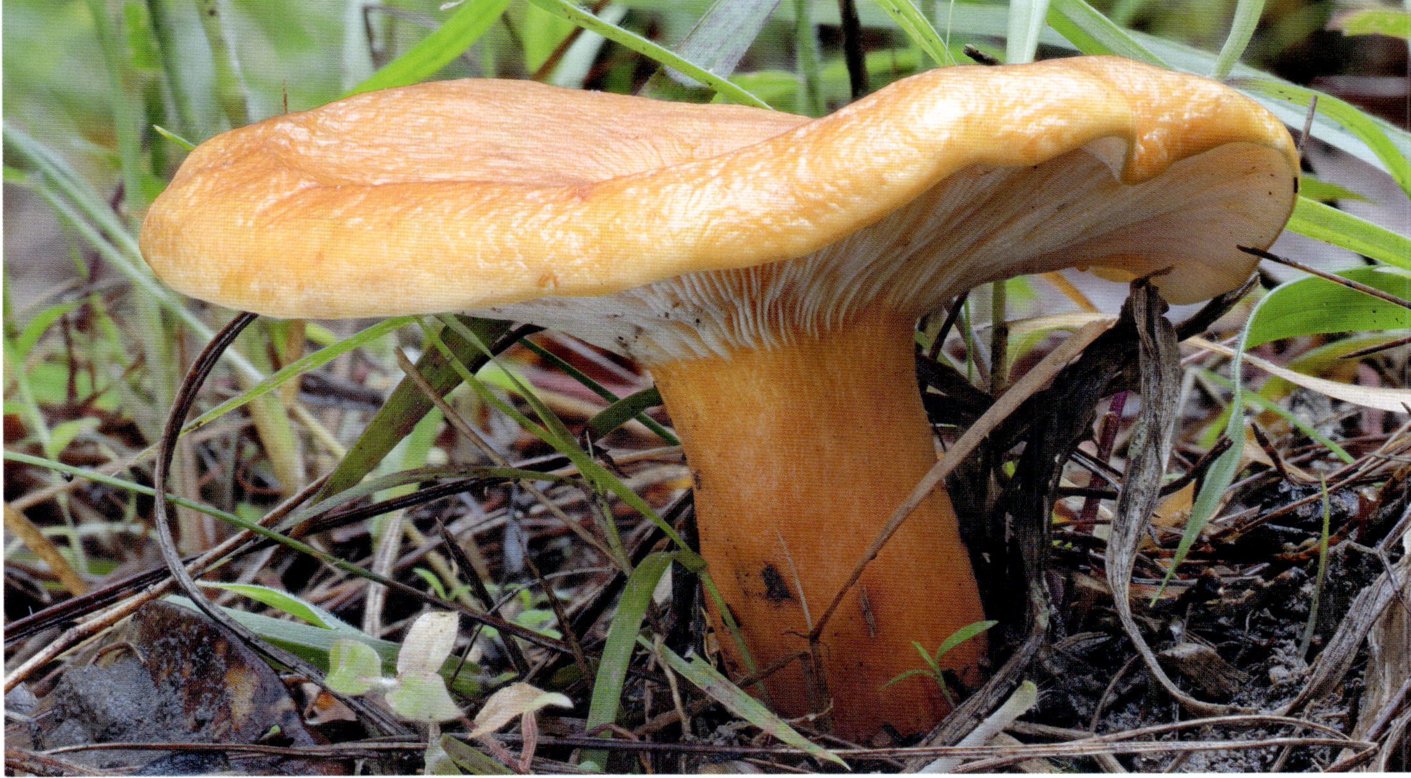

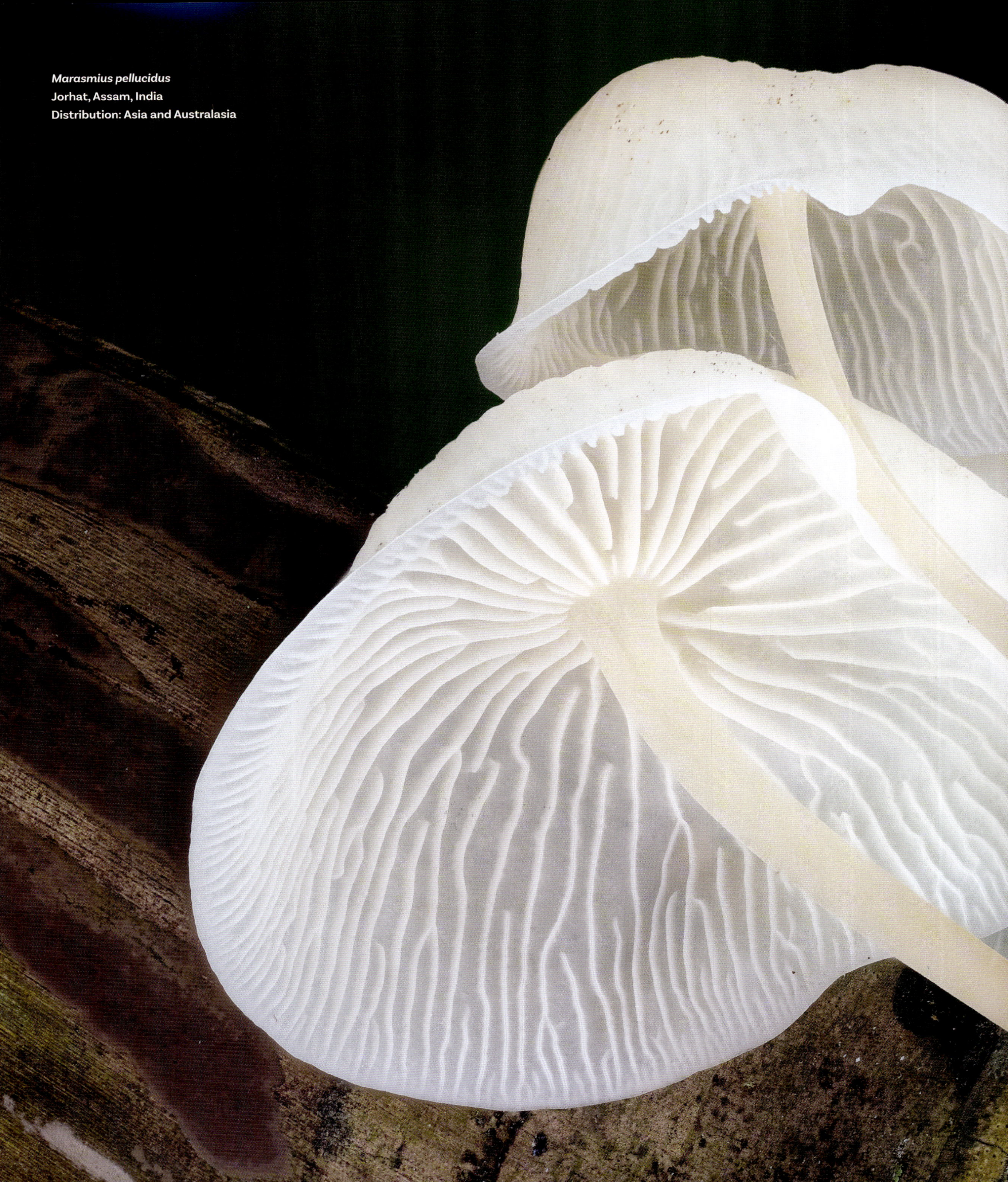

Marasmius pellucidus
Jorhat, Assam, India
Distribution: Asia and Australasia

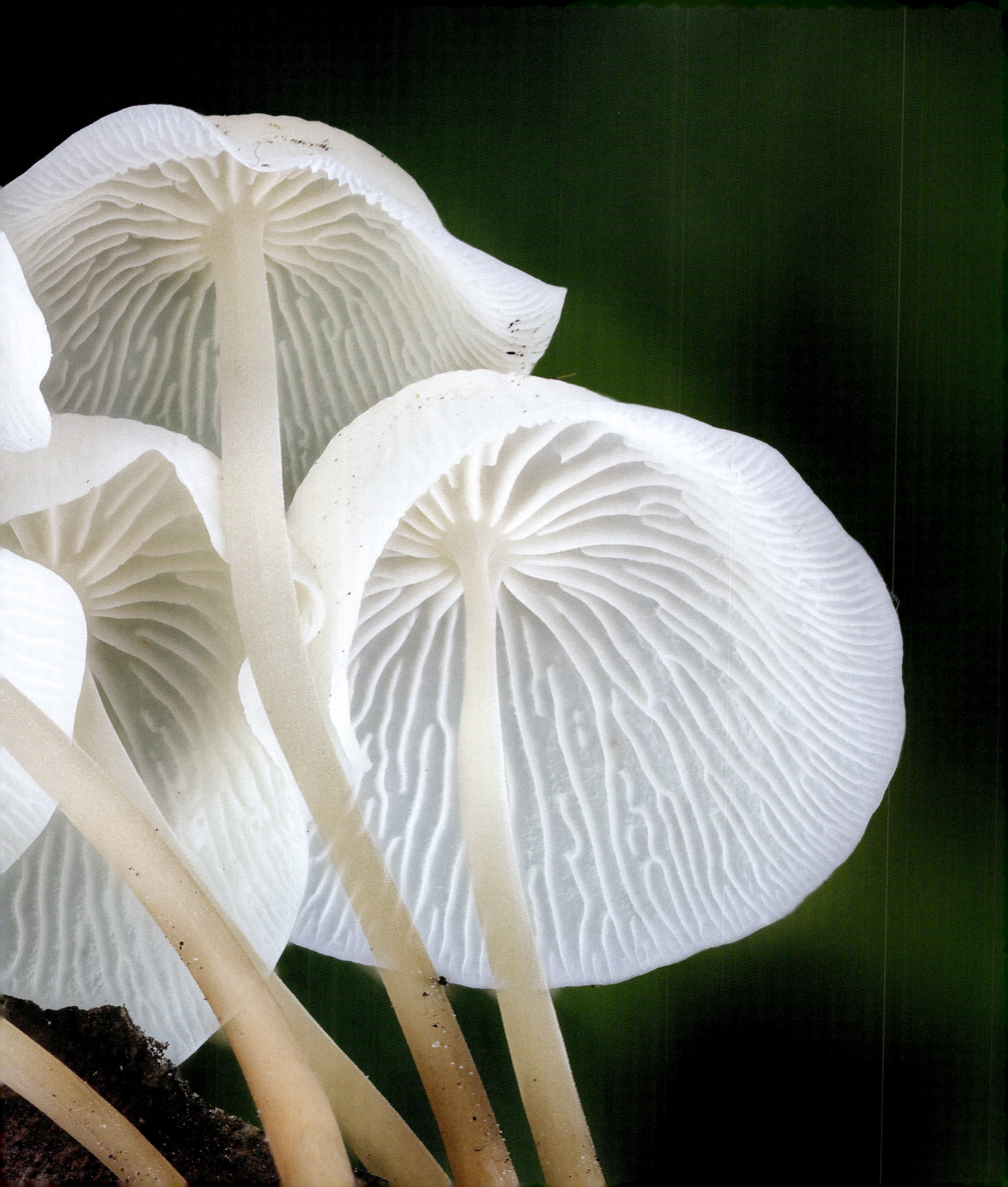

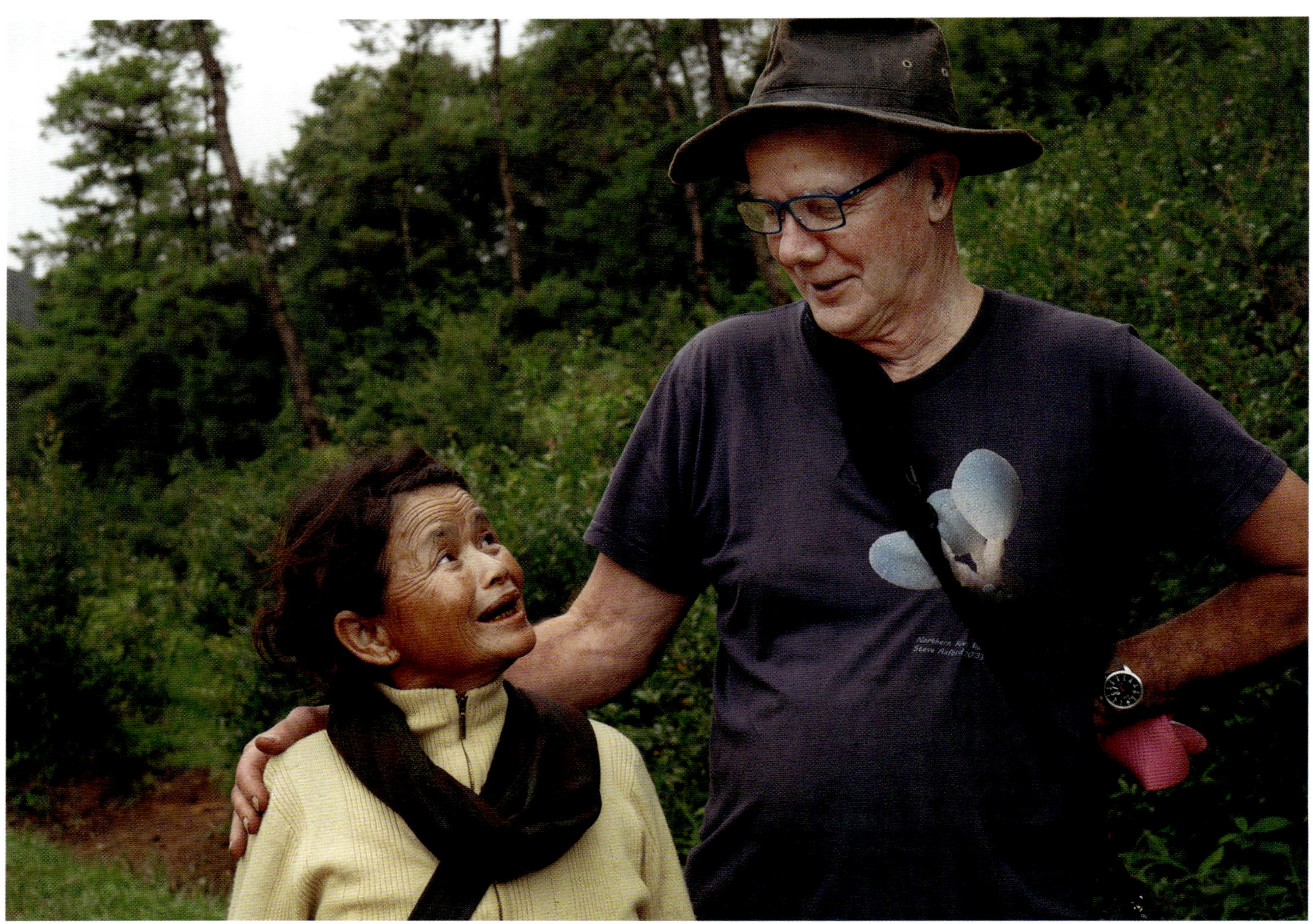

We were fortunate to work with several well-informed guides, who shared the lessons learned by generations of their family.

However, among them, there was one who stood out as an expert in her field – Bilinso Syiemlieh, a Khasi woman introduced to us simply as Kong, which means sister.

Kong's knowledge of the local flora and fungi was truly remarkable.

She had spent years studying these fascinating organisms and had even gone so far as to use herself as a 'guinea-pig' in order to determine which were safe to eat and which were not.

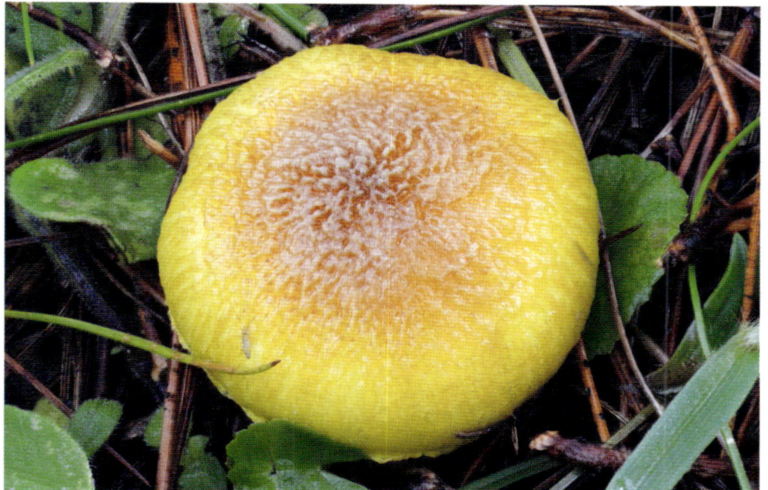

'Kong pointed to a pretty bright yellow mushroom she said was poisonous.

'I asked her how she knew. Although she was telling me in Assamese, her animated description of sweating, shivering, wringing the sweat out of her clothes left me in no doubt that eating this mushroom was too close a brush with death.

'"I just wanted to test it because I liked the look of it," she told me, but over lunch, we discovered there was much more to that story.'

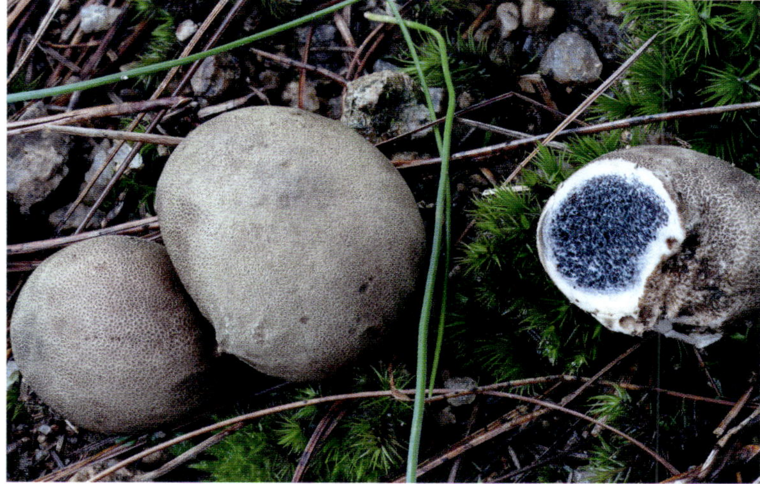

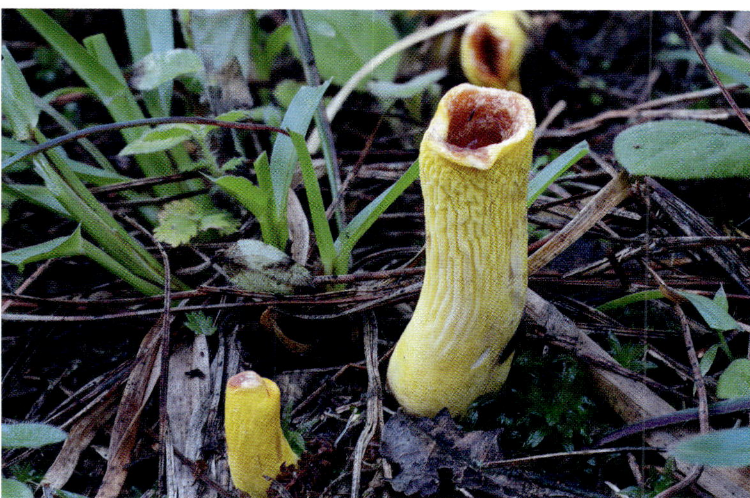

TOP

Tricholoma sp.
Mawkyrwat, Meghalaya, India

This striking golden mushroom is the same species as the one Bilinso Syiemlieh ate, resulting in vomiting and a high fever.

MIDDLE

Scleroderma sp.
Mawkyrwat, Meghalaya, India

BOTTOM

Turbinellus cf. *floccosus*
Mawkyrwat, Meghalaya, India

OPPOSITE

Bilinso Syiemlieh and Stephen share a moment of amusement upon discovering they are the same age.

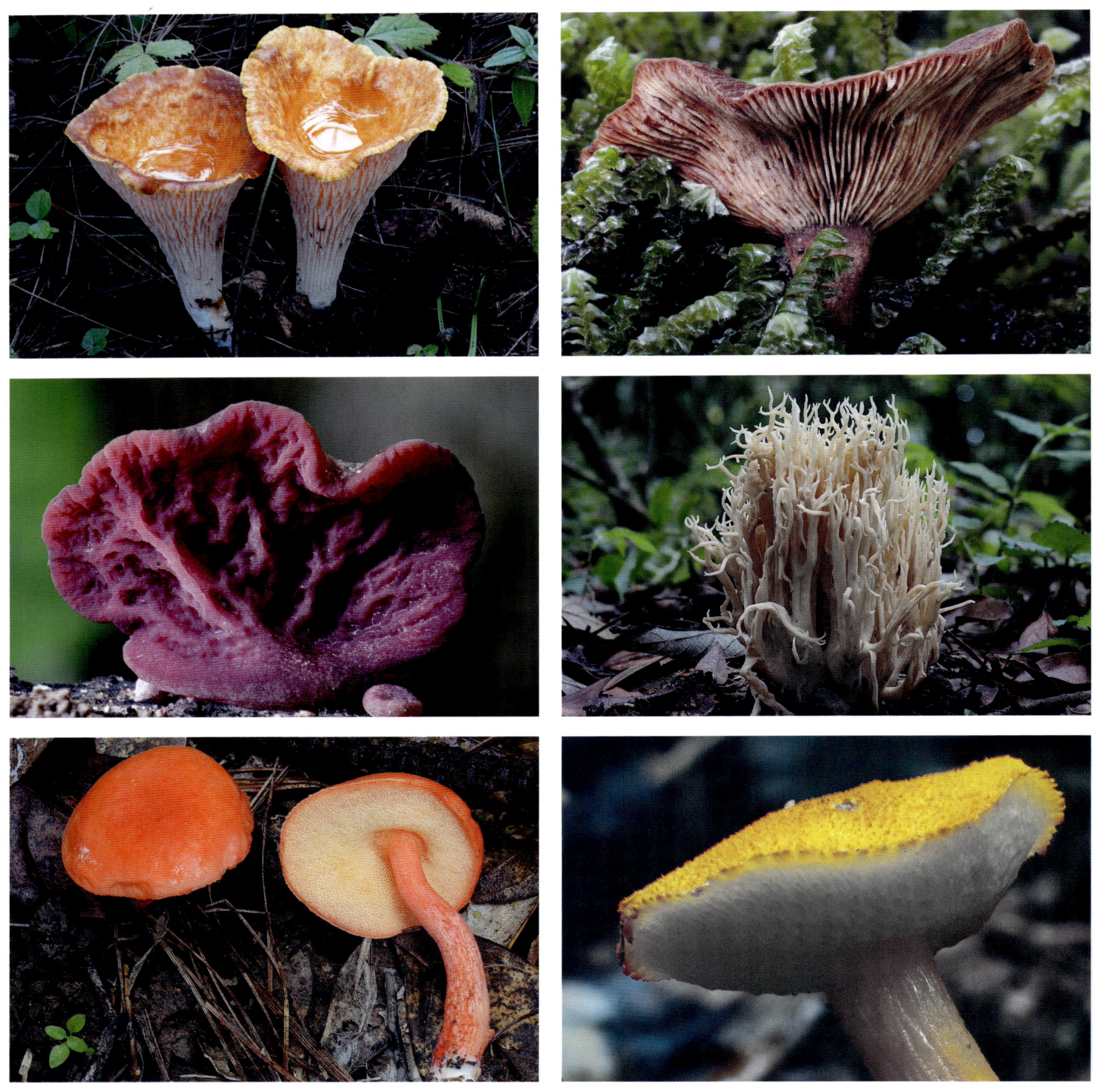

When she was a young mother with five children, Kong's husband, a soldier, unexpectedly became ill and died. In a country without a widow's pension or support for single mothers, Kong turned to the forest to supplement the food she grew. As she foraged for mushrooms, she tested those that were new to her for their edibility.

It was a risky strategy, and one that no mycologist would recommend, but through her experiments, Kong was able to add 57 species to our tally.

They included a wide variety of spongy boletes, blood mushrooms that ooze red juice and impressive coral fungi that grow in the same gully every year.

One of the most surprising edibles was a scaly, leathery vase-shaped mushroom identified from Stephen's photographs as *Turbinellus* cf. *floccosus*. In North America, consumption of *T. floccosus* is known to cause nausea and stomach pain, but in Meghalaya this lookalike is one of the 10 most popular edible mushrooms.[12]

Kong explained its Khasi name is chillum as it resembles the pipes used by villagers for smoking wild tobacco.

OPPOSITE TOP LEFT
Turbinellus cf. *floccosus*
Ialong, Meghalaya, India.
Distribution: Northern Hemisphere

OPPOSITE MIDDLE LEFT
Auricularia delicata, forest ear
Hoollongapar Gibbon Sanctuary, Assam, India
Distribution: worldwide, mainly tropical

OPPOSITE BOTTOM LEFT
Leccinellum sp.
Mawkyrwat, Meghalaya, India

OPPOSITE TOP RIGHT
Lactarius sp.
Mawkyrwat, Meghalaya, India

OPPOSITE MIDDLE RIGHT
Ramaria sp.
Mawphlang, Meghalaya, India

OPPOSITE BOTTOM RIGHT
Aureoboletus sp.
Mawkyrwat, Meghalaya, India

Our four-week expedition in this region of the Eastern Himalayas just scratched the surface of fungal diversity in these unchartered forests.

In addition to the feature documentary we produced, 'Planet Fungi – North-East India', Stephen photographed 232 species.

Of those that were DNA sequenced, 58 species were identified as edible varieties and 64 species identified as those used in traditional medicines. But perhaps most excitingly, we documented 34 species that may be entirely new to science.

It is a reminder of just how much we have yet to learn about fungi in this region of the world.

OPPOSITE
Marasmius haematocephalus
Hoollongapar Gibbon Sanctuary, Assam, India
Distribution: worldwide, mainly tropical

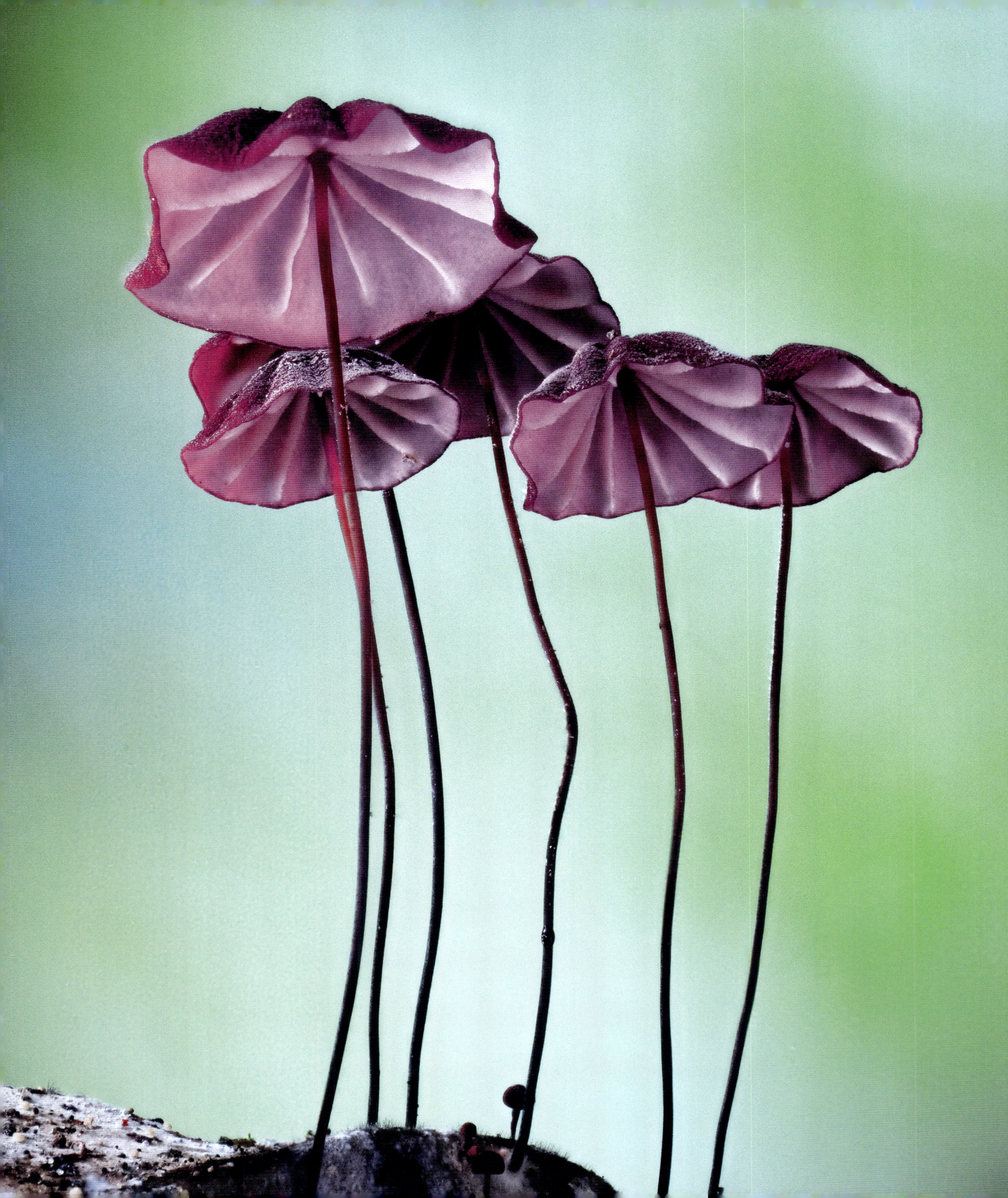

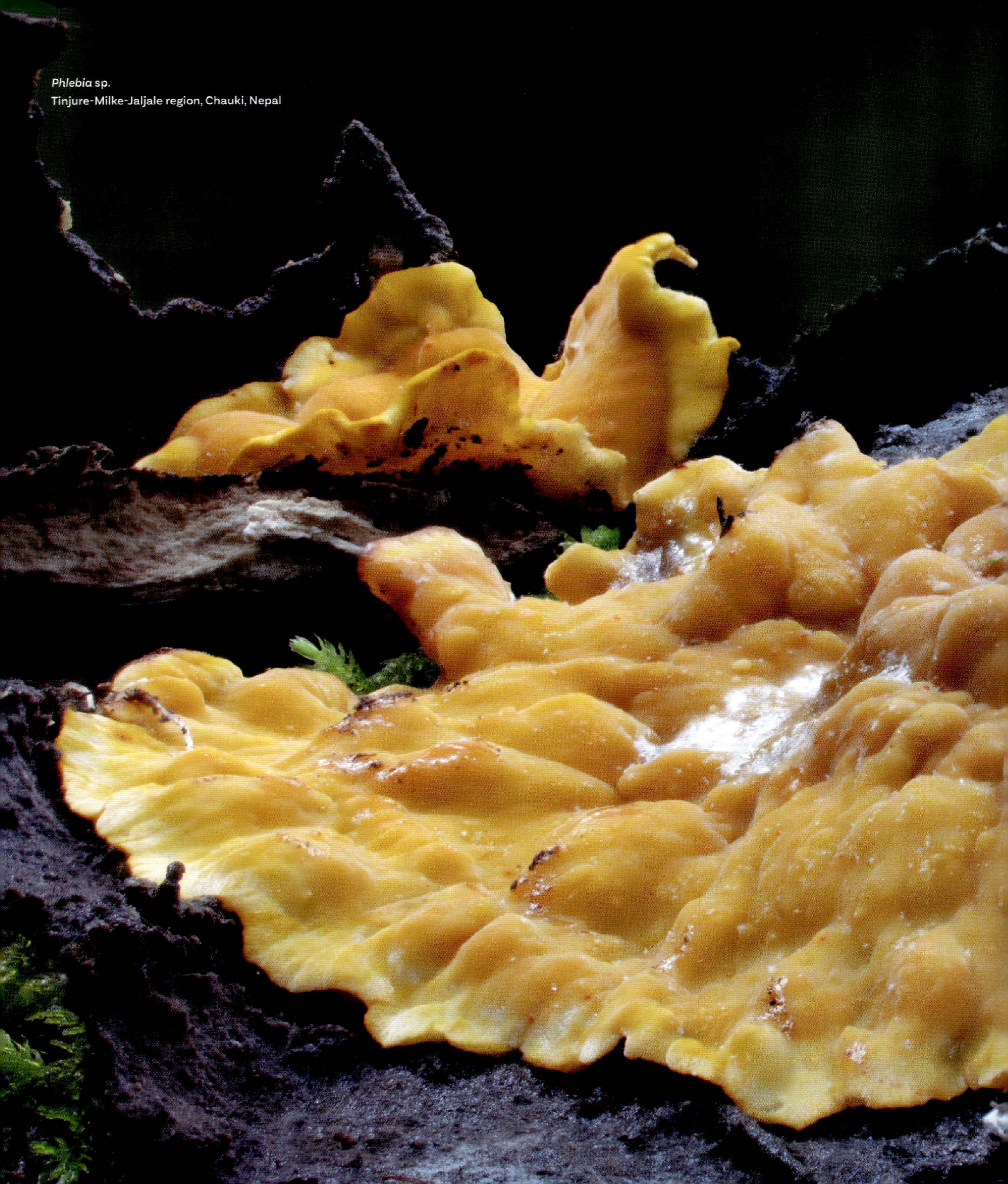

Phlebia sp.
Tinjure-Milke-Jaljale region, Chauki, Nepal

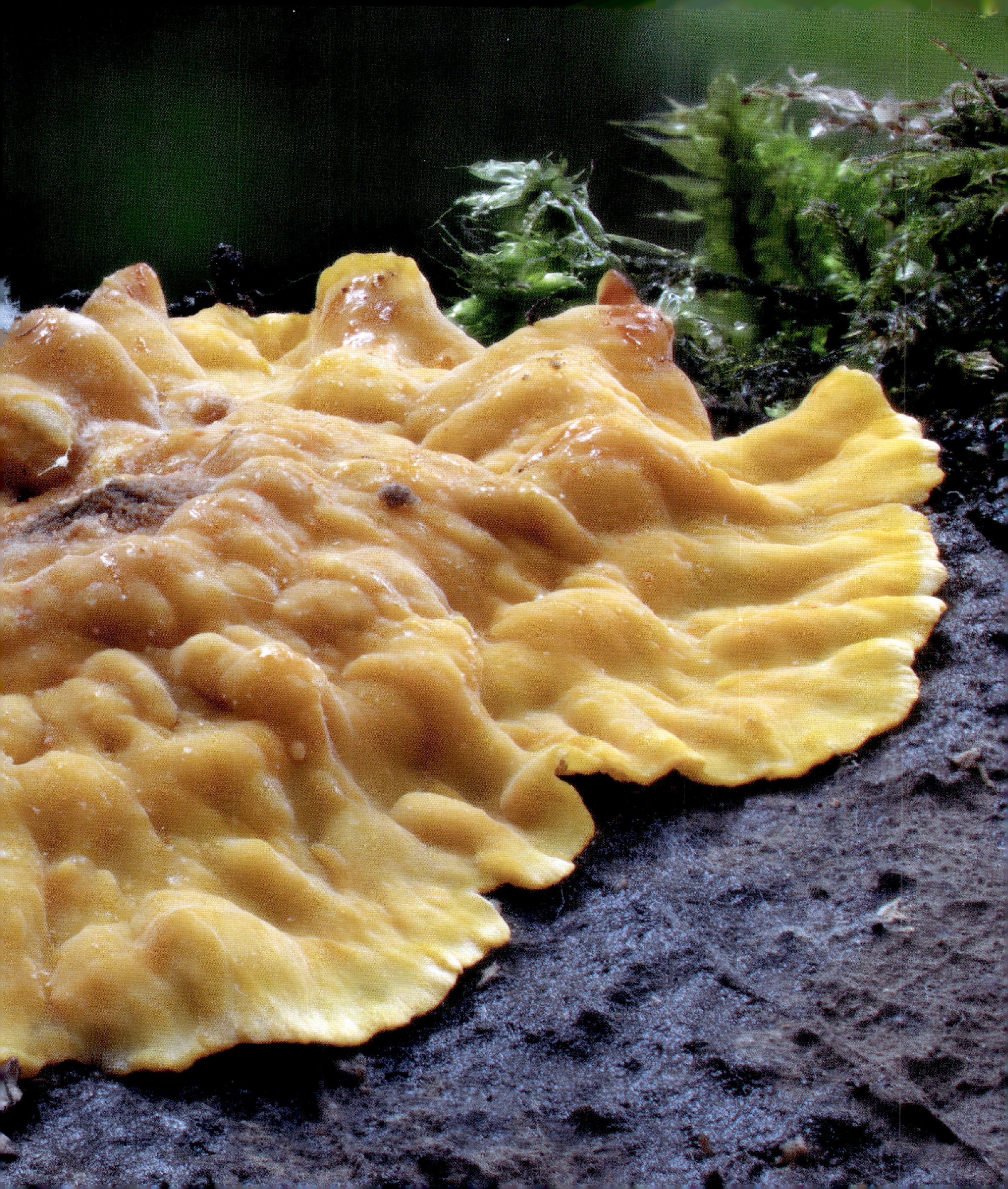

New discoveries

For the intrepid fungi hunter, stumbling upon a novel specimen is a thrill, but for the mycologist, it is just the beginning of unravelling an often-enigmatic identity.

In 2009, Stephen came across what initially appeared to be a blue scrap of paper amid the leafy debris in an endangered subtropical rainforest on Australia's east coast.

Upon closer inspection, it revealed itself as a frosty blue mushroom, unlike any he had ever encountered.

Defying familiar norms, it had a closed, ball-shaped cap partially buried in the soil and tightly packed, paper-thin white gills nestled inside.

As the mushroom grew, it became top-heavy on its fragile stipe, sometimes taking a tumble and rolling around like the ripe fruits of a plum tree, a design possibly evolved to entice creatures into eating it and aiding its spore dispersal.

Coprinopsis pulchricaerulea, frosty blue
Booyong, New South Wales, Australia
Distribution: Australia and New Caledonia

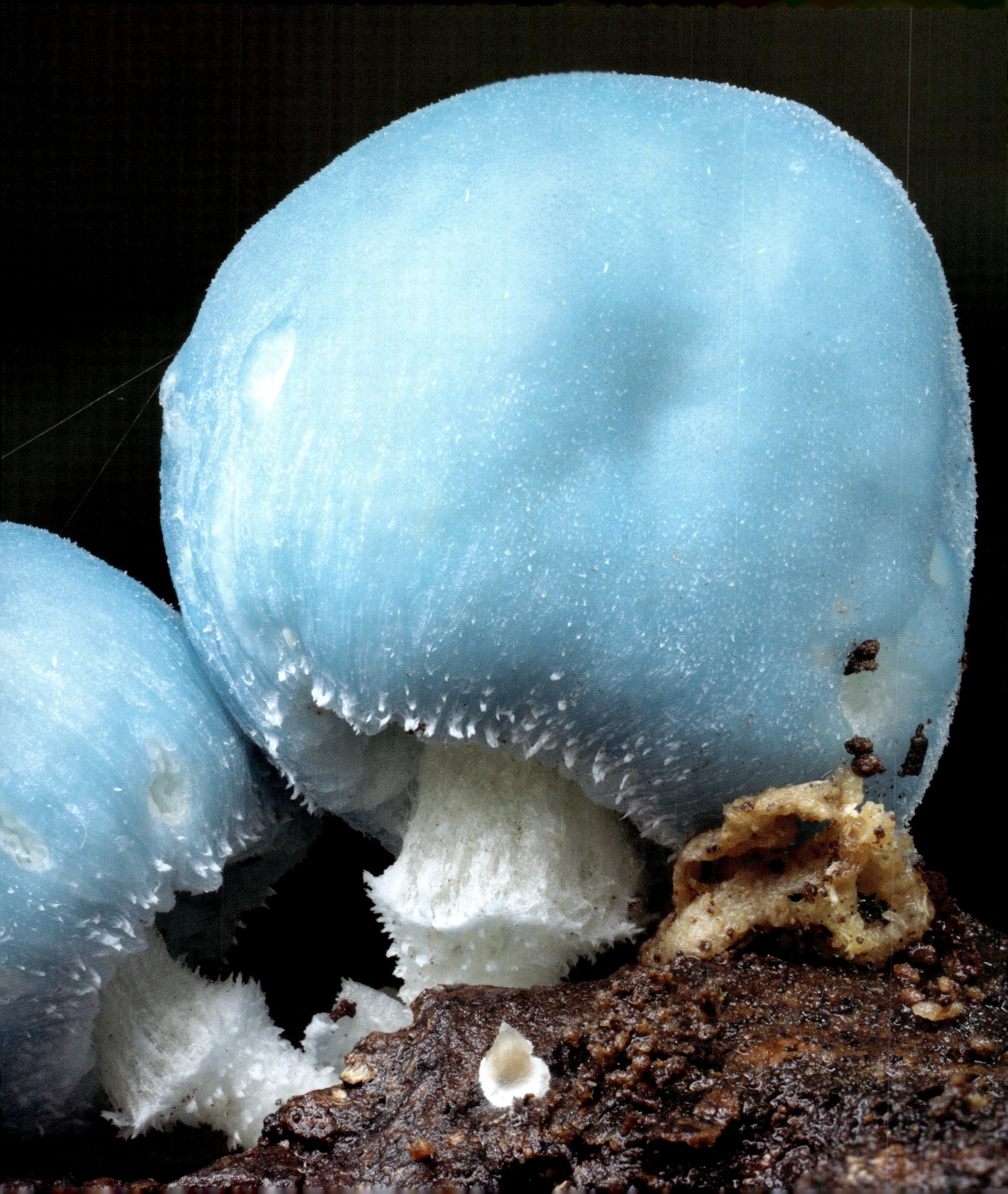

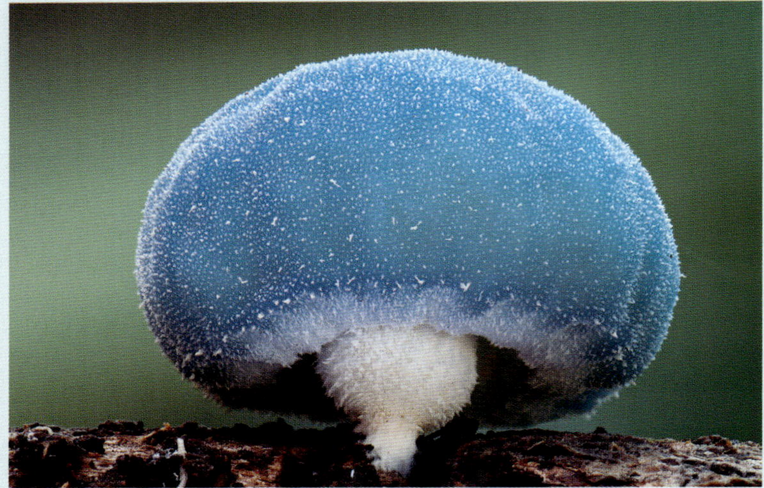

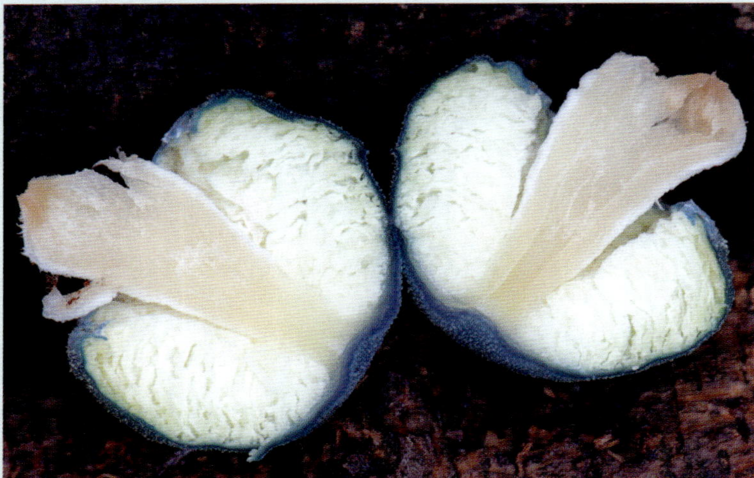

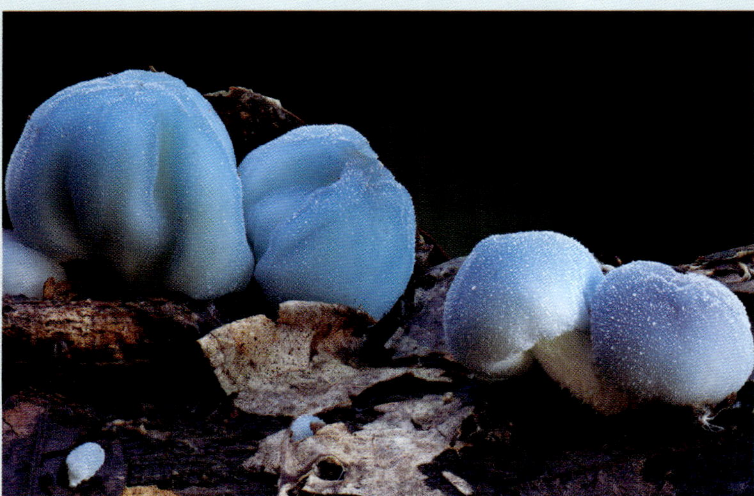

'Photographing and trying to understand this mushroom in all its wonderful and weird forms became an obsession.

'I went on the hunt to find out what it is called, but I couldn't find anything remotely like it.'

The quest for answers led Stephen to our co-writer and mycologist, Dr Tom May, at Royal Botanic Gardens Victoria.

Tom's initial reaction spoke volumes: 'This is something really special.'

'Blue hues are rare in nature, and coupled with the fully enclosed and compacted lamellae, which are the gills, this specimen stood out as truly unique.

'However, identifying a new fungus is like searching for someone in the phone book without knowing their name.'

TOP
The sporing body of *C. pulchricaerulea* has an extremely fragile attachment to its timber substrate.

MIDDLE
Enclosed within the cap (pileus) are tightly packed gills (lamellae).

BOTTOM AND OPPOSITE
Defying our preconceptions about fungal structures, *C. pulchricaerulea* can appear as a small, enclosed classic mushroom, or a blue ball sitting on the surface of the soil, or sometimes weird, contorted shapes that look like mutants.

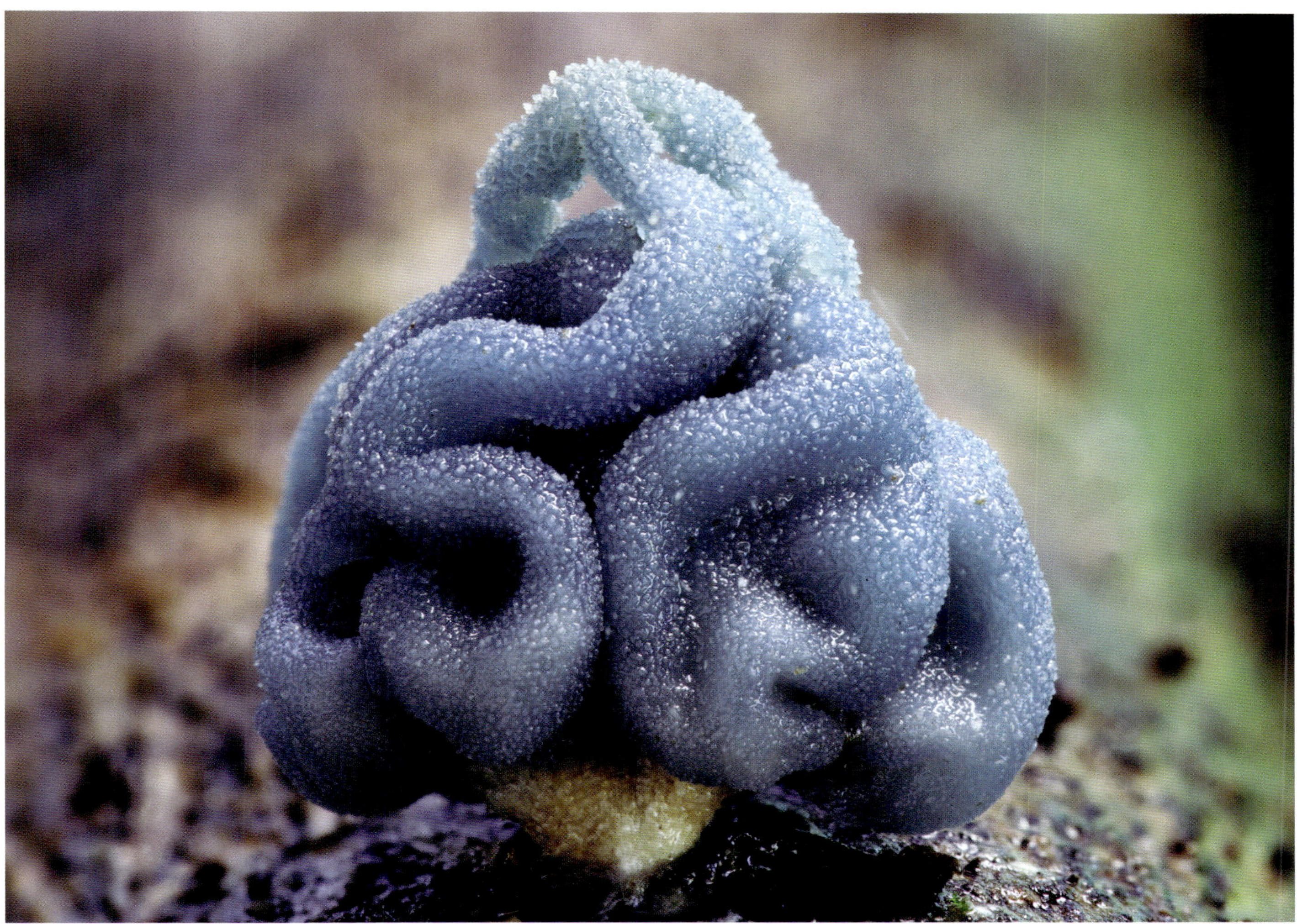

To crack such fungal mysteries, mycologists meticulously compile detailed case files, scrutinising every aspect of the fungus's form and behaviour – from spores, colours and textures to its habitat and interactions.

This traditional approach, akin to assembling an identikit, forms the foundation of fungal taxonomy.

In the quest to unveil the identity of the elusive frosty blue mushroom, Tom initially relied on his mental 'telephone book' of fungal names.

He suspected it could be a pouch fungus similar to blue and red fungi found in New Caledonia, classified at that time as *Leratiomyces*.[1]

'Pouch fungi,' Tom elaborates, *'are a quirky, diverse bunch and share many traits with the frosty blue. They have colourful, fleshy, enclosed sporing bodies, and the lamellae are contorted and compressed inside their pileus. They may be exposed or half-buried and rely on animals to eat them and spread their spores.'*[2]

The initial search among *Leratiomyces* kin hinted at the possibility of a new species.

Since the early 2000s, the powerful new tool of genetic sequencing has been added to the mycologist's arsenal of investigative techniques.

Scientists discovered that a specific region of the genetic code varies between fungal species but remains relatively stable within a species. The region is called a DNA barcode.[3,4]

'This molecular fingerprinting', Tom explains, *'allows us to navigate through vast datasets swiftly. It's like the Eureka! moment in a detective drama when the fingerprint reveals the identity of the mystery person at the crime scene.'*

The next step for Stephen's mystery blue fungus involved gathering additional specimens to rule out the possibility of a mutation or a random discovery.

Then, all specimens would be genetically sequenced.

OPPOSITE TOP LEFT
Coprinopsis pulchricaerulea has many visual traits in common with other pouch fungi but its genetics tell a different story.

OPPOSITE MIDDLE LEFT
Rossbeevera pachydermis, velvet potato fungus
Arthurs Pass, New Zealand
Distribution: probably endemic to New Zealand

OPPOSITE BOTTOM LEFT
Cortinarius porphyroideus, purple pouch fungus
Saint Arnaud, New Zealand
Distribution: New Zealand

OPPOSITE TOP RIGHT
Clavogaster virescens, blue pouch fungus
Tongariro National Park, New Zealand
Distribution: New Zealand

OPPOSITE MIDDLE RIGHT
Leratiomyces erythrocephalus, scarlet pouch fungus
Tongariro National Park, New Zealand
Distribution: New Zealand

OPPOSITE BOTTOM RIGHT
Gallacea scleroderma, velvet potato fungus
Arthurs Pass, New Zealand
Distribution: New Zealand

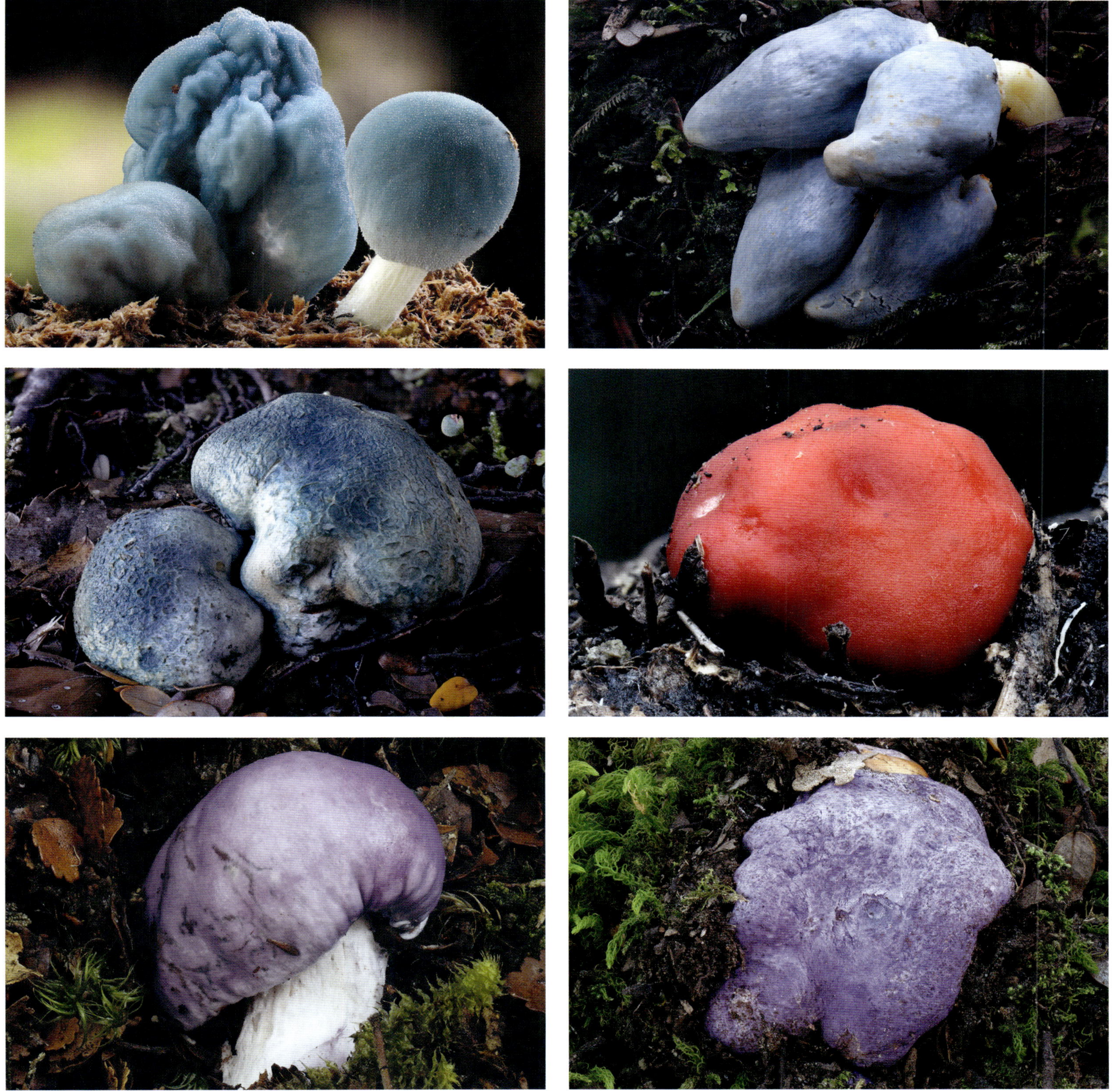

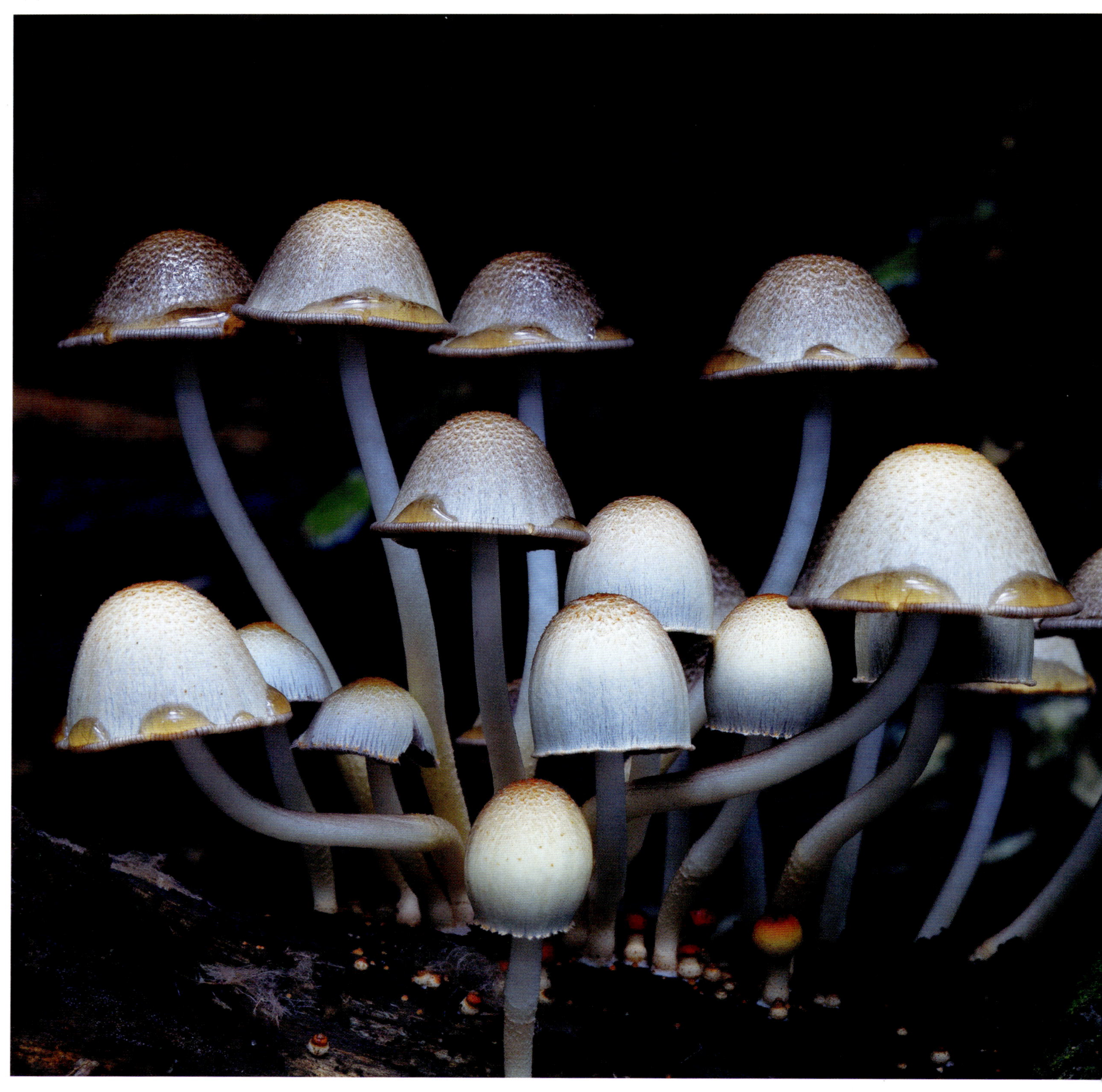

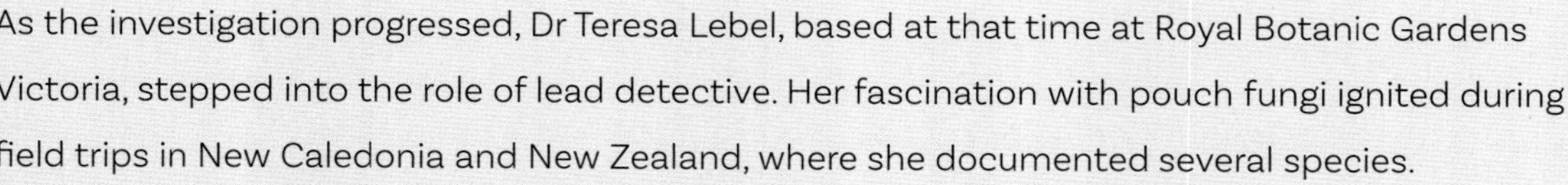

As the investigation progressed, Dr Teresa Lebel, based at that time at Royal Botanic Gardens Victoria, stepped into the role of lead detective. Her fascination with pouch fungi ignited during field trips in New Caledonia and New Zealand, where she documented several species.

When the frosty blue fungus' DNA barcode was sequenced, we held our breath for the big reveal.

But the story took an unexpected turn.

The DNA pointed not to *Leratiomyces* but to inky caps, renowned for their theatrical transformation into black goo as they mature.

After residing in the single genus *Coprinus* for more than 100 years, in 2001, the inky caps were spread across four genera: *Coprinus*, *Coprinopsis*, *Coprinellus* and *Parasola*.

And in another surprise twist, three of the new genera were not even in the same family as *Coprinus*.

All those characteristics we thought were family traits were most likely to be the result of different fungi adapting to similar environmental factors – a classic case of convergent evolution.[5]

Inky caps are no strangers to the fungi hunter.

They are often spotted on our field trips atop animal dung, on roadsides, or exploding into mass fungal displays on decaying stumps and logs.

Their defining characteristics include dark spores and gills.

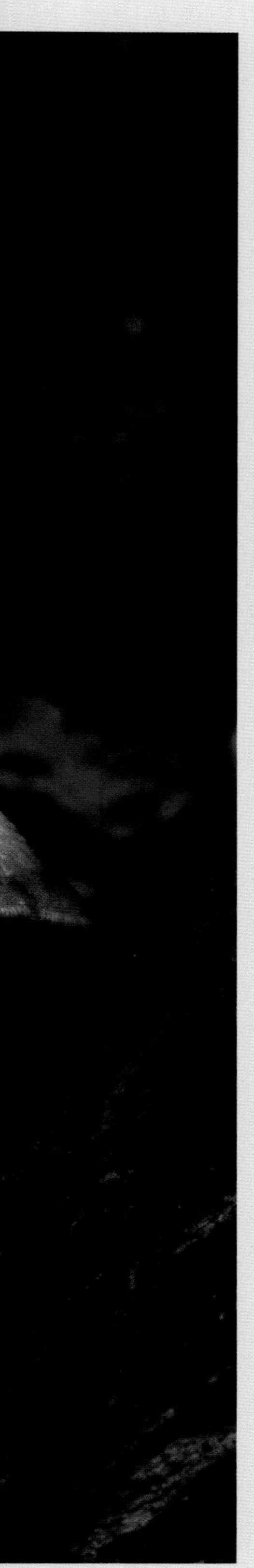

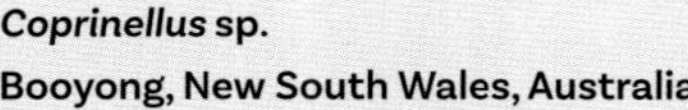

***Coprinellus* sp.**
Booyong, New South Wales, Australia

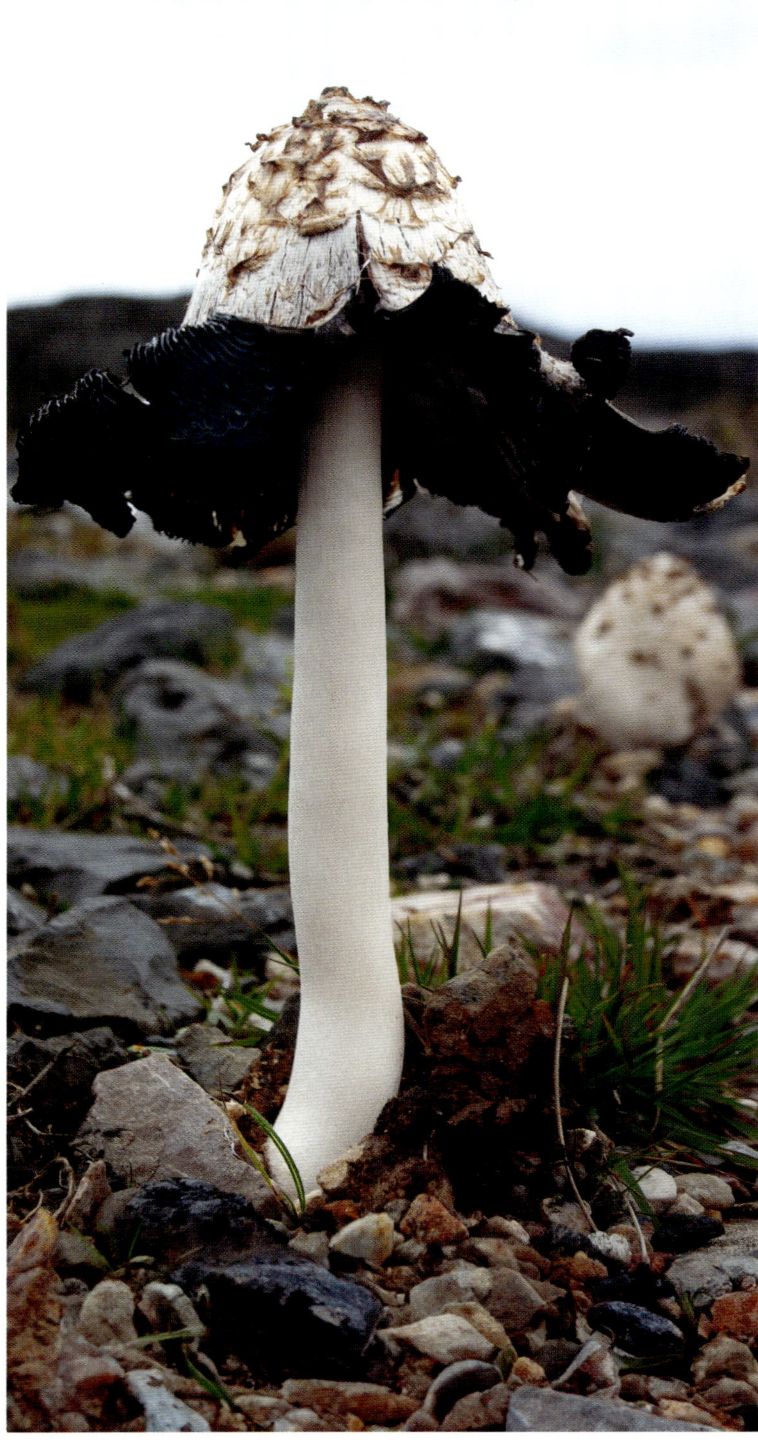

However, our frosty blue was a misfit in this group, with its pink-stained white gills, white spores and a growth pattern akin to a pouch fungus.

Dr Lebel suspected an error in the sequencing process, wondering if an unfortunate contamination of another fungus had slipped into the genetic mix.

DNA sequencing was still a luxury back then. It was expensive, and there was limited access to facilities in Australia.

Undeterred, Dr Lebel embarked on a traditional taxonomic sleuthing journey to crack this cold case.

Her efforts led her to the archived collections of dried fungi in the world's fungaria where she hunted for similar-looking mushrooms, waiting to be matched with their genetic counterparts.

Her focus was on the most promising lead – the pouch fungi.

LEFT
Coprinus comatus, shaggy mane
Pine Lake Walk, Tasmania, Australia
Distribution: worldwide

OPPOSITE
Coprinopsis pulchricaerulea, frosty blue
Booyong, New South Wales, Australia
Distribution: Australia and New Caledonia

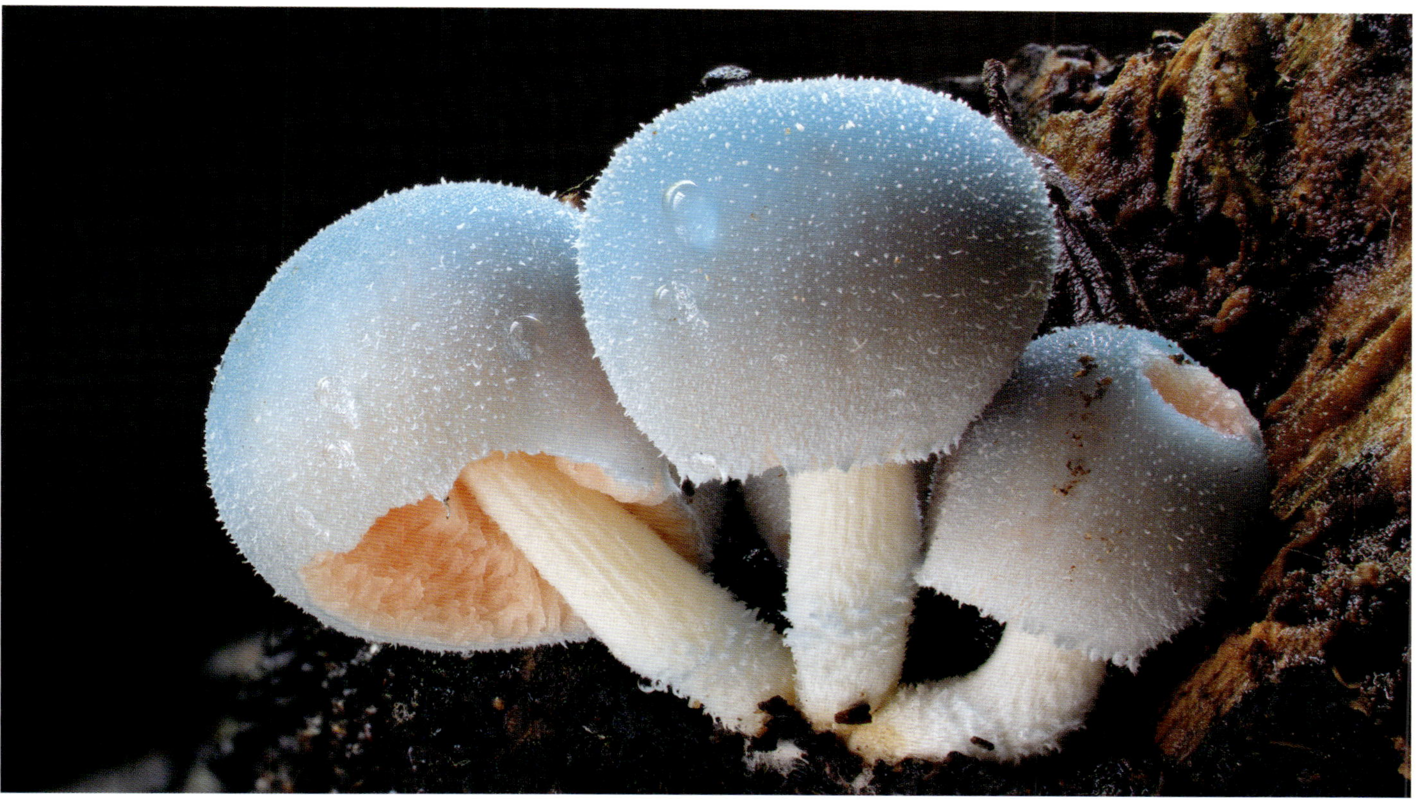

After narrowing her suspects to three promising candidates, the moment of truth arrived with DNA testing.

In a surprise relationship, our frosty blue fungus was very similar to one of the pouch fungi she collected on New Caledonia and a lone, old specimen from Lord Howe Island, a remnant of an ocean volcano off the coast of Eastern Australia.

The most exciting finding was that frosty blue was indeed a new species of *Coprinopsis*, unique in that genus with its combination of structures, colour and growth patterns. It contradicted most of what we knew so far about the morphology of that genus.

This discovery was cause for celebration – the fungus found its home in the tree of life, and it expanded our understanding of the genus *Coprinopsis*.

It was formally christened *Coprinopsis pulchricaerulea*, which means beautiful blue, with the nickname frosty blue, but naming a fungus is merely one chapter in its story.[6]

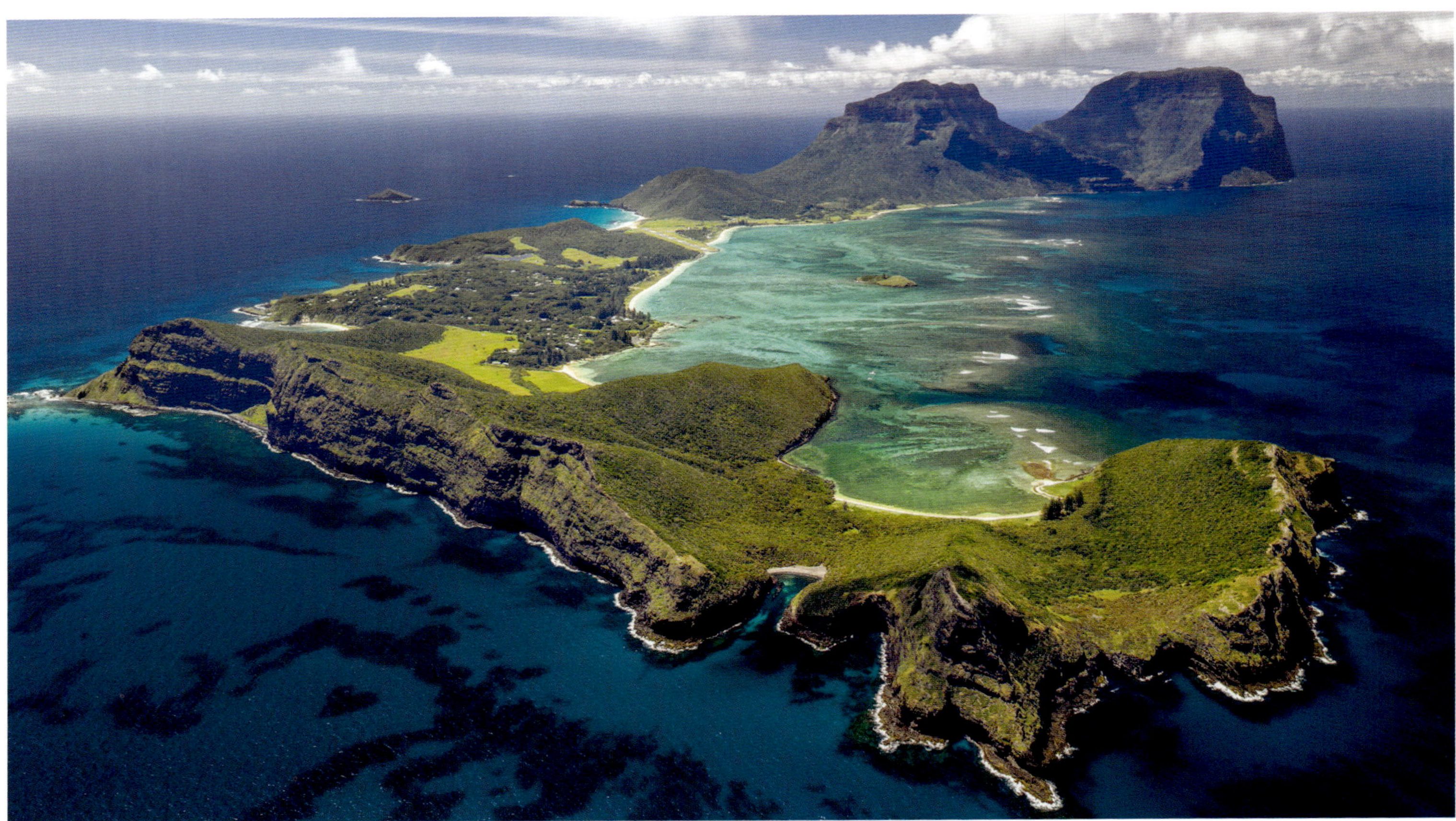

What the DNA didn't tell us is how frosty blue feeds and how it interacts with other organisms. How important is it in the ecosystem? And why has it only been found in three places on the planet?

Driven by curiosity, our investigation led us on an expedition to Lord Howe Island.

Lord Howe is spectacular. In 1982, it was listed as a World Heritage site to honour its rich biodiversity and protect rare species of plants and animals that only exist on this remote atoll.

ABOVE
Lord Howe is an island paradise approximately 700 kilometres north-east of Sydney, Australia.

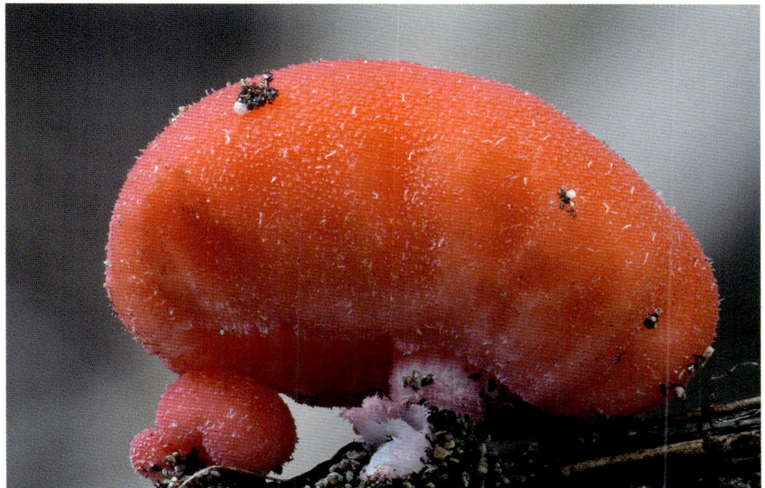

'Searching for the elusive frosty blue mushroom in unfamiliar terrain was daunting, but on our very first hike up a forested hill, luck travelled with us.

'We stumbled upon a treasure trove – seven exquisite frosty blue specimens.

'Sharing the discovery with Ian Hutton, Lord Howe's resident naturalist and botanist, I learned, to my amazement, that he'd been capturing this fungus in his photographs for four decades, unaware of its rarity.

'Then, he introduced us to an even more extraordinary fungus.

'In the lowland forests on Lord Howe, striking pink mushrooms appear on the rotting debris of banyan trees and palm fronds following rain. They look identical to frosty blue except for their stunning pink hue.'

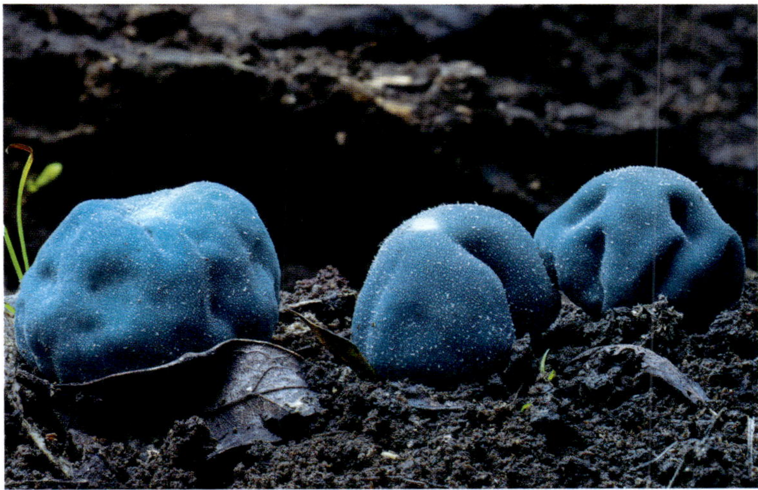

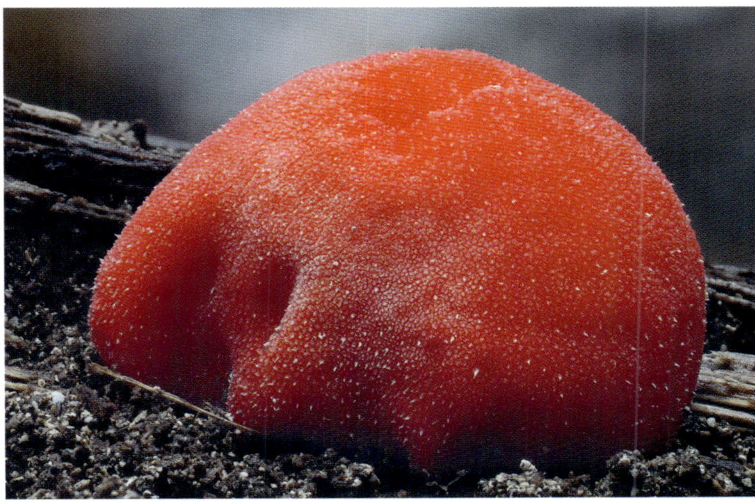

TOP AND BOTTOM
Coprinopsis pulchricaerulea, pink form
Middle Beach Loop Track, Lord Howe Island, Australia

The pink form is known only from Lord Howe Island.

MIDDLE
Coprinopsis pulchricaerulea, frosty blue
Mt Eliza Track, Lord Howe Island, Australia
Distribution: Australia and New Caledonia

With the Lord Howe Island Board's support, we collected samples of the pink mushroom for genetic analysis by Dr Lebel.

In a final twist to the frosty blue saga, the pink mushroom, thought to belong to a different genus, shares an identical genetic fingerprint with *Coprinopsis pulchricaerulea*.

It is but a charming pink variant of our frosty blue.

This revelation highlights the ongoing challenges faced by mycologists in the task of classifying new fungal species.

As DNA sequencing ramps up there are constant surprises.

For instance, some of the most beautiful fungi in the world were thought to belong to the same group, *Marasmius haematocephalus*. But when their genetic makeup is examined, it turns out that despite their similar appearances, they actually belong to several different species.[7]

This taxonomic renaissance is reshaping our perception of fungal diversity, yet it begs the question: is this revolution occurring swiftly enough?

***Marasmius* sp.**
Booyong, New South Wales, Australia
DNA sequencing indicates that this fungus is a novel species.

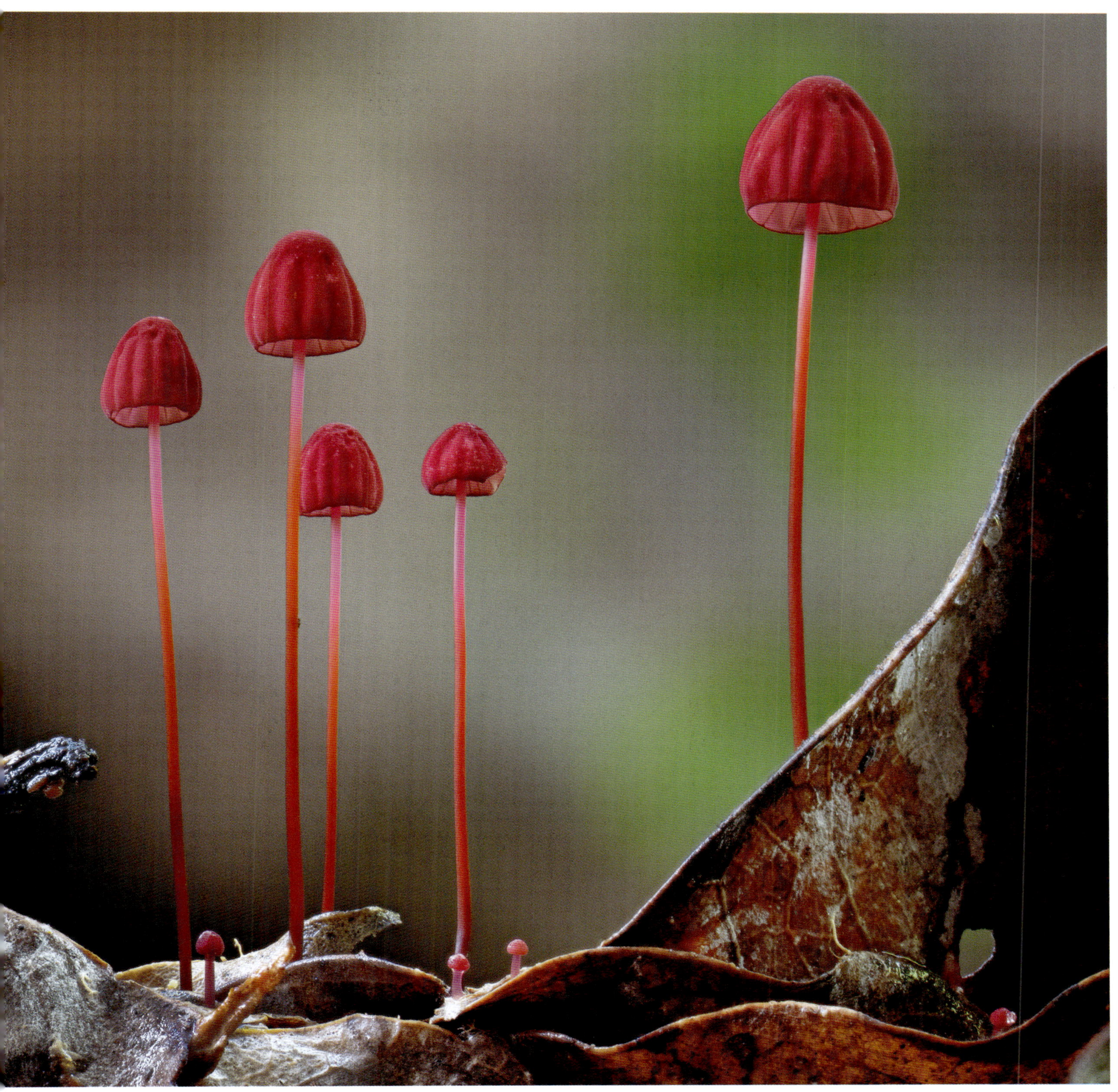

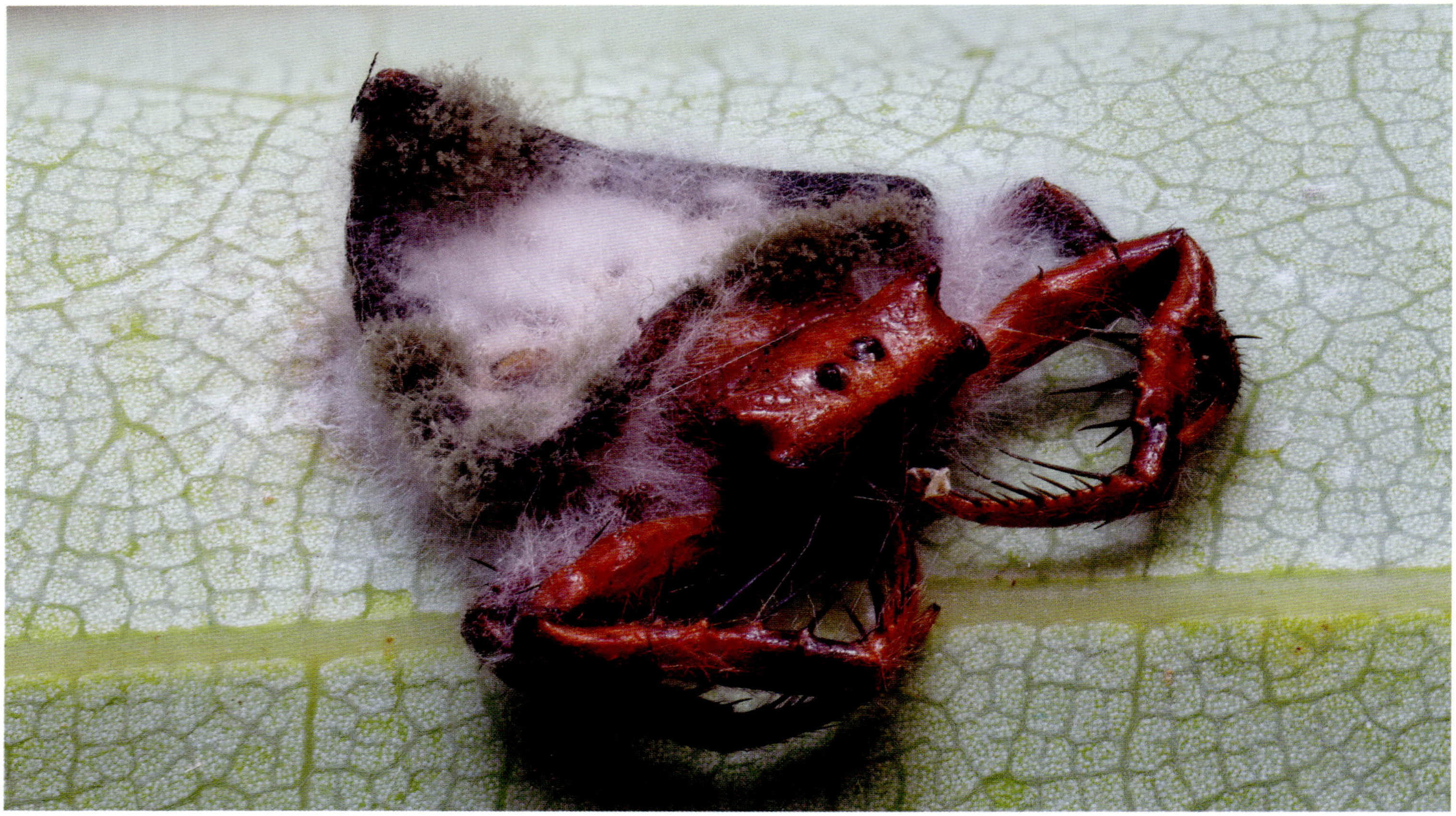

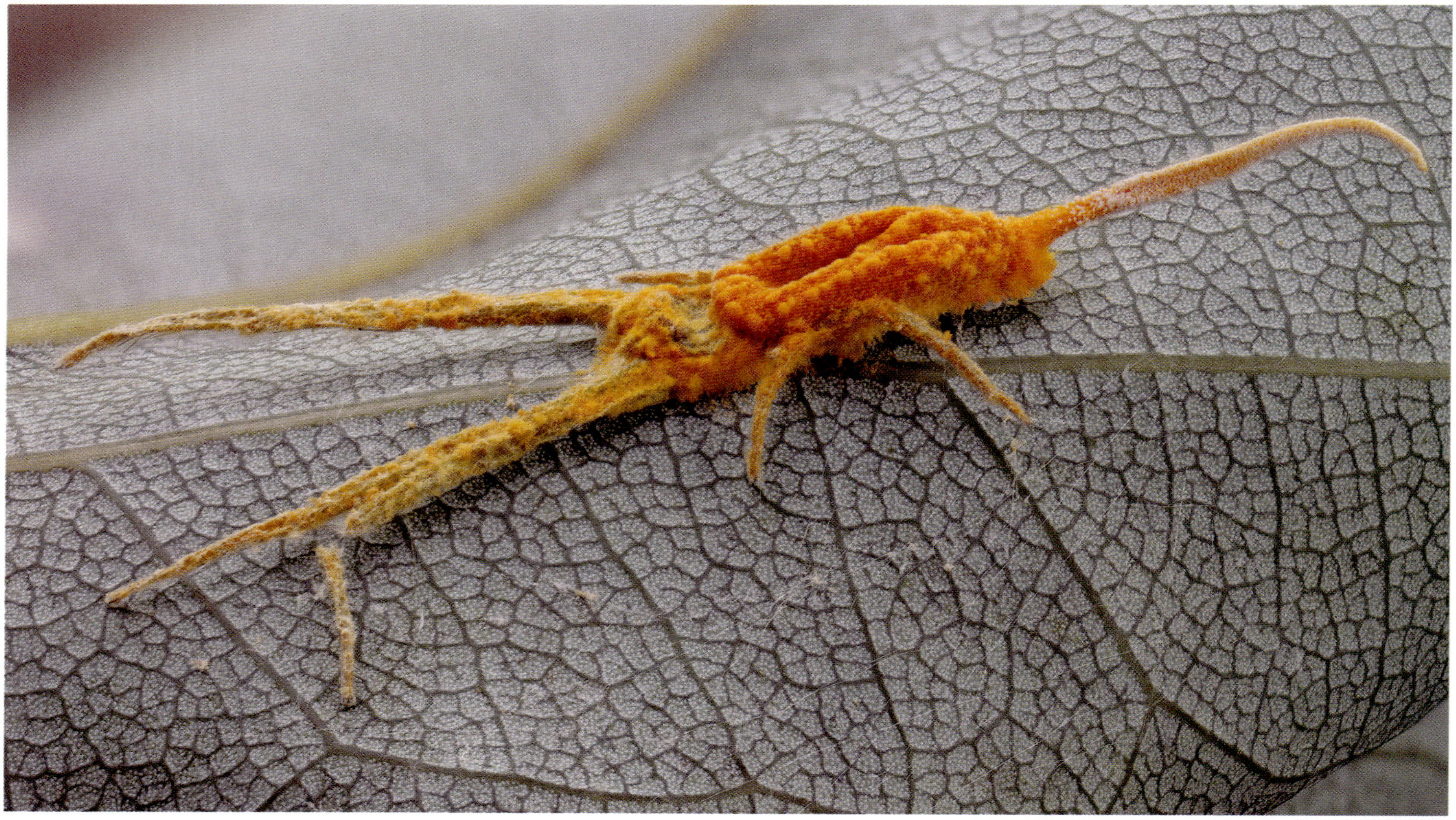

The numbers speak volumes. Since 2020, science has officially recognised 10,000 new fungal species – an impressive feat, yet these are a mere fraction of the estimated 2–5 million fungi awaiting discovery. At our current rate, the journey to unveil and name all these hidden species could span centuries.[8,9]

Mycologists and citizen scientists alike are frustrated at the lack of resourcing for fungal taxonomy and the time it takes to name a new species. They are trialling radical new strategies to shorten the process of collecting, sequencing and analysing, paper writing and peer review, and publishing.

Australian mycologist Professor Roger Shivas is one of them.

We met Professor Shivas through Donovan Teal, a fungi enthusiast we met on the internet. It was Donovan who introduced us to the tiny, creepy, flesh-eating fungi that pin down and inhabit insects on the underside of leaves.

Donovan couriered his specimens of dead spiders, moths, leaf hoppers and ants, covered in fuzzy fungal growths and surreal sporing bodies, to Professor Shivas.

When DNA sequenced, the specimens were found to be laden with hundreds of species of microscopic fungi. There was even a new species of *Penicillium* – a genus famed for penicillin, the antibiotic that has saved millions of human lives.

Professor Shivas and his colleagues named this species *Penicillium tealii* after Donovan. They published it in an initiative called Fungal Planet, where mycologists use a standardised template to publish large numbers of new species in a single issue.

It is an exciting example of how scientists can fast-track the time between collecting and naming a new species so that we can use its identity and description for future observations.

OPPOSITE
Invertebrates can be a mini ecosystem for entomopathogenic microfungi, many of which are proving to be novel species.

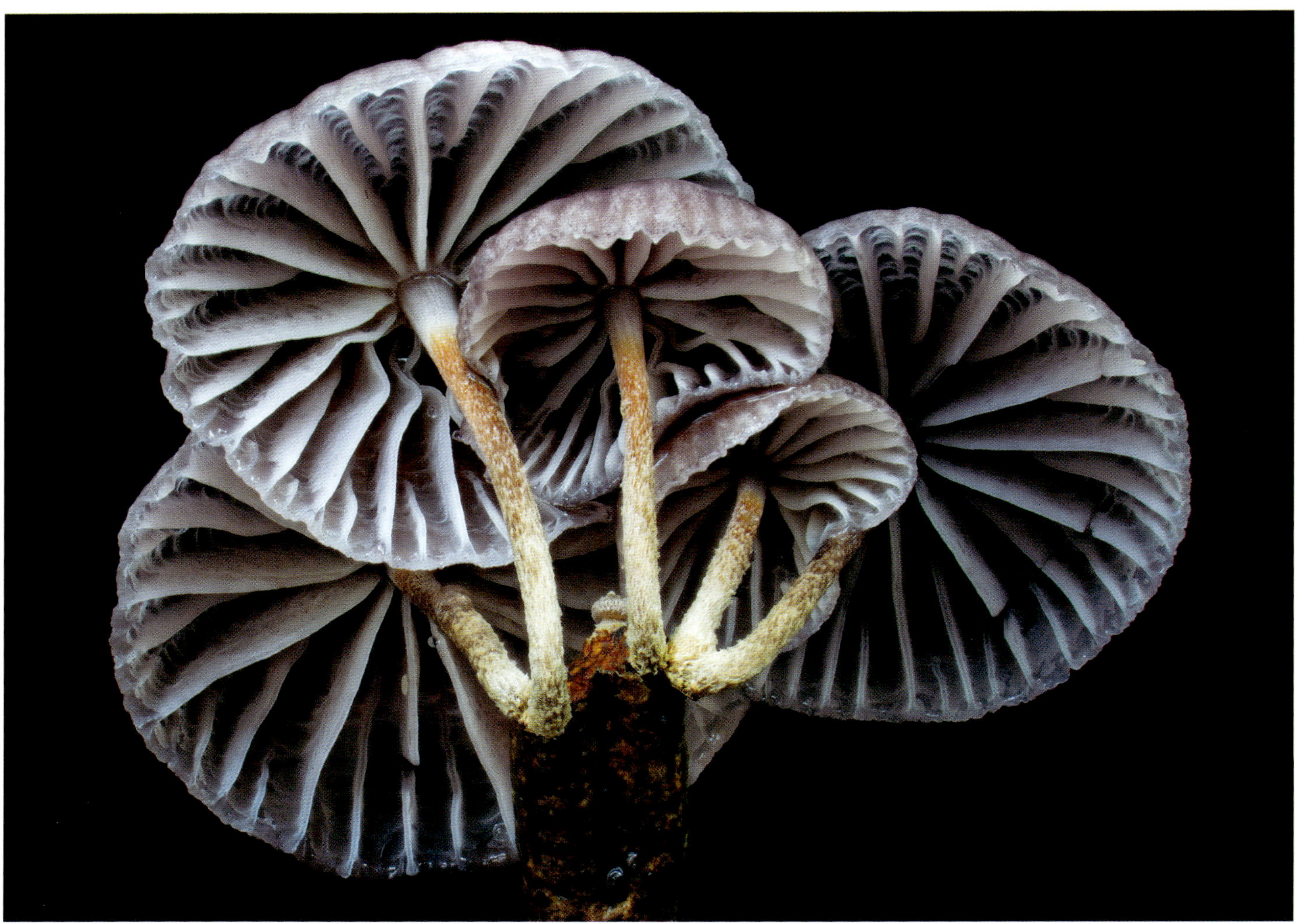

Straddling the traditional and the cutting-edge our current investigations march on with two possible new species from our local subtropical forests under investigation.

The first is exceedingly rare. We have only documented it in one small remnant of an endangered rainforest.

The mushrooms emerge fleetingly after summer's first downpour, painting the forest floor in an unusual switch between shades of mauve, orange and yellow. They vanish as soon as the forest starts to dry up, playing hide-and-seek with us for the rest of the year.

Initial genetic sequencing places this mushroom in the genus *Moniliophthora*, which has only a dozen or so species worldwide, most of which are already in the DNA database.

The stage is set for this fungus to be given a name.

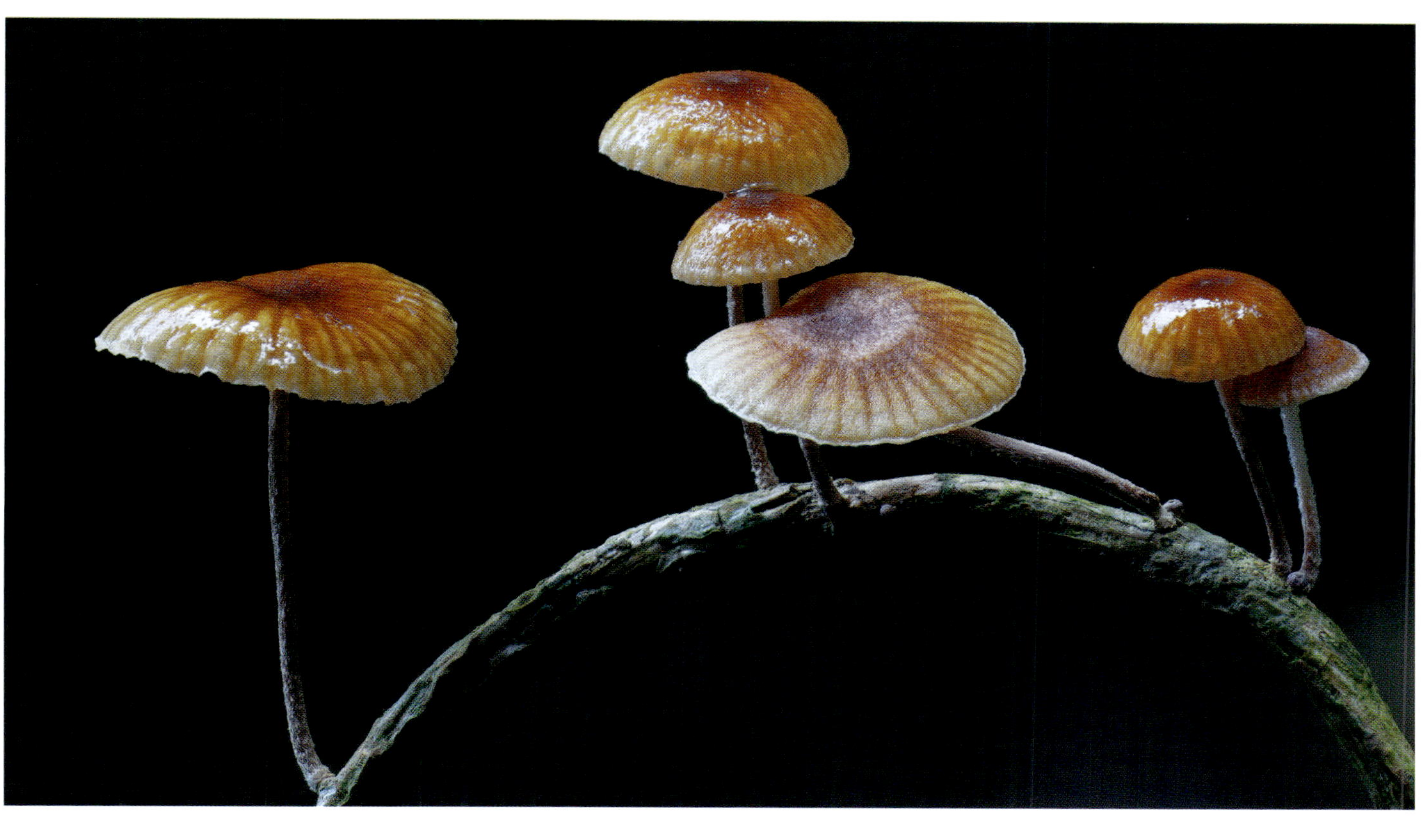

OPPOSITE, ABOVE & RIGHT

Moniliophthora sp.
Booyong, New South Wales, Australia
There are potentially three colour variants of this
fungus: orange, yellow and mauve.

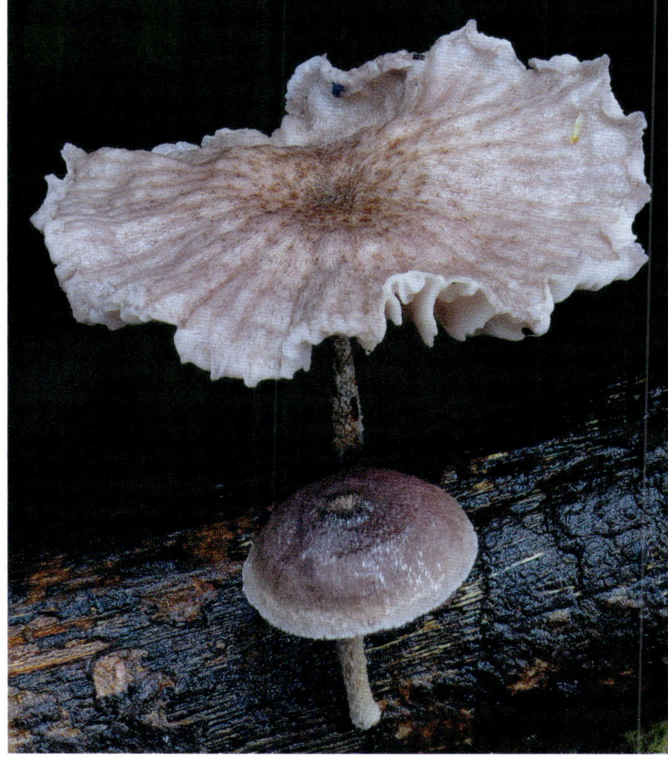

The second candidate is a mushroom that transforms in the most exquisite way.

At first glance, it grows like a bright orange *Coprinellus*, but as it matures, it morphs into a graceful structure with a translucent white cap and gills trimmed with gold.

Unknown to our go-to mycologists, this elusive fungus presents a challenge. It is small, and we have only collected a few specimens.

This scarcity of samples makes it challenging to sequence DNA and determine whether this is a novel species.

To unlock the secrets of this pretty mushroom, we need to gather more specimens.

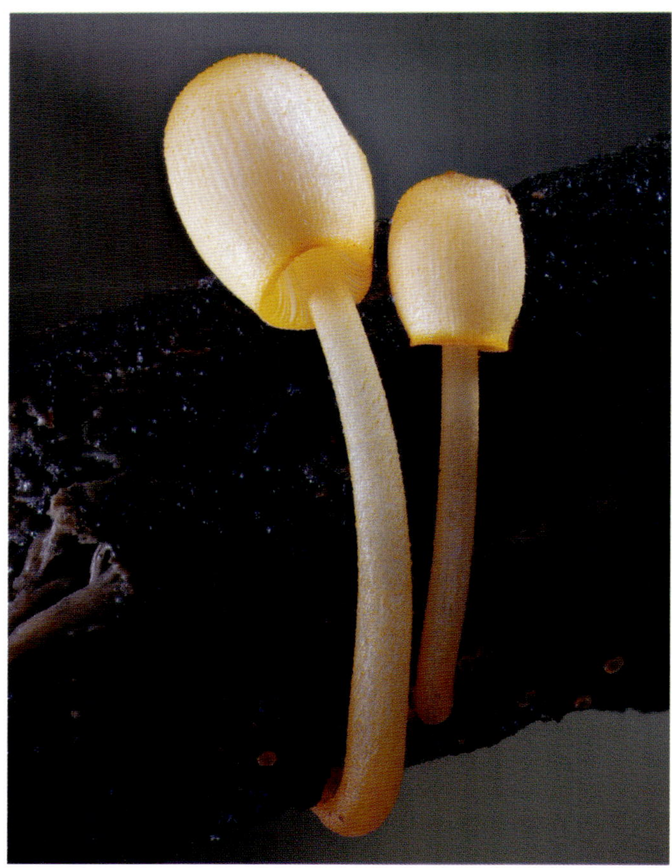

LEFT & OPPOSITE
This mystery mushroom appears, reaches maturity, and vanishes within three days.
On average, the caps are 2 centimetres in diameter. They have been photographed in southern Queensland and north-east New South Wales, Australia.

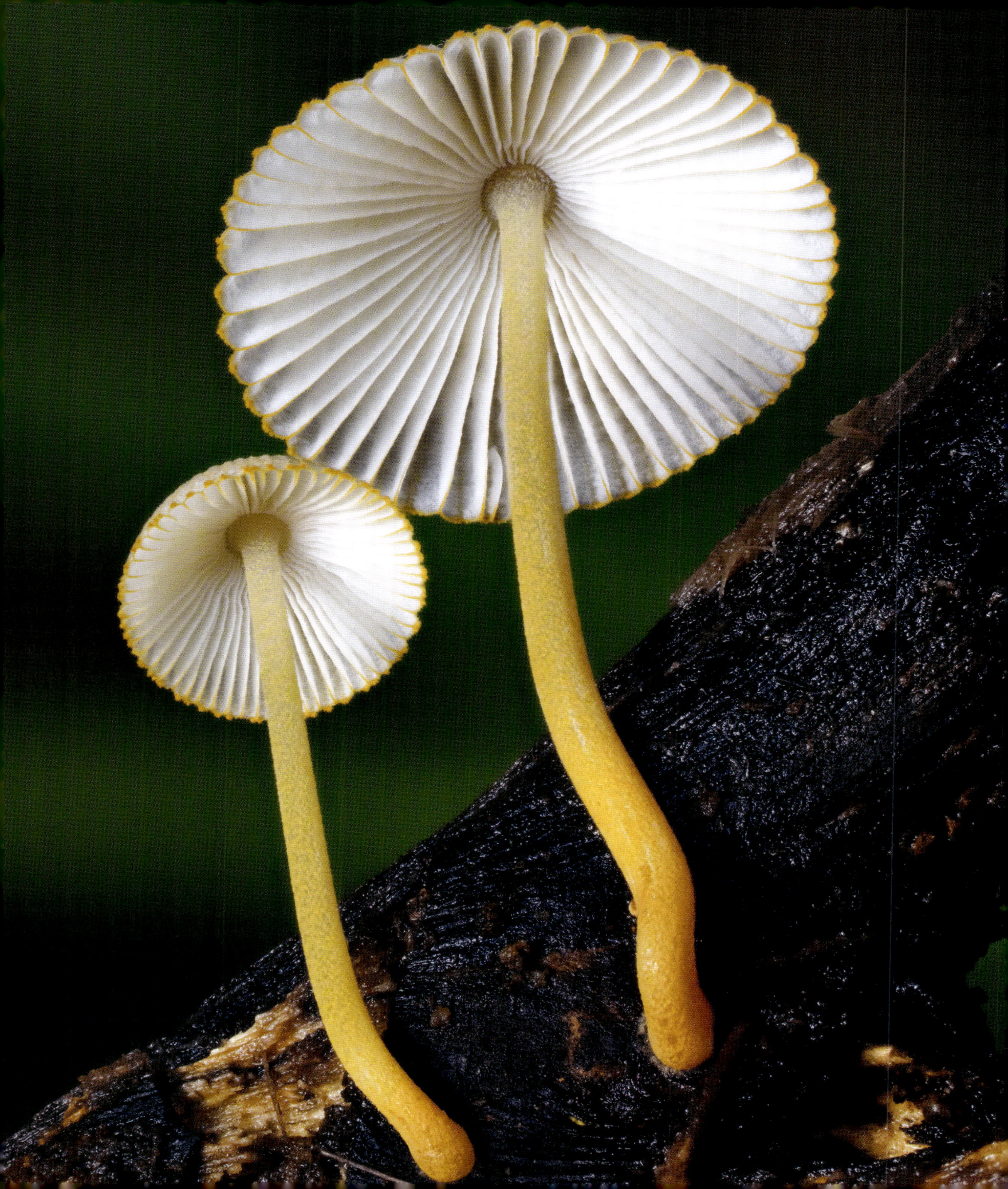

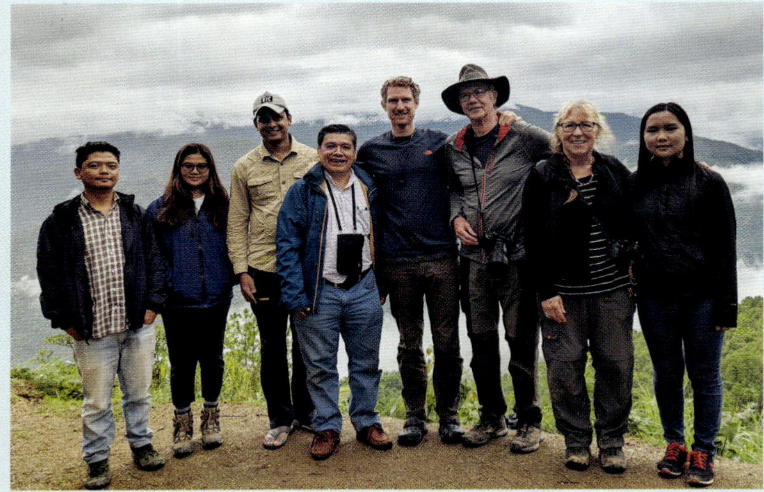

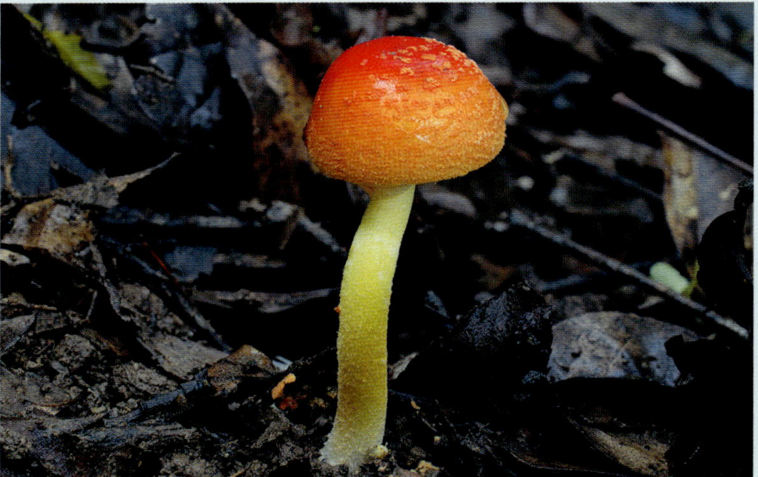

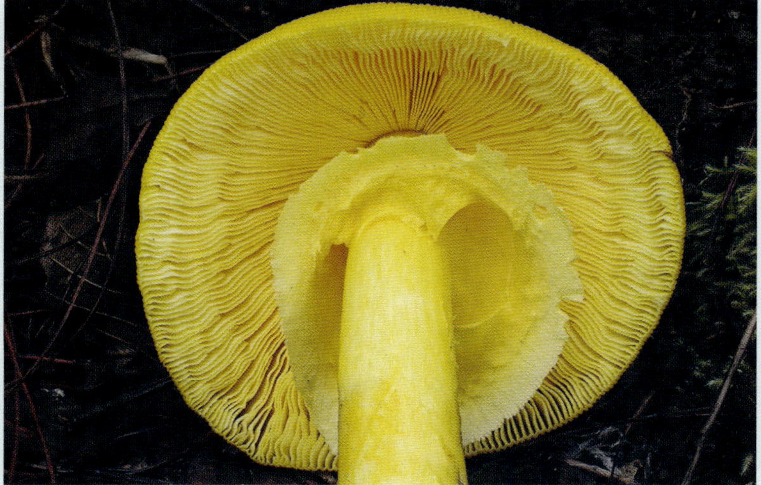

Some of Stephen's recent finds in Asia include a relative of the death cap and a glowing mushroom.

In 2019, fungi hunting led us to Myanmar, where we were invited by a scientific team grappling with reports of fatalities linked to an unknown white mushroom.

We embarked on a mission to find and document the killer.

TOP
We joined a team of mycologists, students and community workers for a fungi field survey in Myanmar, primarily to locate and document a toxic mushroom.

MIDDLE
Amanita rubrovolvata
Chin State, Myanmar
Distribution: Asia

BOTTOM
Amanita sp.
Chin State, Myanmar

OPPOSITE
Amanita pantherina, panthercap
Chin State, Myanmar
Distribution: natural distribution is Northern Hemisphere

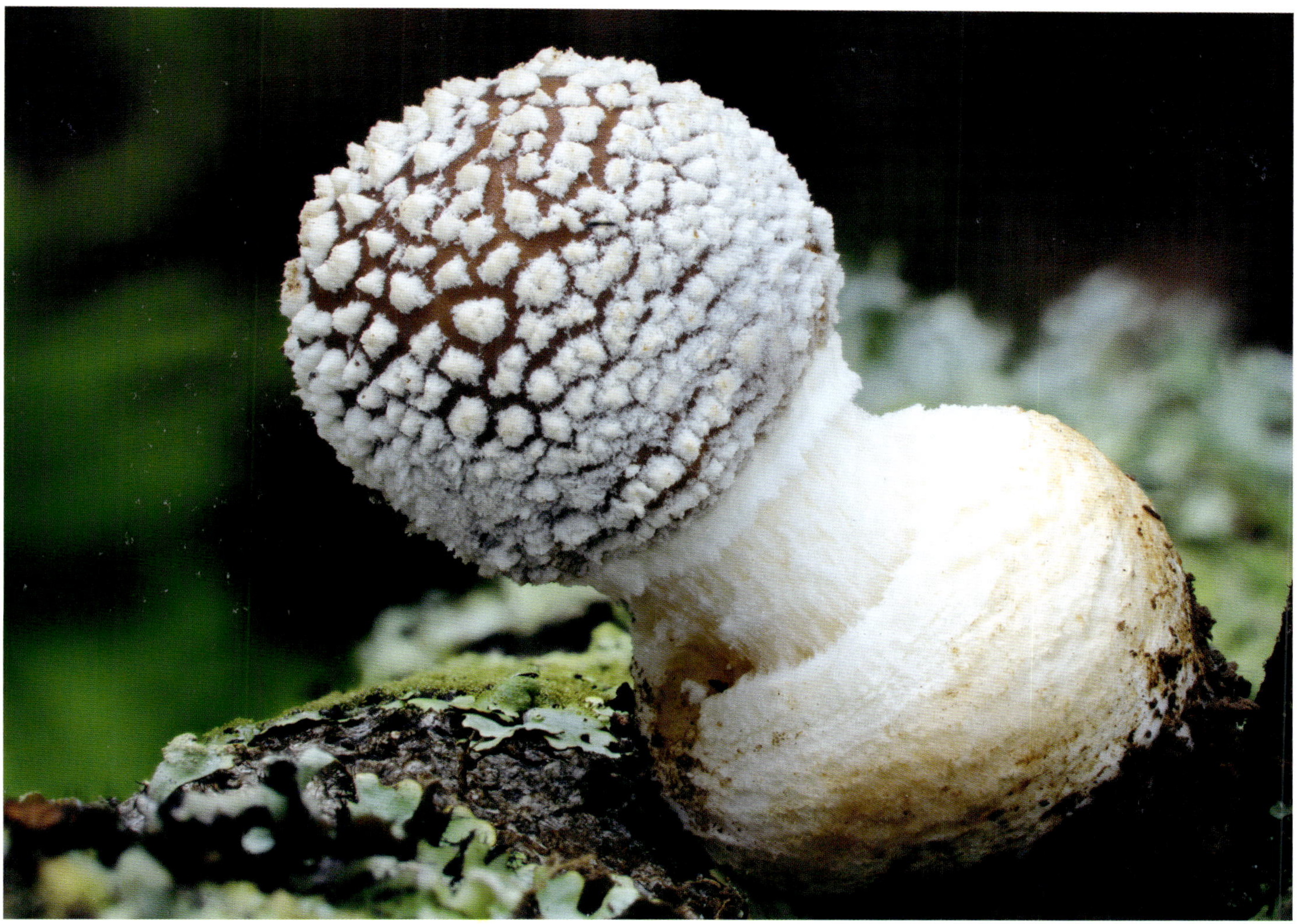

A small window of political calm had settled over the country. Travel restrictions were eased, and permits were secured for a fungi survey in the rugged hills of Chin State, guided by Dr Cin Khan Lian and Rachel Dim of the Ar Yone Oo Social Development Association.

Our team included Professors Peter Mortimer and Samantha Chandranath Karunarathna from China, Professor Jesus Perez Moreno, a visiting mycologist from Mexico, and local postgraduate students and ecologists.

Our mission was urgent. We had only five days to survey forests that had never seen a mycologist and find a poisonous mushroom believed to have caused seven deaths.

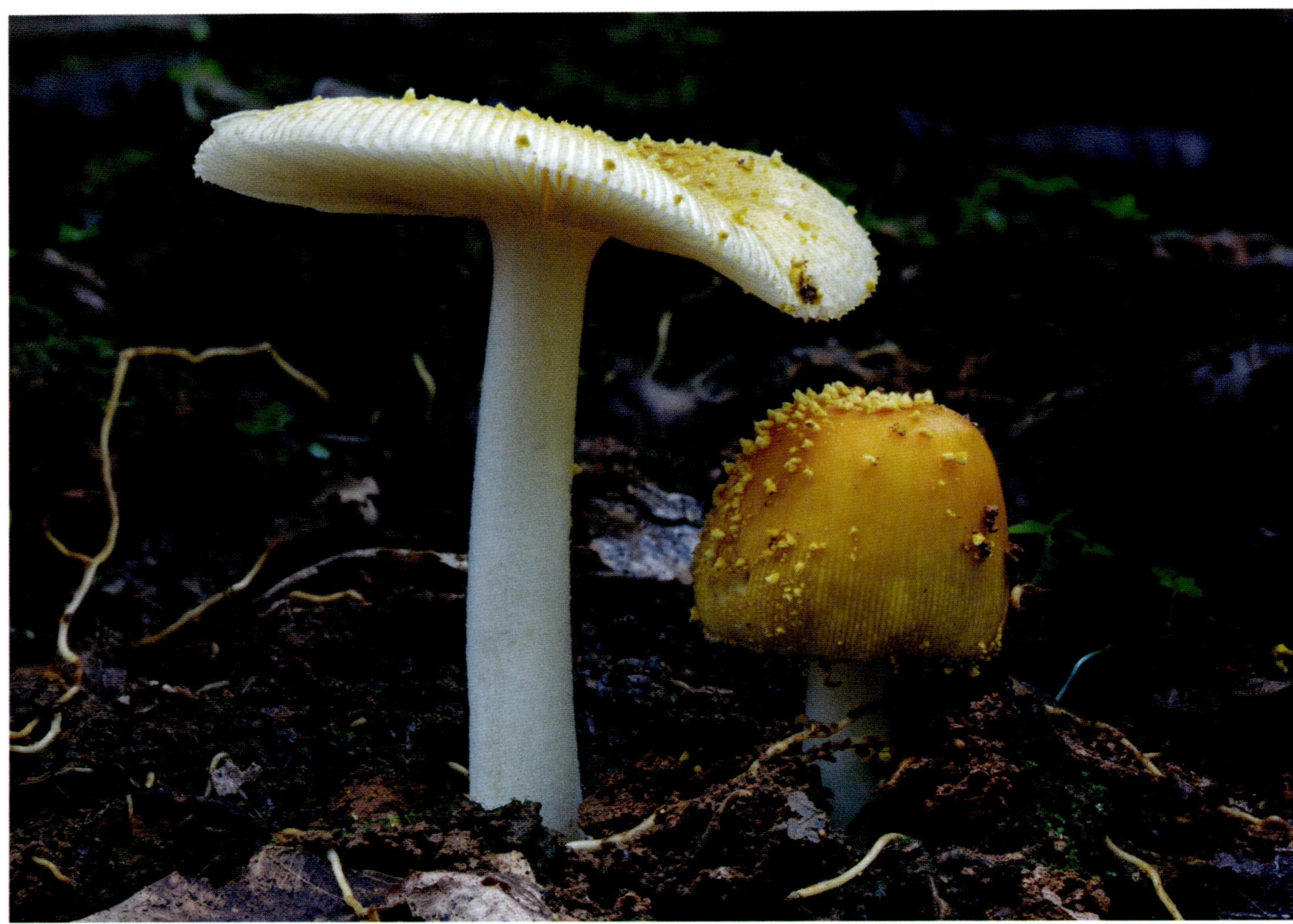

We knew the most likely candidate was an *Amanita* and possibly the infamous death cap *A. phalloides*, the most poisonous of this genus.

Orange, brown, cream, white and red amanitas grew from their distinctive eggs on every forested hilltop. Their elegance is deceptive, distracting foragers from hidden dangers.

ABOVE
***Amanita* sp.**
Chin State, Myanmar

Death from *A. phalloides*, unfolds in three dramatic acts.[10,11]

The first act can start with a period of deceptive calm. The mushroom's taste is often praised and there are no immediate signs of distress.

The second act begins with a sudden onslaught of discomfort. Nausea, abdominal cramps and severe dehydration signal the start of the mushroom's toxic cascade.

This leads to the third act. The individual appears to recover. But beneath this façade, escalating damage to the liver culminates in life-threatening organ destruction.

The final act is a rapid descent into crisis. Liver and renal functions fail. This stage can prove fatal within three to seven days.

These symptoms resembled those reported to our Myanmar colleagues.

TOP
Amanita cf. *rubroflava*
Chin State, Myanmar

BOTTOM
Amanita sp.
Chin State, Myanmar

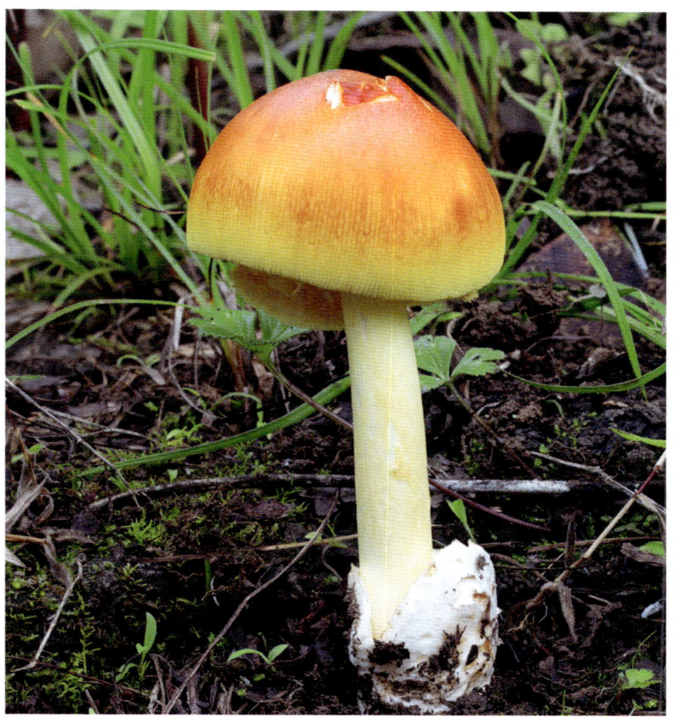

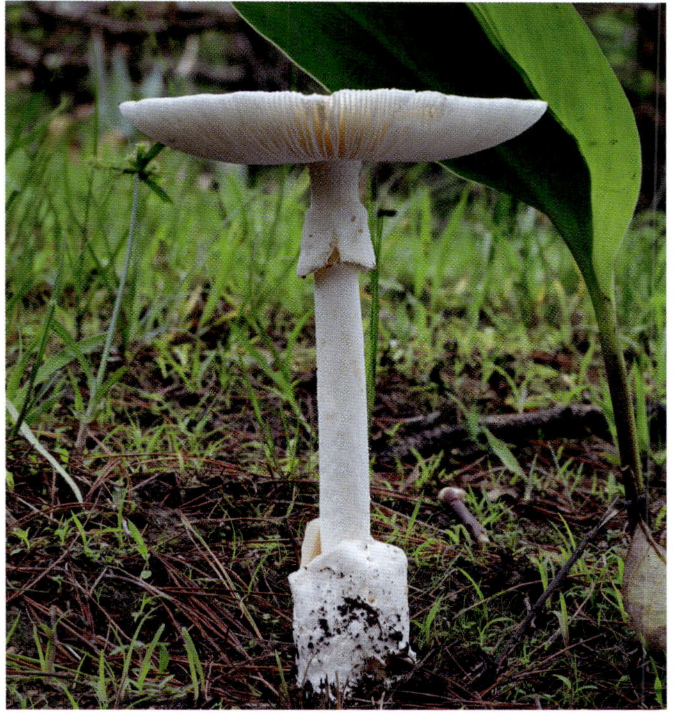

As we travelled, they asked the local villagers to show us the amanitas they were wild-harvesting for food.

The mycologist's sharp eye noticed that one of the species they were collecting had a cap with a subtle yellow tint, similar to the colouring of the cap of an *A. phalloides*.

One of the challenges of international mycology is that countries are reluctant to allow visiting scientists to export scientific specimens. This is well founded due to foreigners' past exploitation of natural resources and intellectual property. However, in this case, transport of a small, dried specimen was permitted.

Professor Karunarathna did some basic DNA sequencing, and the results indicated that it was not only an *Amanita* but potentially a new species in the *A. phalloides* group.

Stephen's photographs bridged science and community. They became tools in a grassroots education campaign to raise awareness.

Posters and field guides detailing edible and poisonous fungi, including the suspected *A. phalloides*, were distributed throughout the region.

Plans were laid for a return to Myanmar in 2020, a collaborative effort with local research facilities to DNA sequence more fully the region's fungi and publish findings about any new species.

However, the world's upheavals intervened. The COVID pandemic and subsequent civil unrest in Myanmar halted scientific progress, leaving this deadly mushroom's true identity shrouded in uncertainty.

DNA sequenced as a new species in the *Amanita phalloides* group, this fungus is believed to be responsible for deaths from mushroom poisoning in Myanmar's Chin State in 2018 and 2019.

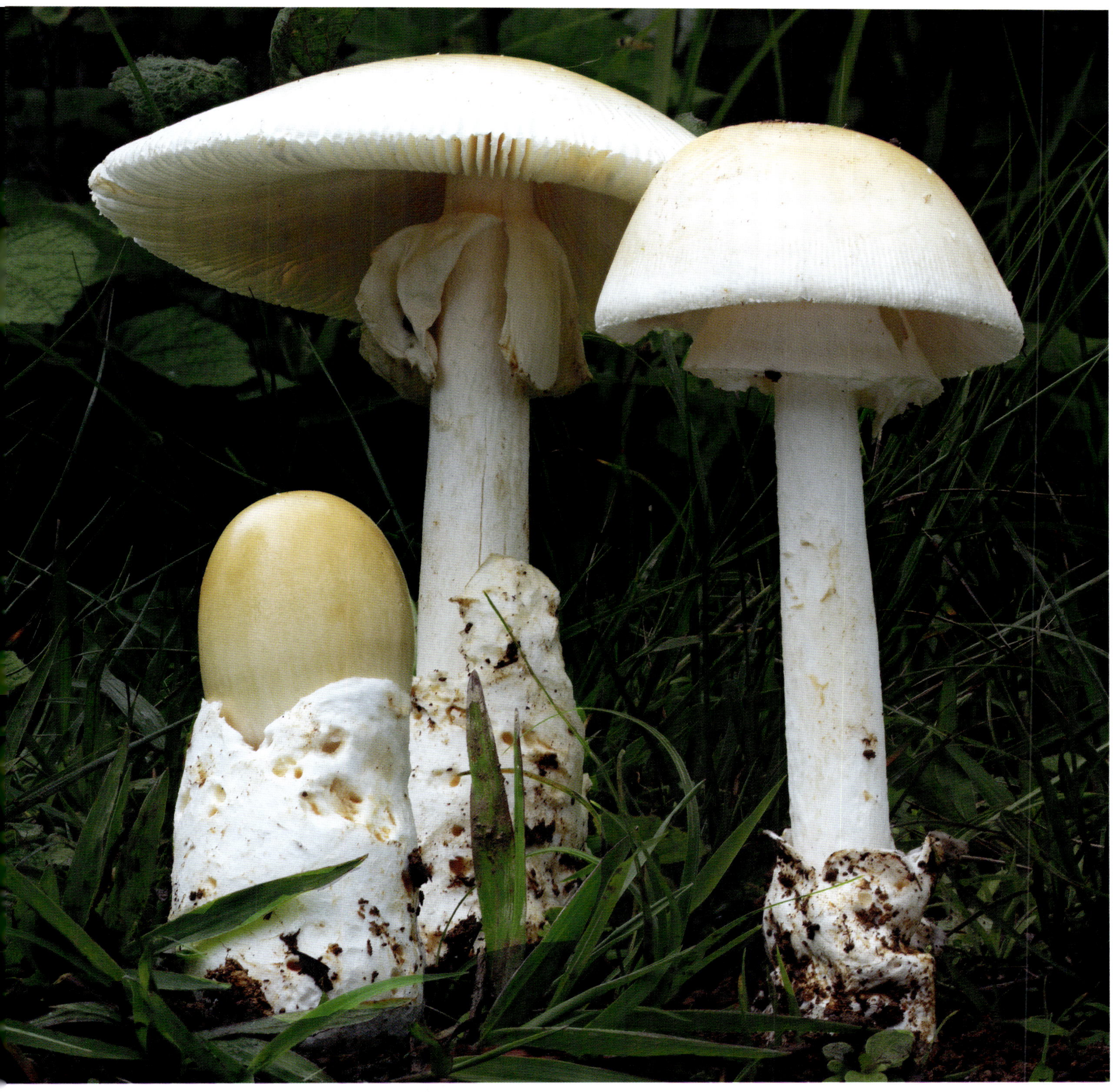

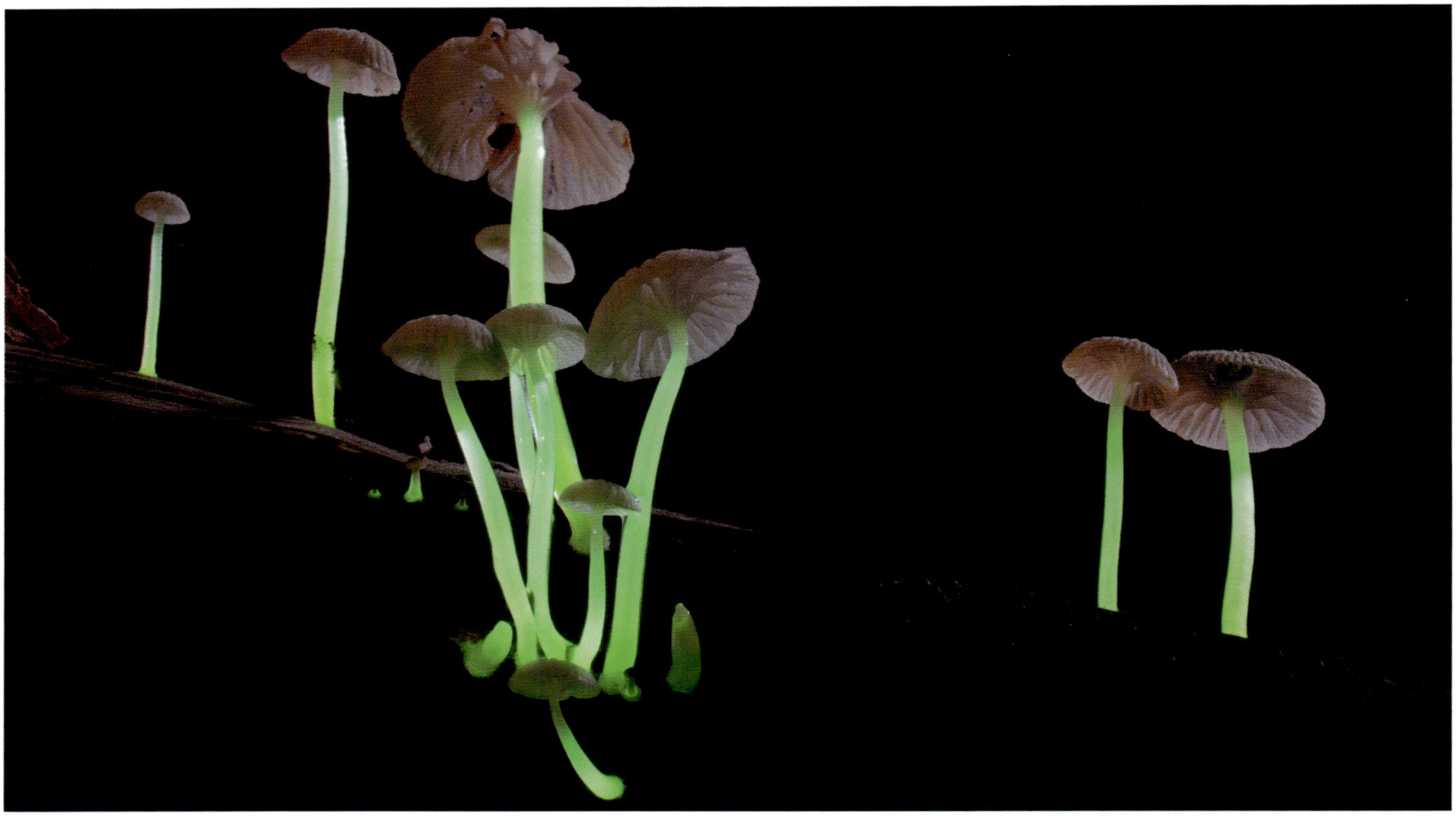

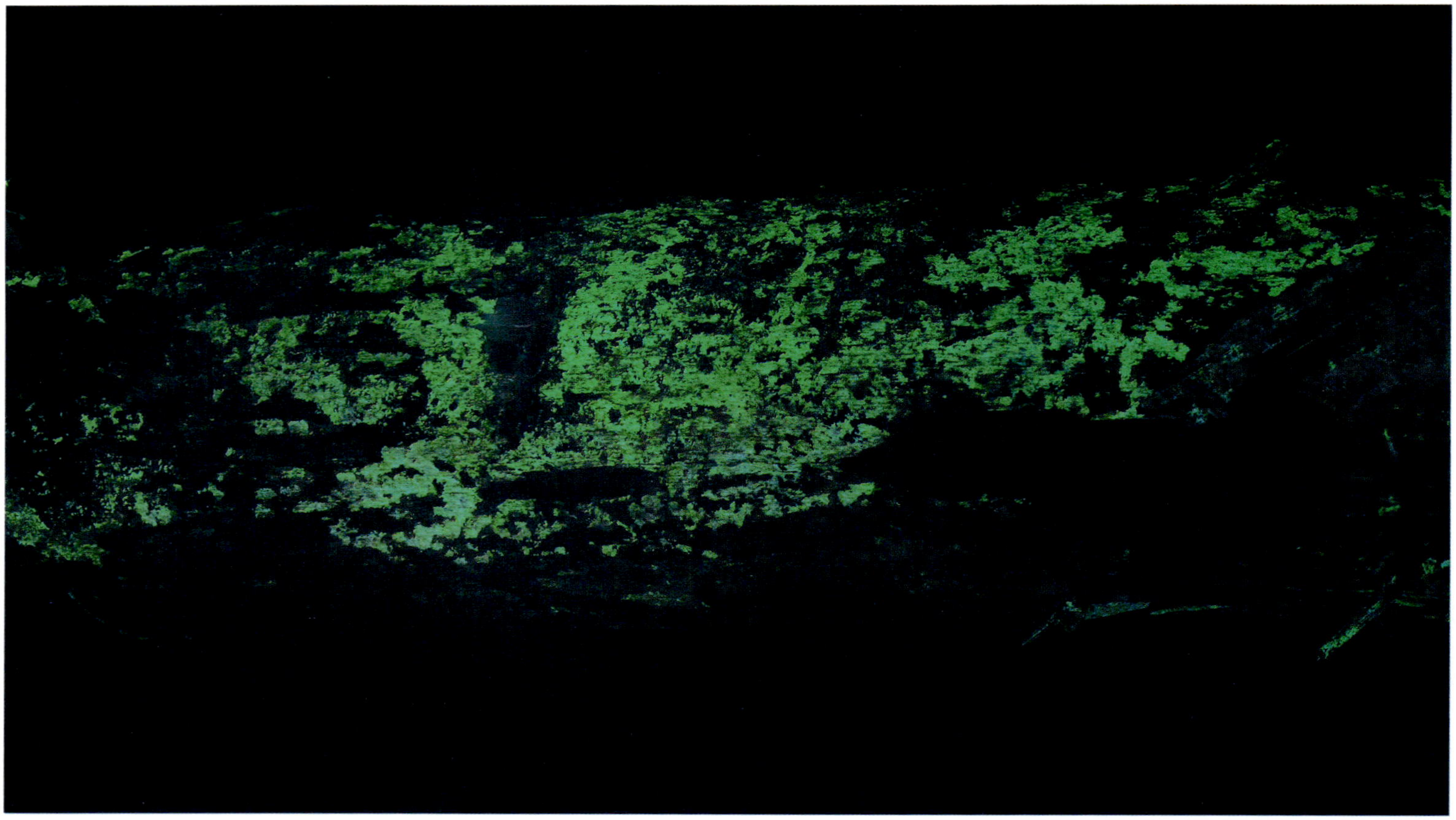

Amid the misty foothills of the Himalayas another new discovery awaited.

During our 2018 expedition in India with the Balipara Foundation, we visited Mawlynnong, a small village near the border of India and Bangladesh.

It is a region renowned for its record-breaking rainfall – around 12 metres annually; in 1985, one village received a mind-boggling 26 metres. The high humidity creates an ideal environment for fungi to thrive.[12]

Wherever we go, we always ask if there is a local variety of bioluminescent fungus. Usually, the answer is no, but in Mawlynnong, our local guide casually responded, 'Of course.'

As twilight wrapped the village, we were led down a dangerously slippery path to a small stream. Growing on the dead debris of bamboo clumps were small glow-in-the-dark mushrooms different to anything we had ever seen.

In most varieties of luminous fungi, it is the caps that tend to glow brightly. However, in this case, it was the stipes and the mycelium that emitted a bright neon green light.

This variety is tiny, on average 10 millimetres tall, with a cap of around 2 millimetres and it has a slimy, glutinous stipe.

We suspected this fungus was something new to science, but without a mycologist in our field crew, we could only speculate.

In 2019, we returned with our collaborators from the Balipara Foundation and Professor Karunarathna from the Kunming Institute of Botany, armed with the necessary permissions to collect and sequence the DNA of this fungus.

Roridomyces phyllostachydis
Mawlynnong, Meghalaya, India
Distribution: India

It is named after the species of bamboo *Phyllostachys mannii*, that appears to be its preferred substrate.

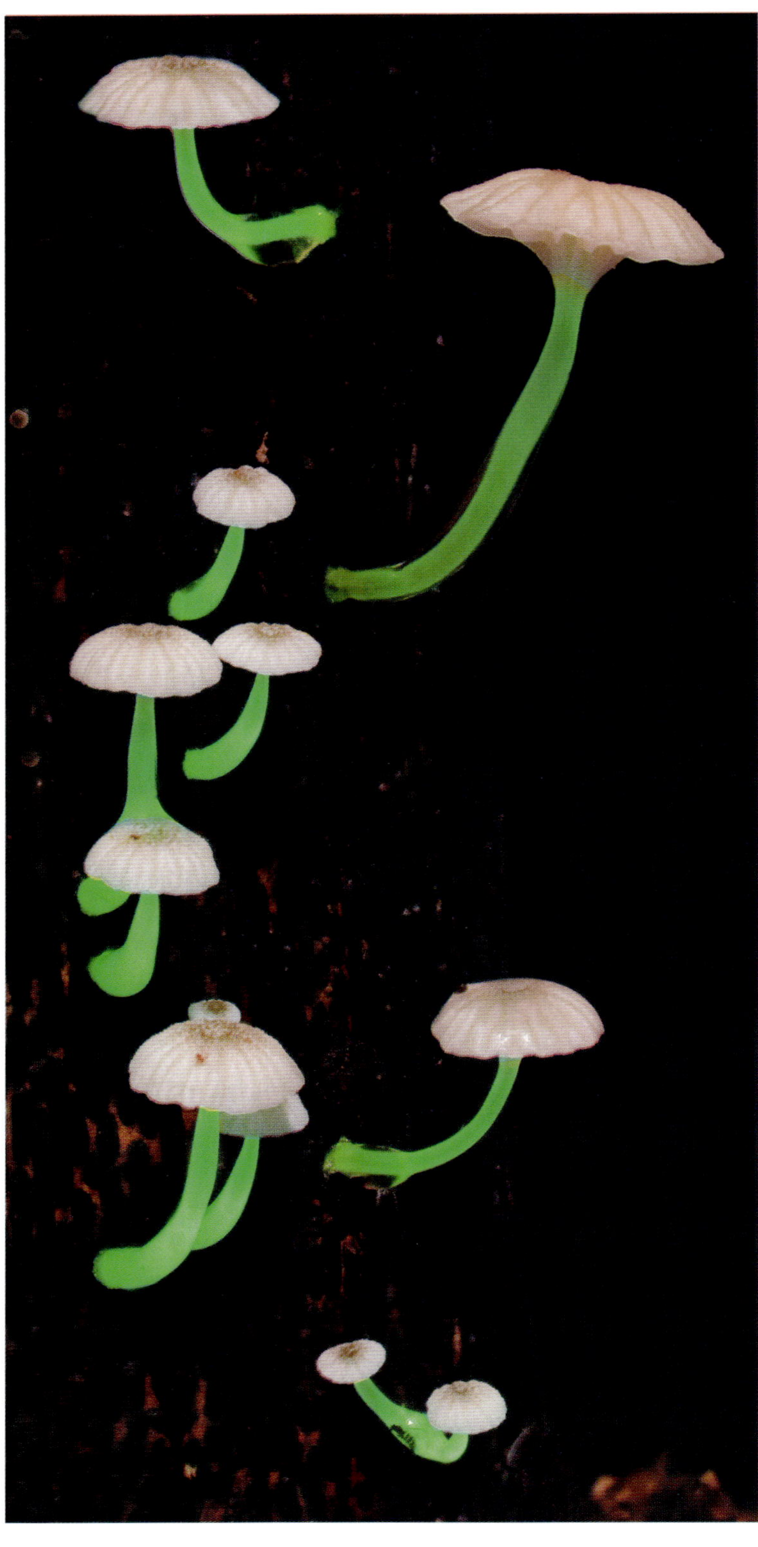

Our hunch was correct.

The world now knows this mushroom as *Roridomyces phyllostachydis*, a name that signifies not just a new species but a milestone – the first of its genus found in India.

Yet, this was just one gem among many. Out of 232 specimens collected on this fungi field trip, DNA sequences suggest 34 are unknown to science.

Only *R. phyllostachydis* has been officially named.

We find ourselves at a pivotal moment in the history of mycology.

For centuries, scientists and naturalists have painstakingly preserved the story of dried fungal specimens in fungaria scattered across the globe.

These collections hold the secrets of the 155,000 known fungal species and countless others that remain unnamed.

Roridomyces phyllostachydis
Mawlynnong, Meghalaya, India
Distribution: India

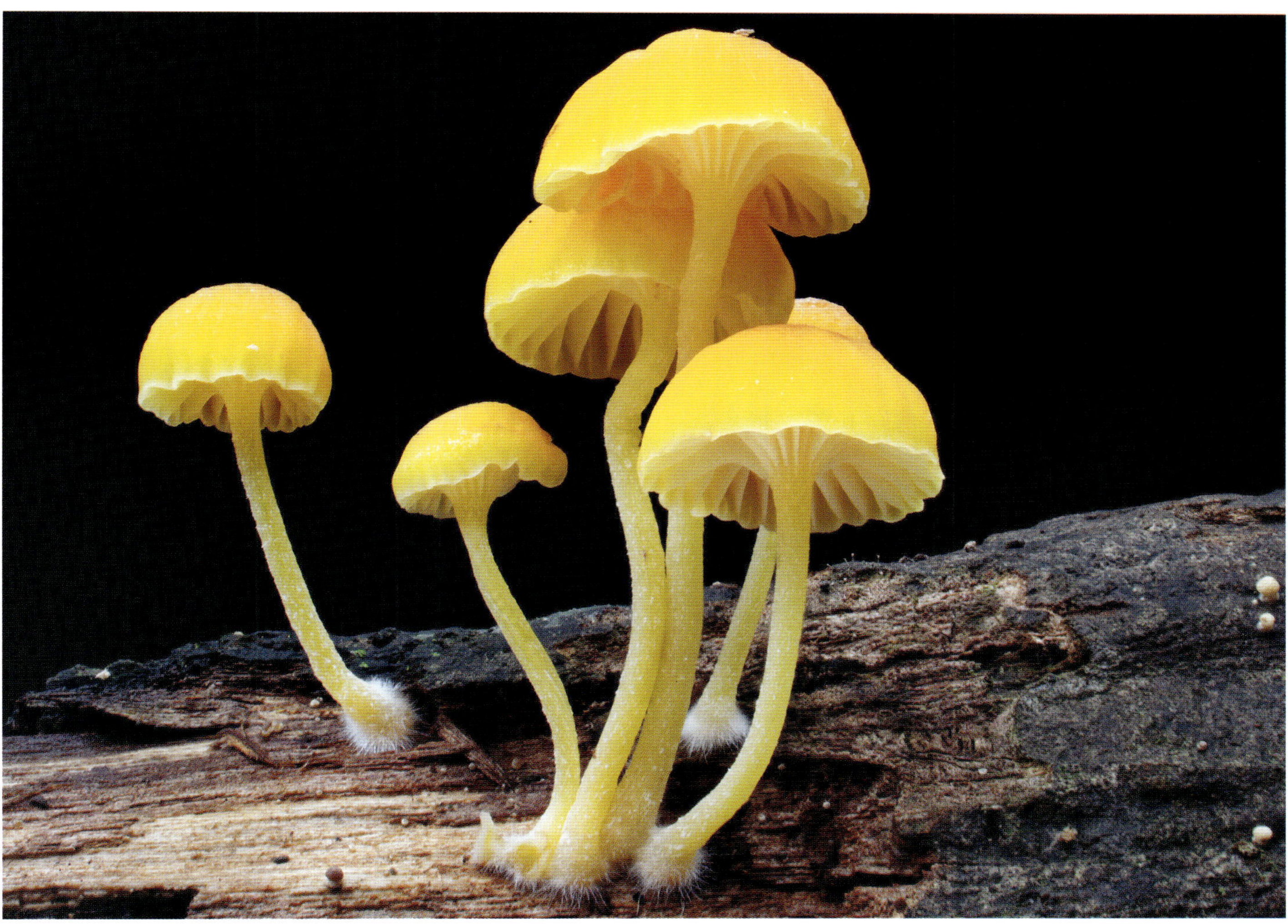

Despite the richness of this precious archive and the advanced technology at our disposal, we have only unlocked the genetic secrets of a mere 10 per cent of known species.

Fungi – the very kingdom connecting life on Earth – remain largely a mystery.

As their habitats and diversity shrink, their names slip through our fingers, and their roles escape our understanding just as we begin to comprehend their importance.

ABOVE
Unidentified mushroom
Nightcap National Park, New South Wales, Australia

This fungus is added to the ever-growing pile of specimens awaiting microscopic examination and sequencing.

There is an urgent need to scan the DNA fingerprints of every known species as swiftly as possible.

This endeavour will be a game changer, turbo-charging the monumental task of naming the millions of unknown fungi that await discovery.

Completing this DNA catalogue will revolutionise fungal taxonomy.

A robust database will emerge, empowering scientists to identify, compare and rapidly detect new discoveries.

It will shed light on the distribution and rarity of species, revealing patterns and connections previously hidden.

Each named fungus is a vital piece in the grand puzzle of biodiversity, bringing us ever closer to understanding the complex world of fungi and their indispensable role in our ecosystem.

'When I'm fungi hunting, knowing the name of a fungus, enables me to immediately recognise what I'm looking at.

'It's like meeting someone and instantly grasping their life story.

'This knowledge is key to unlocking a deeper understanding of the world, transforming a beautiful mushroom tucked away in a damp corner into a profound discovery.'

OPPOSITE
A species of *Ganoderma* documented for the first time in an unsurveyed forest in Nepal. Medicines derived from *Ganoderma* form a multi-billion-dollar global industry.

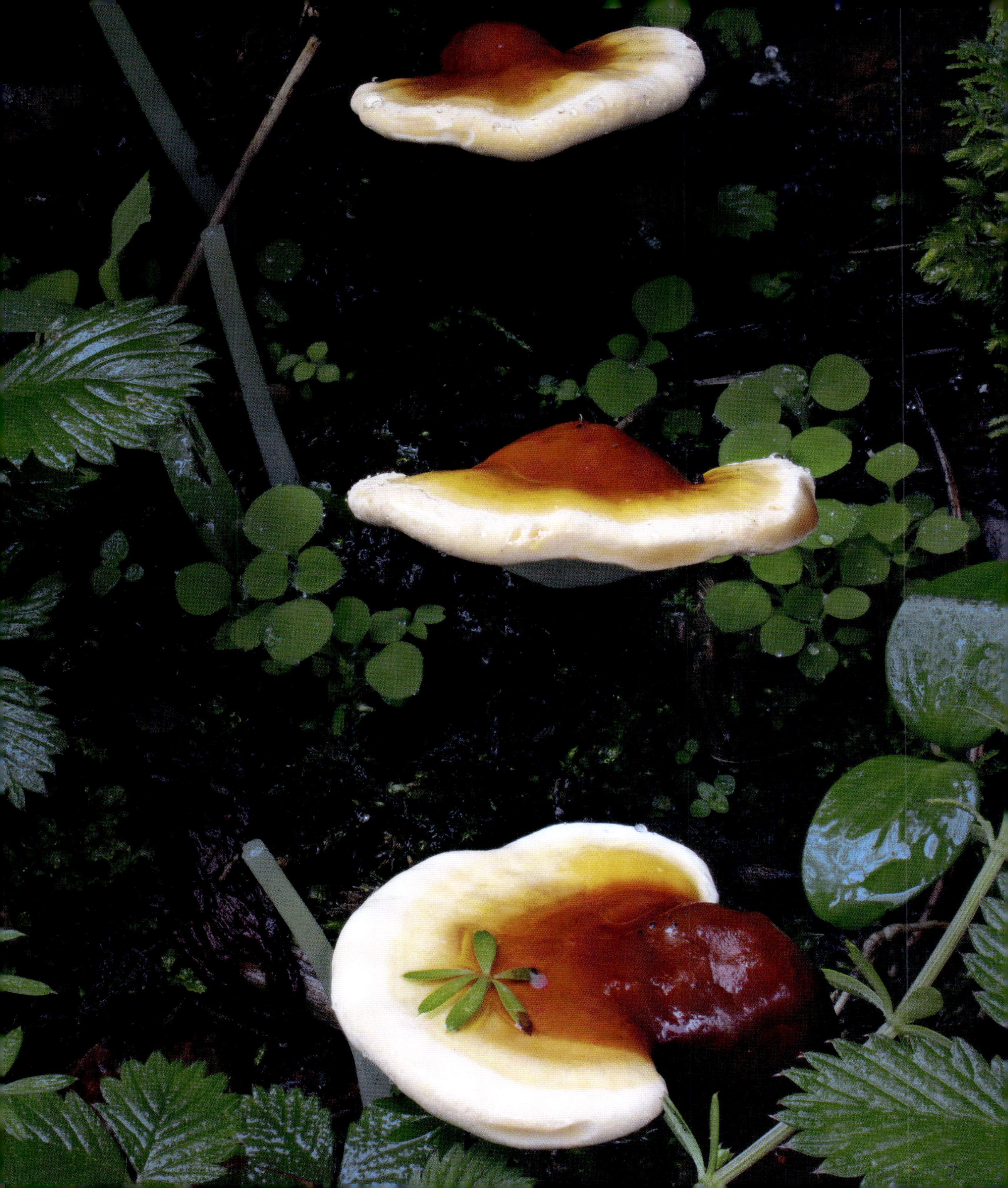

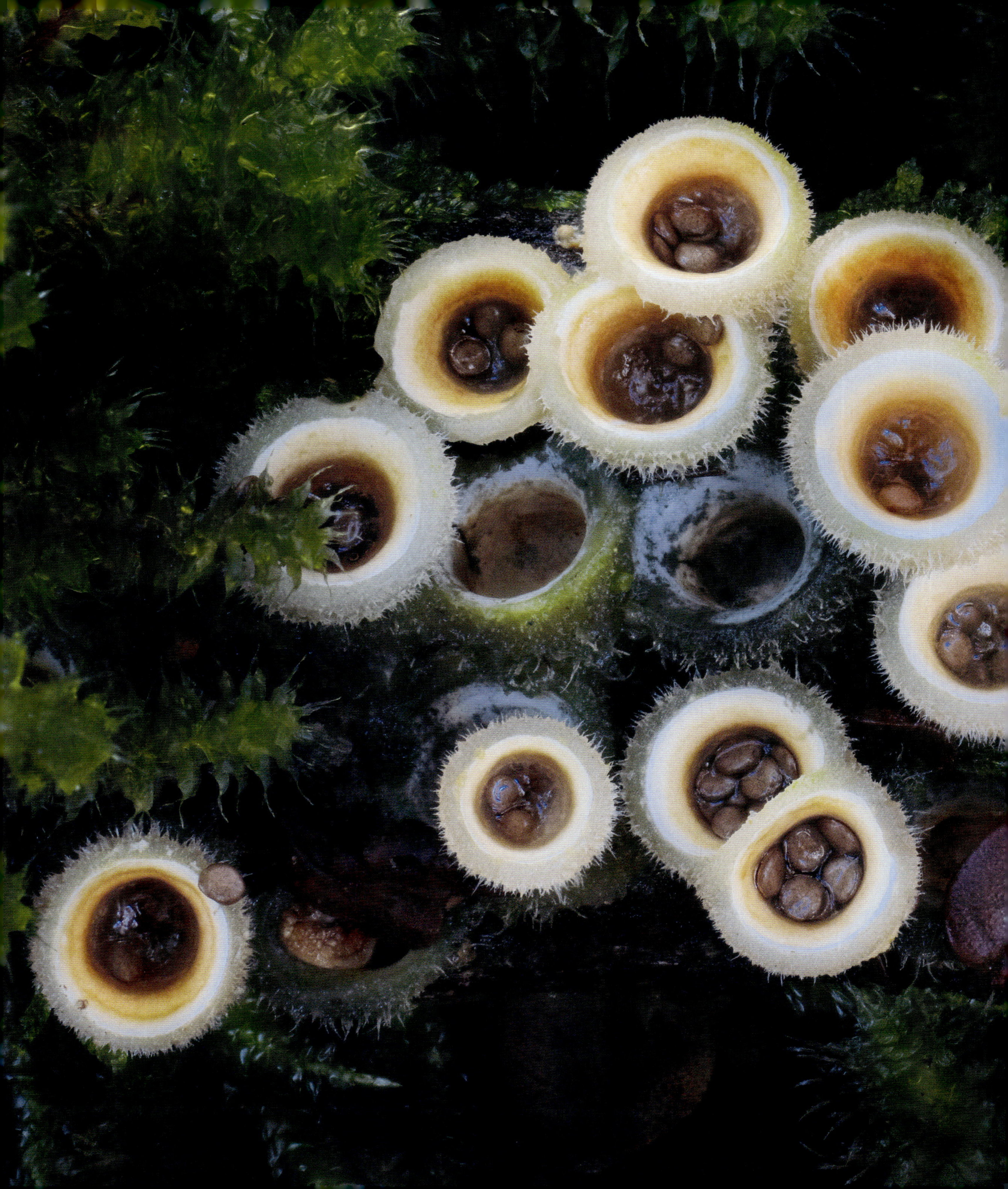

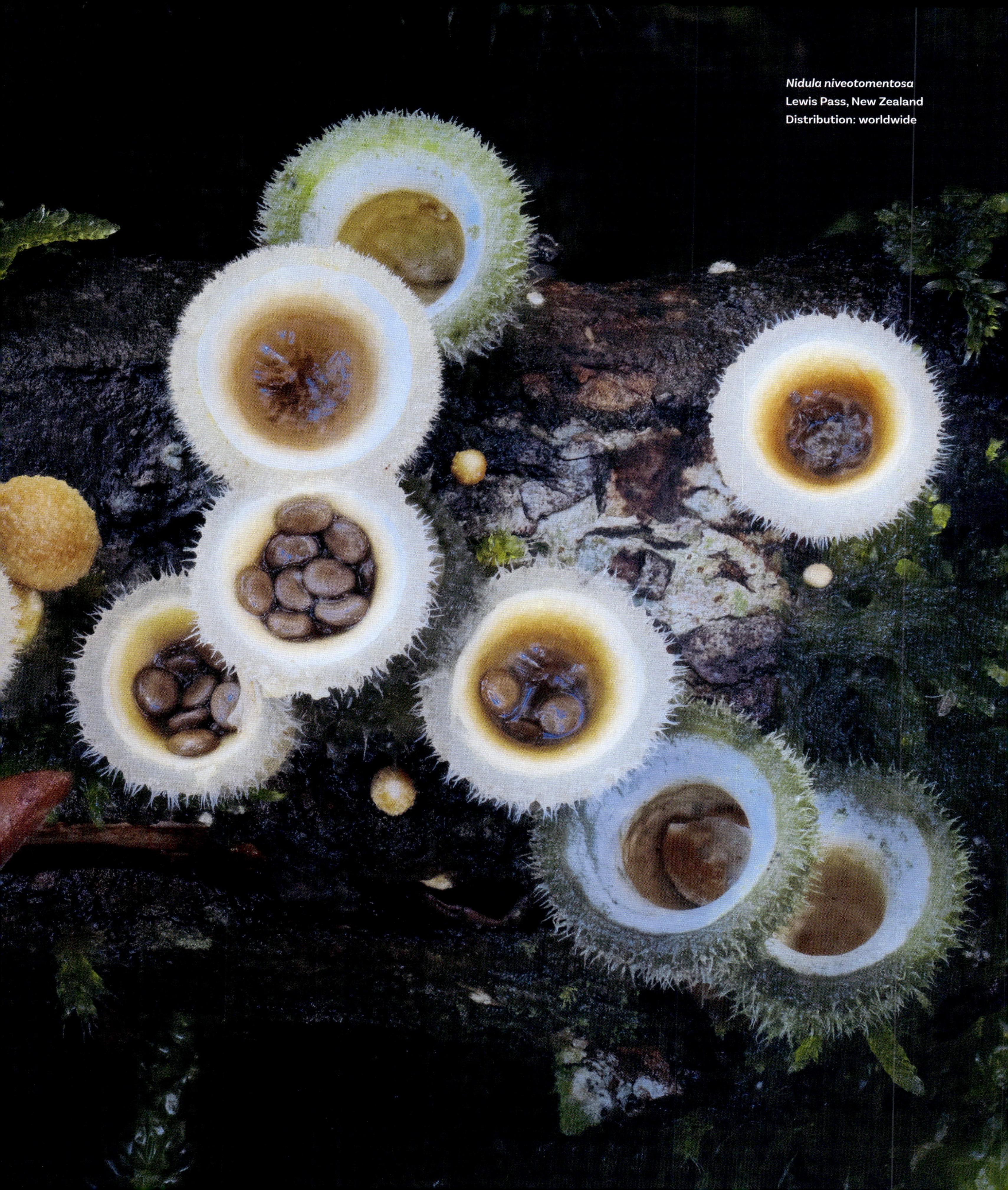

Nidula niveotomentosa
Lewis Pass, New Zealand
Distribution: worldwide

Everything is connected

Without fungi, life as we know it would not exist.

The summer of 2019 in Australia was one of uncertainty and heartbreak.

For months, we awoke to an eerie, apocalyptic orange glow and billowing smoke, accompanied by the constant wail of sirens and the drone of helicopters.

Emergency services worked tirelessly to combat a relentless wall of fire advancing across the country.

'Black Summer' saw a series of mega-blazes fuelled by years of drought and climate change.[1,2]

The environmental devastation was immense. Over 30 million hectares of land were burnt, affecting 3 billion vertebrates and up to 240 trillion invertebrates. Many died. Others were displaced, their habitat destroyed and with it their access to food.[3,4]

A long-overdue thunderstorm finally brought the fires in our region under control.

Within days, Stephen and I began receiving text messages from local firefighters reporting fungi miraculously emerging from the still-smoking ground.

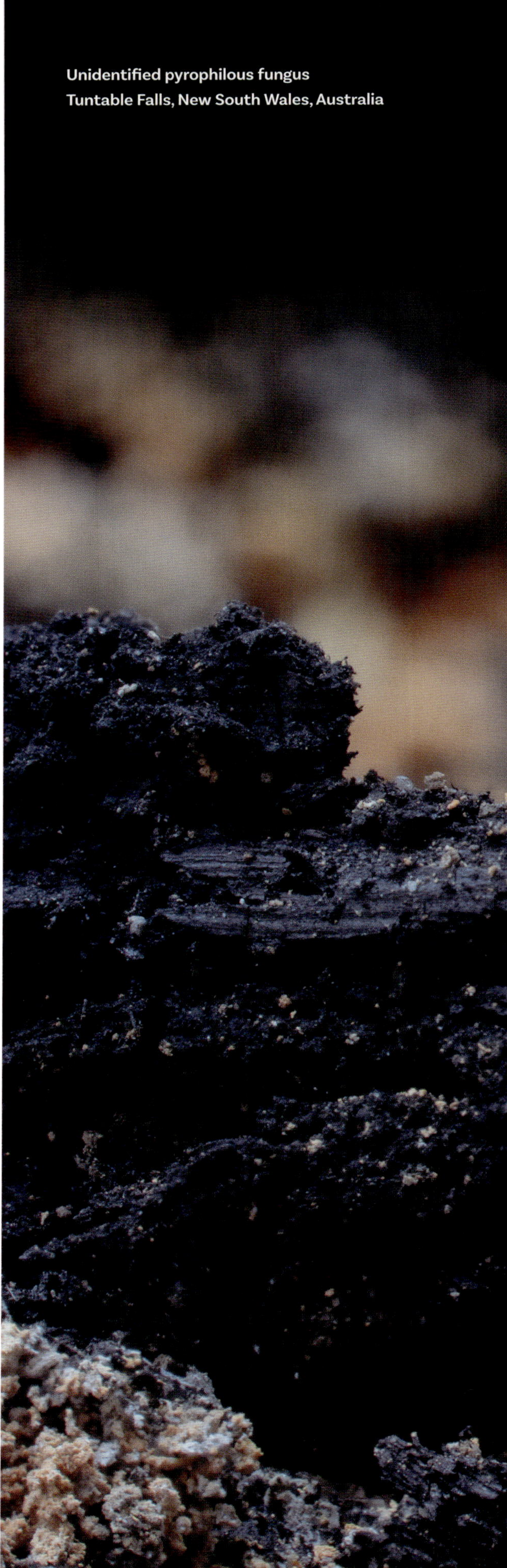

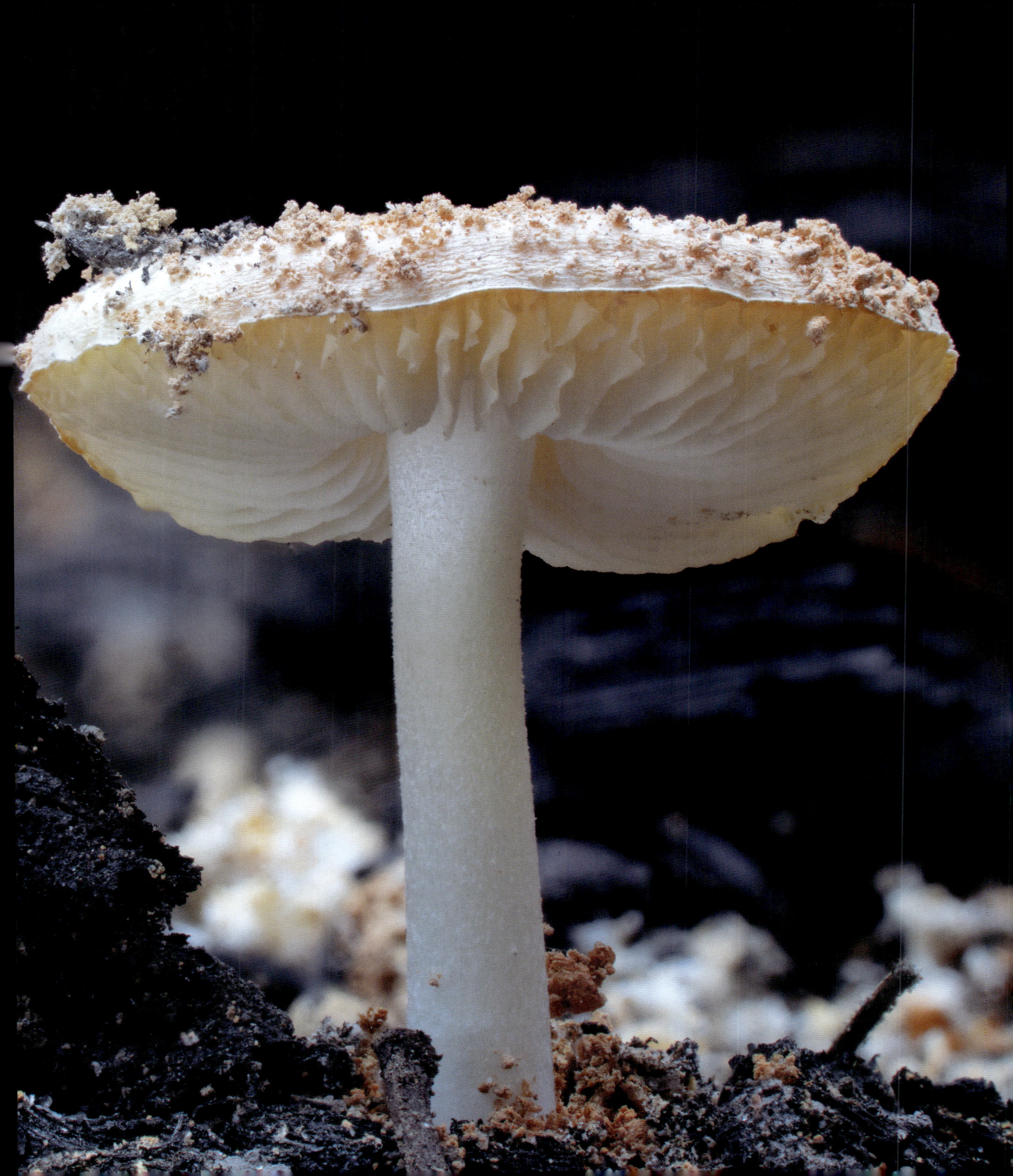

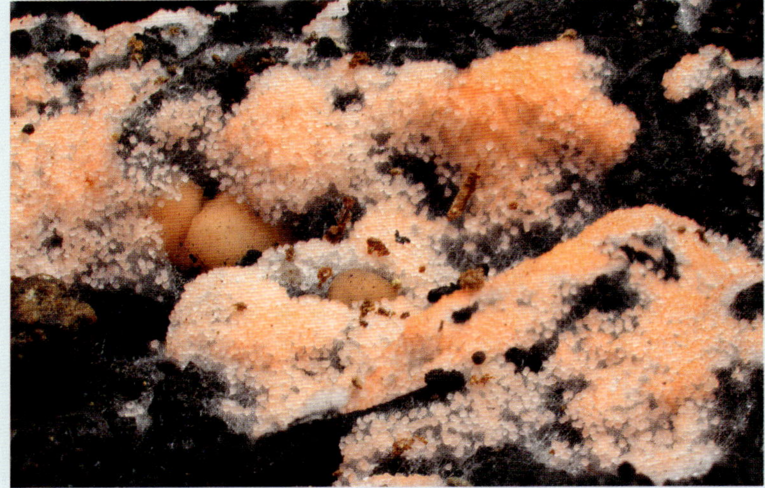

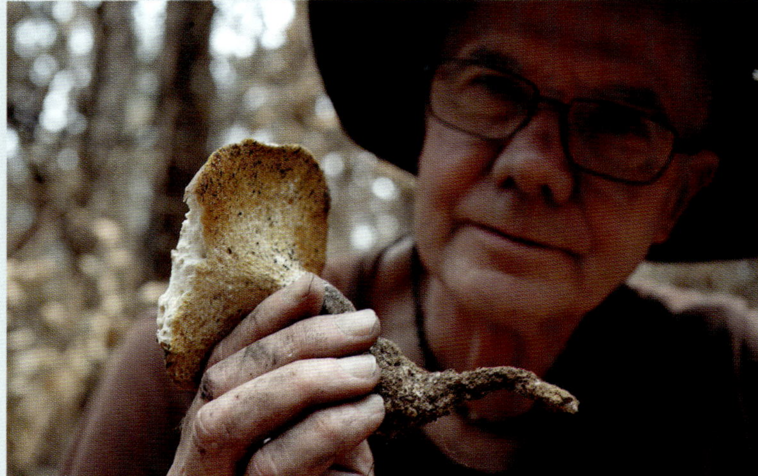

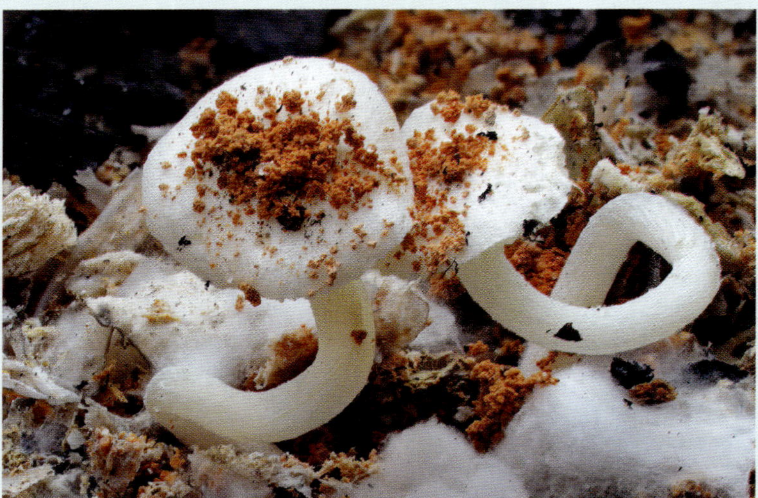

In an incinerated rainforest sanctuary potentially untouched by wildfires for centuries, we witnessed nature's resilience and embarked on a steep learning curve about fire-loving, or pyrophilous, fungi.

Stephen's lens captured the vibrant orange mycelium of a *Pyronema* fungus and the small, eye-lash-rimmed cups of *Anthracobia muelleri*, spreading across the sterilised soil.

These fungi were not only anchoring the ash and soil in place but also preserving moisture just a few millimetres beneath the surface.[5]

TOP
Pyronema sp.
Tuntable Falls, New South Wales, Australia

MIDDLE
Laccocephalum tumulosum
Terania Creek, New South Wales, Australia
Distribution: Australia

BOTTOM
Unidentified pyrophilous fungus
Tuntable Falls, New South Wales, Australia

OPPOSITE
Anthracobia muelleri
Terania Creek, New South Wales, Australia
Distribution: Australia and rarely in New Zealand

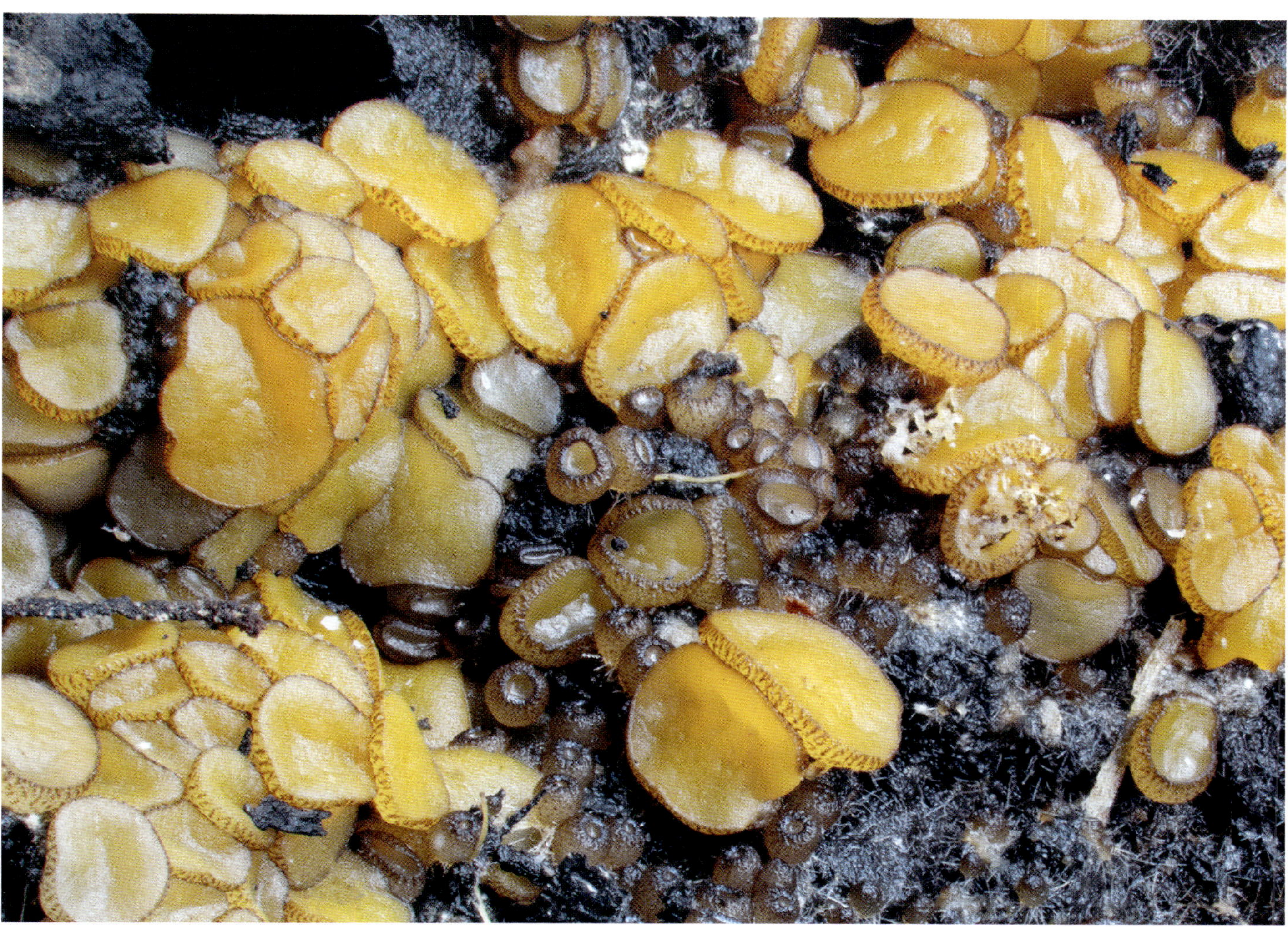

Among the most striking discoveries were the massive sporing bodies of *Laccocephalum tumulosum*, the stonemaker fungus. Its underground sclerotia, brimming with stored energy, can quickly produce a mushroom, possibly triggered by the extreme temperatures of the fire or by smoke. These mushrooms, teeming with insects, provided a rare food source in an otherwise barren landscape.[6]

Three weeks passed, and we returned to the scorched earth to observe any changes. In some of the worst-burned areas, where huge fallen logs had been reduced to charcoal, Stephen lifted the dead leaves to reveal a labyrinth of dainty white mushrooms and their mycelium.

We marvelled at how the spores of these fungi had survived the extreme temperatures on the soil surface as the fire passed.

We learned that soil is an extraordinary insulator, shielding organisms like fungal spores just centimetres below the surface from the firestorm's extreme heat.[5]

Armed with complex enzymes to break down charcoal and the post-fire soil chemistry, these resilient fire-loving fungi seized the moment to fill the niche left by competitors killed by the fire.[7,8]

As they hunted for food, they prepared the soil for plants, infusing it with recycled nutrients, retaining moisture, and nurturing the space for new growth.

Fungi not only benefit plants but their sporing bodies and mycelia provide welcome food for animals in the barren black zone after the fire.

It is a marvel that sporing bodies of some fungi arise only in the aftermath of fire, a phenomenon perhaps not seen in this rainforest for over a hundred years.

When we shared our photos and videos online, people expressed gratitude. They said it was a story of hope, that even in the face of devastation, life persists.

Yet what we couldn't observe was how many species had been lost.

As fires grow in ferocity across our globe, the fate of many fungal species hangs in the balance. Their homes of leaf and log are vanishing, and their plant and animal partners perish.

The full impact these fires have on the fungal kingdom and the ecosystems they support remains a mystery.[6,9]

From the ashes, one truth emerges: in this interconnected web of life, fungi hold a far more pivotal role in the cycle of renewal and resilience than we could have ever imagined.

OPPOSITE
Descolea sp.
Lake Gunn, Fiordland National Park, New Zealand

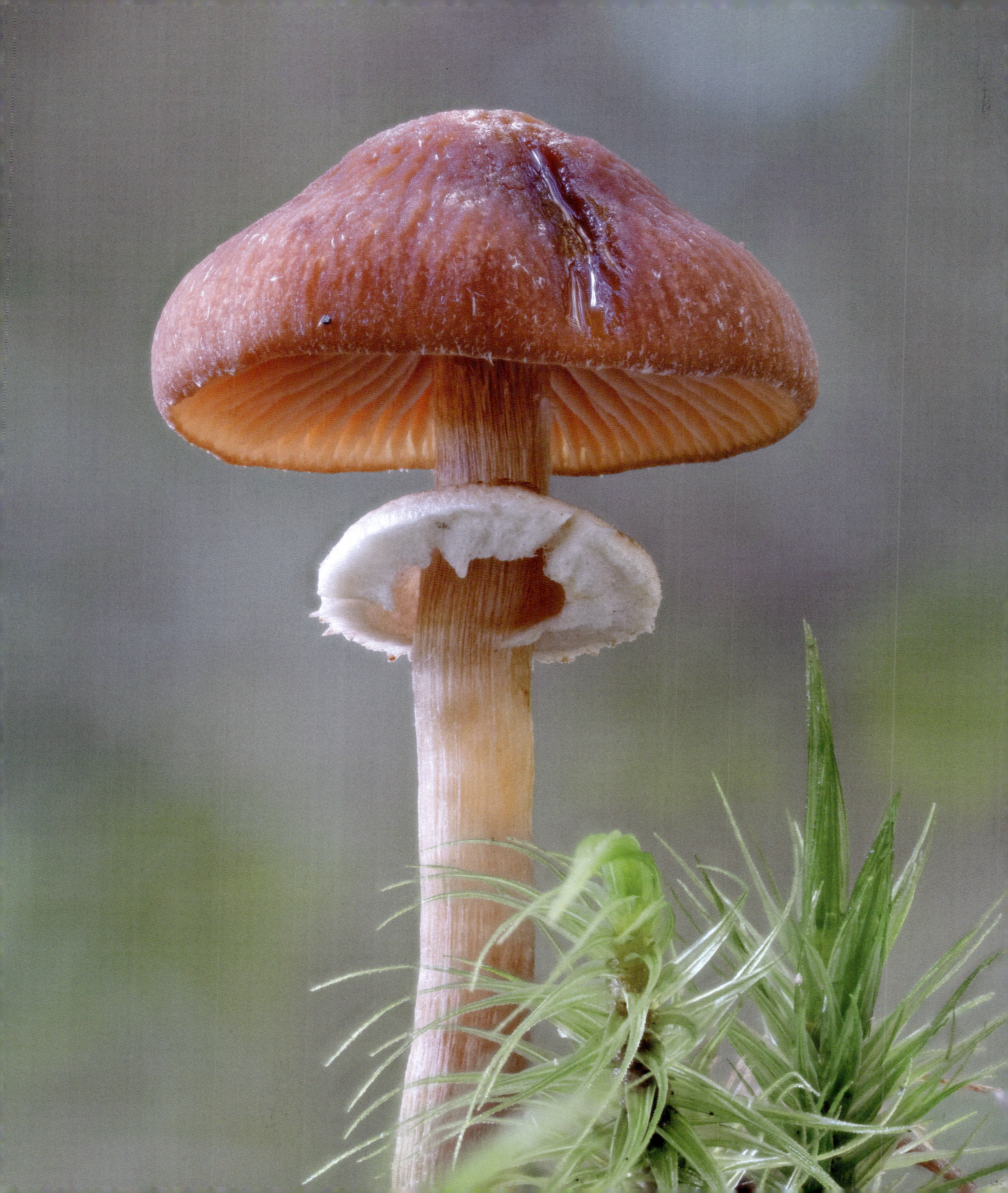

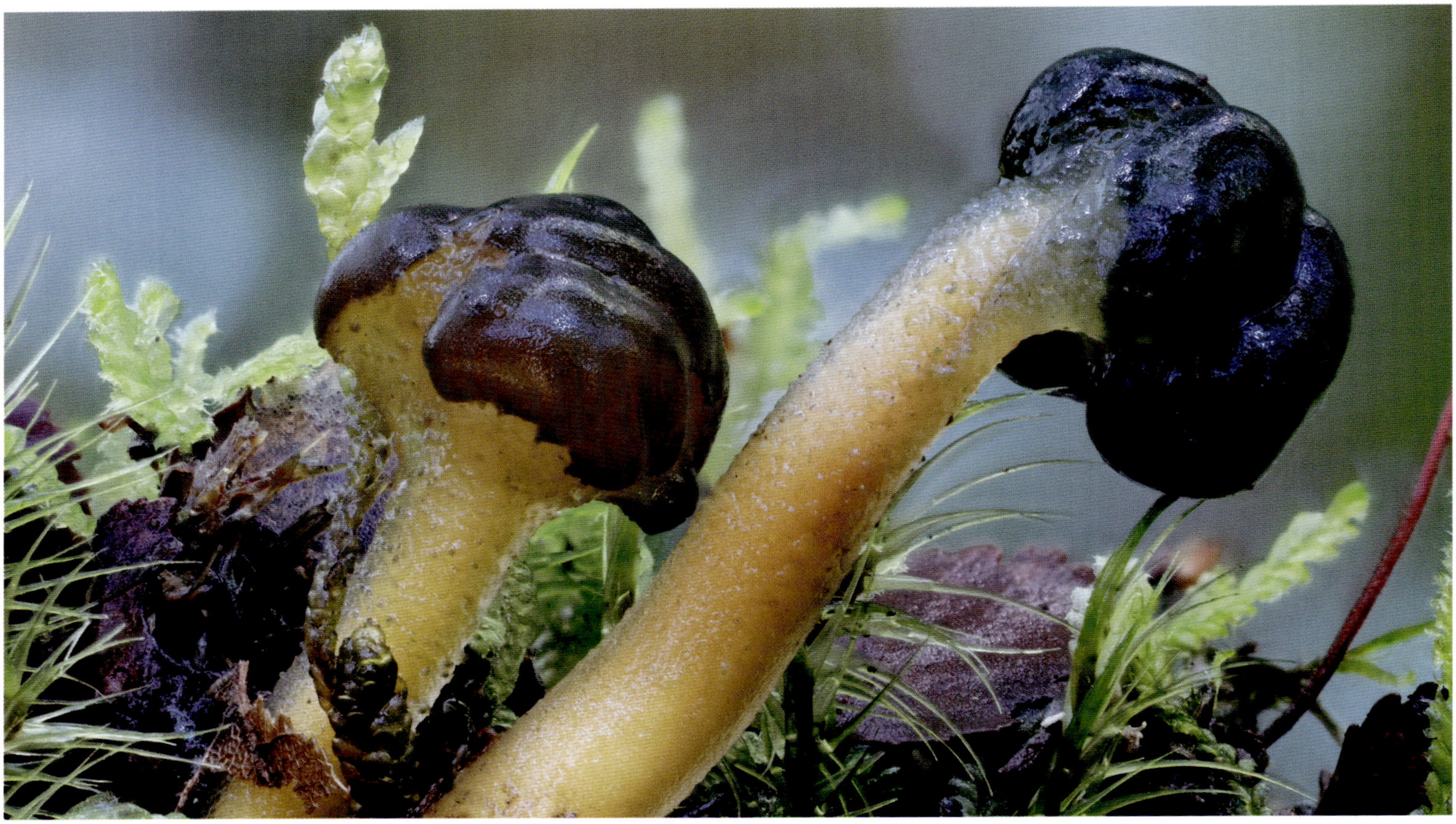

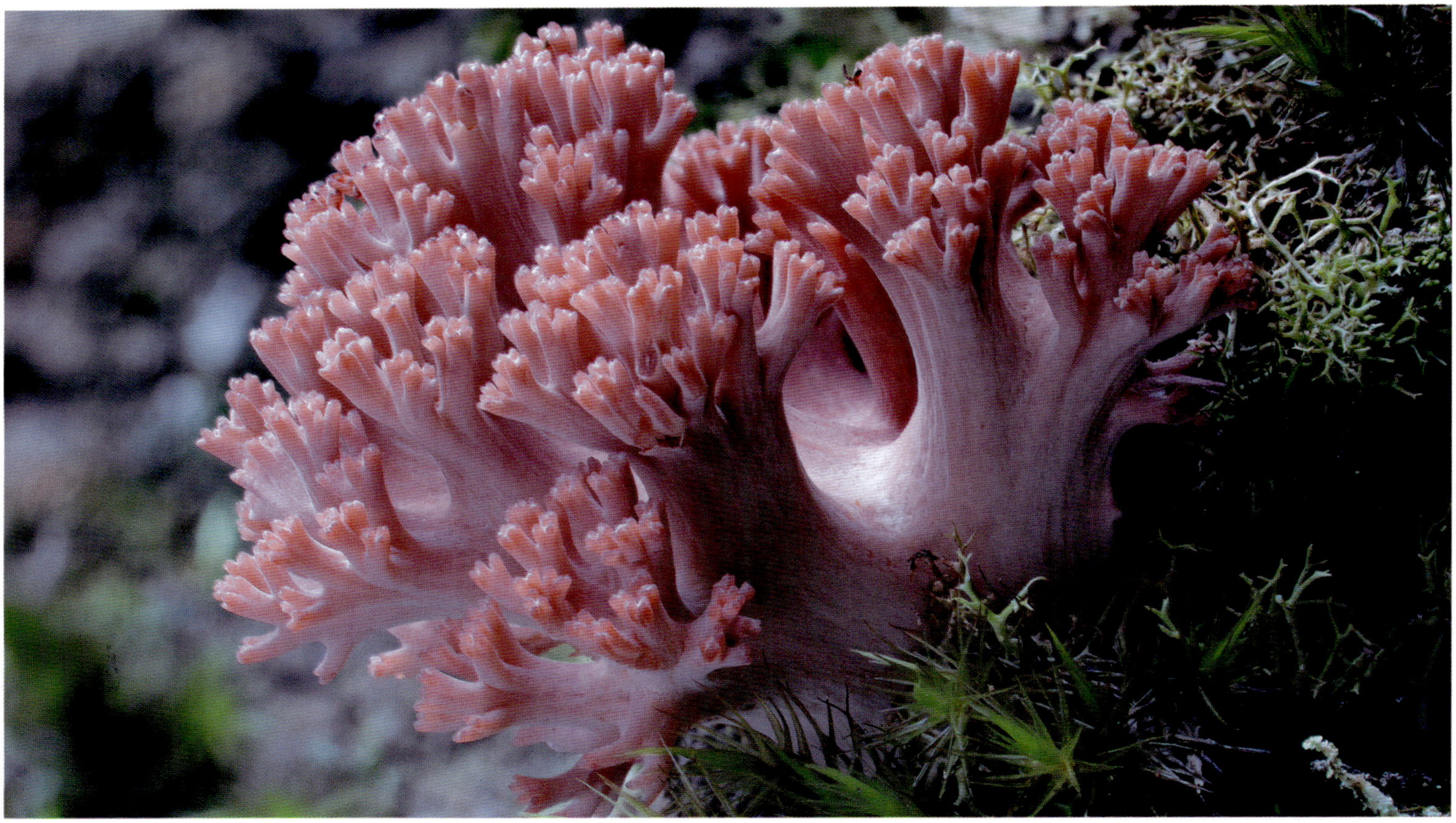

I invite you to take a breath.

Our life-sustaining air comes from a cycle we know well: plants inhale and store carbon dioxide and exhale oxygen, a process essential to our very existence.

But plants do not achieve this alone.

Beneath the surface lies a hidden ally – the ectomycorrhizal fungi. These silent partners are key to locking away carbon.

Within a mere handful of soil, there could be kilometres of fungal threads. As they feast, fungi outcompete other soil microbes, slowing down the decomposition process and the return of carbon to the skies.[10,11]

The biomass of mycelia in soil stores vast amounts of carbon.[12]

In the arid expanse of deserts, the narrative of fungi and carbon is even more striking.

Like their forest kin, desert flora captures carbon through photosynthesis. Then, in their relentless search for moisture, their roots, entwined with fungal partners, burrow deep into the earth.

In this underground realm, they exhale carbon dioxide, which, when united with calcium – a mineral plentiful in these soils – forms caliche, a crystalline vault where carbon is sequestered for millennia.[13,14]

In aiding ecosystems to absorb and store carbon, fungi act as invisible guardians against global warming, safeguarding our planet from the impacts of climate change.

TOP
Leotia lubrica, jelly babies
Lake Gunn, Fiordland National Park, New Zealand
Distribution: worldwide excluding Africa

BOTTOM
Ramaria sp.
Launceston, Tasmania, Australia

But this balance is fragile and easily shattered.

Ectomycorrhizal fungi, along with their plant allies, are acutely vulnerable to the nitrogen we scatter across the globe through the burning of fossil fuels and the use of chemical fertilisers to grow our food.[15]

And, in our pursuit to cut down forests, we sever more than just branches; we snuff out the fungi, disrupting an ancient cycle of life.

The very fungi we endanger, those keepers of carbon, have another critical function in the global climate story – the power to make rain.

One of our photographic quests in chronicling the life of fungi is to capture the magical, sometimes fleeting moment of spore release.

Puffballs can be extraordinarily theatrical, suddenly erupting into a cloud of billions of spores with a raindrop's kiss, or the nudge of a passing animal.

Other fungi employ different mechanisms for spore release, creating one of the most elegant spectacles we have ever witnessed.

OPPOSITE
Lycoperdon perlatum
Speargrass Track, Saint Anaud, New Zealand
Distribution: worldwide

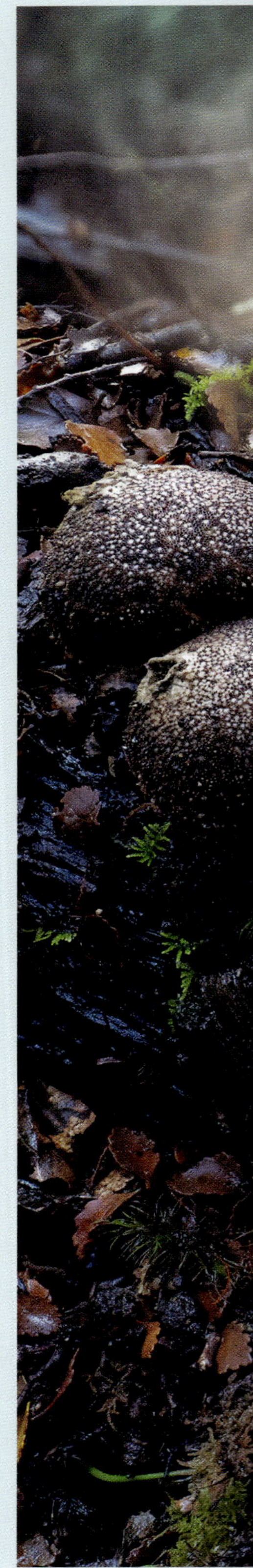

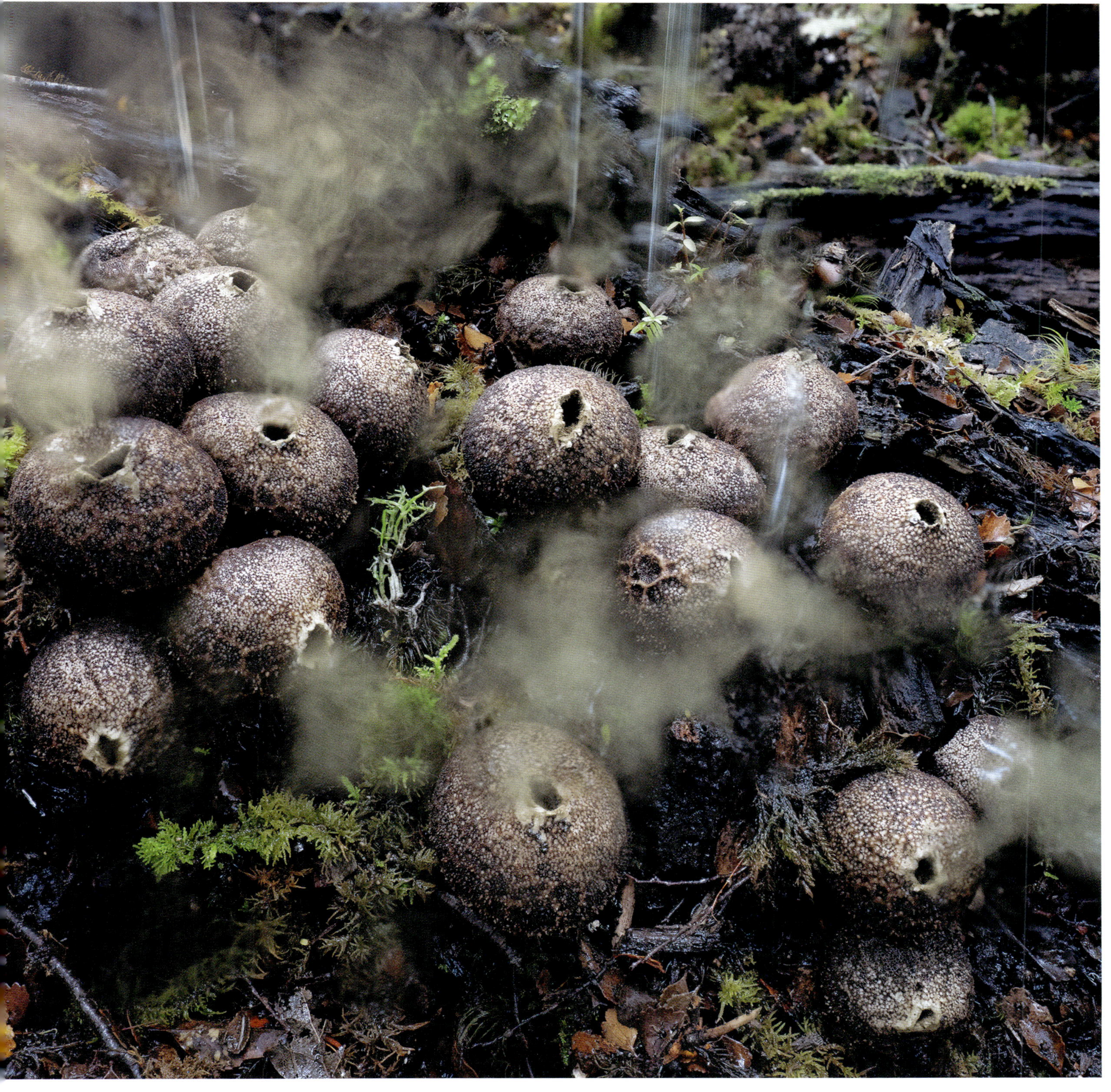

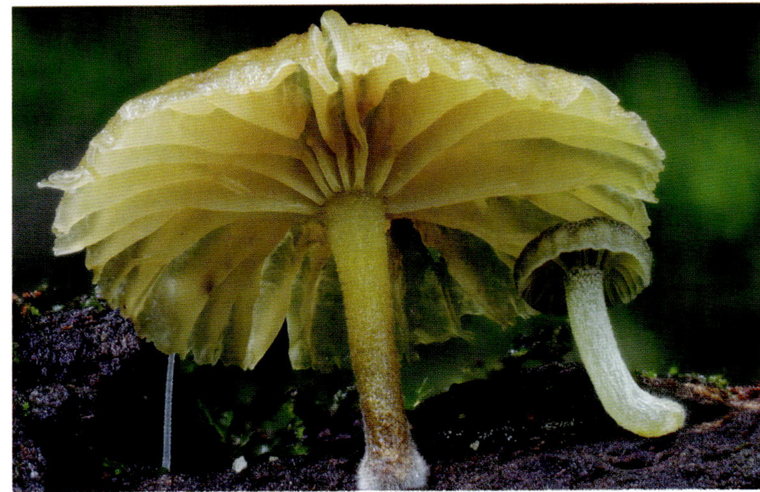

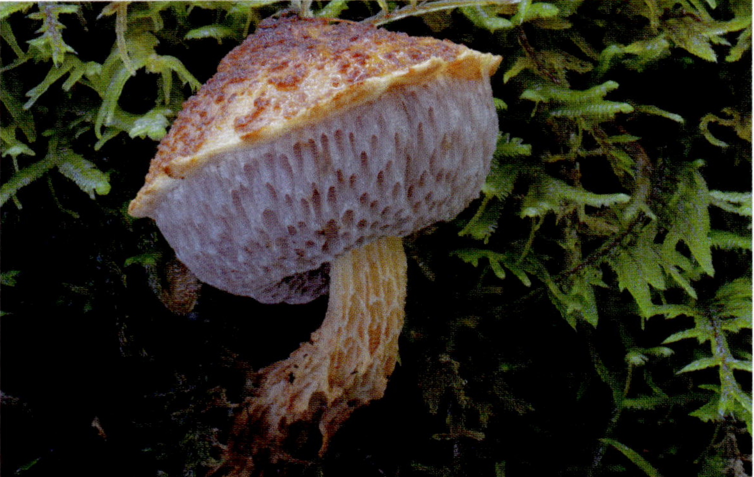

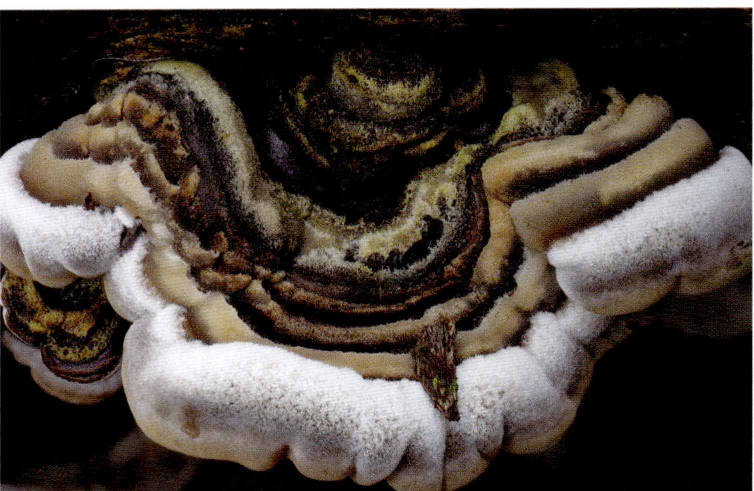

Tucked between the flutes of gills, deep within the tubes of a bolete, or the pinholes of a polypore, an enzyme in the spores attracts moisture. Water droplets form, catapulting millions of tiny spores into the air.[16]

As that moisture evaporates, the spores begin to swirl on miniature air currents, like a murmuration of birds dancing on a breeze of their own making.

Many of the spores continue their ballet as they ride an updraft into the upper atmosphere.

We have only recently understood how the enzymes in these trillions of tiny particles persist in their quest to gather moisture.[16]

TOP
Unidentified mushroom
Booyong, New South Wales, Australia

MIDDLE
***Austroboletus* sp.**
Cradle Mountain, Tasmania, Australia

BOTTOM
***Bjerkandera adusta*, smoky polypore**
Booyong, New South Wales, Australia
Distribution: worldwide

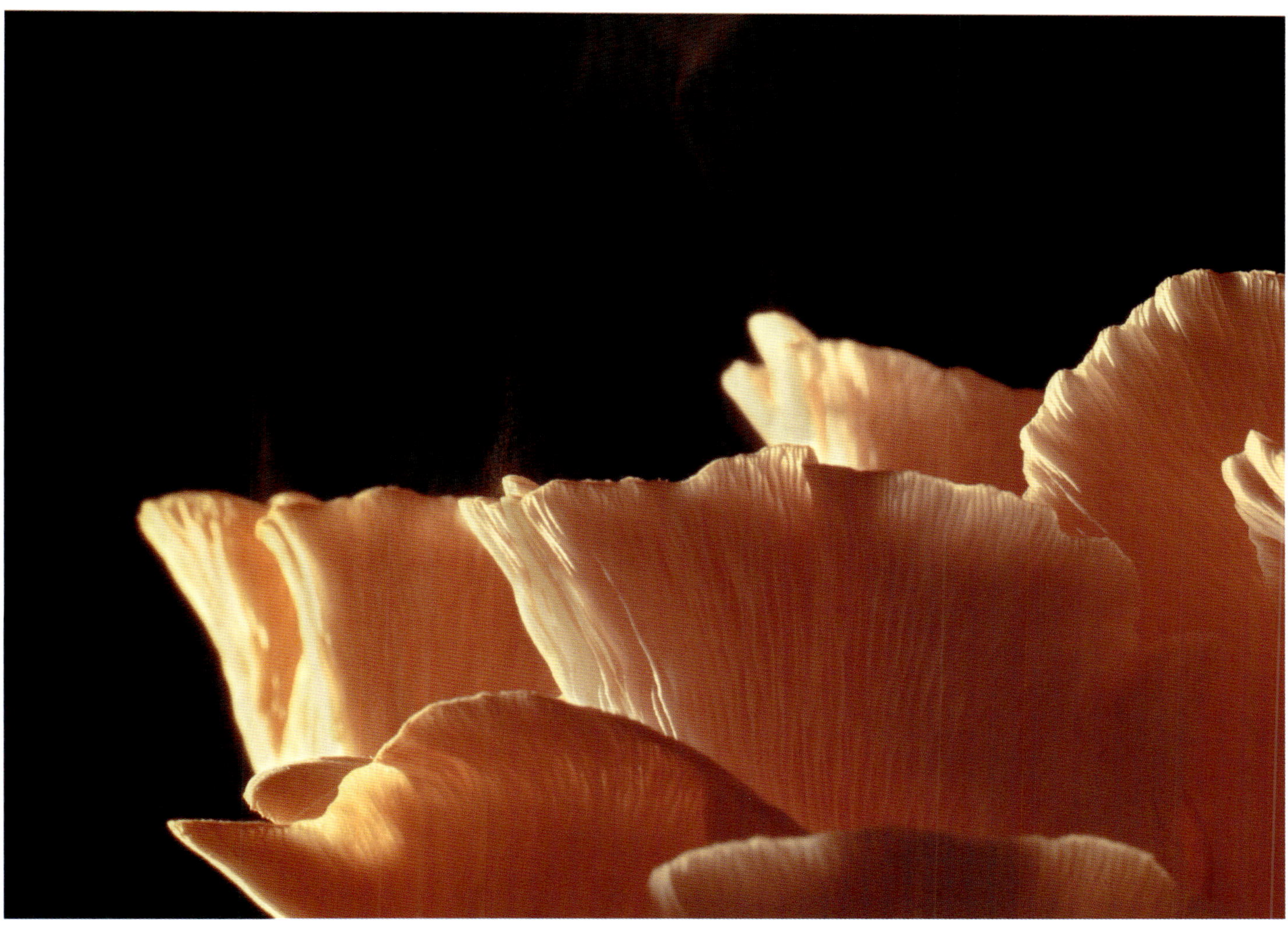

Along with dust, pollen and pollution, spores become scaffolding for raindrops to form.

They are one of nature's cloud seeders, helping to create the rain that is so critical to life on our planet.

As the cycle repeats, the dampened mycelium sends forth its mushrooms, and creatures great and small gather to feast.

ABOVE
Pleurotus djamor, pink oyster
A frame from a slow-motion recording
of a cultivated mushroom sporing.

It isn't just humans who have a taste for fungi; they are entwined with the animal kingdom's food chain.

We have observed Australian marsupial wallabies seek out and devour forest chanterelles, Patagonian birds scratch away to reveal their favourite truffles, and snails methodically chomp away on wood ears.

We have documented fungi gnats weaving lace-like webs beneath the caps of black termite mushrooms, capturing spore-filled meals and the occasional unwary insect.

On a particularly dark and chilly spring morning, fungi hunting found us camped in a tiny alpine remnant in the western plains of New South Wales, Australia. We braved torrential rain in search of a rare sight.

Fellow campers had reported giant neon pink slugs, normally nocturnal, feeding during the day due to the perfect gloomy conditions. We found these magnificent gastropods, up to 15 centimetres long and 2 centimetres wide, grazing upon the lichen, fungi and algae on eucalypt bark and rocks.

This isolated mountaintop, a mere 10 kilometres square, is the only known population on the planet of these endangered slugs.[17]

High-elevation ecosystems, with their unique snails, lichens and fungi, are particularly vulnerable to the effects of climate change.

As the world warms and we experience more drought and wildfires, the organisms in these landscapes are likely to contract even further or vanish entirely.

OPPOSITE TOP LEFT
Triboniophorus cf. *graeffei*, the Mount Kaputar pink slug, has a diet of lichen, microfungi and algae.
Mount Kaputar National Park, New South Wales, Australia

OPPOSITE TOP RIGHT
A rainforest native snail finds *Auricularia delicata*, the forest ear, a tasty meal.
Booyong, New South Wales, Australia

OPPOSITE MIDDLE LEFT
Fungi gnats weave their webs under the caps of *Hymenopellis raphanipes*, black termite mushrooms, to catch a meal of spores.
Kunming, Yunnan, China

OPPOSITE MIDDLE RIGHT
Red springtails enjoying a meal of this succulent *Gliophorus graminicolor*.
Franklin River, Tasmania, Australia

OPPOSITE BOTTOM LEFT
An eastern dwarf tree frog, *Litoria fallax*, lies in wait for a supper of insects it has learnt are attracted to the mushrooms of this *Heimiomyces* sp.
Booyong, New South Wales, Australia

OPPOSITE BOTTOM RIGHT
Bennett's wallabies, *Macropus rufogriseus*, are opportunistic fungi lovers, and seek out their favourites during fungi season.
Lake St Clair, Tasmania, Australia

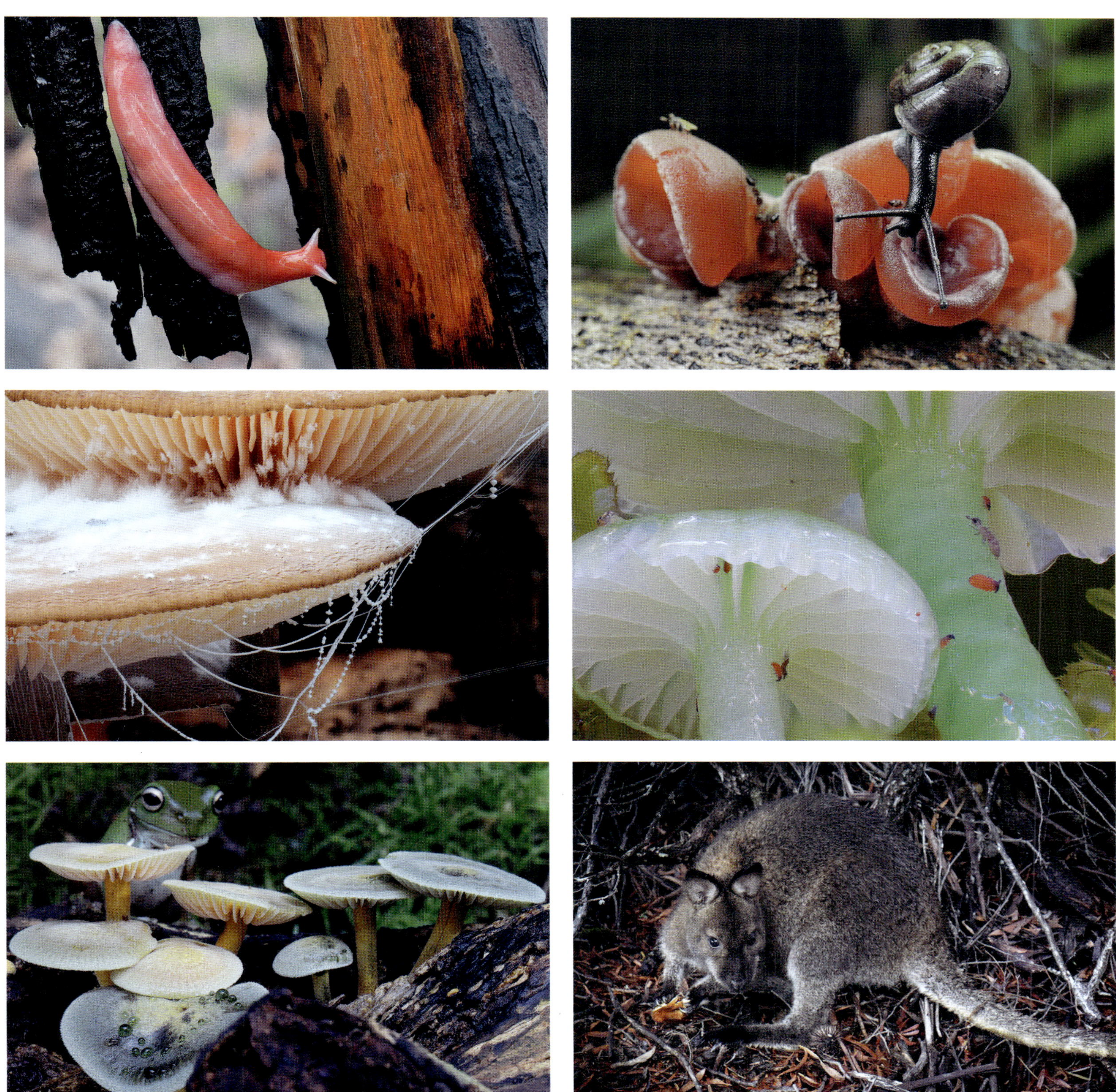

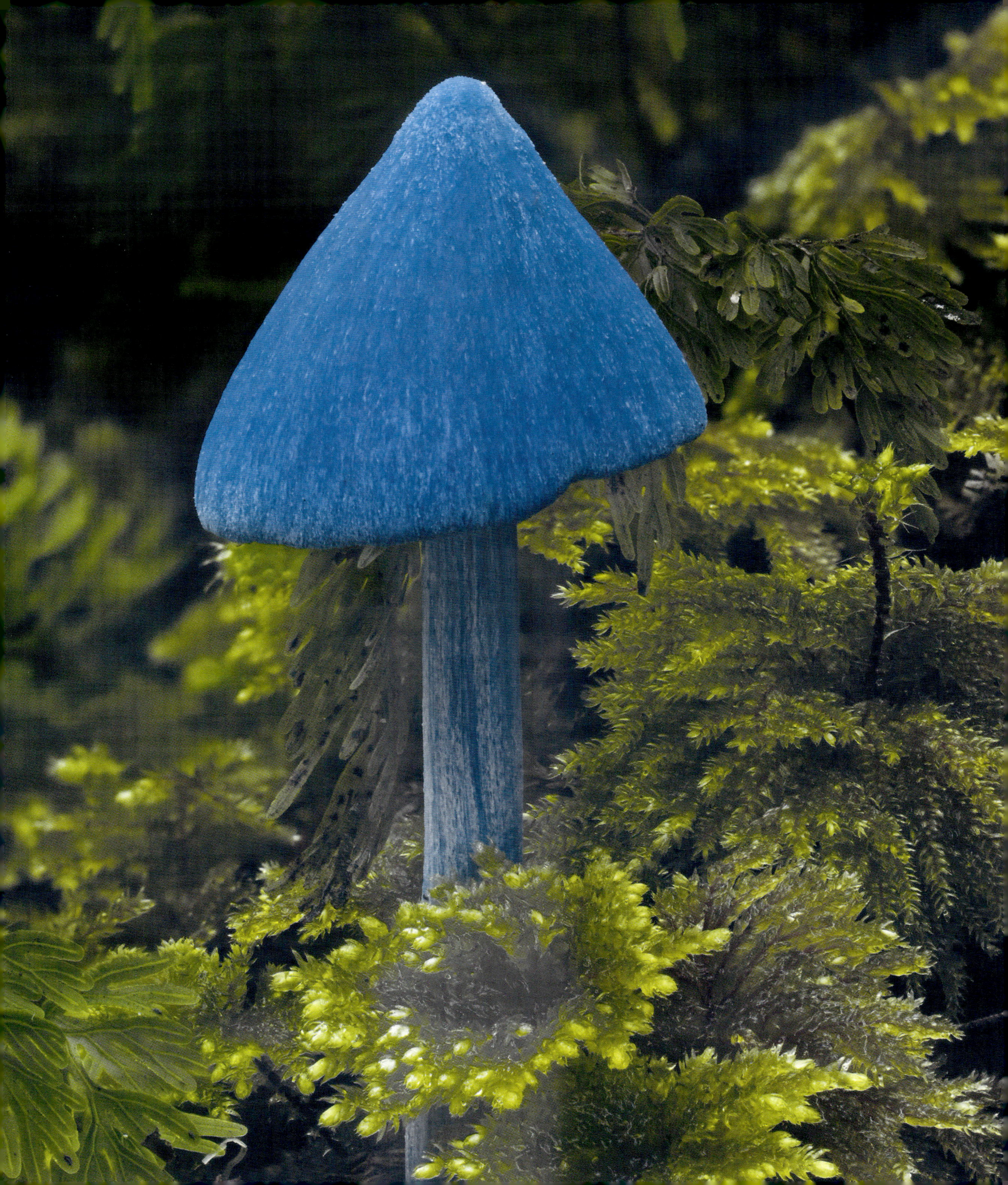

Humans, too, share an ancient bond with fungi.

Beyond the delectable buttery meals of champignons on our dinner plates, fungal fermenters give us wine, beer, cheese, chocolate and bread.

Animals like cows, goats and sheep rely on fungi in their guts to digest plants.[18]

Many of the plants that produce the fruit and vegetables we eat use mycorrhizal partners for their sustenance.

When illness strikes, it is often the fungi that come to our aid with potent antibacterials and drugs to fight cancer, high blood pressure, diabetes, organ transplant rejection and the list goes on.

Fungi are inspiring eco-friendly solutions like mycelium-based packaging and insulation.

Promising research suggests that fungi could also help us clean up our mess, using their insatiable appetite for a huge range of organic compounds to remediate chemical spills and reduce plastic waste.

But beyond the benefits to humans, we are in awe of these organisms, which make plants resilient, sequester carbon, provide food for animals, break down rocks, create soil and even make rain.

As we come to understand the vital roles fungi play in our ecosystems, the imperative to safeguard these unsung heroes becomes ever more pressing.

'The places where we document fungi are just tiny remnants of wilderness. Every year, more and more of these living museums are bulldozed and torn apart. Yet we cannot survive on this planet without them.'

OPPOSITE
***Entoloma hochstetteri*, blue pinkgill**
Arnold River Dam Walk, Dobson, New Zealand
Distribution: New Zealand

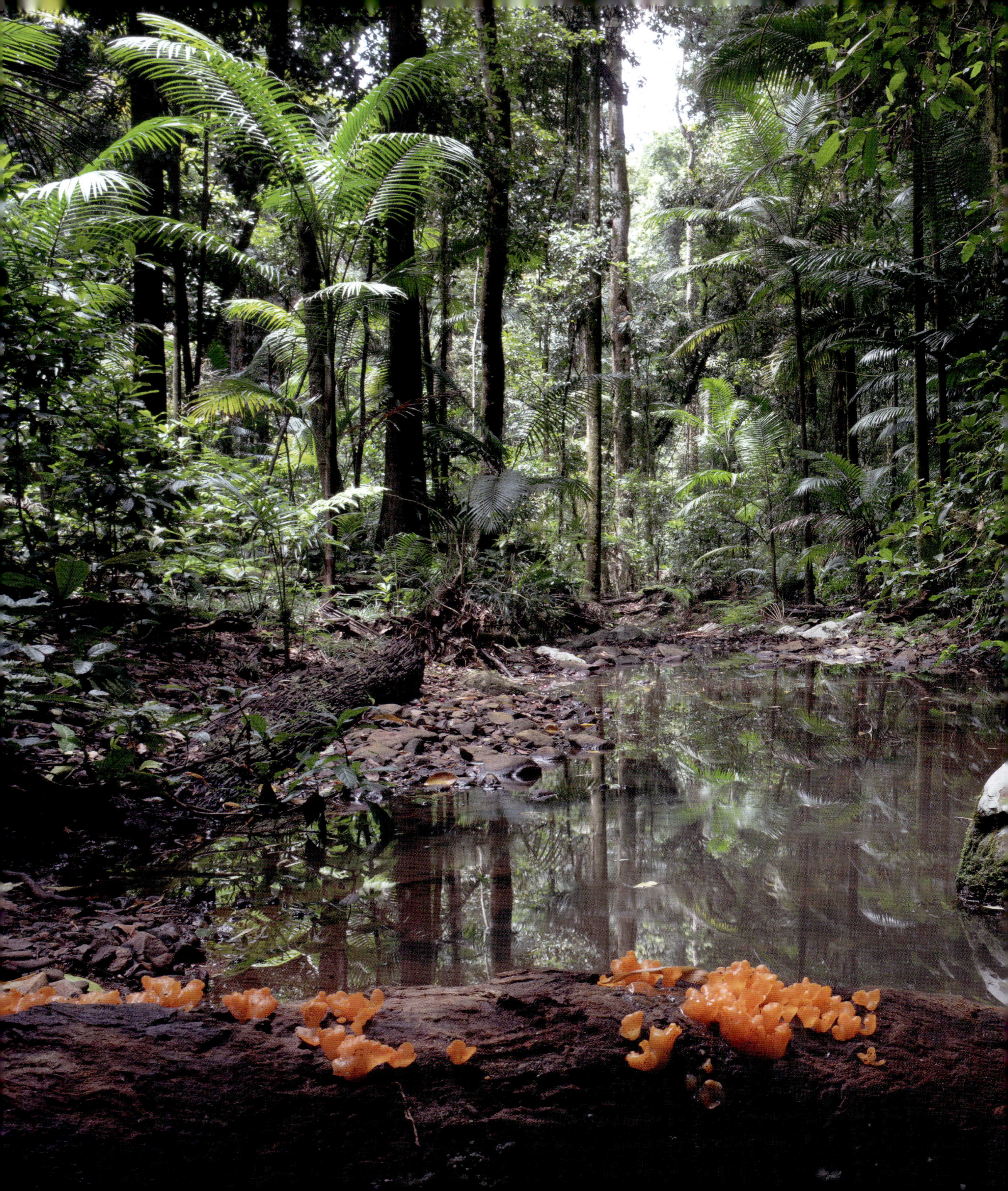

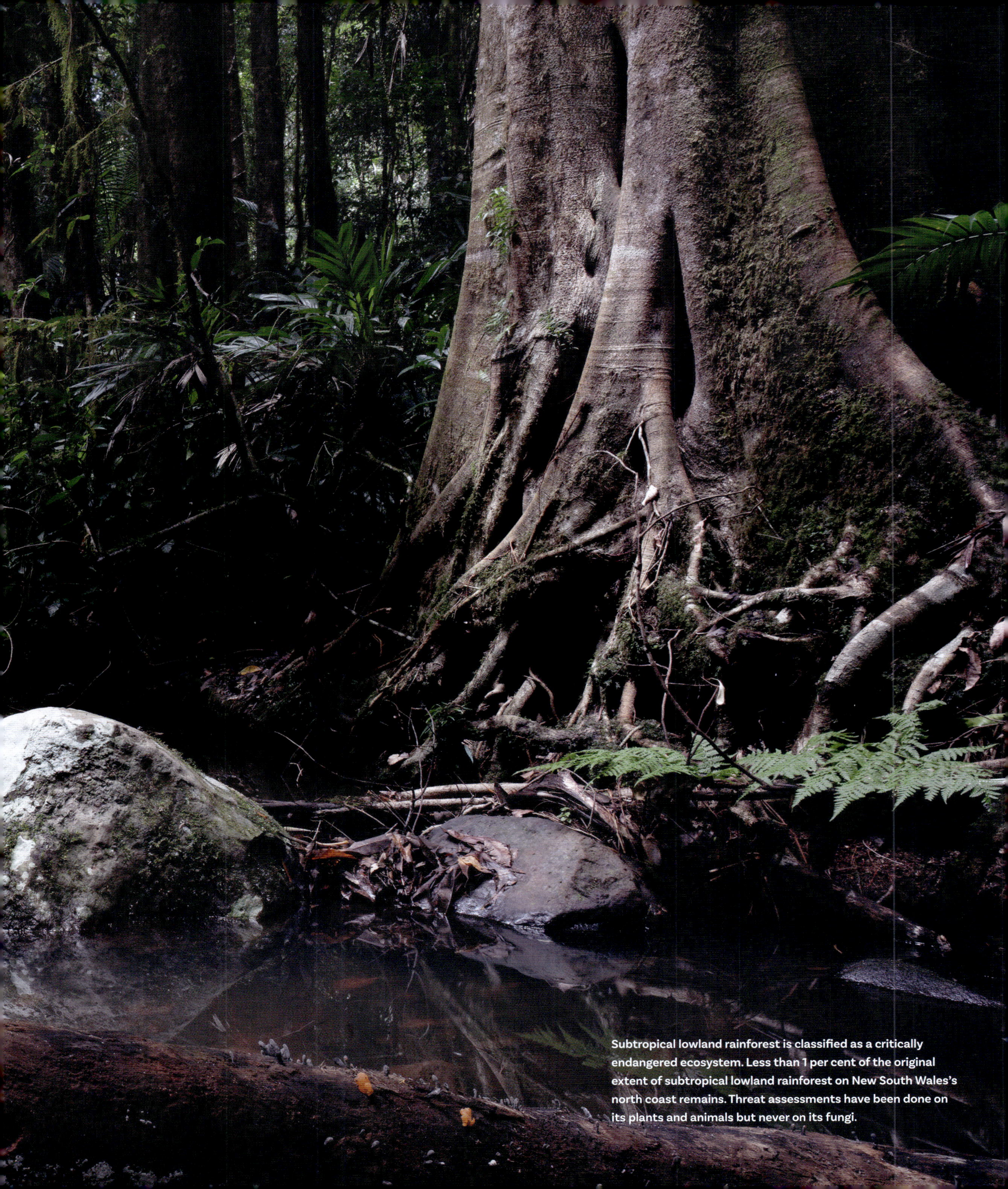

Subtropical lowland rainforest is classified as a critically endangered ecosystem. Less than 1 per cent of the original extent of subtropical lowland rainforest on New South Wales's north coast remains. Threat assessments have been done on its plants and animals but never on its fungi.

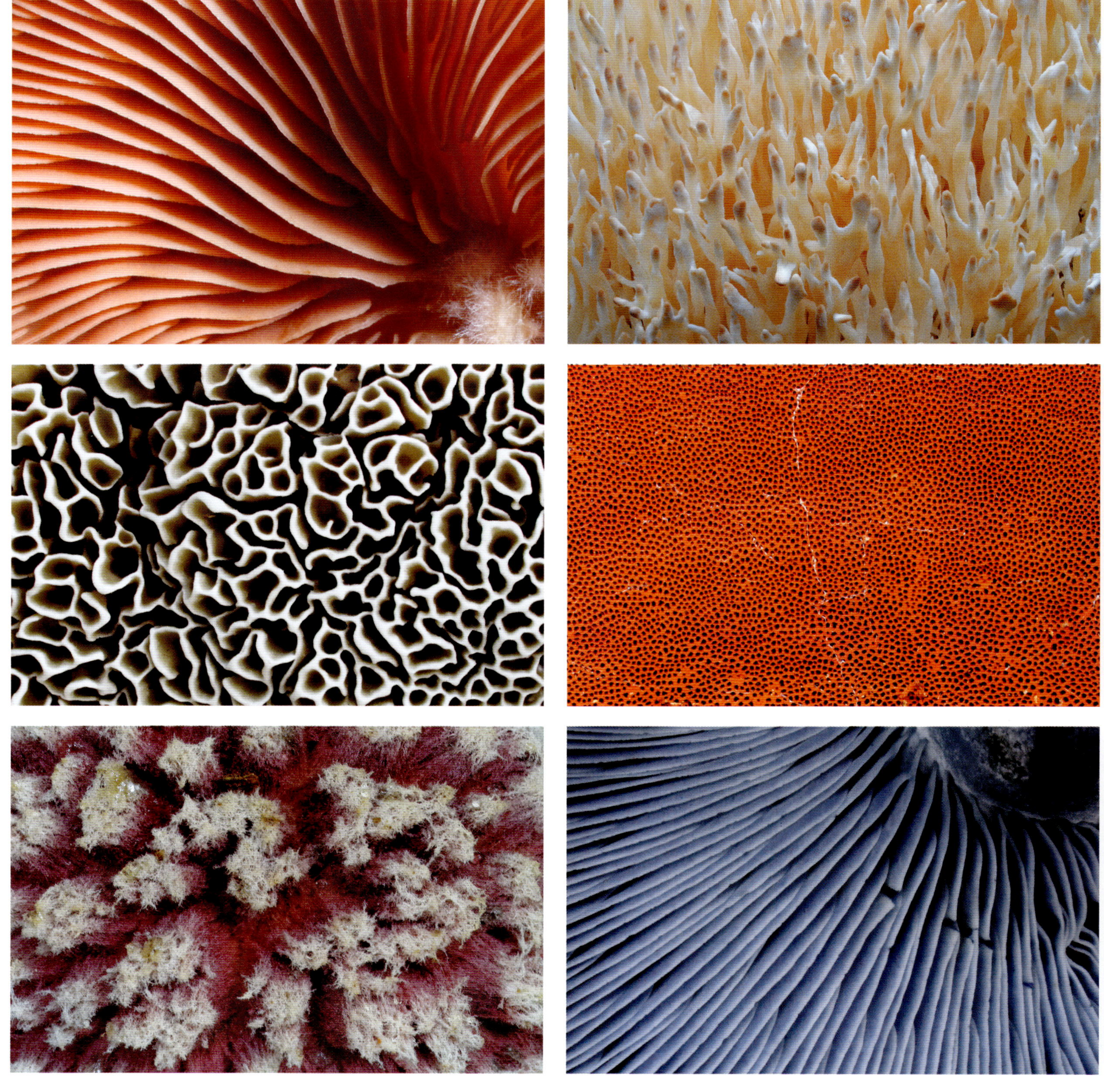

Of the estimated 2–5 million fungal species on Earth, the International Union for Conservation of Nature has scrutinised fewer than 1,000 for potential inclusion on its Red List of Threatened Species. This pales in comparison to the comprehensive assessments of animals such as mammals and birds.

Fungi, a whole kingdom of life, represent less than 1 per cent of all evaluations.

Policy-makers and biodiversity champions agree – fungi are at the core of rich, sustainable ecosystems, yet action to embed fungi in conservation policies remains scarce.

Globally, the 3Fs campaign, 'fauna, flora, and funga', spearheaded by Giuliana Furci and the Fungi Foundation, is gaining traction.

It implores governments and policymakers to enshrine 'Funga' within their research and conservation frameworks and recognise the value of these organisms that create ecosystems.

It is a plea for protection to preserve fungal species before they disappear.

OPPOSITE TOP LEFT
Crepidotus sp.
Booyong, New South Wales, Australia

OPPOSITE MIDDLE LEFT
Aporpium sp.
Booyong, New South Wales, Australia

OPPOSITE BOTTOM LEFT
Boletellus deceptivus
**Rosewood Creek Circuit, Dorrigo National Park,
New South Wales, Australia
Distribution: Australia**

OPPOSITE TOP RIGHT
Aphelaria sp.
Dip Falls, Tasmania, Australia

OPPOSITE MIDDLE RIGHT
Trametes coccinea, **southern cinnabar polypore
Ikara-Flinders Ranges National Park, South Australia,
Australia
Distribution: predominantly Southern Hemisphere**

OPPOSITE BOTTOM RIGHT
Lactarius cf. *indigo*
Mawphlang Sacred Grove, Meghalaya, India

Ultimately, understanding fungi isn't just about conserving a unique kingdom – it is about the stewardship of biodiversity itself – the very foundation of life as we know it.

'When I see a fungus, I see not one organism, but an entire ecosystem.

'In discovering where they fit in the jigsaw, we learn that the three huge kingdoms of visible life – animals, fungi and plants – are interconnected. And often, it's the fungi that are the glue that binds it all together.

'But we have barely scratched the surface in naming and understanding them.

'I have a dream that we can turn this around.

'It's time to turn our research efforts to the underground.'

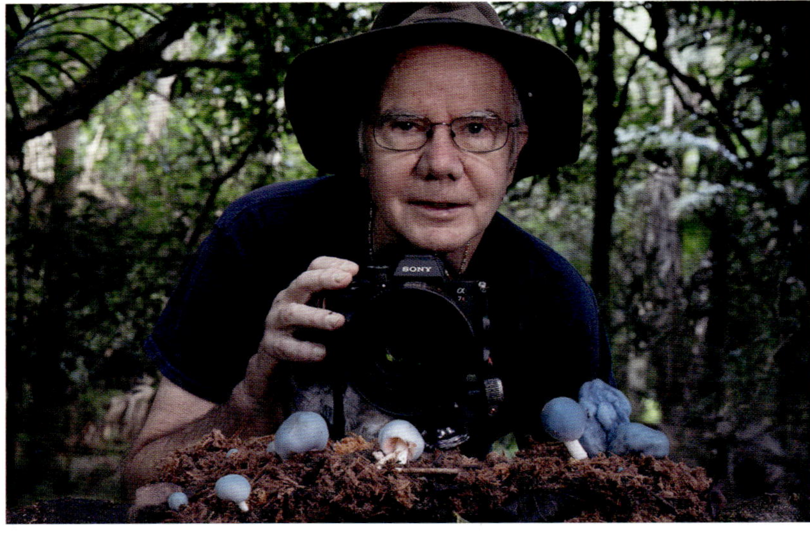

ABOVE
Stephen Axford photographs *Coprinopsis pulchricaerulea*.
Photograph: Catherine Marciniak

OPPOSITE
Cortinarius vitreopileatus
Lake Gunn, Fiordland National Park, New Zealand
Distribution: New Zealand

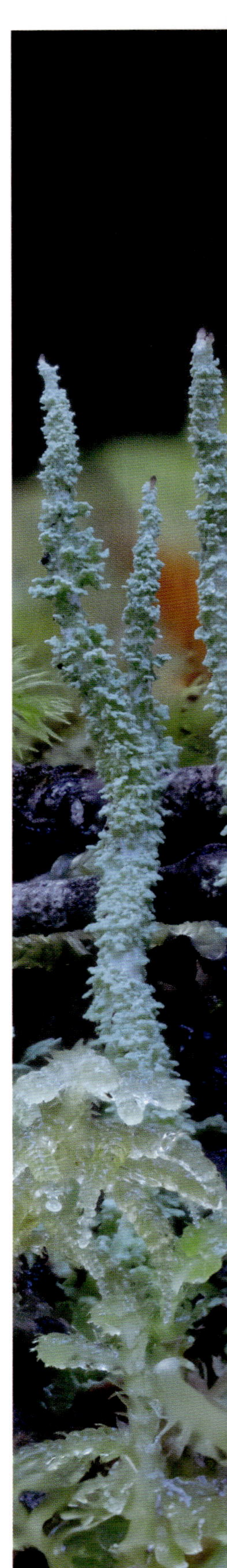

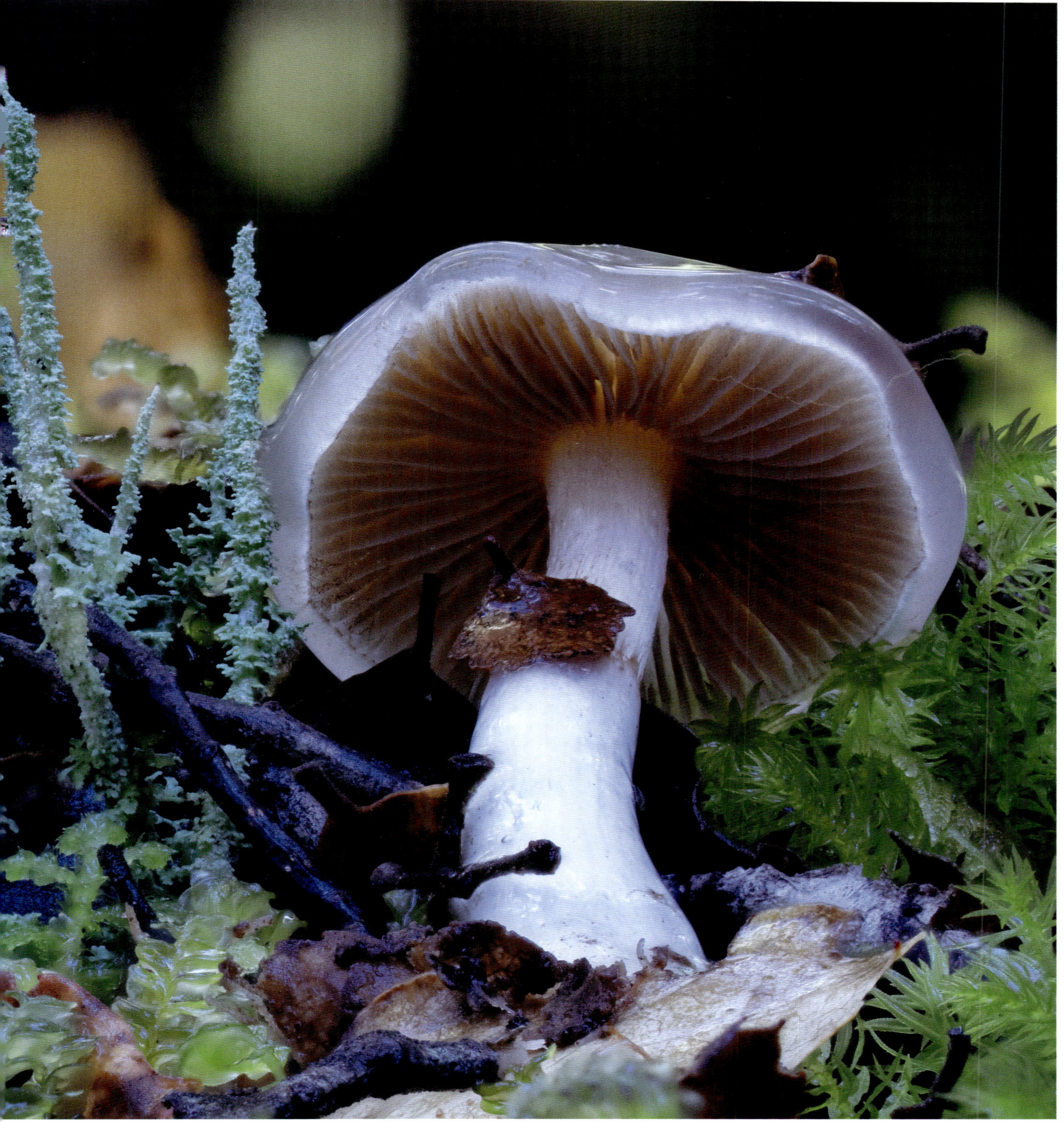

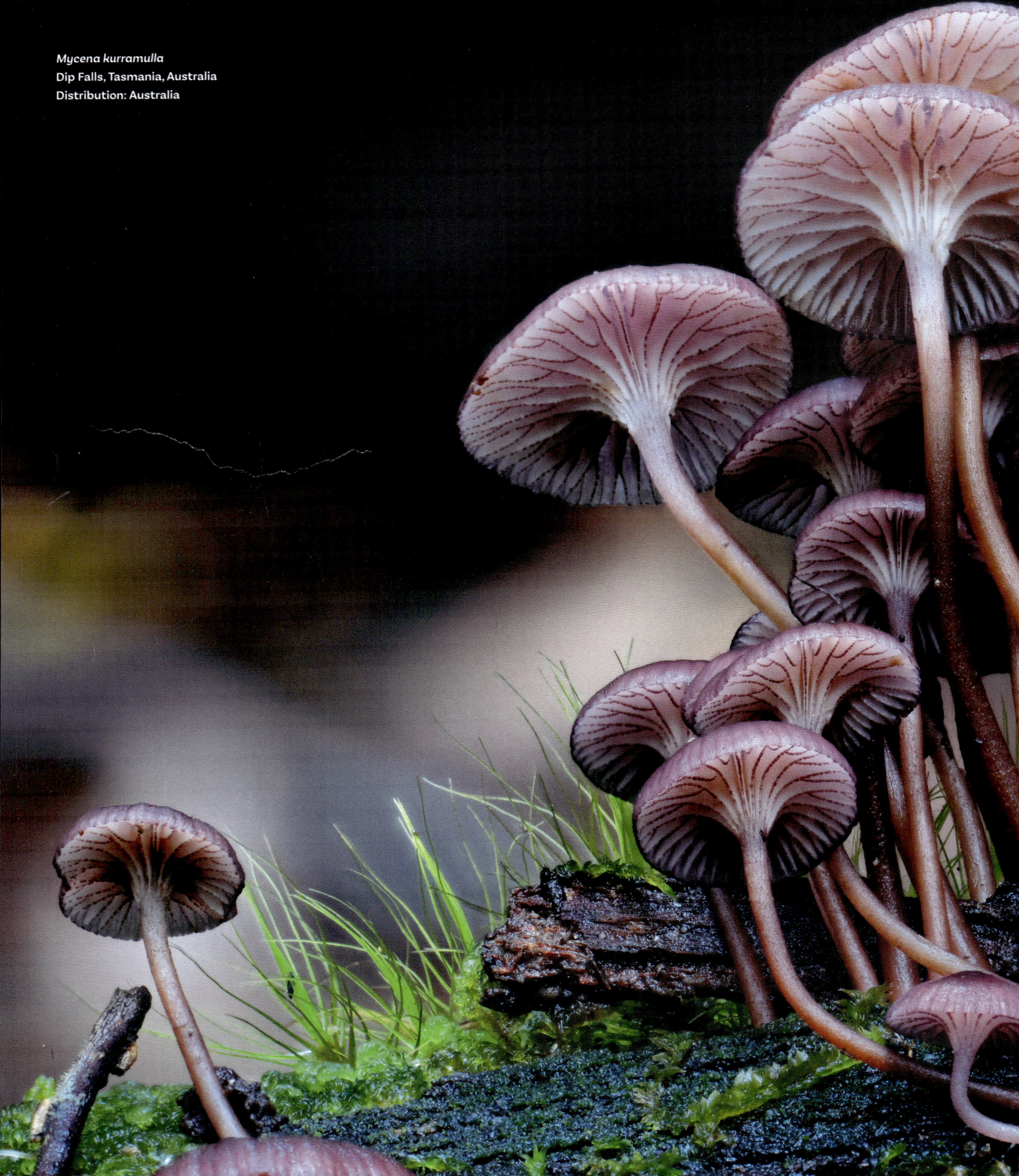

Mycena kurramulla
Dip Falls, Tasmania, Australia
Distribution: Australia

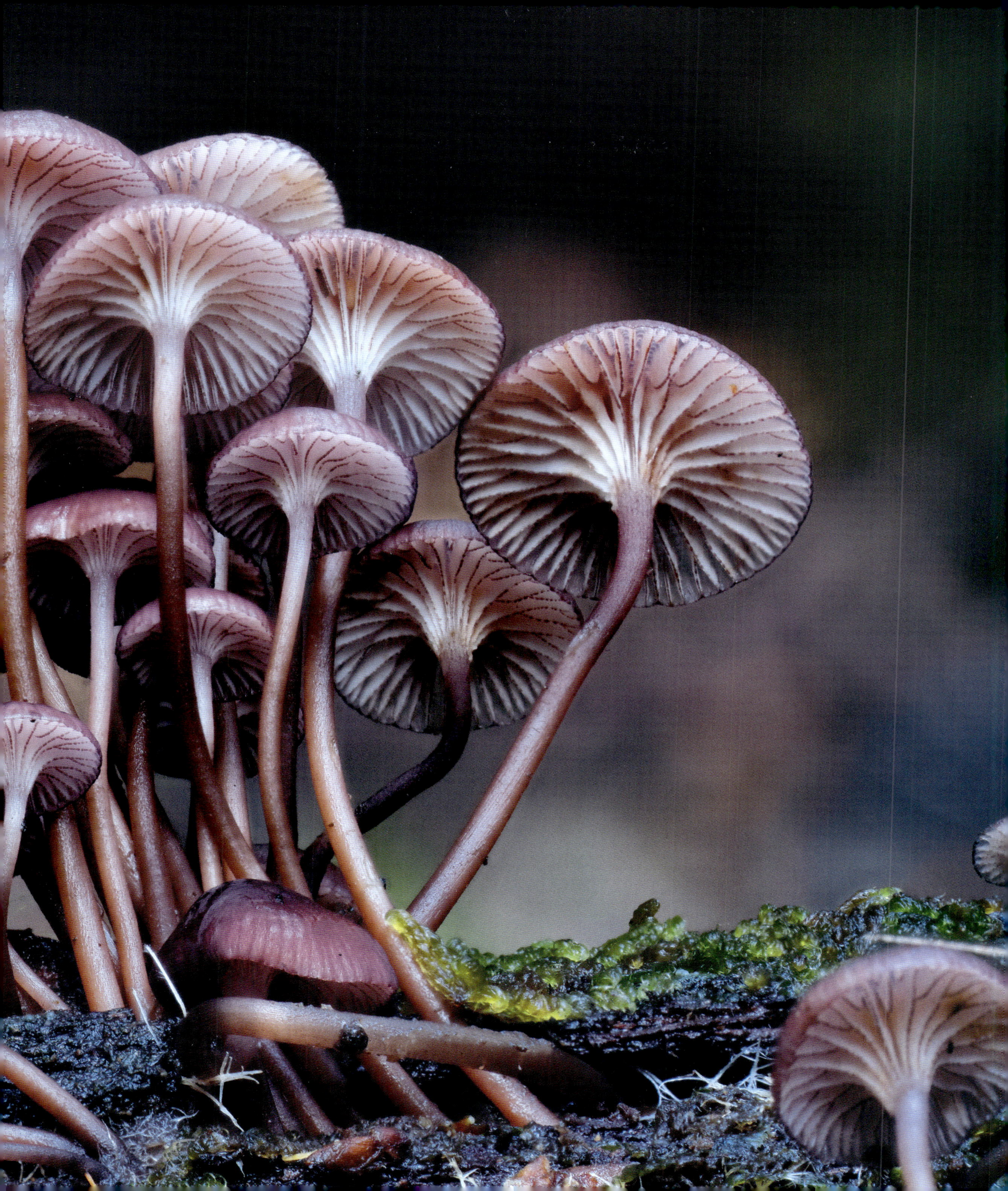

Glossary

bracket fungus (or **bracket**) a **mushroom** with a **cap** that arises laterally, usually from a woody substrate and without or with a rudimentary **stem**

cap the top part of a **mushroom**, under which are the **gills**

confer (abbreviated 'cf.') Latin for compare: used when a fungus is identified as similar to an existing species but not identical, as in *Amanita* cf. *hemibapha*

cyanobacterium (plural **cyanobacteria**, adjective **cyanobacterial**) a blue-green alga, a type of bacterium with the ability to carry out photosynthesis

DNA deoxyribonucleic acid, the molecule that contains the blueprint of each living organism

DNA barcode a short length of **DNA** that has diagnostic features at species level across a group of organisms

ectomycorrhiza (adjective **ectomycorrhizal**) a type of **mycorrhiza** in which a fungus forms a covering of **hyphae** around a plant rootlet

entomopathogen (adjective **entomopathogenic**) a **pathogen** that attacks an insect

funga the fungi of a particular area (as in fauna, flora and funga)

fungarium (plural **fungaria**) a reference collection of preserved fungus specimens (compare *herbarium*, a reference collection of preserved plant specimens)

fungus (plural **fungi**) a member of kingdom *Fungi*, an organism that produces **hyphae** and reproduces by **spores**, such as a **mushroom**

gill a thin plate that hangs below the **cap** of a mushroom and on which spores are produced

hypha (plural **hyphae**) cylindrical cell of a **fungus** that grows at the tip and can branch and anastomose

lamella (plural **lamellae**) see **gill**

lichen a **mutualistic symbiosis** between a fungal **mycobiont** and an algal or **cyanobacterial photobiont**, with the **mycobiont** providing the structure in which the **photobiont** cells live

mating type the genetic programming that determines sexual compatibility in a fungus; compatible matings are between different mating types

microfungus (plural **microfungi**) a fungus whose **sporing body** is not readily visible

mushroom a **sporing body** that has **cap**, **gills** and **stem**, like the familiar field mushroom, also used generally for any **sporing body**

mutualism	(adjective **mutualistic**) a **symbiosis** that is reciprocally beneficial, such as occurs in a **lichen** or a **mycorrhiza**
mycelium	(plural **mycelia**) a network of **hyphae**, the growing and feeding part of a **fungus**
mycobiont	the fungal partner in a **lichen**
mycoheterotroph	(adjective **mycoheterotrophic**) a plant that does not carry out **photosynthesis**, but gains carbon nutrition from fungi
mycological	concerning **fungi**
mycologist	a scientist who studies **fungi**
mycology	the study of **fungi**
mycorrhiza	(plural **mycorrhizas**, adjective **mycorrhizal**) a **mutualistic symbiotic** relationship between a plant and a **fungus**, where there is an exchange of nutrients
nucleus	(plural **nuclei**) the part of a cell that contains genetic information in the form of **DNA**
parasite	(adjective **parasitic**) an organism that takes nutrients from another living organism
parasitism	a **symbiosis** where one kind of organism harms another kind by taking nutrients from it
pathogen	(adjective **pathogenic**) an organism that causes disease in its host, sometimes leading to death of the host
photosynthesis	the process through which atmospheric carbon dioxide and energy from the sun are used to produce sugar and oxygen; as carried out by plants and **cyanobacteria**
pileus	see **cap**
saprotrophic	gaining nutrition by breaking down dead organic matter
spore	the miniscule reproductive propagule of a **fungus**
sporing body	the part of a fungus that produces spores, for example a **mushroom** or **bracket fungus** (used in preference to the term 'fruit body' which has botanical connotations)
sporophore	a technical term for the **sporing body**
stem	the support for the **cap** of a **mushroom** or a **bracket**
stipe	see **stem**
symbiosis	(adjective **symbiotic**) an ongoing interaction between two organisms, such as a **mutualism** or **parasitism**
thallus	the body of a **lichen**

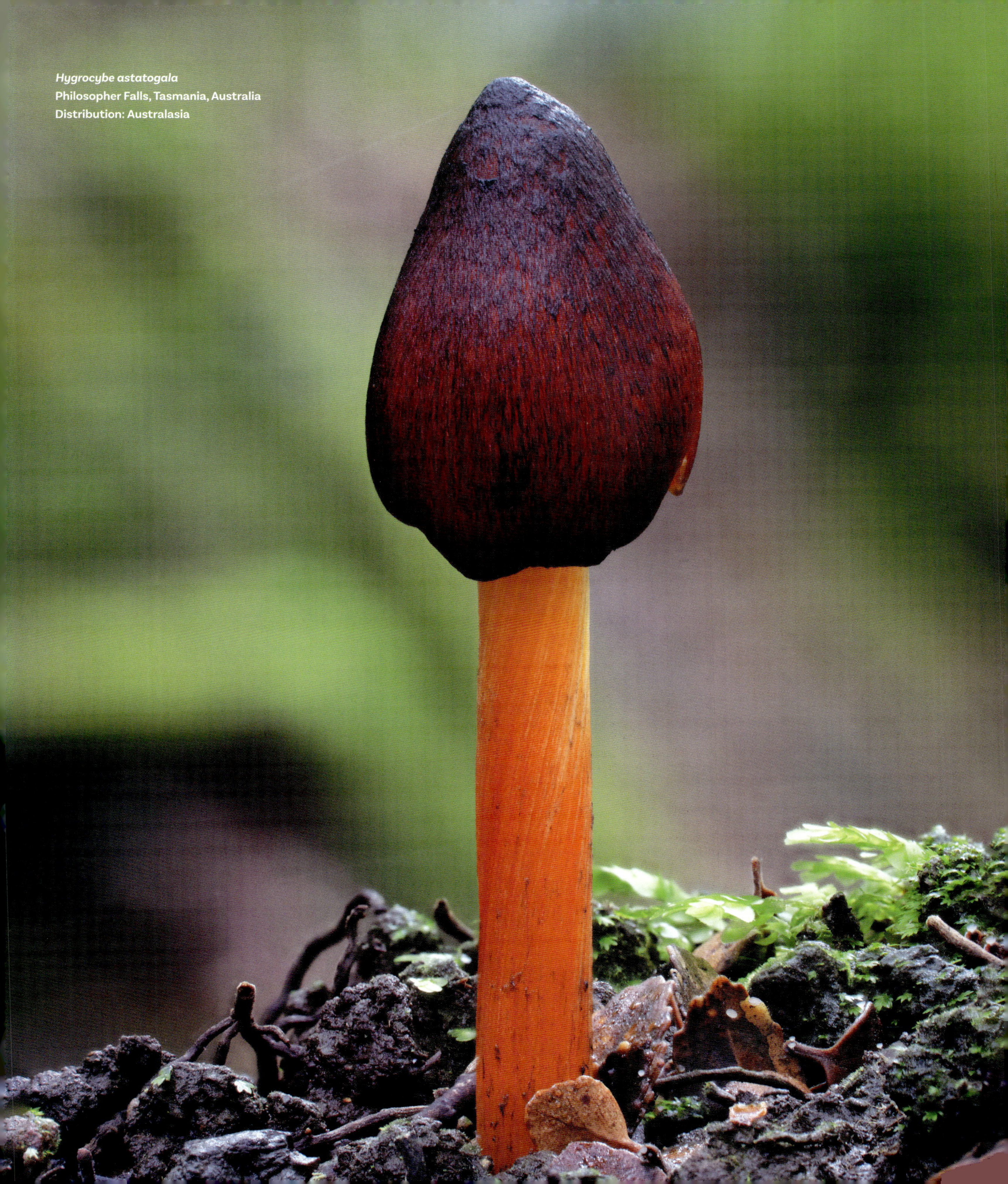

Hygrocybe astatogala
Philosopher Falls, Tasmania, Australia
Distribution: Australasia

Endnotes

HOW FUNGI CHANGED MY VIEW OF THE WORLD

1. Calisher CH (2007) Taxonomy: what's in a name? Doesn't a rose by any other name smell as sweet? *Croatian Medical Journal* **48**, 268–270.

2. Linnaeus C (1753) *Species Plantarum …* Impensis Laurentii Salvii, Holmiae.

3. May TW (2021) Use of target species in citizen science fungi recording schemes. *Biodiversity Information Science and Standards* **5**, e73960. doi:10.3897/biss.5.73960

4. Horak E (1983) Mycogeography in the South Pacific region: Agaricales, Boletales. *Australian Journal of Botany Supplementary Series* **13**, 1–41. doi:10.1071/BT8310001

5. Kothe E (1996) Tetrapolar fungal mating types: sexes by the thousands. *FEMS Microbiology Reviews* **18**, 65–87. doi:10.1111/j.1574-6976.1996.tb00227.x

6. Kothe E, Gola S, Wendland J (2003) Evolution of multispecific mating-type alleles for pheromone perception in the homobasidiomycete fungi. *Current Genetics* **42**, 268–275. doi:10.1007/s00294-002-0352-5

7. Antonelli A, Fry C, Smith RJ, Simmonds MSJ, Kersey PJ, *et al.* (2023) *State of the World's Plants and Fungi 2023*. Royal Botanic Gardens, Kew.

8. Phukhamsakda C, Nilsson RH, Bhunjun CS, Gomes de Farias AR, Sun Y-R, *et al.* (2022) The numbers of fungi: contributions from traditional taxonomic studies and challenges of metabarcoding. *Fungal Diversity* **114**, 327–386. doi:10.1007/s13225-022-00502-3

9. Hyde KD (2022) The numbers of fungi. *Fungal Diversity* **114**, 1. doi:10.1007/s13225-022-00507-y

10. Wu B, Hussain M, Zhang W, Stadler M, Liu X, *et al.* (2019) Current insights into fungal species diversity and perspective on naming the environmental DNA sequences of fungi. *Mycology* **10**, 127–140. doi:10.1080/21501203.2019.1614106

11. Paulus B, Kanowski J, Gadek P, Hyde K (2006) Diversity and distribution of saprobic microfungi in leaf litter of an Australian tropical rainforest. *Mycological Research* **110**, 1441–1454. doi:10.1016/j.mycres.2006.09.002

12. Tedersoo L, Bahram M, Põlme S, Kõljalg U, Yorou NS, *et al.* (2014) Global diversity and geography of soil fungi. *Science* **346**, 1256688. doi:10.1126/science.1256688

TRADERS, UNDERTAKERS AND EXECUTIONERS

1. Lodge D (2000) Ecto or arbuscular mycorrhizas – which are best? *New Phytologist* **146**, 353–354. doi:10.1046/j.1469-8137.2000.00668.x

2. Allen M (1993) *Mycorrhizal Functioning: An Integrative Plant-fungal Process.* Springer, New York.

3. Petersen JH (2013) *The Kingdom of Fungi.* Princeton University Press, Princeton.

4. Begum N, Qin C, Ahanger MA, Raza S, Khan MI, *et al.* (2019) Role of arbuscular mycorrhizal fungi in plant growth regulation: implications in abiotic stress tolerance. *Frontiers in Plant Science* **10**, 1068. doi:10.3389/fpls.2019.01068

5. Lee EH, Eo JK, Ka KH, Eom AH (2013) Diversity of arbuscular mycorrhizal fungi and their roles in ecosystems. *Mycobiology* **41**, 121–125. doi:10.5941/MYCO.2013.41.3.121

6. Truong C (2023) *The Wood-Wide Web. A Story Too Good for its Own Good?* [Video.] <https://www.youtube.com/watch?v=_ix18qNukpk>

7. Marciniak C, Axford S (2023) *Follow the Rain* [Documentary film]. Produced by Planet Fungi.

8. Sheldrake M, Rosenstock NP, Revillini D, Olsson PA, Wright SJ, *et al.* (2017) A phosphorus threshold for mycoheterotrophic plants in tropical forests. *Proceedings. Biological Sciences* **284**, 20162093. doi:10.1098/rspb.2016.2093

9. Floudas D, Binder M, Riley R, Barry K, Blanchette RA, *et al.* (2012) The Paleozoic origin of enzymatic lignin decomposition reconstructed from 31 fungal genomes. *Science* **336**, 1715–1719. doi:10.1126/science.1221748

10. Selosse MA, Strullu-Derrien C, Martin FM, Kamoun S, Kenrick P (2015) Plants, fungi and oomycetes: a 400-million year affair that shapes the biosphere. *New Phytologist* **206**, 501–506. doi:10.1111/nph.13371

11. Riley R, Salamov A, Brown D, Nagy LG, Floudas D, *et al.* (2014) Extensive sampling of basidiomycete genomes demonstrates inadequacy of the white-rot/brown-rot paradigm for wood decay fungi. *Proceedings of the National Academy of Sciences of the United States of America* **111**, 9923–9928. doi:10.1073/pnas.1400592111

12. Marren P (1991) Bracket fungi and their role in nature. *British Wildlife* **3**(1), 1–9.

13. Fogel R, Rogers P (n.d.) Shelf fungi. Intermountain Herbarium, Department of Biology, Utah State University, Fun Facts about Fungi, <https://www.usu.edu/herbarium/education/fun-facts-about-fungi/shelf-fungi>.

14. Gdula AK, Konwerski S, Olejniczak I, Rutkowski T, Skubała P, *et al.* (2021) The role of bracket fungi in creating alpha diversity of invertebrates in the Białowieża National Park, Poland. *Ecology and Evolution* **11**, 6456–6470. doi:10.1002/ece3.7495

15. Money NP (2016) *Fungi: A Very Short Introduction.* Oxford University Press, Oxford.

16. Evans HC, Elliot SL, Hughes DP (2011) *Ophiocordyceps unilateralis*: a keystone species for unraveling ecosystem functioning and biodiversity of fungi in tropical forests? *Communicative & Integrative Biology* **4**, 598–602. doi:10.4161/cib.16721

17. Evans HC, Hywel-Jones NL (1997) Entomopathogenic fungi. In *World Crop Pests: Soft Scale Insects.* (Eds Y Ben-Dov, CJ Hodgson) pp. 3–27. Elsevier, Amsterdam.

FUNGIPHILIA

1. Mantilla C, Zhou L, Wang C, Yang D, Shen S, *et al.* (2021) Favoring your in-group can harm both them and you: ethnicity and public goods provision in China. *Journal of Economic Behavior & Organization* **185**, 211–233. doi:10.1016/j.jebo.2021.02.016

2. Zhang Y, Mo M, Yang L, Mi F, Cao Y, *et al.* (2021) Exploring the species diversity of edible mushrooms in Yunnan, southwestern China, by DNA barcoding. *Journal of Fungi (Basel, Switzerland)* **7**, 310. doi:10.3390/jof7040310

3. Wang R, Herrera M, Xu W, Zhang P, Pérez Moreno J, *et al.* (2022) Ethnomycological study on wild mushrooms in Pu'er Prefecture, Southwest Yunnan, China. *Journal of Ethnobiology and Ethnomedicine* **18**, 55. doi:10.1186/s13002-022-00551-7

4. Anon. (n.d.) Nabanhe National Nature Reserve Administration Bureau inaugurated. Xishuangbanna Tropical Botanical Garden, Chinese Academy of Sciences, <http://english.xtbg.cas.cn/ns/es/200810/t20081007_29700.html> accessed 7 March 2024.

5. Hammond J, Yi Z, McLellan T, Zhao J (2015) 'Situational analysis report: Xishuangbanna Autonomous Dai Prefecture, Yunnan Province, China'. ICRAF Working Paper 194. World Agroforestry Centre East and Central Asia: Kunming, China.

6. Anon. (n.d.) Reflecting on Culture and Everyday Life. Partnerships for Community Development, <https://www.pcd.org.hk/en/content/reflecting-culture-and-everyday-life> accessed 7 March 2024.

7. Goodman J (2017) Buddhist mountaineers: the Bulang of Xishuangbanna. GoKunming, <https://www.gokunming.com/en/blog/item/4022/buddhist-mountaineers-the-bulang-of-xishuangbanna>.

8. Wu Z, Ou X (1995) 'The Xishuangbanna biosphere reserve: a tropical land of natural and cultural diversity, China'. UNESCO South-South Cooperation Programme for Environmentally Sound Socio-Economic Development in the Humid Tropics, Working Papers No. 2. United Nations University, Third World Academy of Sciences: Paris.

9. Chen H, Yi Z-F, Schmidt-Vogt D, Ahrends A, Beckschäfer P, *et al.* (2016) Pushing the limits: the pattern and dynamics of rubber monoculture expansion in Xishuangbanna, SW China. *PLoS One* **11**, e0150062. doi:10.1371/journal.pone.0150062

10. Zeng L, Cao M, Lin L, Peters CM (2022) Tree diversity and regeneration in sacred groves and nature reserves in Xishuangbanna, southwest China. *Journal of Ethnobiology* **42**, 432–460. doi:10.2993/0278-0771-42.4.432

11. Zhang JQ, Corlett RT, Zhai D (2019) After the rubber boom: good news and bad news for biodiversity in Xishuangbanna, Yunnan, China. *Regional Environmental Change* **19**, 1713–1724. doi:10.1007/s10113-019-01509-4

12. Ives M (2013) The rise of rubber takes toll on forests of southwest China. Yale Environment 360, <https://e360.yale.edu/features/the_rise_of_rubber_takes_toll_on_forests_of_southwest_china>.

13. Scally P (2015) Remembering Yunnan's role in World War Two. GoKunming, <https://www.gokunming.com/en/blog/item/3562/remembering-yunnans-role-in-world-war-two>.

14. Shang X, Chen Y (2020) Rural poverty patterns and influencing factors in Yunnan Province, China: based on county level dataset. *Journal of Resources and Ecology* **11**, 366–377. doi:10.5814/j.issn.1674-764x.2020.04.005

15. Wang Y, Ding S (2021) Achievements, experiences and challenges of the battle against poverty in China's ethnic minority areas: focusing on the 'three areas and three prefectures'. *International Journal of Anthropology and Ethnology* **5**, 18. doi:10.1186/s41257-021-00059-0

16. Isaksson E (2020) Ethnic minorities and the fight against poverty in China: the case of Yunnan. The Institute for Security and Development Policy, <https://isdp.eu/ethnic-minorities-and-the-fight-against-poverty-in-china-the-case-of-yunnan/>.

17. Fitra MA, Thomy Z, Samingan, Harnelly E, Kusuma HI (2020) The potency of mushrooms as food alternative in the forest park of Pocut Meurah Intan, Saree, Aceh Besar. The 1st International Conference on Agriculture and Bioindustry 2019 IOP Conference Series: *Earth and Environmental Science* 425, 012058.

18. Moreno RB, Ruthes AC, Baggio CH, Vilaplana F, Komura DL, *et al.* (2016) Structure and antinociceptive effects of β-d-glucans from *Cookeina tricholoma*. *Carbohydrate Polymers* **141**, 220–228. doi:10.1016/j.carbpol.2016.01.001

19. Garcia J, Costa VM, Carvalho ATP, Silvestre R, Duarte JA, *et al.* (2015) A breakthrough on *Amanita phalloides* poisoning: an effective antidotal effect by polymyxin B. *Archives of Toxicology* **89**, 2305–2323. doi:10.1007/s00204-015-1582-x

20. Moor-Smith M, Li R, Ahmad O (2019) The world's most poisonous mushroom, *Amanita phalloides,* is growing in BC. *BC Medical Journal* **61**, 20–24.

21. Leonardo-Silva L, Calaça FJS, Pereira-Silva G, Silva-Neto CM, Xavier-Santos S (2023) A new occurrence of *Gyrodontium sacchari* (Spreng.) Hjortstam Pat. (Boletales, Coniophoraceae) expands the geographic distribution of the genus in Brazil. *Check List* **19**, 27–34. doi:10.15560/19.1.27

22. Robledo G, Giorgio E, Franco CR, Popoff O, Decock C (2014) *Gyrodontium sacchari* (Spreng.: Fr.) Hjortstam (Boletales, Basidiomycota) in America: new records and its geographic distribution. *Check List* **10**, 1514–1519. doi:10.15560/10.6.1514

23. Das SK, Masuda M, Sakurai A, Sakakibara M (2010) Medicinal uses of the mushroom *Cordyceps militaris*: current state and prospects. *Fitoterapia* **81**, 961–968. doi:10.1016/j.fitote.2010.07.010

24. Nguyen TT, Le TN-G, Nguyen TH (2023) First report of emerging fungal pathogens of *Cordyceps militaris* in Vietnam. *Scientific Reports* **13**, 17669. doi:10.1038/s41598-023-43951-9

25. Xu J, Guo J, Mortimer P, Karunarathna S (2016) *Fantastic Mushroom World: Macrofungi of the Nabanhe National Nature Reserve*. Nabanhe National Nature Reserve: Xishuangbanna, China.

LICHEN LOVE

1. Pichler G, Muggia L, Carniel FC, Grube M, Kranner I (2023) How to build a lichen: from metabolite release to symbiotic interplay. *New Phytologist* **238**, 1362–1378. doi:10.1111/nph.18780

2. Honegger R (2000) Simon Schwendener (1829–1919) and the dual hypothesis of lichens. *The Bryologist* **103**, 307–313. doi:10.1639/0007-2745(2000)103[0307:SSATDH]2.0.CO;2

3. Hawksworth DL, Grube M (2020) Lichens redefined as complex ecosystems. *New Phytologist* **227**, 1281–1283. doi:10.1111/nph.16630

4. Lister BC (n.d.) History of ecology. Plant Sciences, Encyclopedia.com, <https://www.encyclopedia.com/science/news-wires-white-papers-and-books/ecology-history> accessed 19 March 2024.

5. Trappe J (2005) A.B. Frank and mycorrhizae: the challenge to evolutionary and ecologic theory. *Mycorrhiza* **15**, 277–281. doi:10.1007/s00572-004-0330-5

6. Robinson G (n.d.) De Bary, (Heinrich) Anton. Complete Dictionary of Scientific Biography, Encyclopedia.com <https://www.encyclopedia.com/science/dictionaries-thesauruses-pictures-and-press-releases/de-bary-heinrich-anton> accessed 19 March 2024.

7. Tipton L, Darcy JL, Hynson NA (2019) A developing symbiosis: enabling cross-talk between ecologists and microbiome scientists. *Frontiers in Microbiology* **10**, 292. doi:10.3389/fmicb.2019.00292

8. Margulis L, Barreno E (2003) Looking at lichens. *Bioscience* **53**, 776–778. doi:10.1641/0006-3568(2003)053[0776:LAL]2.0.CO;2

9. Spribille T, Resl P, Stanton DE, Tagirdzhanova G (2022) Evolutionary biology of lichen symbioses. *New Phytologist* **234**, 1566–1582. doi:10.1111/nph.18048

10. Goward T (2008) Twelve readings on the lichen thallus. I. Face in the mirror. *Evansia* **25**(2), 23–25. doi:10.1639/0747-9859-25.2.24

11. Spribille T, Tuovinen V, Resl P, Vanderpool D, Aime MC, *et al.* (2016) Basidiomycete yeasts in the cortex of ascomycete macrolichens. *Science* **353**, 488–492. doi:10.1126/science.aaf8287 [as reported by Richie M (2017) Overturning 150 Years of Science. UM lichen discovery rocks the research world. Medium, <https://medium.com/vision-2017/overturning-150-years-of-science-b19f776aee54> accessed 22 September 2024]

12. Hawksworth DL, Grube M (2020) Lichens redefined as complex ecosystems. *New Phytologist* **227**, 1281–1283. doi:10.1111/nph.16630

13. Mark K, Laanisto L, Bueno CG, Niinemets Ü, Keller C, Scheidegger C (2020) Contrasting co-occurrence patterns of photobiont and cystobasidiomycete yeast associated with common epiphytic lichen species. *New Phytologist* **227**, 1362–1375. doi:10.1111/nph.16475

14. Eufemio RJ, de Almeida Ribeiro I, Sformo TL, Laursen GA, Molinero V, *et al.* (2023) Lichen species across Alaska produce highly active and stable ice nucleators. *Biogeosciences* **20**, 2805–2812. doi:10.5194/bg-20-2805-2023

15. Kieft TL (1988) Ice nucleation activity in lichens. *Applied and Environmental Microbiology* **54**, 1678–1681. doi:10.1128/aem.54.7.1678-1681.1988

16. Radeka M, Ranogajec J, Kiurski J, Markov S, Marinković-Nedučin R (2007) Influence of lichen biocorrosion on the quality of ceramic roofing tiles. *Journal of the European Ceramic Society* **27**, 1763–1766. doi:10.1016/j.jeurceramsoc.2006.05.001

17. Seaward MRD (2004) Lichens as subversive agents of biodeterioration. In *Biodeterioration of Stone Surfaces*. (Eds LL St. Clair and MRD Seaward) pp. 9–18. Kluwer Academic Publishers, Dordrecht.

18. Anon. (n.d.) Lichens. Australian Antarctic Program, <https://www.antarctica.gov.au/about-antarctica/plants/lichens/> accessed 19 March 2024.

19. Bradwell T, Armstrong R (2007) Growth rates of *Rhizocarpon geographicum* lichens: a review with new data from Iceland. *Journal of Quaternary Science* **22**, 311–320. doi:10.1002/jqs.1058

20. Benítez A, Aragón G, Prieto M (2019) Lichen diversity on tree trunks in tropical dry forests is highly influenced by host tree traits. *Biodiversity and Conservation* **28**, 2909–2929. doi:10.1007/s10531-019-01805-9

21. Ranius T, Johansson P, Berg N, Niklasson M (2008) Influence of tree age and microhabitat quality on the occurrence of crustose lichens associated with old oaks. *Journal of Vegetation Science* **19**, 653–662. doi:10.3170/2008-8-18433

22. Aptroot A, Van Herk K (2007) Further evidence of the effects of global warming on lichens, particularly those with *Trentepohlia* phycobionts. *Environmental Pollution* **146**, 293–298. doi:10.1016/j.envpol.2006.03.018

23. Pinho P, Dias T, Cruz C, Tang YS, Sutton MA, *et al.* (2011) Using lichen functional diversity to assess the effects of atmospheric ammonia in Mediterranean woodlands. *Journal of Applied Ecology* **48**, 1107–1116. doi:10.1111/j.1365-2664.2011.02033.x

24. Devkota S, Dymytrova L, Chaudhary R, Werth S, Scheidegger C (2019) Climate change-induced range shift of the endemic epiphytic lichen *Lobaria pindarensis* in the Hindu Kush Himalayan region. *Lichenologist (London, England)* **51**, 157–173. doi:10.1017/S002428291900001X

SAFARI TO THE END OF THE WORLD

1. Anon. (n.d.) Giuliana Furci. Fungi Foundation, <https://www.ffungi.org/en/meet-giuliana> accessed 7 March 2023.

2. Moens J (2021) Flora, Fauna, and … Funga? The Case for a Third 'F'. Undark, <https://undark.org/2021/08/09/flora-fauna-and-funga/> accessed 29 September 2024.

3. Planet Fungi (2017) *Patagonia Fungi Safari* [Video]. YouTube, <https://youtu.be/NklZ9KkVHxw>.

4. Sanmartín I, Ronquist F (2004) Southern hemisphere biogeography inferred by event-based models: plant versus animal patterns. *Systematic Biology* **53**, 216–243. doi:10.1080/10635150490423430

5. Truong C, Sánchez-Ramírez S, Kuhar F, Kaplan Z, Smith ME (2017) The Gondwanan connection – Southern temperate *Amanita* lineages and the description of the first sequestrate species from the Americas. *Fungal Biology* **121**, 638–651. doi:10.1016/j.funbio.2017.04.006

6. Quijada L, LoBuglio K, Karakehian J, Furci G, Torres D, *et al.* (2020) Exploring the diversity of fungi at 'the end of the world': Leotiomycetes from the forests of Patagonian Chile. *Newsletter of the Friends of the Farlow* **72**, 7–11.

7. Nouhra E, Kuhar F, Truong C, Pastor N, Crespo E, *et al.* (2021) *Thaxterogaster* revisited: a phylogenetic and taxonomic overview of sequestrate *Cortinarius* from Patagonia. *Mycologia* **113**, 1–34. doi:10.1080/00275514.2021.1894535

8. Anon. (n.d.) The Fungarium. Royal Botanic Gardens Kew, <https://www.kew.org/science/collections-and-resources/collections/fungarium> accessed 7 March 2024.

9. Millman L (2019) Darwin's Fungus (*Cyttaria darwinii*). In *Fungipedia: A Brief Compendium of Mushroom Lore.* pp. 51–59. Princeton University Press, Princeton.

10. Wainwright M (2011) Charles Darwin mycologist and refuter of his own myth. *Fungi Magazine* **4**(1), 12–20.

11. Anon. (n.d.) Pan de indio *Cyttaria darwinii*. iNaturalist, Guides, Chile Backroads Trip <https://www.inaturalist.org/guide_taxa/1182867> accessed 7 March 2024.

12. Pfister DH (2009) Fungi at the ends of the earth: Roland Thaxter in southern South America. *ReVista: Harvard Review of Latin America* **8**.

13. Smith ME (2009) Following the rake of Roland Thaxter – unearthing the hypogeous fungi of South America. *Newsletter of the Friends of the Farlow* **53**, 1–8.

14. San-Fabian B, Niskanen T, Liimatainen K, Kooij PW, Mujic AB, *et al.* (2018) New species of *Cortinarius* sect. *Austroamericani*, sect. nov., from South American Nothofagaceae forests. *Mycologia* **110**, 1127–1144. doi:10.1080/00275514.2018.1515449

15. Caiafa MV, Jusino MA, Wilkie AC, Díaz IA, Sieving KE, *et al.* (2021) Discovering the role of Patagonian birds in the dispersal of truffles and other mycorrhizal fungi. *Current Biology* **31**, 5558–5570. doi:10.1016/j.cub.2021.10.024

GLOWING IN THE DARK

1. Andres (2021) Bioluminescent Fungi & a Light-Generating Metabolism. Science and Pandas, <https://www.scienceandpandas.com/bioluminescent-fungi-a-light-generating-metabolism/>.

2. BBC (including Attenborough D and Napper E) (2016) *Planet Earth II: A New World Revealed* [TV documentary series]. British Broadcasting Corporation, Natural History Unit.

3. Oliveira AG, Stevani CV, Waldenmaier HE, Viviani V, Emerson JM, *et al.* (2015) Circadian control sheds light on fungal bioluminescence. *Current Biology* **25**, 964–968. doi:10.1016/j.cub.2015.02.021

4. Wilson T, Hastings JW (2013) *Bioluminescence: Living Lights, Lights for Living.* Harvard University Press, Cambridge, MA.

5. Travouillon KJ, Cooper C, Bouzin JT, Umbrello LS, Lewis SW (2023) All-a-glow: spectral characteristics confirm widespread fluorescence for mammals. *Royal Society Open Science* **10**(10), 230325. doi:10.1098/rsos.230325

6. Evans B, Llewellyn T, Gaya E (2023) The lichen that invented sunscreen. Royal Botanic Gardens, Kew, <https://www.kew.org/read-and-watch/lichen-that-invented-sunscreen>.

7. Veerman J, Vasil'ev S, Paton GD, Ramanauskas J, Bruce D (2007) Photoprotection in the lichen *Parmelia sulcata*: the origins of desiccation-induced fluorescence quenching. *Plant Physiology* **145**, 997–1005. doi:10.1104/pp.107.106872

8. Boch S, Fischer M, Prati D (2015) To eat or not to eat – relationship of lichen herbivory by snails with secondary compounds and field frequency of lichens. *Journal of Plant Ecology* **8**, 642–650. doi:10.1093/jpe/rtv005

9. Galanty A, Popiół J, Paczkowska-Walendowska M, Studzińska-Sroka E, Paśko P, *et al.* (2021) (+)-Usnic Acid as a promising candidate for a safe and stable topical photoprotective agent. *Molecules (Basel, Switzerland)* **26**(17), 5224. doi:10.3390/molecules26175224

10. Lohézic-Le Dévéhat F, Legouin B, Couteau C, Boustie J, Coiffard L (2013) Lichenic extracts and metabolites as UV filters. *Journal of Photochemistry and Photobiology. B, Biology* **120**, 17–28. doi:10.1016/j.jphotobiol.2013.01.009

11. Nguyen K-H, Chollet-Krugler M, Gouault N, Tomasi S (2013) UV-protectant metabolites from lichens and their symbiotic partners. *Natural Product Reports* **30**, 1490–1508. doi:10.1039/c3np70064j

12. Syed AJ, Anderson JC (2021) Applications of bioluminescence in biotechnology and beyond. *Chemical Society Reviews* **50**, 5668–5705. doi:10.1039/D0CS01492C

IN THE SHADOW
OF THE HIMALAYAS

1. Marciniak C (2018) *Planet Fungi. North East India* [Documentary film].

2. Anon. (1979) Flooding in India leaves 13 dead and at least a million homeless. New York Times, 14 October 1979, p. 18, <https://www.nytimes.com/1979/10/14/archives/flooding-in-india-leaves-13-dead-and-at-least-a-million-homeless.html> accessed 22 September 2024.

3. Bhutia DT (2014) Protein energy malnutrition in India: the plight of our under five children. *Journal of Family Medicine and Primary Care* **3**, 63–67. doi:10.4103/2249-4863.130279

4. Singh KT, Chiero V, Kriina M, Alee NT, Chauhan K (2022) Identifying the trend of persistent cluster of stunting, wasting, and underweight among children under five years in northeastern states of India. *Clinical Epidemiology and Global Health* **18**, 101158. doi:10.1016/j.cegh.2022.101158

5. Wang M, Zhao R (2023) A review on nutritional advantages of edible mushrooms and its industrialization development situation in protein meat analogues. *Journal of Future Foods* **3**, 1–7. doi:10.1016/j.jfutfo.2022.09.001

6. González A, Cruz M, Losoya C, Nobre C, Loredo A, *et al.* (2020) Edible mushrooms as a novel protein source for functional foods. *Food & Function* **11**, 7400–7414. doi:10.1039/D0FO01746A

7. Paloi S, Kumla J, Paloi BP, Srinuampan S, Hoijang S, *et al.* (2023) Termite mushrooms (*Termitomyces*), a potential source of nutrients and bioactive compounds exhibiting human health benefits: a review. *Journal of Fungi (Basel, Switzerland)* **9**, 112. doi:10.3390/jof9010112

8. Schalk F, Gostinčar C, Kreuzenbeck NB, Conlon BH, Sommerwek E, *et al.* (2021) The termite fungal cultivar *Termitomyces* combines diverse enzymes and oxidative reactions for plant biomass conversion. *mBio* **12**, e0355120. doi:10.1128/mBio.03551-20

9. Poulsen M (2015) Fungus-growing termite gut microbiota. *Environmental Microbiology* **17**, 2562–2572. doi:10.1111/1462-2920.12765

10. Otani S, Hansen LH, Sørensen SJ, Poulsen M (2016) Bacterial communities in termite fungus combs are comprised of consistent gut deposits and contributions from the environment. *Microbial Ecology* **71**, 207–220. doi:10.1007/s00248-015-0692-6

11. Gover DW (n.d.) Termites and *Termitomyces* sp. an insect – fungus symbiosis. Abstract. Sydney Fungal Studies Group, <https://www.sydneyfungalstudies.org.au/resources/Articles/Termites%20and%20Termitomyces%20(Abstract).pdf> accessed 22 September 2024.

12. Khaund P, Joshi SR (2014) DNA barcoding of wild edible mushrooms consumed by the ethnic tribes of India. *Gene* **550**, 123–130. doi:10.1016/j.gene.2014.08.027

NEW DISCOVERIES

1. Bridge PD, Spooner BM, Beever RE, Park D (2008) Taxonomy of the fungus commonly known as *Stropharia aurantiaca*, with new combinations in *Leratiomyces*. *Mycotaxon* **103**, 109–121.

2. Johns GB (2022) Dispersal and trait evolution: Did New Zealand's sequestrate fungi evolve unique traits to match bird sensory ecologies? Thesis submitted for Master of Science in Biology, University of Auckland, New Zealand.

3. Xu J (2016) Fungal DNA barcoding. *Genome* **59**, 913–932. doi:10.1139/gen-2016-0046

4. Schoch CL, Seifert KA, Huhndorf S, Robert V, Spouge JL, *et al.* (2012) Nuclear ribosomal internal transcribed spacer (ITS) region as a universal DNA barcode marker for Fungi. *Proceedings of the National Academy of Sciences of the United States of America* **109**, 6241–6246. doi:10.1073/pnas.1117018109

5. Redhead SA, Vilgalys R, Moncalvo JM, Johnson J, Hopple JS (2001) *Coprinus* Pers. and the disposition of *Coprinus* species sensu lato. *Taxon* **50**, 203–241. doi:10.2307/1224525

6. Lebel T, Davoodian N, Bloomfield MC, Syme K, May TW, *et al.* (2022) A mixed bag of sequestrate fungi from five different families: Boletaceae, Russulaceae, Psathyrellaceae, Strophariaceae, and Hysterangiaceae. *Swainsona* **36**, 33–65.

7. Guard FE, Dearnaley J, Lebel T, Barrett MD, Bougher NL (2023) *Marasmius australotrichotus* (Marasmiaceae), a new setose species from Australia, and an intriguing range extension for *M. paratrichotus*. *Nuytsia* **34**, 203–219. doi:10.58828/nuy01059

8. Antonelli A, Fry C, Smith RJ, Eden J, Herman R, *et al.* (2023) *State of the World's Plants and Fungi 2023*. Royal Botanic Gardens, Kew.

9. Lücking R (2020) Three challenges to contemporaneous taxonomy from a licheno-mycological perspective. *Megataxa* **1**, 78–103.

10. Santi L, Maggioli C, Mastroroberto M, Tufoni M, Napoli L, *et al.* (2012) Acute liver failure caused by *Amanita phalloides* poisoning. *International Journal of Hepatology* **2012**, 487480. doi:10.1155/2012/487480

11. Horowitz BZ, Moss MJ (2023) Amatoxin mushroom toxicity. StatPearls. StatPearls Publishing, Treasure Island, Florida. Accessed via National Library of Medicine, <https://www.ncbi.nlm.nih.gov/books/NBK431052/> accessed 22 September 2024.

12. Kuttippurath J, Murasingh S, Stott PA, Sarojini BB, Jha MK, *et al.* (2021) Observed rainfall changes in the past century (1901–2019) over the wettest place on Earth. *Environmental Research Letters* **16**, 024018. doi:10.1088/1748-9326/abcf78

EVERYTHING IS CONNECTED

1. Collins L, Bradstock R, Clarke H, Clarke M, Nolan R, *et al.* (2021) The 2019/2020 mega-fires exposed Australian ecosystems to an unprecedented extent of high-severity fire. *Environmental Research Letters* **16**, 044029. doi:10.1088/1748-9326/abeb9e

2. Abram NJ, Henley BJ, Sen Gupta A, Lippman TJR, Clarke H, *et al.* (2021) Connections of climate change and variability to large and extreme forest fires in southeast Australia. *Communications Earth & Environment* **2**, 8. doi:10.1038/s43247-020-00065-8

3. Dickman CR (2021) Ecological consequences of Australia's 'Black Summer' bushfires. *Integrated Environmental Assessment and Management* **17**, 1162–1167. doi:10.1002/ieam.4496

4. Ward M, Tulloch AIT, Radford JQ, Williams BA, Reside AE, *et al.* (2020) Impact of 2019–2020 mega-fires on Australian fauna habitat. *Nature Ecology & Evolution* **4**, 1321–1326. doi:10.1038/s41559-020-1251-1

5. Filialuna O, Cripps C (2021) Evidence that pyrophilous fungi aggregate soil after forest fire. *Forest Ecology and Management* **498**, 119579. doi:10.1016/j.foreco.2021.119579

6. McMullan-Fisher S, May T, Robinson R, Bell TL, Lebel T, *et al.* (2011) Fungi and fire in Australian ecosystems: a review of current knowledge, management implications and future directions. *Australian Journal of Botany* **59**, 70–90. doi:10.1071/BT10059

7. Steindorff A, Carver A, Calhoun S, Stillman K, Liu H, *et al.* (2021) Comparative genomics of pyrophilous fungi reveals a link between fire events and developmental genes. *Environmental Microbiology* **23**, 99–109. doi:10.1111/1462-2920.15273

8. Fischer MS, Stark FG, Berry TD, Zeba N, Whitman T, *et al.* (2021) Pyrolyzed substrates induce aromatic compound metabolism in the post-fire fungus, *Pyronema domesticum*. *Frontiers in Microbiology* **12**, 729289. doi:10.3389/fmicb.2021.729289

9. May TW, Bal P, Hao T, Muscatello A, Bowd E, *et al.* (2023) The impacts of the 2019–20 wildfires on Australian fungi. In *Australia's Megafires: Biodiversity Impacts and Lessons from 2019-2020*. (Eds L Rumpff, S Legge, S van Leeuwen, B Wintle and J Woinarski) pp. 185–201. CSIRO Publishing, Melbourne.

10. Kiers T (2021) Scientists are mapping the world's underground fungi network to fight climate change. CBC Radio, *As it Happens*, <https://www.cbc.ca/radio/asithappens/as-it-happens-the-wednesday-edition-1.6269628/scientists-are-mapping-the-world-s-underground-fungi-network-to-fight-climate-change-1.6269630> accessed 22 September 2024.

11. Averill CA, Hawkes C (2016) Ectomycorrhizal fungi slow soil carbon cycling. *Ecology Letters* **19**, 937–947. doi:10.1111/ele.12631

12. Hawkins HJ, Cargill RIM, Van Nuland ME, Hagen SC, Field KJ, *et al.* (2023) Mycorrhizal mycelium as a global carbon pool. *Current Biology* **33**, R560–R573. doi:10.1016/j.cub.2023.02.027

13. Kobaly R (2022) A Costly omission in planning for climate change. How wide-scale desert soil disturbance releases stored carbon. Desert Report, <https://desertreport.org/a-costly-omission-in-planning-for-climate-change/> accessed 22 September 2024.

14. Boyd S (2022) Carbon sequestration in our desert lands. Desert Report, <https://desertreport.org/carbon-sequestration-in-our-desert-lands/> accessed 22 September 2024.

15. Averill C, Dietze M, Bhatnagar J (2018) Continental-scale nitrogen pollution is shifting forest mycorrhizal associations and soil carbon stocks. *Global Change Biology* **24**, 4544–4553. doi:10.1111/gcb.14368

16. Hassett MO, Fischer MW, Money NP (2015) Mushrooms as rainmakers: how spores act as nuclei for raindrops. *PLoS One* **10**, e0140407. doi:10.1371/journal.pone.0140407

17. Murphy M (2014) *Triboniophorus* sp. nov. 'Kaputar'. The IUCN Red List of Threatened Species 2014, <https://www.iucnredlist.org/species/55242781/55243154> accessed 22 September 2024.

18. Solomon KV, Haitjema CH, Henske JK, Gilmore SP, Borges-Rivera D, *et al.* (2016) Early-branching gut fungi possess a large, comprehensive array of biomass-degrading enzymes. *Science* **351**, 1192–1195. doi:10.1126/science.aad1431

Index

Page numbers in bold refer to photographs.

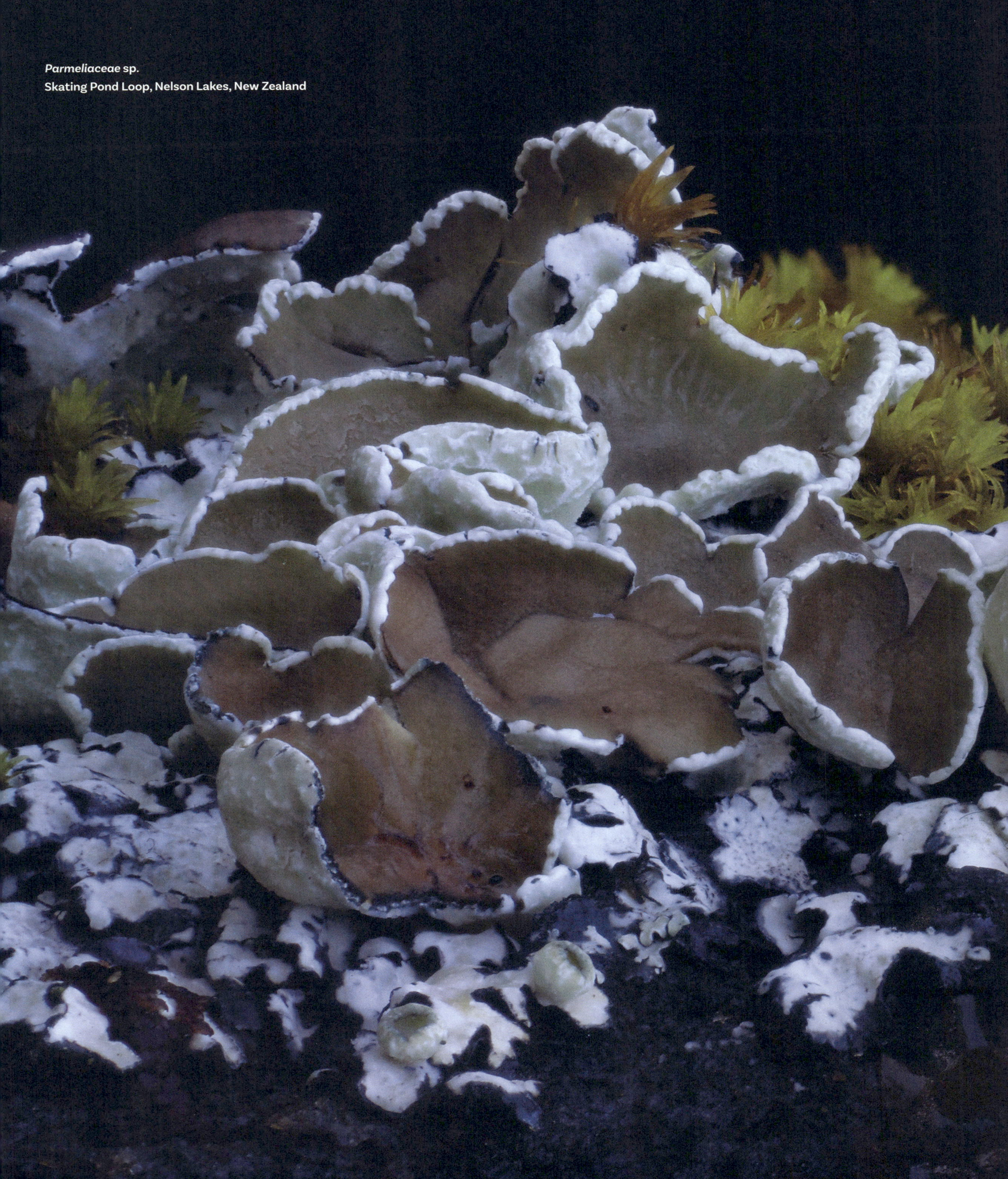

Parmeliaceae sp.
Skating Pond Loop, Nelson Lakes, New Zealand